Student's Solutions Manual

GENERAL CHEMISTRY

THIRD • EDITION

Student's Solutions Manual

GENERAL CHEMISTRY

EBBING

THIRD • EDITION

DARRELL D. EBBING
WAYNE STATE UNIVERSITY

GEORGE H. SCHENK
WAYNE STATE UNIVERSITY

HOUGHTON MIFFLIN COMPANY • BOSTON
Dallas • Geneva, Illinois • Palo Alto • Princeton, New Jersey

Cover Photograph: Laser "spark" spectroscopy of a coal particle at the Combustion Research Facility, located at Sandia National Laboratories in Livermore, California.

Printed in the U.S.A.

ISBN: 0-395-52876-3

IJ-B-998765432

Contents

1 Chemistry and Measurement 1
Solutions to Exercises 1
Answers to Review Questions 2
Solutions to Practice Problems 3
Cumulative-Skills Problems 7

2 Atoms, Molecules, and Ions 8
Solutions to Exercises 8
Answers to Review Questions 9
Solutions to Practice Problems 10
Cumulative-Skills Problems 13

3 Chemical Reactions: An Introduction 14
Solutions to Exercises 14
Answers to Review Questions 16
Solutions to Practice Problems 17
Cumulative-Skills Problems 25

4 Calculations with Chemical Formulas and Equations 26
Solutions to Exercises 26
Answers to Review Questions 32
Solutions to Practice Problems 33
Cumulative-Skills Problems 50

5 The Gaseous State 53
Solutions to Exercises 53
Answers to Review Questions 57
Solutions to Practice Problems 59
Cumulative-Skills Problems 68

6 Thermochemistry 69
Solutions to Exercises 69
Answers to Review Questions 71
Solutions to Practice Problems 73
Cumulative-Skills Problems 79

7 Atomic Structure 82
Solutions to Exercises 82
Answers to Review Questions 84
Solutions to Practice Problems 86
Cumulative-Skills Problems 91

8 Electron Configurations and Periodicity 93
Solutions to Exercises 93
Answers to Review Questions 94
Solutions to Practice Problems 96
Cumulative-Skills Problems 98

9 Ionic and Covalent Bonding 100
Solutions to Exercises 100
Answers to Review Questions 102
Solutions to Practice Problems 104
Cumulative-Skills Problems 116

10 Molecular Geometry and Chemical Bonding Theory 119
Solutions to Exercises 119
Answers to Review Questions 122
Solutions to Practice Problems 124
Cumulative-Skills Problems 130

11 States of Matter; Liquids and Solids 132
Solutions to Exercises 132
Answers to Review Questions 134
Solutions to Practice Problems 136
Cumulative-Skills Problems 143

12 Solutions 145
Solutions to Exercises 145
Answers to Review Questions 149
Solutions to Practice Problems 151
Cumulative-Skills Problems 158

13 Chemical Reactions: Acid–Base and Oxidation–Reduction Concepts 160
Solutions to Exercises 160
Answers to Review Questions 164
Solutions to Practice Problems 166
Cumulative-Skills Problems 180

14 Rates of Reaction 183
Solutions to Exercises 183
Answers to Review Questions 185
Solutions to Practice Problems 189
Cumulative-Skills Problems 197

15 Chemical Equilibrium; Gaseous Reactions 199
Solutions to Exercises 199
Answers to Review Questions 203
Solutions to Practice Problems 205
Cumulative-Skills Problems 214

16 Acid–Base Equilibria 216
Solutions to Exercises 216
Answers to Review Questions 224
Solutions to Practice Problems 227
Cumulative-Skills Problems 246

17 Solubility and Complex-Ion Equilibria 248
Solutions to Exercises 248
Answers to Review Questions 254
Solutions to Practice Problems 256
Cumulative-Skills Problems 270

18 Thermodynamics and Equilibrium 273
Solutions to Exercises 273
Answers to Review Questions 277
Solutions to Practice Problems 279
Cumulative-Skills Problems 287

19 Electrochemistry 290
Solutions to Exercises 290
Answers to Review Questions 294
Solutions to Practice Problems 296
Cumulative-Skills Problems 307

20 Nuclear Chemistry 308
Solutions to Exercises 308
Answers to Review Questions 312
Solutions to Practice Problems 314
Cumulative-Skills Problems 322

21 The Main-Group Elements: Groups IA to IIIA; Metallurgy 324
Solutions to Exercises 324
Answers to Review Questions 324
Solutions to Practice Problems 327

22 The Main-Group Elements: Groups IVA to VIIIA 332
Solutions to Exercises 332
Answers to Review Questions 333
Solutions to Practice Problems 337

23 The Transition Elements 346
Solutions to Exercises 346
Answers to Review Questions 349
Solutions to Practice Problems 352

24 Organic Chemistry 360
Solutions to Exercises 360
Answers to Review Questions 363
Solutions to Practice Problems 366

25 Biochemistry 376
Solutions to Exercises 376
Answers to Review Questions 378
Solutions to Practice Problems 379

Appendix A 385
Solutions to Exercises 385

Preface

This Student's Solutions Manual provides worked-out answers to problems that appear in the third edition of *General Chemistry* by Darrell D. Ebbing. This includes detailed, step-by-step solutions for all in-chapter Exercises. Complete solutions are provided for the odd-numbered Practice Problems, Additional Problems, and Cumulative-Skills Problems that appear at the end of the chapters. Also provided are answers to all of the Review Questions. Please note the following:

Significant Figures The answers are first shown with 1 to 2 nonsignificant figures and no units. The answer is then rounded off to the correct number of significant figures with units. This will help the student to compare their multi-digit calculator readouts to the final answer.

Equilibrium Calculations The answer is first shown at the beginning of each problem. This will help the student gauge whether he or she is on track through the complicated working-out of the solution. The answer is then restated at the end of the problem.

Great effort and care have gone into the preparation of this manual. the solutions have been checked and rechecked for accuracy and completeness several times. We wish to thank the following people who helped check the solutions and who have contributed otherwise to the completion of this manual: John Goodenow, Lawrence Technical University; Dr. Lynne Hitchcock, Wayne State University; Sharyl Majorski, Wayne State University; and Judith Montgomery, Wayne State University.

D.D.E. and G.H.S.

CHAPTER 1

CHEMISTRY AND MEASUREMENT

SOLUTIONS TO EXERCISES

Note on significant figures: The final answer to all mathematical solutions is given first with one nonsignificant figure (last significant figure underlined) and is then rounded to the correct number of figures. Intermediate answers usually also have least one nonsignificant figure.

1.1 In words: mass reacted oxygen = mass red residue - mass metallic mercury

mass reacted oxygen = 2.73 g - 2.53 g = 0.20 g

Thus, the mass of oxygen lost after heating = 0.20 g

1.2 a. $\frac{5.61 \times 7.891}{9.1} = 4.\underline{8}6 = 4.9$ (Two significant figures because 9.1 has only two.)

b. 8.91 - 6.43 = $2.4\underline{7}5$ = 2.48 (Two decimal places because 8.91 has only two decimals.)

c. 6.81 - 6.73 = $0.0\underline{8}0$ = 0.08 (Two decimal places because 6.81 has only two decimals.)

d. 38.91 x (6.81 - 6.730) = 38.91 x $0.0\underline{8}0$ = $\underline{3}.113$ = 3
(The calculated difference of $0.0\underline{8}0$ has two sig figs because 6.81 has two decimal places. When 0.080 is used in multiplying, it has only one significant figure, so the product of 0.080 and 38.91 can have only one significant figure.)

1.3 a. 1.84×10^{-9} m = 1.84 nm
b. 5.67×10^{-12} s = 5.67 ps
c. 7.85×10^{-3} g = 7.85 mg
d. 9.7 x 103 m = 9.7 km
e. 0.000732 s = 732 μs
f. 0.000000000154 = 154 pm

1.4 a. $°C = \frac{°F - 32}{1.8} = \frac{[102.5 - 32]}{1.8} = 39.\underline{1}6 = 39.2$ (3 sig figs from numerator of [70.5])

(Note that 32 and 1.8 are both exact numbers and do not affect significant figures.)

b. K = °C + 273.15 = -78 + 273.15 = $19\underline{5}.15$ = 195 K (78 has no decimal places.)

1.5 $d = \frac{m}{V} = \frac{159\ g}{20.2\ cm^3} = 7.8\underline{7}1 = 7.87\ g/cm^3$ (The metal is iron.)

1.6 $V = \frac{m}{d} = 30.3\ g \times \frac{1\ cm^3}{0.789\ g} = 38.\underline{4}03 = 38.4\ cm^3$

1.7 $121\ pm \times \frac{10^{-12}\ m}{1\ pm} \times \frac{10^3\ mm}{1\ m} = 1.21 \times 10^{-7}\ mm$

1.8 $67.6\ Å \times \frac{10^{-30}\ m}{1\ Å^3} \times \frac{10^3\ m}{1\ m^3} = 6.76 \times 10^{-26}\ dm^3$

1.9 $\frac{36\ in}{1\ yd} \times \frac{2.54\ cm}{1\ in} \times \frac{1\ m}{10^2\ cm} = \frac{9.144 \times 10^{-1}\ m}{1\ yd}$

$3.54\ yd \times \frac{9.144 \times 10^{-1} m}{yd} = 3.2\underline{3}69 = 3.24\ m$

ANSWERS TO REVIEW QUESTIONS

1.1 Matter is any material thing that occupies space. Mass is the quantity of matter in a material. The difference betwen mass and weight is that mass remains the same wherever it is measured but weight varies with the distance from the center of the earth.

1.2 This law states that mass remains constant during a chemical change (reaction). Weighing a flash bulb before and after flashing illustrates this law. The bulb is found to have the same mass after the reaction that occurs during flashing, showing that mass is conserved during chemical changes.

1.3 Mercury reacts with oxygen to form red mercury(II) oxide. Magnesium reacts with oxygen to form white magnesium oxide.

1.4 An experiment is an observation of natural phenomena carried out in a controlled manner so that the results can be duplicated and rational conclusions obtained. A theory is a tested explanation of natural phenomena. Experiments are performed and some regularity is observed; this leads to more experiments and a theory. A hypothesis is a tentative explanation of some regularity.

1.5 Rosenberg conducted controlled experiments, noting a basic relationship, which could be stated as a scientific law. He formed a hypothesis that platinum compounds were responsible for preventing cell division. After more testing, the hypothesis became a theory involving cisplatin.

1.6 The precision of a set of measurements refers to how close they are together. The precision of one measurement is indicated by using significant figures.

1.7 Mul/div rule calculation: 3.00 x 2.0 = 6.0 (2.0 has only two sig figs.)
Add/sub rule calculation: 3.000 + 7.00 = 10.00 (7.00 has two decimal places.)

1.8 An exact number arises when we count a small number of items, or when we define a number rather than measuring it. Examples: absolute: 0 K (absolute zero); measured : 11 K.

1.9 The SI system starts with seven base units and uses prefixes to obtain units of different sizes. Units for other quantities are obtained by deriving them from any of the 7 base units.

1.10 The SI length unit = the meter (39.3 in). The SI mass unit = kilogram (2.2 U.S. pounds).

1.11 An absolute temperature scale is a scale where the lowest temperature that can be attained theoretically is zero degrees. Degrees celsius are related to Kelvins by $^{o}C = (K - 273.15)$.

1.12 Density is the mass of an object per unit volume. It can be used to check purity.

1.13 Units should be carried along in a calculation to obtain units for the answer and avoid errors.

1.14 The conversion factor for converting liters to U.S. gallons is obtained by multiplying the conversion factor for liters to quarts by that for converting quarts to U.S. gallons:

$$\frac{1 \text{ qt}}{0.9464 \text{ L}} \times \frac{1 \text{ U.S. gal}}{4 \text{ qt}} = 0.264\underline{1}58 = \frac{0.2642 \text{ U.S. gal}}{\text{L}}$$

SOLUTIONS TO PRACTICE PROBLEMS

Note on significant figures: The final answer to all mathematical solutions is given first with one nonsignificant figure (last significant figure underlined) and is then rounded to the correct number of digits. Intermediate answers usually also have at least one nonsignificant figure.

1.15 By law of conservation of mass:

Mass(sodium hydrogen carbonate) + mass(acetic acid solution)

= mass(contents of reaction vessel) + mass(carbon dioxide)

8.4 g + 20.0 g = 24.0 g + mass (carbon dioxide)

Mass of carbon dioxide = 20.0 g + 8.4 g - 24.0 g = 4.$\underline{4}$0 = 4.4 g

1.17 By law of conservation of mass:

Mass(zinc) + mass(sulfur) = mass(zinc sulfide)

65.4 g Zn + 32.1 g S must = (65.4 + 32.1) g ZnS

20.0 g Zn + unknown mass of S = X g ZnS

Use this to write a proportion from the masses of zinc and zinc sulfide in both equations:

$$\frac{X}{20.0 \text{ g Zn}} = \frac{(65.4 + 32.1) \text{ g ZnS}}{65.4 \text{ g Zn}}$$

Solving gives X = 29.$\underline{8}$1 = 29.8 g

1.19 All significant figues (s.f.) are underlined:

a. $\underline{73.0000}$ (6 s.f.) c. $\underline{6.300}$ (4 s.f.) e. $\underline{5.10} \times 10^{-7}$ (3 s.f.)
b. 0.0$\underline{503}$ (3 s.f.) d. 0.$\underline{80090}$ (5 s.f.) f. $\underline{2.001}$ (4 s.f.)

1.21 4$\underline{0}$,000 km = 4.$\underline{0} \times 10^4$ km

1.23 a. $(8.71 \times 0.0301) \div 0.056 = 4.\underline{6}8 = 4.7$
b. $0.71 + 81.8 = 82.\underline{5}1 = 82.5$
c. $(934 \times 0.00435) + 107 = 11\underline{1}.06 = 111$
d. $(847.89 - 847.73) \times 14673 = 2.\underline{3}4 \times 10^3 = 2.3 \times 10^3$

1.25 Volume of 5.10 cm $= (4/3) \times 3.1416 \times (5.10\ cm)^3 = 5.5\underline{5}64 \times 10^2\ cm^3$
Volume of 5.00 cm $= (4/3) \times 3.1416 \times (5.00\ cm)^3 = 5.2\underline{3}60 \times 10^2\ cm^3$
Volume difference $= (5.5\underline{5}64 \times 10^2 - 5.2\underline{3}60 \times 10^2) = 0.3\underline{2}04 \times 10^2 = 32\ cm^3$

1.27 a. $5.89 \times 10^{-12}\ s = 5.89\ ps$
b. $2.130 \times 10^{-9}\ m = 2.130\ nm$
c. $0.00721\ g = 7.21\ mg$
d. $6.05 \times 10^3\ m = 6.05\ km$

1.29 a. $6.15\ ps = 6.15 \times 10^{-12}\ s$
b. $3.781\ \mu m = 3.781 \times 10^{-6}\ m$
c. $1.546\ Å = 1.546 \times 10^{-10}\ m$
d. $9.7\ mg = 9.7 \times 10^{-3}\ g$

1.31 a. $\frac{(32°F - 32)}{1.8} = -0.0 = 0°C$
b. $\frac{(-58°F - 32)}{1.8} = -5\underline{0}.0 = -50°C$
c. $\frac{(68°F - 32)}{1.8} = 2\underline{0}.0 = 20°C$
d. $\frac{(-11°F - 32)}{1.8} = -2\underline{3}.8 = -24°C$
e. $(37°C \times 1.8) + 32 = 9\underline{8}.6 = 99°F$
f. $(-70°C \times 1.8) + 32 = -9\underline{4}.0 = -94°F$

1.33 $(-21.1° \times 1.8) + 32°C = -5.\underline{9}8 = -6.0°F$

1.35 $(95° - 32) \div 1.8 = 3\underline{5}°C$; $K = 273.15 + 35 = 30\underline{8}.15 = 308\ K$

1.37 Density = mass ÷ volume $= 12.4\ g \div 1.64\ cm^3 = 7.5\underline{6}09 = 7.56\ g/cm^3$

1.39 Calculate density: density $= 9.42\ g \div 10.7\ mL = 0.88\underline{0}3 = 0.880\ g/mL$
The density is closest to that of benzene ($0.879\ g/cm^3$), so the liquid is benzene.

1.41 Mass (Pt) $= d \times V = \frac{22.5\ g}{cm^3} \times 5.9\ cm^3 = 1.\underline{3}2 \times 10^2 = 1.3 \times 10^2\ g$

1.43 Volume(alc) $= 19.8\ g\ alc \times 1\ cm^3\ alc/0.789\ g\ alc = 25.\underline{0}9 = 25.1\ cm^3\ alc$

1.45 $0.348\ kg \times (10^3\ g/1\ kg) \times (10^3\ mg/1\ g) = 3.48 \times 10^5\ mg$

1.47 $555 \text{ nm} \times \frac{1 \text{ m}}{10^9 \text{ nm}} \times \frac{10^3 \text{ mm}}{1 \text{ m}} = 5.55 \times 10^{-4} \text{ mm}$

1.49 $3.73 \times 10^8 \text{ km}^3 \times \frac{10^9 \text{ m}^3}{1 \text{ km}^3} \times \frac{10^3 \text{ dm}^3}{1 \text{ m}^3} = 3.73 \times 10^{20} \text{ dm}^3\text{(L) or } 3.73 \times 10^{17} \text{ m}^3$

1.51 $47.8 \text{ in} \times 12.5 \text{ in} \times 19.5 \text{ in} \times \frac{1 \text{ gal}}{231 \text{ in}^3} = 50.\underline{4}3 = 50.4 \text{ gal}$

1.53 $3.58 \text{ ton} \times \frac{2000 \text{ lb}}{1 \text{ ton}} \times \frac{16 \text{ oz}}{1 \text{ lb}} \times \frac{1 \text{ g}}{0.03527 \text{ oz}} = 3.2\underline{4}8 \times 10^6 = 3.25 \times 10^6 \text{ g}$

1.55 $2425 \text{ fath} \times \frac{6 \text{ ft}}{1 \text{ fath}} \times \frac{12 \text{ in}}{1 \text{ qt}} \times \frac{1 \text{ L}}{1 \text{ ft}} \times \frac{2.54 \times 10^{-2} \text{ m}}{1 \text{ in}} = 4.43\underline{4}8 \times 10^3 = 4.435 \times 10^3 \text{ m}$

1.57 9.85 g Na + 63.11 g H_2O - 72.53 gal soln = 0.43 g hydrogen

1.59 From the law of conservation of mass:

5.40 g Al + 18.50 g Fe_2O_3 = 11.17 g Fe + 10.20 g Al_2O_3 + g Fe_2O_3 unreacted

g Fe_2O_3 unreacted = 5.40 g + 18.50 g - (11.17 g + 10.20 g) = 2.53 g

1.61 53.10 g + 5.348 g + 56.1 g = $114.\underline{5}4$ = 114.5 g total mass

1.63 $(60.8 \text{ cm})^3 = 2.2\underline{4}7 \times 10^5 = 2.25 \times 10^5 \text{ cm}^3$

1.65 5.85 cm sphere volume = $(4/3)(3.1416)(5.85 \text{ cm})^3 = 8.3\underline{8}60 \times 10^2 \text{ cm}^3$

5.61 cm sphere volume = $(4/3)(3.1416)(5.61 \text{ cm})^3 = 7.3\underline{9}56 \times 10^2 \text{ cm}^3$

Difference = $0.9\underline{9}04 \times 10^2 = 9.9 \times 10^1 \text{ cm}^3$

1.67 a. $0.7\underline{5}98 = 0.76$ b. $16.\underline{2}9 = 16.3$ c. $47\underline{6}.3 = 476$ d. $0.11\underline{1}9 = 0.112$

1.69 a. 9.12 cg/mL b. 66 pm c. 7.1 µm d. 56 nm

1.71 a. 1.07×10^{-12} s b. 5.8×10^{-6} m c. 3.19×10^{-7} m d. 1.53×10^{-2} s

1.73 $(3410^\circ\text{C} \times 1.8) + 32 = 6170^\circ\text{F}$

1.75 $(825^\circ\text{C} \times 1.8) + 32 = 15\underline{1}7^\circ\text{F} = 1.52 \times 10^3 \; ^\circ\text{F}$

1.77 K = 273.15 + 29.8 = $302.\underline{9}5$ = 303.0; (29.8^{o}C x 1.8) + 32 = $85.\underline{6}4$ = 85.6^{o}F

1.79 (1666°F - 32) ÷ 1.8 = $907.\underline{7}7$ = 907.8^{o}C; 273.15 + 907.77 = $1180.\underline{9}2$ = 1180.9 K

1.81 Density = (22.5 g/cm^3)(1 kg/10^3 g)(10^2 cm/1 m)^3 = 2.25 x 10^4 kg/m^3

1.83 Density = 38.4 g ÷ (65.7 mL - 51.2 mL) = $2.6\underline{4}8$ = 2.65 g/mL

1.85 Density(unknown) = 22.3 g/15.0 mL = $1.4\underline{8}6$ g/mL = 1.49 g/cm^3

This density is closest to that of chloroform (1.489 g/cm^3), so the unknown is chloroform.

1.87 Mass = 21.4 g/cm^3 x (2.20 cm)^3 = $2.2\underline{7}8$ x 10^2 = 2.28 x 10^2 g (0.228 kg)

1.89 Volume = (1 mL/1.053 g) x 35.00 g = $33.2\underline{3}8$ = 33.24 mL (33.24 cm^3)

1.91 a. 8.45 kg x (10^3 g/1 kg) x (10^{3}mg/1 g) = 8.45 x 10^6 mg

b. 318 μs x (1 s/l0^6 μs) x (10^3 ms/1 s) = 0.318 ms

c. 93 km x (10^{3}m/1 km) x (10^9 nm/1 m) = 9.3 x 10^{13} nm

d. 37.1 mm x (1 m/10^3 mm) x (10^2 cm/1 mm) = 3.71 cm

1.93 a. 5.91 kg x (10^3 g/1 kg) x (10^3 mg/1 g) = 5.91 x 10^6 mg

b. 753 mg x (1 g/10^3 mg) x (10^{6}μg/1 g) = 7.53 x 10^5 μg

c. 90.1 mHz x (1 Hz/10^6 mHz) x (10^3 kHz/1 Hz) = 9.01 x 10^4 kHz

d. 498 mJ x (1 J/10^3 mJ) x (1 kJ/10^3 J) = 4.98 x l0^{-4} kJ

1.95 Volume = 12,230 km^3 x [(10^4 dm)/(1 km)]^3 x (1 L/1 dm^3) = 1.2230 x 10^{16} L
(assuming final zero in 12,230 km^3 is significant)

1.97 Volume = 10.0 ft x 12.0 ft x 9.0 ft x (12 in/1 ft)^3 x (2.54 cm/1 in)^3 x (1 dm/10 cm)^3 x (1 L/1 dm^3)

= $3.\underline{0}58$ x 10^4 = 3.1 x 10^4 L

1.99 Mass (wt) = 563 carats x (200 mg/1 carat) x (1 g/10^3 mg) = $11\underline{2}.6$ = 113 g

Cumulative-Skills Problems

1.101 From the law of conservation of mass:

10.0 g marble + (50.0 mL x 1.096 g/mL) HCl = 60.4 g soln + mass CO_2

Mass CO_2 = 10.0 g + 54.8 g - 60.4g = 4.4 g; Vol CO_2 = 4.4g/(1.798 g/L) = 2.4 L

1.103 Volume of sphere = $(4/3)(3.1416)(1.58 \text{ in})^3(2.54 \text{ cm/in})^3 = 27\underline{0}.746 \text{ cm}^3$
Mass of sphere = $270.746 \text{ cm}^3 \times 7.88 \text{ g/cm}^3 = 2.1\underline{3}3 \times 10^3 = 2.13 \times 10^3$ g

1.105 Mass = $(840{,}000 \text{ mi}^2 - 132{,}000 \text{ mi}^2) \times (5280 \text{ ft/1 mi})^2 \times 5000 \text{ ft} \times (12 \text{ in/1 ft})^3 \times (2.54 \text{ cm/1 in})^3 \times (.917 \text{ g/cm}^3)$

$= 2.\underline{5}6 \times 10^{21} = 2.6 \times 10^{21}$ g

1.107 Let x = mass of ethanol and y = mass of water. From the law of conservation of mass:

total mass = x + y = 49.6 g
so y = 49.6 g - x, and
total volume = volume of ethanol + volume of water

Express the volumes of ethanol and water using both densities (given) and masses x and y:

$54.2 \text{ cm}^3 = (x/0.789 \text{ cm}^3) + [(49.6 \text{ g} - x)/(0.998 \text{ g/cm}^3)]$

Solve the above equation for x:

$54.2 \text{ cm}^3 = (x/0.789 \text{ cm}^3) + [49.6 \text{ g}/(0.998 \text{ g/cm}^3)] - (x/0.998 \text{ g/cm}^3)$
$(x/0.789) - (x/0.998) = [54.2 - (49.6/0.998)]$ g

Multiply both sides by (0.789)(0.998):

$0.998x - 0.789x = (0.789)(0.998) \times [54.2 - (49.6/0.998)]$ g
$0.2090x = 3.54$ g
$x = 16.\underline{9}4$ g

Answer 1: mass % ethanol = (16.94 g/49.6 g soln) x 100% = $34.\underline{1}4$ = 34.1%

To calculate the proof, first find the volume of ethanol from its mass of 16.94 g:

Volume = $16.94 \text{ g}/(0.789 \text{ g/cm}^3) = 21.\underline{4}7 \text{ cm}^3$
Volume % = $(21.47 \text{ cm}^3/54.2 \text{ cm}^3) \times 100\% = 39.\underline{6}1\%$

Answer 2: Proof = 2 x volume % = 2 x 39.61 = $79.\underline{2}2$ = 79.2 proof

1.109 Mass water displaced = 18.49 - 16.21 = 2.28 g

Volume mineral = $2.28 \text{ g} \times (1 \text{ cm}^3/0.9982 \text{ g}) = 2.2\underline{8}4 \text{ cm}^3$

Mass air displaced = $2.284 \text{ cm}^3 \times 1.205 \times 10^{-3} \text{ g/cm}^3 = 2.7\underline{5}2 \times 10^{-3}$ g

$$\text{Density} = \frac{18.49 + 2.752 \times 10^{-3} \text{ g}}{2.284 \text{ cm}^3} = 8.096 = 8.10 \text{ g/cm}^3$$

CHAPTER 2

ATOMS, MOLECULES, AND IONS

SOLUTIONS TO EXERCISES

Note on significant figures: The final answer to all mathematical solutions is given first with one nonsignificant figure (last significant figure underlined) and is then rounded to the correct number of figures. Intermediate answers usually also have at least one nonsignificant figure.

2.1 Physical properties: soft, silver color, melts at 64°C, and density of 0.86 g/cm^3.

Chemical properties: metal, reacts with water, reacts with oxygen, and reacts with chlorine.

2.2 The fraction of potassium in each sample is as follows:

$$\text{A: } \frac{1.400 \text{ g K}}{2.502 \text{ g}} = 0.5596 \quad \text{B: } \frac{1.217 \text{ g K}}{1.819 \text{ g}} = 0.6690 \quad \text{C: } \frac{1.832 \text{ g K}}{2.761 \text{ g}} = 0.6635$$

Sample A has a markedly different fraction of K (and obviously O) than B or C; it's a mixture.

2.3 a. Se: Group VIA, period 4; nonmetal
b. Cs: Group IA, period 6; metal
c. Fe: Group VIIIB, period 4; metal
d. Cu: Group IB, period 4; metal
e. Br: Group VIIA, period 4; nonmetal

2.4 The formula is K_2CrO_4.

2.5 a. CaO: calcium oxide [group IIA forms only +2 cations]

b. $PbCrO_4$: lead(II) chromate [CrO_4^{2-} is the chromate anion (Table 2.5), so this is Pb^{2+}]

2.6 The formula is $Tl(NO_3)_3$.

2.7 a. KF is ionic b. P_4O_6 is molecular c. B_2O_3 is molecular

2.8 a. Nitrogen monoxide b. Phosphorus trichloride c. Phosphorous pentachloride

2.9 a. CS_2 b. SO_3

2.10 Perbromate ion: BrO_4^-

2.11 Sodium carbonate decahydrate

2.12 $Na_2S_2O_3 \cdot 5H_2O$

ANSWERS TO REVIEW QUESTIONS

2.1 Gases: very compressible, fluid, low density. Liquids: not very compressible, fluid, moderate density. Solids: not very compressible, not fluid, high density.

2.2 An example is the element sodium. Physical properties: soft, silvery metal melting at 98°C. Chemical properties: reacts vigorously with water and with chlorine (produces NaCl).

2.3 An element: oxygen, O_2. A compound: water, H_2O. A heterogeneous mixture: sand and water. A homogeneous mixture: air.

2.4 a. Two phases: liquid Na and solid Na
b. Phases: liquid water, solid quartz, and solid seashells

2.5 If the material is a compound, all samples will have the same melting point, density, color, and elemental composition. If it is a mixture, most, if not all, of these properties will vary.

2.6 Atomic theory states that matter is composed of small particles called atoms. There are over 100 different kinds of atoms, from which an infinite number of different kinds of compounds can be made. A chemical reaction consists of rearranging the combinations of atoms present in certain kinds of matter to form new kinds of matter.

2.7 Apply the law of multiple proportions: the masses of chlorine for a fixed mass of iron are in the ratio of small whole numbers. Divide each amount of chlorine (1.270 g and 1.904 g) by 1.270 (the lower amount): this gives 1.000 g and 1.499 g, respectively. Convert these to whole numbers by multiplying by the appropriate number (2), giving 2.000 g and 2.998 g. The ratio of these amounts of chlorine is essentially 2:3.

2.8 Oxygen consists of three different <u>isotopes</u>.

2.9 Dalton obtained the relative atomic masses in a compound by measuring the relative masses of elements required to form the compound. From this he deduced the relative <u>atomic</u> masses. To do this, he had to know the formula of the compound.

2.10 The atomic weight of an element is the <u>average</u> atomic mass for the naturally occurring elements. The atomic weight would be different elsewhere in the universe if the percentages of isotopes in the element were different from those on earth.

2.11 The element is tin.

2.12 Characteristic metal properties: lustrous appearance, good conductor of heat and electricity.

2.13 The formula is C_2H_6.

2.14 A molecular formula indicates how many of certain kinds of atoms are in the molecule. A structural formula also shows how the atoms are bonded together.

2.15 The mixture and water molecules have the same number of atoms but the atoms are combined in different ways.

2.16 An ionic binary compound: NaCl; a molecular binary compound: H_2O.

2.17 The two hydrogen atoms and one oxygen atom in water form bonds with a definite spatial arrangement. The calcium and chloride ions in calcium chloride form ions, and exist in a crystal with a regular repeating arrangment of many ions, not just three ions.

2.18 CuCl: copper(I) chloride; $CuCl_2$: copper(II) chloride.
Advantages of the Stock system: more than two different ions of the same metal can be named with the Stock system. In the former (older) system, a new suffix other than "-ic" and "-ous" must be established and/or memorized.

SOLUTIONS TO PRACTICE PROBLEMS

Note on significant figures: The final answer to all mathematical solutions is given first with one nonsignificant figure (last significant figure underlined) and is then rounded to the correct number of digits. Intermediate answers usually also have at least one nonsignificant figure.

2.19 a. Solid b. Liquid c. Gas d. Solid

2.21 a. Physical change b. Physical change c. Chemical change d. Physical change

2.23 Physical change: liquid mercury is cooled to solid mercury.
Chemical changes: solid HgO forms liquid mercury and gaseous O_2, and glowing wood plus O_2 transforms into burning wood.

2.25 a. Physical property
b. Chemical property
c. Physical property
d. Physical property
e. Chemical property

2.27 Physical properties: solid (room temperature); lustrous, blue-black color; crystals vaporize to a gas (without forming a liquid); and the gas is violet colored.
Chemical properties: combines with many metals, and iodine combines with aluminum to form aluminum iodide.

2.29 a. Physical process
b. Chemical reaction
c. Physical process
d. Chemical reaction
e. Physical process

2.31 The fraction of Fe in each sample is calculated as follows:

$$\text{A: } \frac{1.094 \text{ g Fe}}{1.518 \text{ g}} = 0.7207 \qquad \text{B: } \frac{1.449 \text{ g Fe}}{2.056 \text{ g}} = 0.7047 \qquad \text{C: } \frac{1.335 \text{ g Fe}}{1.873 \text{ g}} = 0.7128$$

Each sample has a slightly different composition, so this is a mixture, not a compound.

2.33 a. Solution b. Substance c. Substance d. Heterogeneous mixture

2.35 a. Pure substance; liquid and gas
b. Mixture; only solids
c. Mixture; liquid and solids
d. Pure substance; liquid and solid

2.37 Calculate the ratio of oxygen for 1 g (fixed amount) of nitrogen in both compounds:

A: $\frac{2.755 \text{ g O}}{1.206 \text{ g N}} = \frac{2.28\underline{4}4 \text{ g O}}{1 \text{ g N}}$ B: $\frac{4.714 \text{ g O}}{1.651 \text{ g N}} = \frac{2.85\underline{5}2 \text{ g O}}{1 \text{ g N}}$

$$\frac{\text{g O in B/1 g N}}{\text{g O in A/1 g N}} = \frac{2.85\underline{5}2 \text{ g O}}{2.28\underline{4}4 \text{ g O}} = \frac{1.24\underline{9}8 \text{ g O}}{1 \text{ g O}}$$

B contains 1.25 times as many O atoms as A (there are 5 O's in B for every 4 O's in B).

2.39 Isotope of A: atom C; atom with same mass number: atom B.

2.41 a. Neon
b. Zinc
c. Silver
d. Magnesium

2.43 a. K   b. S   c. Fe   d. Mn

2.45 Divide the mass of N by one third of the mass of hydrogen to find the relative mass of N:

$$\frac{\underline{\text{Atomic mass of N}}}{\text{Atomic mass of H}} = \frac{7.933 \text{ g N}}{1/3 \times 1.712 \text{ g H}} = \frac{13.9\underline{0}1 \text{ g N}}{1 \text{ g H}} = \frac{13.90}{1}$$

2.47 a. C: IVA, period 2; nonmetal
b Po: VIA, period 6; metal
c. Cr: VIB, period 4; metal
d. Mg: IIA, period 3; metal
e. B: IIIA, period 2; metalloid

2.49 a. Tellurium   b. Aluminum

2.51 a. O (oxygen)   b. Na (sodium)   c. Fe (iron)   d. Ce (cerium)

2.53 The solid sulfur consists of all S_8 molecules, which are four times as heavy as S_2 molecules. Hot sulfur is a mixture of S_2 and S_8 just above the boiling point, but at high temperatures is all S_2 molecules. Both hot sulfur and solid sulfur consist of molecules with <u>all</u> S-S bonds.

2.55 $2.05 \times 10^{22} \text{ } N_2O \text{ molec} \times \frac{2 \text{ N atoms}}{1 \text{ } N_2O \text{ molec}} = 4.10 \times 10^{22} \text{ N atoms}$

$$1.00 \text{ g} \times \frac{2.05 \times 10^{22} \text{ } N_2O \text{ molec}}{1.50 \text{ g}} \times \frac{2 \text{ N atoms}}{1 \text{ } N_2O \text{ molec}} = 2.7\underline{3}3 \times 10^{22}$$

$$= 2.73 \times 10^{22} \text{ N atoms}$$

2.57 $3.3 \times 10^{21} \text{ H atoms} \times \frac{1 \text{ } NH_3 \text{ molec}}{3 \text{ H atoms}} = 1.1 \times 10^{21} \text{ } NH_3 \text{ molec}$

2.59 a. N_2H_4 b. H_2O_2 c. C_3H_8O d. PCl_3

2.61 $\frac{2Fe}{1Fe_2(SO_4)_3} \times \frac{1Fe_2(SO_4)_3}{3SO_4^{2-}} \times \frac{1SO_4^{2-}}{4\ O} = \frac{2Fe}{12\ O} = \frac{1Fe}{6\ O}$

The ratio is 2 Fe to 12 O, as written; this simpifies to 1 Fe to 6 O.

2.63 a. $Fe(CN)_3$ b. K_2SO_4 c. Li_3N d. $SrCl_2$

2.65 a. Na_2SO_4: sodium sulfate (group IA forms only +1 cations)
b. CaO: calcium oxide (group IIA forms only +2 cations)
c CuCl: copper(I) chloride (group IB forms +1 and +2 cations)
d. Cr_2O_3: chromium(III) oxide (group VIB forms numerous oxidation states)

2.67 a. Lead(II) dichromate: $PbCr_2O_7$ (dichromate is in Table 2.5)
b. Barium hydrogen carbonate: $Ba(HCO_3)_2$ (the HCO_3^- ion is in Table 2.5)
c. Cesium oxide: Cs_2O (group 1A ions form +1 cations)
d. Iron(II) acetate: $Fe(C_2H_3O_2)_2$ (the acetate ion = -1(Table 2.5); for the sum of charges to be 0, the Fe must be +2)

2.69 a. Molecular b. Ionic c. Molecular d. Ionic

2.71 a. Dinitrogen monoxide
b. Tetraphosphorus dec(a)oxide
c. Arsenic trichloride
d. Dichlorinehept(a)oxide

2.73 a. NBr_3 b. XeO_4 c. OF_2 d. Cl_2O_5

2.75 a. Bromate ion: BrO_3^-
b. Hyponitrite ion: $N_2O_2^{2-}$
c. Thiosulfate ion: $S_2O_3^{2-}$
d. Arsenate ion: AsO_4^{3-}

2.77 $Na_2SO_4 \cdot 10H_2O$ is sodium sulfate decahydrate.

2.79 Iron(II) sulfate heptahydrate is $FeSO_4 \cdot 7H_2O$.

2.81 a. Element = bromine, Br (Table 2.2)
b. Element = phosphorus, P (Table 2.2)
c. Element = gold, Au
d. Element = carbon (graphite), C

2.83 A: Fraction P $= \frac{1.156\ g}{(1.156 + 3.971)\ g} = 0.225\underline{4}7$

B: Fraction P $= \frac{1.542\ g}{(1.542 + 5.297)\ g} = 0.225\underline{4}7$

A and B are the same compound because the fractions of P (and Cl) are the same.

2.85 An ion with 36 electrons has a charge of 1+, the formula is Rb^{+}; an ion with 35 electrons has a charge of 2+, the formula is Rb^{2+}.

2.87 a. Bromine, Br b. Oxygen, O c. Niobium, Nb d. Fluorine, F

2.89 a. Chromium(III) ion b. Chromium(II) ion c. Copper(I) ion d. Copper(II) ion

2.91 All possible ionic compounds: Na_2SO_4, NaCl, $NiSO_4$, and $NiCl_2$.

2.93 a. Tin(II) phosphate
b. Ammonium nitrite
c. Magnesium hydroxide
d. Chromium(II) sulfate

2.95 a. Hg_2Cl_2 [Mercury(I) exists as the polyatomic Hg_2^{2+} ion (Table 2.5)]
b. CuO c. $(NH_4)_2Cr_2O_7$ d. ZnS

2.97 a. Arsenic trichloride
b. Selenium dioxide
c. Dinitrogen trioxide
d. Silicon tetrafluoride

Cumulative-Skills Problems (require skills from Chapters 1 and 2)

2.99 The spheres occupy a diameter of 2 x 1.86 Å = 3.72 Å.

$$\text{The line of Na atoms} = \frac{3.72\ \text{Å}}{1\ \text{Na atom}} \times 2.619 \times 10^{22}\ \text{Na atoms} = 9.7\underline{4}27 \times 10^{22}\ \text{Å}$$

$$9.7\underline{4}27 \times 10^{22}\ \text{Å} \times \frac{1 \times 10^{-10}\ \text{m}}{1\ \text{Å}} \times \frac{1\ \text{mile}}{1.609 \times 10^{3}\ \text{m}} = 6.0\underline{5}5 \times 10^{9} = 6.06 \times 10^{9}\ \text{miles}$$

2.101 $NiSO_4 \cdot 7H_2O(s) \rightarrow NiSO_4 \cdot 6H_2O(s) + H_2O(g)$
[8.753] = [8.192 g + (8.753 - 8.192 = 0.561 g)]

The 8.192 g of $NiSO_4 \cdot 6H_2O$ must contain 6 x 0.561 = 3.366 g H_2O.

Mass of anhydrous $NiSO_4$ = 8.192 g $NiSO_4 \cdot 6H_2O$ - 3.366 g $6H_2O$ = 4.826 g

2.103
$$\text{Mass of O} = 0.6015\ \text{L} \times \frac{1.330\ \text{g O}}{1\ \text{L}} = 0.799\underline{9}95\ \text{g O}$$

$$15.9994\ \text{amu O} \times \frac{3.177\ \text{g X}}{0.799995\ \text{g O}} = 63.5\underline{3}8\ \text{amu X} = 63.54\ \text{amu} = \text{atomic wt of X}$$

X is copper, which has an atomic number of 29.

CHAPTER 3

CHEMICAL REACTIONS: AN INTRODUCTION

SOLUTIONS TO EXERCISES

Note on significant figures: The final answer to calculations is given first with one nonsignificant figure (last significant figure underlined); it is then rounded to the correct number of figures. Intermediate answers have one nonsignificant figure.

3.1 a. Write a "2" in front of $POCl_3$ to balance O first because it occurs in only one reactant and one product. The "$2POCl_3$" then requires a "2" in front of PCl_3 for the final balancing:

$$O_2 + 2PCl_3 \rightarrow 2POCl_3$$

b. Write a "6" in front of N_2O to balance O first because it occurs in only one reactant and one product. The "$6N_2O$" then requires a "6" in front of N_2 for the final balancing:

$$P_4 + 6N_2O \rightarrow P_4O_6 + 6N_2$$

c. Balance S atoms first by writing "$2As_2S_3$" and "$6SO_2$"; ultimately, this will achieve an even number of oxygens on the right for balancing the O_2's on the left. The "2" in front of the As_2S_3 then requires a "2" in front of the As_2O_3. Finally, to balance (6 + 12) O atoms on the right, write a "9" in front of the O_2:

$$2As_2S_3 + 9O_2 \rightarrow 2As_2O_3 + 6SO_2$$

d. Balance H atoms first by writing a "4" in front of H_3PO_4 and a "3" in front of the $Ca(H_2PO_4)_2$ to give 12 H's on each side. This also balances all the other atoms:

$$Ca_3(PO_4)_2 + 4H_3PO_4 \rightarrow 3Ca(H_2PO_4)_2$$

3.2 a. Combustion reaction
b. Combination reaction
c. Decomposition reaction
d. Metathesis reaction
e. Displacement reaction

3.3 a. Weak acid
b. Weak acid
c. Strong acid
d. Strong base

3.4 a. Ionic equation:

$$2H^+(aq) + 2NO_3^-(aq) + Mg(OH)_2(s) \rightarrow 2H_2O(l) + Mg^{2+}(aq) + 2NO_3^-(aq)$$

Net ionic equation:

$$2H^+(aq) + Mg(OH)_2(s) \rightarrow 2H_2O(l) + Mg^{2+}(aq)$$

b. Ionic equation:

$$Pb^{2+}(aq) + 2NO_3^-(aq) + 2Na^+(aq) + SO_4^{2-}(aq) \rightarrow PbSO_4(s) + 2Na^+(aq) + 2NO_3^-(aq)$$

Net ionic equation:

$$Pb^{2+}(aq) + SO_4^{2-}(aq) \rightarrow PbSO_4(s)$$

3.5 Molecular equation showing the formation of insoluble PbI_2:

$$Pb(C_2H_3O_2)_2(aq) + 2NaI(aq) \rightarrow PbI_2(s) + 2NaC_2H_3O_2(aq)$$

Ionic equation:

$$Pb^{2+}(aq) + 2I^-(aq) \rightarrow PbI_2(s)$$

3.6 Molecular equation showing the formation of the water molecule:

$$HCN(aq) + LiOH(aq) \rightarrow H_2O(l) + LiCN(aq)$$

Net Ionic equation (HCN is a weak acid, but K^+ is omitted since KOH is a strong base):

$$HCN(aq) + OH^-(aq) \rightarrow H_2O(l) + CN^-(aq)$$

3.7 Molecular equations:

$$H_2SO_4(aq) + KOH(aq) \rightarrow KHSO_4(aq) + H_2O(l)$$

$$KHSO_4(aq) + KOH(aq) \rightarrow K_2SO_4(aq) + H_2O(l)$$

Net ionic equations:

$$H^+(aq) + OH^-(aq) \rightarrow H_2O(l)$$

$$HSO_4^-(aq) + OH^-(aq) \rightarrow H_2O(l) + SO_4^{2-}(aq)$$

3.8 Molecular equation showing insoluble $CaCO_3$ and formation of CO_2 gas and molecular water:

$$CaCO_3(s) + 2HNO_3(aq) \rightarrow CO_2(g) + H_2O(l) + Ca(NO_3)_2(aq)$$

Net ionic equation:

$$CaCO_3(s) + 2H^+(aq) \rightarrow CO_2(g) + H_2O(l) + Ca^{2+}(aq)$$

ANSWERS TO REVIEW QUESTIONS

3.1 From experiment, we determine the substances that are reactants and products. We also use experiments to determine the formulas of these substances. Using pencil and paper, we can determine the coefficients of these substances for a balanced equation and write the equation.

3.2 In words, one N_2 molecule (= two N atoms) reacts with three H_2 molecules (= six H atoms) to form two NH_3 molecules (= two N atoms and six H atoms). There are the same number of N atoms and H atoms on both sides of the equation.

3.3 The traditional scheme depends on how atoms or groups of atoms are rearranged during the reaction. The types of reactions are: combination reactions, decomposition reactions, displacement reactions, metathesis reactions, and combustion reactions.

3.4 The electrical conductivity of an electrolyte solution refers to the conduction of electrical charge (or electric current) through a solution by means of the movement of the positive and negative ions of the electrolyte. In a sodium chloride solution, for example, the negative chloride ions carry a negative charge to the positively charged electrode, and the positive sodium ions carry a positive charge to the negatively charged electrode.

3.5 Chemical equilibrium is a dynamic equilibrium, because the forward and reverse reactions are occurring at the same rate. An example is the ammonia-water reaction in solution:

$$NH_3(aq) + H_2O(l) \rightleftharpoons NH_4^+(aq) + OH^-(aq)$$

where the rate of NH_3 reacting with H_2O is the same as the rate of NH_4^+ reacting with OH^-.

3.6 A strong electrolyte is a chemical species existing in solution almost entirely as ions (nearly100% ionized). A weak electrolyte is a chemical species existing in water mainly in the molecular form (only a small percentage exists as ions). An example of a strong electrolyte is hydrochloric acid, HCl. An example of a weak electrolyte is ammonia, NH_3.

3.7 An acid is a substance that, when dissolved in water, increases the concentration of the hydronium ion, H_3O^+ (= hydrogen ion, H^+). More generally it is a species (molecule or ion) that donates a proton to another species in a proton-transfer reaction. A base is a substance that, when dissolved in water, increases the concentration of hydroxide ion, OH-. More generally, it is a species that accepts a proton in a proton-transfer reaction. An example of an acid is HCl. An example of a base is NaOH.

3.8 The hydronium ion may be represented as $H_3O^+(aq)$ or as $H^+(aq)$. Ionization of water:

$$H_2O(l) \rightleftharpoons H^+(aq) + OH^-(aq), \text{ or } 2H_2O(l) \rightleftharpoons H_3O^+(aq) + OH^-(aq)$$

3.9 The NH_3 molecule forms basic solutions because it produces hydroxide ions when it reacts with water. The pair of electrons on the nitrogen in ammonia (H_3N:) accepts a proton from water to form the H_3NH^+ ion and the OH^- ion.

3.10 The advantage of using a molecular equation to represent an ionic equation is that it states explicitly what chemical species have been added and what chemical species are obtained as products. It also makes stoichiometric calculations easy to perform. The disadvantages are: 1) the molecular equation does not represent the fact that the reaction actually involves ions, and 2) the molecular equation does not indicate which species exist as ions and which exist as molecular solids or molecular gases.

3.11 A spectator ion is an ion that does not take part in the reaction; it remains unchanged after the reaction is over. In the following ionic reaction, the Na^+ and Cl^- are spectator ions:

$$Na^+(aq) + OH^-(aq) + H^+(aq) + Cl^-(aq) \rightarrow Na^+(aq) + Cl^-(aq) + H_2O(l)$$

3.12 Metathesis reactions are: precipitation, acid-base, and gas-formation reactions.

3.13 An example of a neutralization reaction is:

$$HBr(acid) + KOH(base) \rightarrow KBr(salt) + H_2O(l)$$

3.14 An example of polyprotic acid is carbonic acid, H_2CO_3. The successive neutralization is given by the following molecular equations:

$$H_2CO_3(aq) + NaOH(aq) \rightarrow NaHCO_3(aq) + H_2O(l)$$

$$NaHCO_3(aq) + NaOH(aq) \rightarrow Na_2CO_3(aq) + H_2O(l)$$

3.15 Magnesite must evolve carbon dioxide, the only odorless gas in Table 3.4. Thus magnesite is probably a metal(M) carbonate salt. It is only possible to write a net ionic equation from the observations, using $MCO_3(s)$ to represent the mineral (which may be insoluble in water):

$$MCO_3(s) + 2H^+(aq) \rightarrow CO_2(g) + H_2O(l) + M^{2+}(aq)$$

(In fact, magnesite is $MgCO_3$, so it would be insoluble in water.)

3.16 Precipitation (molecular equation): $FeCl_3(aq) + 3NaOH(aq) \rightarrow Fe(OH)_3(s) + 3NaCl(aq)$

Neutralization (molecular equation): $2HBr(aq) + Ba(OH)_2(aq) \rightarrow BaBr_2(aq) + 2H_2O(l)$

Gas formation (molecular equation): $K_2CO_3(aq) + 2HCl(aq) \rightarrow 2KCl(aq) + CO_2(g)$

3.17 To prepare AgCl and $NaNO_3$, first make solutions of $AgNO_3$ and NaCl by weighing the stoichiometric amounts of both solid compounds. Then mix the two solutions together, forming a precipitate of silver chloride and a solution of soluble sodium nitrate. Filter off the silver chloride and wash it with water to remove the sodium nitrate solution. Then allow it to dry to obtain pure crystalline silver chloride. Finally, take the filtrate containing the sodium nitrate and evaporate it, leaving pure crystalline sodium nitrate.

3.18 To prepare $CaCl_2$, first add the solid $CaCO_3$ to the HCl solution, forming a solution of calcium chloride and carbon dioxide gas. Warm the solution to evolve all of the dissolved carbon dioxide. Then take the solution containing the calcium chloride and evaporate it, leaving pure crystalline $CaCl_2$.

SOLUTIONS TO PRACTICE PROBLEMS

Note on significant figures: The final answer to all mathematical solutions is given first with one nonsignificant figure (last significant figure underlined) and is then rounded to the correct number of digits. Intermediate answers usually also have at least one nonsignificant figure.

3.19 $1\ As_4O_6 \times \frac{6\ O\ atoms}{1\ As_4O_6} + 6\ H_2O \times \frac{1\ O\ atom}{1\ H_2O} = 12\ O\ atoms$

3.21 a. Balance: $Sn + NaOH \rightarrow Na_2SnO_2 + H_2$

If Na is balanced first by writing a "2" in front of NaOH, the entire equation is balanced.

$$Sn + 2NaOH \rightarrow Na_2SnO_2 + H_2$$

b. Balance: $Al + Fe_3O_4 \rightarrow Al_2O_3 + Fe$

First balance O (it appears once on each side) by writing a "3" in front of Fe_3O_4 and a "4" in front of Al_2O_3:

$$Al + 3Fe_3O_4 \rightarrow 4Al_2O_3 + Fe$$

Now balance Al against the 8 Al's on the right and Fe against the 9 Fe's on the left:

$$8Al + 3Fe_3O_4 \rightarrow 4Al_2O_3 + 9Fe$$

c. Balance: $CH_3OH + O_2 \rightarrow CO_2 + H_2O$

First balance H (it appears once on each side) by writing a "2" in front of H_2O:

$$CH_3OH + O_2 \rightarrow CO_2 + 2H_2O$$

To avoid fractional coefficients for O, multiply the equation by two:

$$2CH_3OH + 2O_2 \rightarrow 2CO_2 + 4H_2O$$

Finally balance O by changing "$2O_2$" to "$3O_2$"; this balances the entire equation:

$$2CH_3OH + 3O_2 \rightarrow 2CO_2 + 4H_2O$$

d. Balance: $P_4O_{10} + H_2O \rightarrow H_3PO_4$

First balance P (it appears once on each side) by writing a "4" in front of H_3PO_4:

$$P_4O_{10} + H_2O \rightarrow 4H_3PO_4$$

Finally balance H by writing a "6" in front of H_2O; this balances the entire equation:

$$P_4O_{10} + 6H_2O \rightarrow 4H_3PO_4$$

e. Balance: $PCl_5 + H_2O \rightarrow H_3PO_4 + HCl$

First balance Cl (it appears once on each side) by writing a "5" in front of HCl:

$$PCl_5 + H_2O \rightarrow H_3PO_4 + 5HCl$$

Finally balance H by writing a "4" in front of H_2O; this balances the entire equation:

$$PCl_5 + 4H_2O \rightarrow H_3PO_4 + 5HCl$$

3.23 a. Balance Cl with a "2" in front of HCl; this balances the entire equation.

$$SbCl_5 + H_2O \rightarrow SbOCl_3 + 2HCl$$

b. Balance O with a "2" in front of MgO:

$$Mg + SiO_2 \rightarrow 2MgO + Si$$

Finally balance Mg with a "2" in front of Mg; this balance the entire equation.

$$2Mg + SiO_2 \rightarrow 2MgO + Si$$

c. Balance Cl with a "2" in front of NaCl; this balances the entire equation.

$$CaCl_2 + Na_2CO_3 \rightarrow CaCO_3 + 2NaCl$$

d. Tentatively balance C with a "6" in front of CO_2:

$$C_6H_6 + O_2 \rightarrow 6CO_2 + H_2O$$

Next balance H; use "6" (not 3) H_2O and double the C_6H_6 and CO_2 coefficients to avoid an odd number of O's.

$$2C_6H_6 + O_2 \rightarrow 12CO_2 + 6H_2O$$

Finally balance O with a "15" in front of O_2 to balance the entire equation:

$$2C_6H_6 + 15O_2 \rightarrow 12CO_2 + 6H_2O$$

e. Balance Al by writing a "2" in front of $Al(OH)_3$:

$$Al_2S_3 + H_2O \rightarrow 2Al(OH)_3 + H_2S$$

Next balance S by writing a "3" in front of H_2S:

$$Al_2S_3 + H_2O \rightarrow 2Al(OH)_3 + 3H_2S$$

Finally balance O and H by writing a "6" in front of H_2O to balance the entire equation.

$$Al_2S_3 + 6H_2O \rightarrow 2Al(OH)_3 + 3H_2S$$

3.25 Balance: $Ca_3(PO_4)_2(s) + H_2SO_4(aq) \rightarrow CaSO_4(s) + H_3PO_4(aq)$

Balance Ca first with a "3" in front of the $CaSO_4$:

$$Ca_3(PO_4)_2 + H_2SO_4 \rightarrow 3CaSO_4 + H_3PO_4$$

Next balance the P with a "2" in front of H_3PO_4:

$$Ca_3(PO_4)_2 + H_2SO_4 \rightarrow 3CaSO_4 + 2H_3PO_4$$

Finally balance the S with a "3" in front of H_2SO_4; this balances the equation.

$$Ca_3(PO_4)_2(s) + 3H_2SO_4(aq) \rightarrow 3CaSO_4(s) + 2H_3PO_4(aq)$$

3.27 Balance: $NH_4Cl(aq) + Ba(OH)_2(aq) \rightarrow NH_3(g) + BaCl_2(aq) + H_2O(l)$

Balance O first with a "2" in front of H_2O:

$$NH_4Cl + Ba(OH)_2 \rightarrow NH_3 + BaCl_2 + 2H_2O(l)$$

Balance H with a "2" in front of NH_4Cl and a "2" in front of NH_3; this balances the equation.

$$2NH_4Cl(aq) + Ba(OH)_2(aq) \xrightarrow{\Delta} 2NH_3(g) + BaCl_2(aq) + 2H_2O(l)$$

3.29 a. Displacement
b. Decomposition
c. Combustion
d. Metathesis
e. Combination

3.31 Decomposition reaction: $C_{12}H_{22}O_{11}(s) \xrightarrow{\Delta} 12C(s) + 11H_2O(g)$

3.33 Metathesis reaction: $Mg(OH)_2(s) + 2HCl(aq) \rightarrow MgCl_2(aq) + 2H_2O(l)$

3.35 a. Weak acid b. Strong base c. Strong acid d. Weak acid

3.37 a. $HF(aq) + OH^-(aq) \rightarrow F^-(aq) + H_2O(l)$

b. $Ag^+(aq) + Br^-(aq) \rightarrow AgBr(s)$

c. $S^{2-}(aq) + 2H^+(aq) \rightarrow H_2S(g)$

d. $OH^-(aq) + NH_4^+(aq) \rightarrow NH_3(g) + H_2O(l)$

3.39 Molecular equation: $Pb(NO_3)_2(aq) + Na_2SO_4(aq) \rightarrow PbSO_4(s) + 2NaNO_3(aq)$

Net ionic equation: $Pb^{2+}(aq) + SO_4^{2-}(aq) \rightarrow PbSO_4(s)$

3.41 a. AgBr is insoluble (silver halides are insoluble).

b. $Pb(NO_3)_2$ is soluble (all nitrates are soluble).

c. $SrSO_4$ is insoluble (in Group IIA, the sulfates of Ba and Sr are insoluble).

d. Na_2CO_3 is soluble (all sodium salts are soluble).

3.43 a. $FeSO_4(aq) + NaCl(aq) \rightarrow NR$

b. $Na_2CO_3(aq) + MgBr_2(aq) \rightarrow MgCO_3(s) + 2NaBr(aq)$

$CO_3^{2-}(aq) + Mg^{2+}(aq) \rightarrow MgCO_3(s)$

c. $MgSO_4(aq) + 2NaOH(aq) \rightarrow Mg(OH)_2(s) + Na_2SO_4(aq)$

$Mg^{2+}(aq) + 2OH^-(aq) \rightarrow Mg(OH)_2(s)$

d. $NiCl_2(aq) + NaBr(aq) \rightarrow NR$

3.45 a. $Ba(NO_3)_2(aq) + LiSO_4(aq) \rightarrow BaSO_4(s) + 2LiNO_3(aq)$

$Ba^{2+}(aq) + SO_4^{2-}(aq) \rightarrow BaSO_4(s)$

b. $Ca(NO_3)_2 + NaBr(aq) \rightarrow NR$

c. $Al_2(SO_4)_3 + 6NaOH(aq) \rightarrow 2Al(OH)_3(s) + 3Na_2SO_4(aq)$

$2Al^{3+}(aq) + 6OH^-(aq) \rightarrow 2Al(OH)_3(s)$

d. $3CaBr_2(aq) + 2Na_3PO_4(aq) \rightarrow Ca_3(PO_4)_2(s) + 6NaBr(aq)$

$3Ca^{2+}(aq) + 2PO_4^{3-}(aq) \rightarrow Ca_3(PO_4)_2(s)$

3.47 a. $NaOH(aq) + HNO_3(aq) \rightarrow H_2O(l) + NaNO_3(aq)$

$H^+(aq) + OH^-(aq) \rightarrow H_2O(l)$

b. $2HCl(aq) + Ba(OH)_2(aq) \rightarrow 2H_2O(l) + BaCl_2(aq)$

$H^+(aq) + OH^-(aq) \rightarrow H_2O(l)$

c. $2HC_2H_3O_2(aq) + Ca(OH)_2(s) \rightarrow 2H_2O(l) + Ca(C_2H_3O_2)_2(aq)$

$2HC_2H_3O_2(aq) + Ca(OH)_2(s) \rightarrow 2H_2O(l) + Ca^{2+}(aq) + 2C_2H_3O_2^-(aq)$

d. $HNO_3(aq) + NH_3(aq) \rightarrow NH_4NO_3(aq)$

$H^+(aq) + NH_3(aq) \rightarrow NH_4^+(aq)$

3.49 a. $2HBr(aq) + Ca(OH)_2(aq) \rightarrow 2H_2O(l) + CaBr_2(aq)$

$H^+(aq) + OH^-(aq) \rightarrow H_2O(l)$

b. $3HNO_3(aq) + Al(OH)_3(s) \rightarrow 3H_2O(l) + Al(NO_3)_3(aq)$

$3H^+(aq) + Al(OH)_3(s) \rightarrow 3H_2O(l) + Al^{3+}(aq)$

c. $2HCN(aq) + Ca(OH)_2(aq) \rightarrow 2H_2O(l) + Ca(CN)_2(aq)$

$HCN(aq) + OH^-(aq) \rightarrow H_2O(l) + CN^-(aq)$

d. $HCN(aq) + LiOH(aq) \rightarrow H_2O(l) + LiCN(aq)$

$HCN(aq) + OH^-(aq) \rightarrow H_2O(l) + CN^-(aq)$

3.51 a. $H_3PO_4(aq) + 2KOH(aq) \rightarrow 2H_2O(l) + K_2HPO_4(aq)$

$H_3PO_4(aq) + 2OH^-(aq) \rightarrow 2H_2O(l) + HPO_4^{2-}(aq)$

b. $3H_2SO_4(aq) + 2Al(OH)_3(s) \rightarrow 6H_2O(l) + Al_2(SO_4)_3(aq)$

$3H^+(aq) + Al(OH)_3(s) \rightarrow 3H_2O(l) + Al^{3+}(aq)$

c. $2HC_2H_3O_2(aq) + Ca(OH)_2(aq) \rightarrow 2H_2O(l) + Ca(C_2H_3O_2)_2(aq)$

$HC_2H_3O_2(aq) + OH^-(aq) \rightarrow H_2O(l) + C_2H_3O_2^-(aq)$

d. $H_2SO_3(aq) + NaOH(aq) \rightarrow H_2O(l) + NaHSO_3(aq)$

$H_2SO_3(aq) + OH^-(aq) \rightarrow HSO_3^-(aq) + H_2O(l)$

3.53 Molecular equations: $2H_2SO_3(aq) + Ca(OH)_2(aq) \rightarrow 2H_2O(l) + Ca(HSO_3)_2(aq)$

$Ca(HSO_3)_2(aq) + Ca(OH)_2(aq) \rightarrow 2H_2O(l) + 2CaSO_3(s)$

Ionic equations: $2H_2SO_3(aq) + 2OH^-(aq) \rightarrow 2H_2O(l) + 2HSO_3^-(aq)$

$Ca^{2+}(aq) + HSO_3^-(aq) + OH^-(aq) \rightarrow CaSO_3(s) + H_2O(l)$

3.55 a. Molecular equation: $2HBr(aq) + CaS(aq) \rightarrow H_2S(g) + CaBr_2(aq)$

Ionic equation: $2H^+(aq) + S^{2-}(aq) \rightarrow H_2S(g)$

b. Molecular equation: $2HNO_3(aq) + MgCO_3(s) \rightarrow CO_2(g) + H_2O(l) + Mg(NO_3)_2(aq)$

Ionic equation: $2H^+(aq) + MgCO_3(s) \rightarrow CO_2(g) + H_2O(l) + Mg^{2+}(aq)$

c. Molecular equation: $H_2SO_4(aq) + K_2SO_3(aq) \rightarrow SO_2(g) + H_2O(l) + K_2SO_4(aq)$

Ionic equation: $2H^+(aq) + SO_3^{2-}(aq) \rightarrow SO_2(g) + H_2O(l)$

3.57 Molecular equation: $FeS(s) + 2HCl(aq) \rightarrow H_2S(g) + FeCl_2(aq)$

Ionic equation: $FeS(s) + 2H^+(aq) \rightarrow H_2S(g) + Fe^{2+}(aq)$

3.59 a. Balance the C and H first:

$C_2H_6 + O_2 \rightarrow 2CO_2 + 3H_2O$

Avoid a fractional coefficient for O on the left by doubling all coefficients except O_2's, and then balance the O's:

$2C_2H_6 + 7O_2 \rightarrow 4CO_2 + 6H_2O$

b. Balance the P first:

$P_4O_6 + H_2O \rightarrow 4H_3PO_3$

Then balance the O (or H), which also gives the H (or O) balance:

$P_4O_6 + 6H_2O \rightarrow 4H_3PO_3$

c. Balancing the O first is the simplest approach. (Starting with K and Cl and then O will cause the initial coefficient for $KClO_3$ to be changed in balancing O last.)

$4KClO_3 \rightarrow 3KClO_4 + KCl$

d. Balance the N first:

$$(NH_4)_2SO_4 + NaOH \rightarrow 2NH_3 + H_2O + Na_2SO_4$$

Then balance the Na, followed by O; this also balances the H:

$$(NH_4)_2SO_4 + 2NaOH \rightarrow 2NH_3 + 2H_2O + Na_2SO_4$$

e. Balance the N first:

$$2NBr_3 + NaOH \rightarrow N_2 + NaBr + HOBr$$

Note that NaOH and HOBr each have 1 O, and that NaOH and NaBr each have 1 Na; thus the coefficients of all three must be equal; from "$2NBr_3$" this coefficient must = 6Br/2 = 3:

$$2NBr_3 + 3NaOH \rightarrow N_2 + 3NaBr + 3HOBr$$

3.61 When metallic iron is added to a solution of copper(II) sulfate, insoluble copper metal and a solution of iron(II) sulfate are formed.

3.63 When ammonia gas is burned in oxygen in the presence of a plantinum catalyst, the balanced equation for the formation of nitric oxide gas and water vapor is:

$$4NH_3(g) + 5O_2(g) \xrightarrow{Pt} 4NO(g) + 6H_2O(g)$$

3.65 a. The equation for heating ammonium dichromate is:

$$(NH_4)_2Cr_2O_7(s) \xrightarrow{\Delta} Cr_2O_3(s) + N_2(g) + 4H_2O(g)$$

b. The equation for heating ammonium nitrite is:

$$NH_4NO_2(aq) \xrightarrow{\Delta} N_2(g) + 2H_2O(g)$$

c. The equation for the reaction of potassium chromate and lead nitrate is:

$$K_2CrO_4(aq) + Pb(NO_3)_2(aq) \rightarrow PbCrO_4(s) + 2KNO_3(aq)$$

d. The equation for the reaction of hydrogen chloride gas and gaseous ammonia is:

$$NH_3(g) + HCl(g) \rightarrow NH_4Cl(s)$$

e. The equation for the reaction of aluminum and sulfuric acid is:

$$2Al(s) + 3H_2SO_4(aq) \rightarrow Al_2(SO_4)_3(aq) + 3H_2(g)$$

3.67 a. Decomposition
b. Decompositon
c. Metathesis
d. Combination
e. Displacement

3.69 For the reaction of magnesium metal with hydrobromic acid, the equations are:

Molecular equation: $Mg(s) + 2HBr(aq) \rightarrow H_2(g) + MgBr_2(aq)$

Ionic equation: $Mg(s) + 2H^+(aq) \rightarrow H_2(g) + Mg^{2+}(aq)$

3.71 For the reaction of nickel(II) sulfate and lithium hydroxide, the equations are:

Molecular equation: $NiSO_4(aq) + 2LiOH(aq) \rightarrow Ni(OH)_2(s) + LiSO_4(aq)$

Ionic equation: $Ni^{2+}(aq) + 2OH^-(aq) \rightarrow Ni(OH)_2(s)$

3.73 a. Molecular equation: $HCN(aq) + LiOH(aq) \rightarrow H_2O(l) + LiCN(aq)$

Ionic equation: $HCN(aq) + OH^-(aq) \rightarrow H_2O(l) + CN^-(aq)$

b. Molecular equation: $2HNO_3(aq) + Li_2CO_3(aq) \rightarrow CO_2(g) + H_2O(l) + 2LiNO_3(aq)$

Ionic equation: $2H^+(aq) + CO_3^{2-}(aq) \rightarrow CO_2(g) + H_2O(l)$

c. Molecular equation: $AgNO_3(aq) + LiCl(aq) \rightarrow AgCl(s) + LiNO_3(aq)$

Ionic equation: $Ag^+(aq) + Cl^-(aq) \rightarrow AgC(s)$

d. Molecular equation: $LiCl(aq) + MgSO_4(aq) \rightarrow$ NR [Li_2SO_4 & $MgCl_2$ are soluble.]

3.75 a. Molecular equation: $2HC_2H_3O_2(aq) + Sr(OH)_2(aq) \rightarrow H_2O(l) + Sr(C_2H_3O_2)_2(aq)$

Ionic equation: $HC_2H_3O_2(aq) + OH^-(aq) \rightarrow H_2O(l) + C_2H_3O_2^-(aq)$

b. Molecular equation: $NH_4I(aq) + CsCl(aq) \rightarrow$ NR

Ionic equation: none (NR)

c. Molecular equation: $NaNO_3(aq) + CsCl(aq) \rightarrow$ NR [NaCl & $CsNO_3$ are soluble]

d. Molecular equation: $AgNO_3(aq) + NH_4I(aq) \rightarrow AgI(s) + NH_4NO_3(aq)$

Ionic equation: $Ag^+(aq) + I^-(aq) \rightarrow AgI(s)$

3.77 For each preparation, the compound to be prepared is given first, followed by the compound from which it is to be prepared. Then the method of preparation is given, followed by the molecular equation for the preparation reaction. Steps such as evaporation, etc. are not given in the molecular equation.

a. To prepare $CuCl_2$ from $CuSO_4$, add a solution of $BaCl_2$ to a solution of the $CuSO_4$, precipitating $BaSO_4$. The $BaSO_4$ can be filtered off, leaving aqueous $CuCl_2$, which can be obtained in solid form by evaporation of the solution. Molecular equation:

$$CuSO_4(aq) + BaCl_2(aq) \rightarrow BaSO_4(s) + CuCl_2(aq)$$

b. To prepare $Ca(C_2H_3O_2)_2$ from $CaCO_3$, add a solution of acetic acid, $HC_2H_3O_2$, to the solid $CaCO_3$, forming CO_2, H_2O, and aqueous $Ca(C_2H_3O_2)_2$. The aqueous $Ca(C_2H_3O_2)_2$ can be converted to the solid form by evaporation of the solution, which also removes the CO_2 and H_2O products. Molecular equation:

$$CaCO_3(s) + 2HC_2H_3O_2(aq) \rightarrow Ca(C_2H_3O_2)_2(aq) + CO_2 + H_2O(l)$$

c. To prepare $NaNO_3$ from Na_2SO_3, add a solution of nitric acid, HNO_3, to the solid Na_2SO_3, forming SO_2, H_2O, and aqueous $NaNO_3$. The aqueous $NaNO_3$ can be converted to the solid by evaporation of the solution, which also removes the SO_2 and H_2O products. Molecular equation:

$$Na_2SO_3(s) + 2HNO_3(aq) \rightarrow 2NaNO_3(aq) + SO_2(g) + H_2O(l)$$

d. To prepare $MgCl_2$ from $Mg(OH)_2$, add a solution of hydrochloric acid (HCl) to the solid $Mg(OH)_2$, forming H_2O and aqueous $MgCl_2$. The aqueous $MgCl_2$ can be converted to the solid form by evaporation of the solution. Molecular equation:

$$Mg(OH)_2(s) + 2HCl(aq) \rightarrow MgCl_2(aq) + 2H_2O(l)$$

Cumulative-Skills Problems (require skills from chapters 1, 2, and 3)

3.79 Lead(II) nitrate reacts with cesium sulfate in a precipitation reaction. For this reaction, the formulas are listed first, followed by the molecular and net ionic equations, the names of the products, and the molecular equation for another reaction giving the same precipitate.

Lead(II) nitrate is $Pb(NO_3)_2$, and cesium sulfate is Cs_2SO_4.

Molecular equation: $Pb(NO_3)_2(aq) + Cs_2SO_4(aq) \rightarrow PbSO_4(s) + 2CsNO_3(aq)$

Net ionic equation: $Pb^{2+}(aq) + SO_4^{2-}(aq) \rightarrow PbSO_4(s)$

$PbSO_4$ is lead(II) sulfate, and $CsNO_3$ is cesium nitrate.

Molecular equation: $Pb(NO_2)_2(aq) + Na_2SO_4(aq) \rightarrow PbSO_4(s) + 2NaNO_2(aq)$

3.81 Net ionic equation: $2Br^-(aq) + Cl_2(g) \rightarrow 2Cl^-(aq) + Br_2(l)$

Molecular equation:	$CaBr_2(aq)$ +	$Cl_2(g)$ →	$CaCl_2(aq)$ +	$Br_2(l)$
Masses:	40.0 g	14.2 g	22.2 g	(40.0 + 14.2 - 22.2) g

Combining the three known masses gives the unknown mass of Br_2. Now use a ratio of the known masses of $CaBr_2$ to Br_2 to convert pounds of Br_2 to grams of $CaBr_2$:

$$10.0 \text{ lb } Br_2 \times \frac{40.0 \text{ g } CaBr_2}{(40.0 + 14.2 - 22.2) \text{ g } Br_2} \times \frac{453.6 \text{ g}}{1 \text{ lb}} = 56\underline{7}0 = 5.67 \times 10^3 \text{ g } CaBr_2$$

CHAPTER 4

CALCULATIONS WITH CHEMICAL FORMULAS AND EQUATIONS

SOLUTIONS TO EXERCISES

Note on significant figures: The final answer to all mathematical solutions is given first with one nonsignificant figure (last significant figure underlined) and is then rounded to the correct number of figures. Intermediate answers usually also have at least one nonsignificant figure.

4.1 a. NO_2:

1 x AW OF N			=	14.0067 amu
2 x AW of O	=	2 x 15.9994 amu	=	31.9988 amu
MW of NO_2			=	46.0055 = 46.0 amu (3 sf)

b. $C_6H_{12}O_6$

6 x AW of C	=	6 x 12.011	=	72.066 amu
12 x AW of H	=	12 x 1.0079	=	12.0948 amu
6 x AW of O	=	6 x 15.9994	=	95.9964 amu
MW of $C_6H_{12}O_6$			=	180.1572 amu = 180 amu (3 sf)

c. NaOH

1 x AW of Na	=	22.98977 amu
1 x AW of O	=	15.9994 amu
1 x AW of H	=	1.0079 amu
MW of NaOH	=	39.9971 amu = 40.0 amu (3 sf)

d. $Mg(OH)_2$

1 x AW of Mg			=	24.305 amu
2 x AW of O	=	2 x 15.9994	=	31.9988 amu
2 x AW of H	=	2 x 1.0079	=	2.0158 amu
MW of $Mg(OH)_2$			=	58.3196 amu = 58.3 amu (3 sf)

4.2 a. The atomic weight of Ca = 40.08 amu; thus the molar mass = 40.08 g/mol, and 1 mol of Ca = 6.02×10^{23} Ca atoms.

$$\text{Mass of 1 Ca} = \frac{40.08 \text{ g}}{1 \text{ mol}} \times \frac{1 \text{ mol}}{6.02 \times 10^{23} \text{ atom}} = 6.6\underline{5}7 \times 10^{-23} = \frac{6.66 \times 10^{-23} \text{ g}}{\text{atom}}$$

b. The molecular weight of C_2H_5OH, or C_2H_6O, = (2 x 12.01) + (6 x 1.01) + 16.00 = 46.08. Its molar mass = 46.08 g/mol, and 1 mol = 6.02×10^{23} molecules of C_2H_6O.

$$\text{Mass of 1 } C_2H_6O = \frac{46.08 \text{ g}}{1 \text{ mol}} \times \frac{1 \text{ mol}}{6.02 \times 10^{23} \text{ molec}} = 7.6\underline{5}4 \times 10^{-23} = \frac{7.65 \times 10^{-23} \text{ g}}{\text{molec}}$$

4.3 The molar mass of H_2O_2 is 34.0 g/mol. Therefore,

$$0.909 \text{ mol } H_2O_2 \times \frac{34.0 \text{ g } H_2O_2}{1 \text{ mol } H_2O_2} = 30.\underline{9}06 = 30.9 \text{ g } H_2O_2$$

4.4 The molar mass of HNO_3 is 63.0 g/mol. Therefore,

$$28.5 \text{ g } HNO_3 \times \frac{1 \text{ mol } HNO_3}{63.0 \text{ g } HNO_3} = 0.45\underline{2}3 = 0.452 \text{ mol } HNO_3$$

4.5 Convert the mass of HCN from milligrams to grams. Then convert grams of HCN to moles of HCN. Finally convert moles of HCN to the number of HCN molecules.

$$56 \text{ mg HCN} \times \frac{1 \text{ g}}{1000 \text{ mg}} \times \frac{1 \text{ mol HCN}}{27.0 \text{ g HCN}} \times \frac{6.02 \times 10^{23} \text{ HCN molec}}{1 \text{ mol HCN}}$$

$$= 1.\underline{2}4 \times 10^{21} = 1.2 \times 10^{21} \text{ HCN molec}$$

4.6 The molecular weight of NH_4NO_3 = 80.04; thus its molar mass = 80.04 g/mol. Hence:

$$\%\text{N} = \frac{28.0 \text{ g}}{80.0 \text{ g}} \times 100\% = 35.\underline{0}0 = 35.0\%$$

$$\%\text{H} = \frac{4.0 \text{ g}}{80.0 \text{ g}} \times 100\% = 5.\underline{0}0 = 5.0\%$$

$$\%\text{O} = \frac{48.0 \text{ g}}{80.0 \text{ g}} \times 100\% = 60.\underline{0}0 = 60.0\%$$

4.7 From the previous exercise, NH_4NO_3 is 35.0% N (fraction N = 0.350), so the mass of N in 48.5 g of NH_4NO_3 is:

$$48.5 \text{ g } NH_4NO_3 \times (0.350 \text{ g N/1 g } NH_4NO_3) = 16.\underline{9}75 = 17.0 \text{ g N}$$

4.8 First convert the mass of CO_2 to moles of CO_2. Next convert this to moles of C (1 mol of CO_2 is equivalent to 1 mol C). Finally convert to mass of carbon, changing mg to g first:

$$5.80 \times 10^{-3} \text{ g } CO_2 \times \frac{1 \text{ mol } CO_2}{44.0 \text{ g } CO_2} \times \frac{1 \text{ mol C}}{1 \text{ mol } CO_2} \times \frac{12.0 \text{ g C}}{1 \text{ mol C}} = 1.5\underline{8}3 \times 10^{-3} \text{ g C}$$

Do the same series of calculations for water, noting that 1 mol H_2O contains 2 mol H.

$$1.58 \times 10^{-3} \text{ g } H_2O \times \frac{1 \text{ mol } H_2O}{18.02 \text{ g } H_2O} \times \frac{2 \text{ mol H}}{1 \text{ mol } H_2O} \times \frac{1.008 \text{ g H}}{1 \text{ mol H}} = 1.7\underline{6}7 \times 10^{-4} \text{ g H}$$

The mass percentages of C and H can be calculated using the masses from above:

$$\%\,C = \frac{1.583\ mg}{3.87\ mg} \times 100\% = 40.\underline{9}0 = 40.9\%\ C$$

$$\%\,H = \frac{0.1767\ mg}{3.87\ mg} \times 100\% = 4.5\underline{6}58 = 4.57\%\ H$$

The mass percentage of O can be determined by subtracting the sum of the above percentages from 100%:

$$\%\,O = 100.000\% - (40.90 + 4.5658) = 54.\underline{5}342 = 54.5\%\ O$$

4.9 Convert the masses to moles, which are proportional to subscripts in the empirical formula:

$$33.4\ g\ S \times \frac{1\ mol\ S}{32.06\ g\ S} = 1.0\underline{4}18\ mol\ S$$

$$(83.5 - 33.4)\ g\ O \times \frac{1\ mol\ O}{16.00\ g\ O} = 3.1\underline{3}12\ mol\ O$$

Next obtain the smallest integers from the moles by dividing each by the smallest number of moles:

$$\text{For O: } \frac{3.1312}{1.0418} = 3.01 \qquad \text{For S: } \frac{1.0418}{1.0418} = 1.00$$

The empirical formula is SO_3.

4.10 For a 100.0 g sample of benzoic acid, 68.8 g are C, 5.0 g are H, and 26.2 g are O. Using the molar masses, convert these masses to moles:

$$68.8\ g\ C = 68.8\ g\ C \times \frac{1\ mol\ C}{12.01\ g\ C} = 5.7\underline{2}9\ mol\ C$$

$$5.0\ g\ H = 5.0\ g\ H \times \frac{1\ mol\ H}{1.008\ g\ H} = 4.\underline{9}6\ mol\ H$$

$$26.2\ g\ O = 26.2\ g\ O \times \frac{1\ mol\ O}{16.00\ g\ O} = 1.6\underline{3}8\ mol\ O$$

These numbers are in the same ratio as the subscripts in the empirical formula. They must be changed to integers. First divide each one by the smallest mol number:

$$\text{For C: } \frac{5.729}{1.638} = 3.497 \qquad \text{For H: } \frac{4.96}{1.638} = 3.03 \qquad \text{For O: } \frac{1.638}{1.638} = 1.000$$

Rounding off, we obtain $C_{3.5}H_{3.0}O_{1.0}$. Multiplying the numbers by 2 gives whole numbers for an empirical formula of $C_7H_6O_2$.

4.11 For a 100.0 g sample of acetaldehye, 54.5 g are C, 9.2 g are H, and 36.3 g are O. Using the molar masses, convert these masses to moles:

$$54.5 \text{ g C} = 54.5 \text{ g C} \times \frac{1 \text{ mol C}}{12.01 \text{ g C}} = 4.5\underline{3}7 \text{ mol C}$$

$$9.2 \text{ g H} = 9.2 \text{ g H} \times \frac{1 \text{ mol H}}{1.008 \text{ g H}} = 9.\underline{1}2 \text{ mol H}$$

$$36.3 \text{ g O} = 36.3 \text{ g O} \times \frac{1 \text{ mol O}}{16.00 \text{ g O}} = 2.2\underline{6}8 \text{ mol O}$$

These numbers are in the same ratio as the subscripts in the empirical formula. They must be changed to integers. First divide each one by the smallest mol number:

$$\text{For C: } \frac{4.537}{2.268} = 2.000 \quad \text{For H: } \frac{9.12}{2.268} = 4.02 \quad \text{For O: } \frac{2.268}{2.268} = 1.000$$

Rounding off, we obtain C_2H_4O, the empirical formula.

4.12

H_2	+	Cl_2	→	$2HCl$
1 molec(mol) H_2	+	1 molec(mol) Cl_2	→	2 molecs (mol) HCl (molec, mole interp.)
2.018 g H_2	+	70.9 g Cl_2	→	2 x 36.5 g HCl (mass interp.)

4.13 Equation: $Na + H_2O \rightarrow 1/2H_2 + NaOH$, or $2Na + H_2O \rightarrow H_2 + 2NaOH$

From the above equation, 1 mol of Na corresponds to 1/2 mol of H_2, or 2 mol of Na corresponds to 1 mol of H_2. Therefore:

$$7.81 \text{ g } H_2 \times \frac{1 \text{ mol } H_2}{2.016 \text{ g } H_2} \times \frac{2 \text{ mol Na}}{1 \text{ mol } H_2} \times \frac{22.99 \text{ g Na}}{1 \text{ mol Na}} = 17\underline{8}.1 = 178 \text{ g Na}$$

4.14 Balanced equation: $2ZnS + 3O_2 \rightarrow 2ZnO + 2SO_2$

Convert grams of ZnS to moles of ZnS. Then determine the relationship between ZnS and O_2 (2ZnS is equivalent to $3O_2$). Finally, convert to mass O_2.

$$5.00 \times 10^3 \text{g ZnS} \times \frac{1 \text{ mol ZnS}}{97.44 \text{ g ZnS}} \times \frac{3 \text{ mol } O_2}{2 \text{ mol ZnS}} \times \frac{32.00 \text{ g } O_2}{1 \text{ mol } O_2} \times \frac{1 \text{ kg}}{10^3 \text{ g}}$$

$$= 2.4\underline{6}3 = 2.46 \text{ kg } O_2$$

4.15 Balanced equation: $2HgO \rightarrow 2Hg + O_2$

Convert the mass of O_2 to mol of O_2. Using the fact that 1 mol of O_2 is equivalent to 2 mol of Hg, determine the number of mol of Hg, and convert to mass of Hg.

$$6.47 \text{ g } O_2 \times \frac{1 \text{ mol } O_2}{32.00 \text{ g } O_2} \times \frac{2 \text{ mol Hg}}{1 \text{ mol } O_2} \times \frac{200.59 \text{ g Hg}}{1 \text{ mol Hg}} = 81.\underline{1}1 = 81.1 \text{ g Hg}$$

4.16 First determine the limiting reactant by calculating the moles of $AlCl_3$ that would be obtained if Al and HCl were totally consumed:

$$0.15 \text{ mol Al} \times \frac{2 \text{ mol } AlCl_3}{2 \text{ mol Al}} = 0.1\underline{5}0 \text{ mol } AlCl_3$$

$$0.35 \text{ mol HCl} \times \frac{2 \text{ mol } AlCl_3}{6 \text{ mol HCl}} = 0.1\underline{1}66 \text{ mol } AlCl_3$$

Since the HCl produces the smaller amount of $AlCl_3$, the reaction will stop when it is totally consumed but before the Al is consumed. The limiting reagent is therefore the HCl. The amount of $AlCl_3$ produced must be $0.1\underline{1}66$, or 0.12 mol.

4.17 First, determine the limiting reactant by calculating the moles of ZnS produced by totally consuming Zn and S_8:

$$7.36 \text{ g Zn} \times \frac{1 \text{ mol Zn}}{65.38 \text{ g Zn}} \times \frac{8 \text{ mol ZnS}}{8 \text{ mol Zn}} = 0.11\underline{2}57 \text{ mol ZnS}$$

$$6.45 \text{ g } S_8 \times \frac{1 \text{ mol } S_8}{256.5 \text{ g } S_8} \times \frac{8 \text{ mol ZnS}}{1 \text{ mol } S_8} = 0.20\underline{1}1 \text{ mol ZnS}$$

The reaction will stop when Zn is totally consumed; S_8 is in excess and not all of it is converted to ZnS. The limiting reactant is therefore the Zn. Now convert the moles of ZnS obtained from the Zn to grams of ZnS:

$$0.11257 \text{ mol Zn} \times \frac{97.44 \text{ g ZnS}}{1 \text{ mol ZnS}} = 10.\underline{9}6 = 11.0 \text{ g ZnS}$$

4.18 First write the balanced equation:

$$CH_3OH + CO \rightarrow HC_2H_3O_2$$

Convert grams of each reactant to moles of acetic acid:

$$15.0 \text{ g } CH_3OH \times \frac{1 \text{ mol } CH_3OH}{32.03 \text{ g } CH_3OH} \times \frac{1 \text{ mol } HC_2H_3O_2}{1 \text{ mol } CH_3OH} = 0.46\underline{8}3 \text{ mol } HC_2H_3O_2$$

$$10.0 \text{ g CO} \times \frac{1 \text{ mol CO}}{28.0 \text{ g CO}} \times \frac{1 \text{ mol } HC_2H_3O_2}{1 \text{ mol } CH_3OH} = 0.35\underline{7}1 \text{ mol } HC_2H_3O_2$$

Thus CO is the limiting reactant, and 0.03571 mol $HC_2H_3O_2$ are obtained. The mass of product is:

$$0.3571 \text{ mol } HC_2H_3O_2 \times \frac{60.03 \text{ g } HC_2H_3O_2}{1 \text{ mol } HC_2H_3O_2} = 21.\underline{4}4 \text{ g } HC_2H_3O_2$$

The percentage yield is:

$$\frac{19.1 \text{ g actual yield}}{21.44 \text{ g theo yield}} \times 100\% = 89.\underline{0}8 = 89.1\%$$

4.19 Convert mass of NaCl to moles of NaCl. Then divide moles of solute by liters of solution.

$$0.0678 \text{ g NaCl} \times \frac{1 \text{ mol NaCl}}{58.4 \text{ g NaCl}} = 1.1\underline{6}1 \times 10^{-3} \text{ mol NaCl}$$

$$25.0 \text{ mL} \times \frac{1 \text{ L}}{1 \times 10^3 \text{ mL}} = 0.0250 \text{ L}$$

$$\text{Molarity} = \frac{1.161 \times 10^{-3} \text{ mol NaCl}}{0.0250 \text{ L soln}} = 0.046\underline{4}4 = 0.0464 \text{ M}$$

4.20 Convert grams of NaCl to moles NaCl to volume of NaCl solution.

$$0.0958 \text{ g NaCl} \times \frac{1 \text{ mol NaCl}}{58.4 \text{ g NaCl}} \times \frac{1 \text{ L soln}}{0.163 \text{ mol NaCl}} \times \frac{1000 \text{ mL}}{1 \text{ L}} = 10.\underline{0}6 = 10.1 \text{ mL NaCl}$$

4.21 One (1) liter of solution is equivalent to 0.15 mol of NaCl. The amount of NaCl in 50.0 mL of solution is:

$$50.0 \text{ mL} \times \frac{1 \text{ L}}{1000 \text{ mL}} \times \frac{0.15 \text{ mol NaCl}}{1 \text{ L soln}} = 0.007\underline{5}0 \text{ mol NaCl}$$

$$0.00750 \text{ mol NaCl} \times \frac{58.4 \text{ g NaCl}}{1 \text{ mol NaCl}} = 0.4\underline{3}8 = 0.44 \text{ g NaCl}$$

4.22 Rearrange the equation $M_iV_i = M_fV_f$, where i = initial and f = final:

$$V_f = \frac{M_iV_i}{M_f} = \frac{(1.5 \text{ M})(0.048 \text{ L})}{0.18 \text{ M}} = 0.4\underline{0}0\text{L} \ \ (400 \text{ mL})$$

4.23 Rearrange the equation $M_iV_i = M_fV_f$, where i = initial and f = final.

$$V_i = \frac{M_fV_f}{M_i} = \frac{(1.00 \text{ M})(0.0500 \text{ L})}{14.8 \text{ M}} = 0.033\underline{7}8 \text{ L} \ \ (3.38 \text{ mL})$$

4.24 Calculate the molarity of 95% H_2SO_4. Assume 100.0 g of solution, which then contains 95.0 g of H_2SO_4.

$$95 \text{ g } H_2SO_4 \times \frac{1 \text{ mol } H_2SO_4}{98.08 \text{ g } H_2SO_4} = 0.9\underline{6}9 \text{ mol } H_2SO_4$$

Now calculate the volume of 100.0 g H_2SO_4 solution from the density:

$$V \text{ of } H_2SO_4 = \frac{m}{d} = \frac{100.0 \text{ g soln}}{1.84 \text{ g } H_2SO_4/\text{mL soln}} = 54.\underline{3}4 \text{ mL sol}$$

$$\text{Molarity} = \frac{0.9\underline{6}9 \text{ mol } H_2SO_4}{0.05434 \text{ L}} = 1\underline{7}.8 \text{ M } H_2SO_4$$

To determine the volume of 17.8 M H_2SO_4 necessary for dilution, rearrange the formula $M_iV_i = M_fV_f$:

$$V_i = \frac{M_fV_f}{M_i} = \frac{(0.15 \text{ M})(1.0 \text{ L})}{17.8 \text{ M}} = 8.\underline{4}2 \times 10^{-3} \text{ L} \ \ (8.4 \text{ mL})$$

4.25 Convert the volume of Na_3PO_4 to moles using the molarity of Na_3PO_4.

$$45.7 \times 10^{-3}\ \text{L}\ Na_3PO_4 \times \frac{0.265\ \text{mol}\ Na_3PO_4}{1\ \text{L}} = 0.012\underline{1}1\ \text{mol}\ Na_3PO_4$$

Finally, calculate the amount of $NiSO_4$ requried to react with this amount of Na_3PO_4:

$$0.1211\ \text{mol}\ Na_3PO_4 \times \frac{3\ \text{mol}\ NiSO_4}{2\ \text{mol}\ Na_3PO_4} \times \frac{1\ \text{L}\ NiSO_4}{0.375\ \text{M}\ NiSO_4} = 0.048\underline{4}4\ \text{L}\ \ (48.4\ \text{mL}\ NiSO_4)$$

4.26 Convert the mass of Na_2CO_3 to moles. Then use the balanced equation to determine the moles of $Ca(OH)_2$. Finally determine the volume of $Ca(OH)_2$ using its molarity.

$$2.55\ \text{g}\ Na_2CO_3 \times \frac{1\ \text{mol}\ Na_2CO_3}{106.0\ \text{g}\ Na_2CO_3} \times \frac{1\ \text{mol}\ Ca(OH)_2}{1\ \text{mol}\ Na_2CO_3} \times \frac{1\ \text{L}}{0.150\ \text{mol}\ Ca(OH)_2}$$

$$= 0.16\underline{0}3\ \text{L}\ (160\ \text{mL})$$

4.27 Determine the moles of NaOH needed for the titration.

$$0.0391\ \text{L NaOH} \times \frac{0.108\ \text{mol NaOH}}{1\ \text{L}} = 0.0042\underline{2}3\ \text{mol NaOH} = 0.0042\underline{2}3\ \text{mol}\ HC_2H_3O_2$$

Since the moles of NaOH are equal to the moles of $HC_2H_3O_2$, the latter can be converted to grams of $HC_2H_3O_2$, and then to the mass percentage.

$$0.004223\ \text{mol}\ HC_2H_3O_2 \times \frac{60.05\ \text{g}\ HC_2H_3O_2}{1\ \text{mol}\ HC_2H_3O_2} = 0.25\underline{3}59\ \text{g}\ \ HC_2H_3O_2$$

$$\text{Mass percentage} = \frac{0.25359\ \text{g}\ HC_2H_3O_2}{5.00\ \text{g vinegar}} \times 100\% = 5.0\underline{7}1 = 5.07\%$$

ANSWERS TO REVIEW QUESTIONS

4.1 The molecular weight is the sum of the atomic weights of all the atoms in a molecule, whereas the formula weight is the sum of the atomic weights of all the atoms in one formula unit of the compound, whether the compound is molecular or not. A given substance could have both a molecular weight and a formula weight if it existed as discrete molecules.

4.2 A formula weight equals the sum of the atomic weights of all atoms in a formula.

4.3 A mole of N_2 contains Avogadro's number of N_2 molecules and 2 x Avogadro's number of N atoms. One mole of $Fe_2(SO_4)_3$ contains 3 moles of SO_4^{2-} ions; it contains 12 moles of O.

4.4 N_2O_4 will have the same percentage composition as NO_2: 30.5% N and 69.5% O.

4.5 A sample of the compound of known mass is burned. The percentage composition of the compound is determined from the masses of CO_2 and H_2O obtained.

4.6 Assume 100 g of substance and convert the masses of the elements to moles of the elements, which are proportional to the numbers of atoms in the formula. Reduce these numbers of moles to the smallest whole numbers.

4.7 The empirical formula is C_3H_6O.

4.8 The molecular formula is 34.0 ÷ 17.0 x HO = H_2O_2.

4.9 The equation means that one molecule of CH_4 reacts with two molecules of O_2 to give one molecule of CO_2 plus two molecules of H_2O. It also means that one mole of CH_4 reacts with two moles of O_2 to give one mole of CO_2 plus two moles of H_2O. In terms of mass, 16.0 g CH_4 reacts with 64.0 g O_2 to give 44.0 g CO_2 plus 36.0 g H_2O.

4.10 A chemical equation directly relates the moles of substances involved in the reaction. Since moles of a substance can be related to mass, the chemical equation indirectly relates masses of substances.

4.11 The limiting reactant is the reactant that is entirely consumed when the reaction is complete. Therefore, the amount of product is directly related to the amount of the limiting reactant.

4.12 By definition,

$$\text{molar concentration} = \frac{\text{moles of solute}}{\text{liters of solution}}$$

Therefore,

$$\text{moles of solute} = \text{molar concentration} \times \text{liters of solution}$$

4.13 During dilution, the moles of solute remain constant. Therefore, the product of molar concentration and liters of solution remains constant. (See the answer to the previous question.)

4.14 The number of moles of hydrochloric acid reacted equals the product of molar concentration and volume of acid used in the titration. From the chemical equation, we see that the number of moles of hydrochloric acid equals the number of moles of sodium hydroxide.

SOLUTIONS TO PRACTICE PROBLEMS

Note on significant figures: The final answer is given first with one nonsignificant figure (rightmost significant figure underlined), and is then rounded to the correct number of significant figures. Intermediate answers usually also have at least one nonsignificant figure. Atomic weights are rounded to two decimal places, except for that of hydrogen.

4.15 a. Formula weight of CH_3OH = AW of C + 4(AW of H) + AW of O. Using the values of atomic weights in Table 3.3 rounded to 4 significant figures:

$$\text{FW} = 12.01 \text{ amu} + 4 \times 1.008 \text{ amu} + 16.00 \text{ amu} = 32.0\underline{4}2 = 32.0 \text{ amu (3 sf)}$$

b. FW of PCL_3 = AW of P + 3(AW of Cl)

$$= 30.97 \text{ amu} + 3 \times 35.45 \text{ amu} = 137.3\underline{2}0 = 137 \text{ amu (3 sf)}$$

c. FW of K_2CO_3 = 2(AW of K) + AW of C + 3(AW of O)

$$= 2 \times 39.10 \text{ amu} + 12.01 \text{ amu} + 3 \times 16.00 \text{ amu}$$

$$= 138.2\underline{1}0 = 138 \text{ amu (3sf)}$$

d. FW of $Ni_3(PO_4)_2$ = 3(AW of Ni) + 2(AW of P) + 8(AW of O)

= 3 x 58.70 amu + 2 x 30.97 amu + 8 x 16.00 amu

= 366.0$\underline{4}$0 = 366 amu (3 sf)

4.17 First find the formula weight of NH_4NO_3 by adding the respective atomic weights. Then convert it to the molar mass:

FW of NH_4NO_3 = 2(AW of N) + 4(AW of H) + 3(AW of O)

= 2 x 14.07 amu + 4 x 1.008 amu + 3 x 16.00 amu

= 80.1$\underline{7}$2 amu

The molar mass of NH_4NO_3 = 80.17 g/mol.

4.19 a. The atomic weight of Na = 22.99 amu; thus the molar mass = 22.99 g/mol. Since 1 mol of Na atoms = 6.02 x 10^{23} Na atoms, we calculate:

$$\text{Mass of one Na atom} = \frac{22.99 \text{ g/mol}}{6.02 \times 10^{23} \text{ atom/mol}} = 3.8\underline{2} \times 10^{-23} \text{ g/atom}$$

b. The atomic weight of S = 32.06 amu; thus the molar mass = 32.06 g/mol. Since 1 mol of S atoms = 6.02 x 10^{23} S atoms, we calculate:

$$\text{Mass of one S atom} = \frac{32.06 \text{ g/mol}}{6.02 \times 10^{23} \text{ atom/mol}} = 5.3\underline{3} \times 10^{-23} \text{ g/atom}$$

c. The formula weight of CH_3Cl = [12.01 + (3 x 1.008) + 35.45] = 50.48 amu; thus the molar mass = 50.4$\underline{8}$4 g/mol. Since 1 mol of CH_3Cl molecules = 6.02 x 10^{23} CH_3Cl molecules, we calculate:

$$\text{Mass of one CH}_3\text{Cl molec} = \frac{50.48 \text{ g/mol}}{6.02 \times 10^{23} \text{ molec/mol}} = 8.3\underline{9} \times 10^{-23} \text{ g/molec}$$

d. The formula weight of Na_2SO_3 = [(2 x 22.99) + 32.06 + (3 x 16.00)] = 126.0 amu; thus the molar mass = 126.0 g/mol. Since 1 formula weight of Na_2SO_3 = 6.02 x 10^{23} Na_2SO_3 formula units, we calculate:

$$\text{Mass of one Na}_2\text{SO}_3 = \frac{126.0 \text{ g/mol}}{6.02 \times 10^{23} \text{ unit/mol}} = 2.0\underline{9} \times 10^{-22} \text{ g/unit}$$

4.21 First find the formula weight (in amu) using Table 4.3:

FW of $(CH_3CH_2)_2O$ = 4 x 12.01 amu + 10 x 1.008 amu + 16.00 amu

= 74.12 amu

$$\text{Mass of (CH}_3\text{CH}_2)_2\text{O molec} = \frac{74.12 \text{ g/mol}}{6.02 \times 10^{23} \text{ molec/mol}} = 1.2\underline{3} \times 10^{-22} \text{ g/molec}$$

4.23 From the table of atomic weights, we obtain the following molar masses for parts a through d: Na = 22.99 g/mol; S = 32.06 g/mol; C = 12.01 g/mol; H = 1.008 g/mol; Cl = 35.45 g/mol; and finally, O = 16.00 g/mol.

a. $$0.15 \text{ mol Na} \times \frac{22.99 \text{ g Na}}{1 \text{ mol Na}} = 3.\underline{4}48 = 3.4 \text{ g Na}$$

b. $$0.594 \text{ mol S} \times \frac{32.06 \text{ g S}}{1 \text{ mol S}} = 19.\underline{0}4 = 19.0 \text{ g S}$$

c. Using molar mass = 50.48 g/mol for CH_3Cl, we obtain:

$$2.78 \text{ mol } CH_3Cl \times \frac{50.48 \text{ g } CH_3Cl}{1 \text{ mol } CH_3Cl} = 14\underline{0}.3 = 140 \text{ g } CH_3Cl$$

d. Using molar mass = 126.04 g/mol for Na_2SO_3, we obtain:

$$38 \text{ mol } Na_2SO_3 \times \frac{126.04 \text{ g } Na_2SO_3}{1 \text{ mol } Na_2SO_3} = 4.\underline{7}89 \times 10^3 = 4.8 \times 10^3 \text{ g } Na_2SO_3$$

4.25 First find the molar mass of H_3BO_3: 3 x 1.008 amu + 10.81 amu + 3 x 16.00 amu = 61.83. Therefore the molar mass of H_3BO_3 = 61.83 g/mol. The mass of H_3BO_3 is calculated:

$$0.543 \text{ mol } H_3BO_3 \times \frac{61.83 \text{ g } H_3BO_3}{1 \text{ mol } H_3BO_3} = 33.\underline{5}7 = 33.6 \text{ g } H_3BO_3$$

4.27 From the table of atomic weights, we obtain the following rounded molar masses for parts a to d: C = 12.01 g/mol; Br = 79.90 g/mol; H = 1.008 g/mol, Li = 6.94 g/mol; and O = 16.00 g/mol.

a. $$3.43 \text{ g C} \times \frac{1 \text{ mol C}}{12.01 \text{ g C}} = 0.28\underline{5}59 = 0.286 \text{ mol C}$$

b. $$7.05 \text{ g } Br_2 \times \frac{1 \text{ mol } Br_2}{159.80 \text{ g } Br_2} = 0.044\underline{1}1 = 0.0441 \text{ mol } Br_2$$

c. The molar mass of C_4H_{10} = 4 x 12.01 + 10 x 1.008 = 58.12 g C_4H_{10}/mol C_4H_{10}. The mass of C_4H_{10} is calculated:

$$76 \text{ g } C_4H_{10} \times \frac{1 \text{ mol } C_4H_{10}}{58.12 \text{ g } C_4H_{10}} = 1.\underline{3}07 = 1.3 \text{ mol } C_4H_{10}$$

d. The molar mass of Li_2CO_3 = 2 x 6.94 + 12.01 + 3 x 16.00 g = 73.89 g Li_2CO_3/mol Li_2CO_3. The mass of Li_2CO_3 is calculated:

$$35.4 \text{ g } C_4H_{10} \times \frac{1 \text{ mol } Li_2CO_3}{73.89 \text{ g } Li_2CO_3} = 0.47\underline{9}09 = 0.479 \text{ mol } Li_2CO_3$$

4.29 Calculate the formula weight of calcium sulfate: 40.08 amu + 32.006 amu + 4 x 16.00 amu = 136.1 amu. Therefore the molar mass of $CaSO_4$ is 136.1 g/mol. Use this to convert the mass of $CaSO_4$ to moles:

$$0.791 \text{ g } CaSO_4 \times \frac{1 \text{ mol } CaSO_4}{136.1 \text{ g } CaSO_4} = 0.0058\underline{1}1 \text{ mol } CaSO_4$$

Calculate the molecular weight of water: 2 x 1.008 amu + 16.00 amu = 18.02 amu. Therefore the molar mass of H_2O = 18.02 g/mol. Use this to convert the rest of the sample to moles of water:

$$0.209 \text{ g } H_2O \times \frac{1 \text{ mol } H_2O}{18.02 \text{ g } H_2O} = 0.011\underline{5}9 \text{ mol } H_2O$$

Since 0.01159 mol is about twice 0.005811 mol, both numbers of moles are consistent with the formula $CaSO_4{\cdot}2H_2O$.

4.31 The following rounded atomic weights are used: Li = 6.94 g/N_A; Br = 79.90 g/N_A; N = 14.01 g/N_A; H = 1.008 g/N_A; Pb = 207.2 g/N_A; Cr = 52.00 g/N_A; O = 16.00 g/N_A; S = 32.06 g/N_A.

a. $$\text{No. Li atoms} = 7.46 \text{ g Li} \times \frac{6.02 \times 10^{23} \text{ atoms}}{6.94 \text{ g Li}} = 6.4\underline{7} \times 10^{23} \text{ atoms}$$

b. $$\text{No. Br atoms} = 32.0 \text{ g } Br_2 \times \frac{2 \times 6.02 \times 10^{23} \text{ atoms}}{2 \times 79.90 \text{ g } Br_2} = 2.4\underline{1} \times 10^{23} \text{ atoms}$$

c. $$\text{No. } NH_3 \text{ molec} = 43 \text{ g } NH_3 \times \frac{6.02 \times 10^{23} \text{ atoms}}{17.03 \text{ g } NH_3} = 1.\underline{5}2 \times 10^{24} \text{ atoms}$$

d. $$\text{No. } PbCrO_4 \text{ units} = 159 \text{ g } PbCrO_4 \times \frac{6.02 \times 10^{23} \text{ units}}{323.2 \text{ g } PbCrO_4} = 2.9\underline{6} \times 10^{23} \text{ units}$$

e. $$\text{No. } SO_4^{2-} \text{ ions} = 14.3 \text{ g } Cr_2(SO_4)_3 \times \frac{3 \times 6.02 \times 10^{23} \text{ ions}}{392.16 \text{ g } Cr_2(SO_4)_3} = 6.5\underline{8} \times 10^{22} \text{ ions}$$

4.33 Calculate the molecular weight of CCl_4: 12.01 amu + 4 x 35.45 amu = 153.81 amu; use this and Avogadro's number to express it as 153.81 g/N_A to calculate the number of molecules:

$$7.58 \text{ mg } CCl_4 \times \frac{1 \text{ g}}{10^3 \text{ mg}} \times \frac{6.02 \times 10^{23} \text{ molec}}{153.81 \text{ g } CCl_4} = 2.9\underline{6}6 \times 10^{19} = 2.97 \times 10^{19} \text{ molec}$$

4.35 $$\text{Mass \% carbon} = \frac{\text{mass of C in sample}}{\text{mass of sample}} \times 100\%$$

$$\text{\% carbon} = \frac{1.584 \text{ g}}{1.836 \text{ g}} \times 100\% = 86.2\underline{7}4 = 86.27\%$$

4.37 $$\text{Mass \% phosphorus oxychloride} = \frac{\text{mass of } POCl_3 \text{ in sample}}{\text{mass of sample}} \times 100\%$$

$$\text{\% } POCl_3 = \frac{1.72 \text{ mg}}{8.53 \text{ mg}} \times 100\% = 20.\underline{1}6 = 20.2\%$$

4.39 Start with the definition for percentage nitrogen and rearrange this equation to find the mass of N in the fertilizer.

$$\text{Mass \% N} = \frac{\text{mass of N in fert.}}{\text{mass of fert.}} \times 100\%$$

$$\text{Mass N} = \frac{\text{mass \% N}}{100\%} \times \text{mass of fert.} = \frac{15.8\%}{100\%} \times 4.15 \text{ kg} = 0.65\underline{5}7 = 0.656 \text{ kg N}$$

4.41 Convert moles to mass, using the molar masses from the respective atomic weights. Then calculate the mass percentages from the respective masses.

$$0.0972 \text{ mol Al} \times \frac{26.98 \text{ g Al}}{1 \text{ mol Al}} = 2.6\underline{2}2 \text{ g Al}$$

$$0.0381 \text{ mol Mg} \times \frac{24.31 \text{ g Mg}}{1 \text{ mol Mg}} = 0.92\underline{6}2 \text{ g Mg}$$

$$\text{\% Al} = \frac{\text{mass of Al}}{\text{mass of alloy}} = \frac{2.622 \text{ g Al}}{3.548 \text{ g alloy}} \times 100\% = 73.\underline{9}007 = 73.9\% \text{ Al}$$

$$\text{\% Mg} = \frac{\text{mass of Mg}}{\text{mass of alloy}} = \frac{0.9262 \text{ g Mg}}{3.548 \text{ g alloy}} \times 100\% = 26.\underline{1}06 = 26.1\% \text{ Mg}$$

4.43 In each part, the numerator consists of the mass of the element in one mole of the compound; the denominator is the mass of one mole of the compound. Use the atomic weights of: C = 12.01 g/mol; O = 16.00 g/mol; K = 39.10 g/mol; Mn = 54.94 g/mol; Co = 58.93 g/mol; N = 14.01 g/mol. (100% is written as 100.000% to emphasize that it is an exact number.)

a. $$\text{\% C} = \frac{\text{mass of C}}{\text{mass of CO}} = \frac{12.01 \text{ g C}}{28.01 \text{ g CO}} \times 100\% = 42.878 = 42.9\%$$

$$\text{\% O} = 100.000\% - 42.8\underline{7}8\%\text{C} = 57.122 = 57.1\%$$

b. $$\text{\% C} = \frac{\text{mass of C}}{\text{mass of } CO_2} = \frac{12.01 \text{ g C}}{44.01 \text{ g } CO_2} \times 100\% = 27.289 = 27.3\%$$

$$\text{\% O} = 100.000\% - 27.2\underline{8}9\% \text{ C} = 72.7\underline{1}1 = 72.7\%$$

c. $$\text{\% K} = \frac{\text{mass of K}}{\text{mass of } KMnO_4} = \frac{39.10 \text{ g K}}{158.04 \text{ g } KMnO_4} \times 100\% = 24.741 = 24.7\%$$

$$\text{\% Mn} = \frac{\text{mass of Mn}}{\text{mass of } KMnO_4} = \frac{54.94 \text{ g Mn}}{158.04 \text{ g } KMnO_4} \times 100\% = 34.763 = 34.8\%$$

$$\text{\% O} = 100.000\% - (24.7\underline{4}1 + 34.7\underline{6}3) = 40.496 = 40.5\%$$

d. $$\text{\% Co} = \frac{\text{mass of Co}}{\text{mass of } Co(NO_3)_2} = \frac{58.93 \text{ g Co}}{182.95 \text{ g } Co(NO_3)_2} \times 100\% = 32.211 = 32.2\%$$

$$\text{\% N} = \frac{\text{mass of N}}{\text{mass of } Co(NO_3)_2} = \frac{2 \times 14.01 \text{ g N}}{182.95 \text{ g } Co(NO_3)_2} \times 100\% = 15.316 = 15.3\%$$

$$\text{\% O} = 100.000\% - (32.2\underline{1}1 + 15.3\underline{1}6) = 52.473 = 52.5\%$$

4.45 One mol of $CF_3CHBrCl$ contains 2 mol of C (at wt = 12.01), 3 mol of F (at wt = 19.00), one mol of H (at wt = 1.008), one mol of Br (at wt = 79.90), and one mol of Cl (at wt = 35.45).

$$\% C = \frac{\text{mass of C}}{\text{mass of } C_4F_3HBrCl} = \frac{2 \times 12.01 \text{ g}}{197.38 \text{ g}} \times 100\% = 12.169 = 12.2\%$$

$$\% F = \frac{\text{mass of F}}{\text{mass of } C_4F_3HBrCl} = \frac{3 \times 19.0 \text{ g}}{197.38 \text{ g}} \times 100\% = 28.878 = 28.9\%$$

$$\% Br = \frac{\text{mass of Br}}{\text{mass of } C_4F_3HBrCl} = \frac{79.90 \text{ g}}{197.38 \text{ g}} \times 100\% = 40.4802 = 40.5\%$$

$$\% Cl = \frac{\text{mass of Cl}}{\text{mass of } C_4F_3HBrCl} = \frac{35.45 \text{ g}}{197.38 \text{ g}} \times 100\% = 17.9602 = 18.0\%$$

$$\% H = \frac{\text{mass of H}}{\text{mass of } C_4F_3HBrCl} = \frac{1.008 \text{ g}}{197.38 \text{ g}} \times 100\% = 0.51069 = 0.511\%$$

4.47 Find the moles of C in each amount in one-step operations: calculate the moles of each compound using the molar mass; then multiply by the number of moles of C per mole of compound:

$$\text{Mol C (glucose)} = 4.71 \text{ g} \times \frac{1 \text{ mol}}{180.2 \text{ g}} \times \frac{6 \text{ mol C}}{1 \text{ mol glucose}} = 0.15\underline{7} \text{ mol}$$

$$\text{Mol C (ethanol)} = 5.85 \text{ g} \times \frac{1 \text{ mol}}{46.07 \text{ g}} \times \frac{2 \text{ mol C}}{1 \text{ mol ethanol}} = 0.25\underline{4} \text{ mol (more C)}$$

4.49 First calculate the mass of C in the glycol by multiplying the mass of CO_2 by the molar mass of C and the reciprocal of the molar mass of CO_2. Then calculate the mass of H in the glycol by multiplying the mass of H_2O by the molar mass of 2H and the reciprocal of the molar mass of H_2O. Then use the masses to calculate the mass percentages. Calculate O by difference.

$$9.06 \text{ mg } CO_2 \times \frac{1 \text{ mol } CO_2}{44.01 \text{ g } CO_2} \times \frac{12.01 \text{ g C}}{1 \text{ mol C}} = 2.4\underline{7}2 \text{ mg C}$$

$$5.58 \text{ mg } H_2O \times \frac{1 \text{ mol } H_2O}{18.016 \text{ mg } H_2O} \times \frac{2 \text{ H}}{1 \text{ } H_2O} \times \frac{1.008 \text{ g H}}{1 \text{ mol H}} = 0.62\underline{4}4 \text{ mg H}$$

$$\text{mg O} = 6.38 \text{ mg} - (2.472 + 0.6244) = 3.2\underline{8}3 \text{ mg O}$$

$$\% C = [2.472 \text{ mg C}/6.38 \text{ mg glycol}] \times 100\% = 38.\underline{7}4 = 38.7\%$$

$$\% H = [0.6244 \text{ mg H}/6.38 \text{ mg glycol}] \times 100\% = 9.7\underline{8}6 = 9.79\%$$

$$\% O = [3.283 \text{ mg O}/6.38 \text{ mg glycol}] \times 100\% = 51.\underline{4}6 = 51.5\%$$

4.51 Start by calculating the moles of Os and O; then divide each by the smaller number of moles to obtain integers for the empirical formula.

$$\text{Mol Os} = 2.16 \text{ g Os} \times \frac{1 \text{ mol}}{190.2 \text{ g Os}} = 0.011\underline{3}6 \text{ mol (smaller no.)}$$

$$\text{Mol O} = (2.89 - 2.16)\text{ g O} \times \frac{1\text{ mol}}{16.00\text{ g Os}} = 0.04\underline{5}6\text{ mol}$$

$$\text{Integer for Os} = 0.01136 \div 0.01136 = 1.0\underline{0}0$$

$$\text{Integer for O} = 0.0456 \div 0.01136 = 4.0\underline{1}$$

Since 4.01 = 4.0 within experimental error, the empirical formula is OsO_4.

4.53 Assume a sample of 100.0 g of potassium manganate. By multiplying this by the percentage composition, one obtains 39.7 g of K, 27.9 g of Mn, and 32.5 g of O. Convert each of these masses to moles by dividing by molar mass.

$$\text{Mol K} = 39.7\text{ g K} \times \frac{1\text{ mol}}{39.10\text{ g K}} = 1.0\underline{1}5\text{ mol}$$

$$\text{Mol Mn} = 27.9\text{ g Mn} \times \frac{1\text{ mol}}{54.94\text{ g Mn}} = 0.50\underline{7}8\text{ mol (smallest no.)}$$

$$\text{Mol O} = 32.5\text{ g O} \times \frac{1\text{ mol}}{16.00\text{ g Mn}} = 2.0\underline{3}1\text{ mol}$$

Now divide each number of moles by the smallest number to obtain the smallest set of integers for the empirical formula.

$$\text{Integer for K} = 1.015 \div 0.5078 = 2.00\text{, or }2$$

$$\text{Integer for Mn} = 0.5078 \div 0.5078 = 1.00\text{, or }1$$

$$\text{Integer for O} = 2.031 \div 0.5078 = 4.00\text{, or }4$$

The empirical formula is thus K_2MnO_4.

4.55 Assume a sample of 100.0 g of acrylic acid. By multiplying this by the percentage composition, one obtains 50.0 g C, 5.6 g H, and 44.4 g O. Convert each of these masses to moles by dividing by the molar mass.

$$\text{Mol C} = 50.0\text{ g C} \times \frac{1\text{ mol}}{12.01\text{ g C}} = 4.1\underline{6}3\text{ mol}$$

$$\text{Mol H} = 5.6\text{ g H} \times \frac{1\text{ mol}}{1.008\text{ g H}} = 5.\underline{5}6\text{ mol}$$

$$\text{Mol O} = 44.4\text{ g O} \times \frac{1\text{ mol}}{16.00\text{ g O}} = 2.7\underline{7}5\text{ mol (smallest no.)}$$

Now divide each number of moles by the smallest number to obtain the smallest number of moles, and tentative integers for the empirical formula.

$$\text{Tentative integer for C} = 4.163 \div 2.775 = 1.50\text{, or }1.5$$

$$\text{Tentative integer for H} = 5.56 \div 2.775 = 2.00\text{, or }2$$

$$\text{Tentative integer for O} = 2.775 \div 2.775 = 1.00\text{, or }1$$

Since 1.5 is not a whole number, multiply each tentative integer by 2 to give the final integer for the empirical formula:

C: 2 x 1.5 = 3
H: 2 x 2 = 4
O: 2 x 1 = 2

The empirical formula is thus $C_3H_4O_2$.

4.57 The formula weight corresponding to the empirical formula C_2H_6N may be found by adding the respective atomic weights.

Formula weight = 2 x 12.01 amu + 6 x 1.008 amu + 14.01 amu = 44.08 amu

Dividing the molecular weight by the formula weight gives the number of times the C_2H_6N unit occurs in the molecule. Since the molecular weight is an average of 88.5 [(90 + 87) ÷ 2], this quotient is:

88.5 amu ÷ 44.1 amu = 2.0̲06, or 2

Therefore the molecular formula is $(C_2H_6N)_2$, or $C_4H_{12}N_2$.

4.59 Assume a sample of 100.0 g of oxalic acid. By multiplying this by the percentage composition, one obtains 26.7 g C, 2.2 g H, and 71.1 g O. Convert each of these masses to moles by dividing by the molar mass.

$$\text{Mol C} = 26.7\text{ g C} \times \frac{1\text{ mol}}{12.01\text{ g C}} = 2.2\underline{2}3\text{ mol}$$

$$\text{Mol H} = 2.2\text{ g H} \times \frac{1\text{ mol}}{1.008\text{ g H}} = 2.\underline{1}8\text{ mol (smallest no.)}$$

$$\text{Mol O} = 71.1\text{ g O} \times \frac{1\text{ mol}}{16.00\text{ g O}} = 4.4\underline{4}3\text{ mol}$$

Now divide each number of moles by the smallest number to obtain the smallest set of integers for the empirical formula.

Integer for C = 2.223 ÷ 2.18 = 1.02, or 1

Integer for H = 2.18 ÷ 2.18 = 1.00, or 1

Integer for O = 4.443 ÷ 2.18 = 2.0̲3̲8, or 2

The empirical formula is thus CHO_2. The formula weight corresponding to this formula may be found by adding the respective atomic weights:

Formula weight = 12.01 amu + 1.008 amu + 2 x 16.00 amu = 45.02 amu

Dividing the molecular weight by the formula weight gives the number of times the CHO_2 unit occurs in the molecule. Since the molecular weight is 90 amu, this quotient is:

90 amu ÷ 45.02 amu = 2.0̲0, or 2

The molecular formula is thus $(CHO_2)_2$, or $C_2H_2O_4$.

4.61

C_2H_4	+	$3O_2$	→	$2CO_2$	+	$2H_2O$
1 molecule C_2H_4	+	3 molecules O_2	→	2 molecules CO_2	+	2 molecules H_2O
1 mole C_2H_4	+	3 moles O_2	→	2 moles CO_2	+	2 moles H_2O
28.054 g C_2H_4	+	3 x 31.999 g O_2	→	2 x 44.01 g CO_2	+	2 x 18.015 g H_2O

4.63 $3NO_2 + H_2O \rightarrow 2HNO_3 + NO$

3 mol of NO_2 is equivalent to 2 mol of HNO_3 (from equation)

1 mol of NO_2 is equivalent to 46.01 g NO_2 (from molecular weight of NO_2)

1 mol of HNO_3 is equivalent to 63.01 g HNO_3 (from molecular weight of HNO_3)

$$5.00 \text{ g } HNO_3 \times \frac{1 \text{ mol } HNO_3}{63.01 \text{ g } HNO_3} \times \frac{3 \text{ mol } NO_2}{2 \text{ mol } HNO_3} \times \frac{46.01 \text{ g } NO_2}{1 \text{ mol } NO_2} = 5.4\underline{76} = 5.48 \text{ g } NO_2$$

4.65 $WO_3 + 3H_2 \rightarrow W + 3H_2O$

1 mol of W is equivalent to 3 moles of H_2 (from equation)

1 mol of H_2 is equivalent to 2.016 g H_2 (from molecular weight of H_2)

1 mol of W is equivalent to 183.8 g W (from atomic weight of W)

4.81 kg of H_2 is equivalent to 4.81 x 10^3 g of H_2

$$4.81 \times 10^3 \text{ g } H_2 \times \frac{1 \text{ mol } H_2}{2.016 \text{ g } H_2} \times \frac{1 \text{ mol W}}{3 \text{ mol } H_2} \times \frac{183.8 \text{ g W}}{1 \text{ mol W}} = 1.4\underline{6}1 \times 10^5 = 1.46 \times 10^5 \text{ g W}$$

4.67 Using the approach of the previous problem, we write the equation and set up the calculation below the equation (after calculating the two molecular weights):

$CS_2 + 3Cl_2 \rightarrow CCl_4 + S_2Cl_2$

$$62.7 \text{ g } Cl_2 \times \frac{1 \text{ mol } Cl_2}{70.91 \text{ g } Cl_2} \times \frac{1 \text{ mol } CS_2}{3 \text{ mol } Cl_2} \times \frac{76.13 \text{ g } CS_2}{1 \text{ mol } CS_2} = 22.\underline{4}3 = 22.4 \; CS_2$$

4.69 Using the approach of the problems before 4.67, we write the equation and set up the calculation below the equation (after calculating the two molecular weights):

$2N_2O_5 \rightarrow 4NO_2 + O_2$

$$1.618 \text{ g } O_2 \times \frac{1 \text{ mol } O_2}{32.00 \text{ g } O_2} \times \frac{4 \text{ molNO}_2}{1 \text{ mol } O_2} \times \frac{46.01 \text{ g } NO_2}{1 \text{ mol } NO_2} = 9.30\underline{5}52 = 9.306 \text{ g } NO_2$$

4.71 First determine whether KO_2 or H_2O is the limiting reagent by calculating the moles of O_2 that each would form, if each were the limiting reagent. Identify the limiting reactant by the smaller number of moles of O_2 formed.

$$0.10 \text{ mol } H_2O \times \frac{3 \text{ mol } O_2}{2 \text{ mol } H_2O} = 0.1\underline{5}0 \text{ mol } O_2$$

$$0.15 \text{ mol } KO_2 \times \frac{3 \text{ mol } O_2}{4 \text{ mol } KO_2} = 0.1\underline{1}25 \text{ mol } O_2 \quad (KO_2 \text{ is limiting reactant})$$

The moles of O_2 produced = 0.11 mol.

4.73 First determine whether CO or H_2 is the limiting reagent by calculating the moles of CH_3OH that each would form, if each were the limiting reagent. Identify the limiting reactant by the smaller number of moles of CH_3OH formed. Use the molar mass of CH_3OH to calculate the mass of CH_3OH formed. Then calculate the mass of the unused reactant left.

$$CO + 2H_2 \rightarrow CH_3OH$$

$$10.2 \text{ g } H_2 \times \frac{1 \text{ mol } H_2}{2.016 \text{ g } H_2} \times \frac{1 \text{ mol } CH_3OH}{2 \text{ mol } H_2} = 2.5\underline{2}9 \text{ mol } CH_3OH$$

$$35.4 \text{ g CO} \times \frac{1 \text{ mol CO}}{28.01 \text{ g CO}} \times \frac{1 \text{ mol } CH_3OH}{1 \text{ mol CO}} = 1.2\underline{6}3 \text{ mol } CH_3OH \quad (\text{CO = limiting reagent})$$

$$\text{Mass } CH_3OH \text{ formed} = 1.2\underline{6}3 \text{ mol } CH_3OH \times \frac{32.042 \text{ g } CH_3OH}{1 \text{ mol } CH_3OH} = 40.\underline{4}6 \text{ g } CH_3OH$$

Hydrogen is left unconsumed at the end of the reaction. The mass of H_2 that reacts can be obtained from the moles of product obtained:

$$1.2\underline{6}3 \text{ mol } CH_3OH \times \frac{2 \text{ mol } H_2}{1 \text{ mol } CH_3OH} \times \frac{2.016 \text{ g } H_2}{1 \text{ mol } H_2} = 5.0\underline{9}2 \text{ g } H_2$$

The unreacted H_2 = 10.2 g total H_2 - 5.092 g reacted H_2 = 5.$\underline{1}$08 = 5.1 g H_2.

4.75 First determine which of the three reactants is the limiting reagent by calculating the moles of $TiCl_4$ that each would form, if each were the limiting reagent. Identify the limiting reagent by the smallest number of moles of $TiCl_4$ formed. Use the molar mass of $TiCl_4$ to calculate the mass of $TiCl_4$ formed.

$$3TiO_2 + 4C + 6Cl_2 \rightarrow 3TiCl_4 + 2CO_2 + 2CO$$

$$4.15 \text{ g } TiO_2 \times \frac{1 \text{ mol } TiO_2}{79.88 \text{ g } TiO_2} \times \frac{3 \text{ mol } TiCl_4}{3 \text{ mol } TiO_2} = 0.051\underline{9}5 \text{ mol } TiCl_4$$

$$5.67 \text{ g C} \times \frac{1 \text{ mol C}}{12.01 \text{ g C}} \times \frac{3 \text{ mol } TiCl_4}{4 \text{ mol C}} = 0.35\underline{4}07 \text{ mol } TiCl_4$$

$$6.78 \text{ g } Cl_2 \times \frac{1 \text{ mol } Cl_2}{70.90 \text{ g } Cl_2} \times \frac{3 \text{ mol } TiCl_4}{6 \text{ mol } Cl_2} = 0.047\underline{8}1 \text{ mol } TiCl_4 \quad (Cl_2 = \text{limiting reagent})$$

$$\text{Mass } TiCl_4 \text{ formed} = 0.047\underline{8}1 \text{ mol } TiCl_4 \times \frac{189.68 \text{ g } TiCl_4}{1 \text{ mol } TiCl_4} = 9.0\underline{6}8 = 9.07 \text{ g } TiCl_4$$

4.77 First determine which of the two reactants is the limiting reagent by calculating the moles of aspirin that each would form, if each were the limiting reagent. Identify the limiting reagent by the smallest number of moles of aspirin formed. Use the molar mass of aspirin to calculate the theoretical yield in grams of aspirin. Then calculate the percentage yield.

$$C_7H_6O_3 + C_4H_6O_3 \rightarrow C_9H_8O_4 + C_2H_4O_2$$

$$4.00 \text{ g } C_4H_6O_3 \times \frac{1 \text{ mol } C_4H_6O_3}{102.1 \text{ g } C_4H_6O_3} \times \frac{1 \text{ mol } C_9H_6O_4}{1 \text{ mol } C_4H_6O_3} = 0.039\underline{1}8 \text{ mol } C_9H_8O_4$$

$$2.00 \text{ g } C_7H_6O_3 \times \frac{1 \text{ mol } C_7H_6O_3}{138.1 \text{ g } C_7H_6O_3} \times \frac{1 \text{ mol } C_9H_8O_4}{1 \text{ mol } C_7H_6O_3} = 0.014\underline{4}8 \text{ mol } C_9H_8O_4$$

Thus, $C_7H_6O_3$ is the limiting reagent. The theoretical yield of $C_9H_8O_4$ is:

$$0.014\underline{4}8 \text{ mol } C_9H_8O_4 \times \frac{180.2 \text{ g } C_9H_8O_4}{1 \text{ mol } C_9H_8O_4} = 2.6\underline{0}9 \text{ g } C_9H_8O_4$$

The percentage yield is:

$$\% \text{ yield} = \frac{\text{actual yield}}{\text{theoretical yield}} \times 100\% = \frac{2.10 \text{ g}}{2.609 \text{ g}} \times 100\% = 80.\underline{4}9 = 80.5\%$$

4.79 $$\text{Molarity} = \frac{\text{moles solute}}{\text{liters of solution}} = \frac{0.0341 \text{ mol}}{0.0250 \text{ L}} = 1.3\underline{6}4 = 1.36 \text{ M}$$

4.81 Find the number of moles of solute ($KMnO_4$), using the molar mass of 158.0 g $KMnO_4$ per 1 mol of $KMnO_4$:

$$0.798 \text{ g } KMnO_4 \times \frac{1 \text{ mol } KMnO_4}{158.0 \text{ g } KMnO_4} = 5.0\underline{5}06 \times 10^{-3} \text{ mol } KMnO_4$$

$$\text{Molarity} = \frac{\text{moles solute}}{\text{liters of solution}} = \frac{5.0506 \times 10^{-3} \text{ mol}}{0.0500 \text{ L}} = 0.10\underline{1}01 = 0.101 \text{ M}$$

4.83 $$0.150 \text{ mol } CuSO_4 \times \frac{1 \text{ L soln}}{0.120 \text{ mol } CuSO_4} = 1.2\underline{5}0 = 1.25 \text{ L soln}$$

4.85 $$0.0353 \text{ g KOH} \times \frac{1 \text{ mol KOH}}{56.10 \text{ g KOH}} \times \frac{1 \text{ L soln}}{0.0176 \text{ mol KOH}} = 0.035\underline{7}51 \text{ L (35.8 mL soln)}$$

4.87 From the molarity, 1 L of heme solution is equivalent to 0.0019 mol of heme solute. Before starting the calculation, note that 25 mL of soln is equivalent to 25×10^{-3} L of soln:

$$25 \times 10^{-3} \text{ L soln} \times \frac{0.0019 \text{ mol heme}}{1 \text{ L soln}} = 4.\underline{7}50 \times 10^{-5} = 4.8 \times 10^{-5} \text{ mol heme}$$

4.89 Multiply the volume of solution by molarity to convert it to moles; then convert to mass of solute by multiplying by the molar mass:

$$50 \times 10^{-3} \text{ L soln} \times \frac{0.025 \text{ mol } Na_2Cr_2O_7}{1 \text{ L soln}} \times \frac{262.0 \text{ g } Na_2Cr_2O_7}{1 \text{ mol } Na_2Cr_2O_7} = 0.3\underline{2}75$$

$$= 0.33 \text{ g } Na_2Cr_2O_7$$

4.91 Start with the equation: $M_i \times V_i = M_f \times V_f$. Rearrange to solve for V_f:

$$V_f = \frac{M_i \times V_i}{M_f} = \frac{34 \text{ mL} \times 1.32 \text{ M}}{0.28 \text{ M}} = 1.\underline{6}02 \times 10^2 = 1.6 \times 10^2 \text{ mL}$$

4.93 Start with the equation: $M_i \times V_i = M_f \times V_f$. Rearrange to solve for V_i:

$$V_i = \frac{M_f \times V_f}{M_i} = \frac{0.18 \text{ M} \times 45.0 \text{ mL}}{2.00 \text{ M}} = 4.\underline{0}5 = 4.0 \text{ mL}$$

4.95 Start by calculating the molarity of the 68% HNO_3. Assume 100 g of 68% HNO_3 solution. First find the volume of this solution by dividing by the density; then calculate the number of moles in the 100 g. Finally divide moles by volume to find molarity:

$$100 \text{ g soln} \times \frac{1.00 \times 10^{-3} \text{ L}}{1.41 \text{ g soln}} = 0.070\underline{9}2 \text{ L soln}$$

$$68.0 \text{ g } HNO_3 \times \frac{1 \text{ mol } HNO_3}{63.02 \text{ g } HNO_3} = 1.0\underline{7}9 \text{ mol } HNO_3$$

$$\text{Molarity} = \frac{1.079 \text{ mol } HNO_3}{0.07092 \text{ L soln}} = 15.\underline{2}1 \text{ M}$$

$$V_i = \frac{M_f \times V_f}{M_i} = \frac{0.150 \text{ M} \times 425 \text{ mL}}{15.21 \text{ M}} = 4.1\underline{9}1 = 4.19 \text{ mL}$$

Measure 4.19 mL of 68% HNO_3 and dilute it to a final volume of 425 mL.

4.97 Using molarity, convert volume of Na_2CO_3 to moles of Na_2CO_3 ; then use the equation to convert to moles of HNO_3, and finally to volume:

$$2HNO_3 + Na_2CO_3 \rightarrow 2NaNO_3 + H_2O + CO_2(g)$$

$$42.4 \times 10^{-3} \text{ L } Na_2CO_3 \times \frac{0.150 \text{ mol } Na_2CO_3}{1 \text{ L soln}} \times \frac{2 \text{ mol } HNO_3}{1 \text{ mol } Na_2CO_3} \times \frac{1 \text{ L } HNO_3}{0.250 \text{ mol } HNO_3}$$

$$= 0.050\underline{8}8 \text{ L (50.9 mL) of } HNO_3$$

4.99 The reaction is: $H_2SO_4 + 2NaHCO_3 \rightarrow Na_2SO_4 + 2H_2O + CO_2(g)$.

$$2.05 \text{ g } NaHCO_3 \times \frac{1 \text{ mol } NaHCO_3}{84.00 \text{ g } NaHCO_3} \times \frac{1 \text{ mol } H_2SO_4}{2 \text{ mol } NaHCO_3} \times \frac{1 \text{ L soln}}{0.150 \text{ mol } H_2SO_4}$$

$= 0.081\underline{3}49$ L soln (81.3 mL)

4.101 First find the mass of H_2O_2 required to react with $KMnO_4$.

$$5H_2O_2 + 2KMnO_4 + 3H_2SO_4 \rightarrow 5O_2(g) + 2MnSO_4 + K_2SO_4 + 8H_2O$$

$$46.9 \times 10^{-3} \text{ L } KMnO_4 \times \frac{0.145 \text{ mol } KMnO_4}{1 \text{ L soln}} \times \frac{5 \text{ mol } H_2O_2}{2 \text{ mol } KMnO_4} \times \frac{34.01 \text{ g } H_2O_2}{1 \text{ mol } H_2O_2}$$

$= 0.57\underline{8}2$ g H_2O_2

% H_2O_2 = (mass H_2O_2 ÷ mass sample) x 100 = (0.5782 g ÷ 20.0 g) x 100

$= 2.8\underline{9}1 = 2.89\%$

4.103 For 1 mol of caffeine, there are 8 mol of C, 10 mol of H, 4 mol of N, and 2 mol of O. Convert these amounts to masses by multipying by the respective molar masses:

8 mol C x 12.01 g C/1 mol C = 96.08 g C
10 mol H x 1.008 g H/1 mol H = 10.08 g H
4 mol N x 14.01 g N/1 mol N = 56.04 g N
2 mol O x 16.00 g O/1 mol O = $\underline{32.00 \text{ g O}}$

1 mol of caffeine (total) = 194.2 g (molar mass)

Each mass % is calculated by dividing the mass of the element by the molar mass of caffeine, and multiplying by 100%: mass % = (mass element ÷ mass caffeine) x 100%

Mass % C = (96.08 g ÷ 194.2 g) x 100% = 49.5% (3 sf)

Mass % H = (10.08 g ÷ 194.2 g) x 100% = 5.19% (3 sf)

Mass % N = (56.04 g ÷ 194.2 g) x 100% = 28.9% (3 sf)

Mass % O = (32.00 g ÷ 194.2 g) x 100% = 16.5% (3 sf)

4.105 Assume a sample of 100.0 g of dichlorobenzene. By multiplying this by the percentage composition, one obtains 49.0 g C, 2.7 g of H, and 48.2 g of Cl. Convert each mass to moles by dividing by the molar mass:

$$49.0 \text{ g C} \times \frac{1 \text{ mol C}}{12.01 \text{ g C}} = 4.0\underline{8}0 \text{ mol C}$$

$$2.7 \text{ g H} \times \frac{1 \text{ mol H}}{1.008 \text{ g H}} = 2.\underline{6}8 \text{ mol H}$$

$$48.2 \text{ g Cl} \times \frac{1 \text{ mol Cl}}{35.45 \text{ g Cl}} = 1.3\underline{6}0 \text{ mol Cl}$$

Divide each number of moles by the smallest number to obtain the smallest set of integers for the empirical formula.

Integer for C = 4.080 mol $\div$ 1.360 mol = 3.00, or 3

Integer for H = 2.68 mol $\div$ 1.360 mol = 1.97, or 2

Integer for Cl = 1.360 mol $\div$ 1.360 mol = 1.00, or 1

The empirical formula is thus C_3H_2Cl. Find the formula weight by adding the atomic weights:

Formula weight = 3 x 12.01 amu + 2 x 1.008 amu + 35.45 amu = $73.4\underline{9}6$ = 73.50 amu

Divide the molecular weight by the formula weight to find the number of times the C_3H_2Cl unit occurs in the molecule. Since the molecular weight is 147 amu, this quotient is:

147 amu $\div$ 73.50 amu = 2.00, or 2

The molecular formula is thus $(C_3H_2Cl)_2$, or $C_6H_4Cl_2$.

4.107 For these calculations, the relative numbers of moles of gold and chlorine must be determined. These can be found from the masses of the two elements in the sample:

Total mass = mass of Au + mass of Cl = 328 mg

The mass of chlorine in the precipitated AgCl = the mass of chlorine in the compound of gold and chlorine. The mass of Cl in the 0.464 g of AgCl is:

$$0.464 \text{ g AgCl} \times \frac{1 \text{ mol AgCl}}{143.32 \text{ g AgCl}} \times \frac{1 \text{ mol Cl}}{1 \text{ mol AgCl}} \times \frac{35.45 \text{ g Cl}}{1 \text{ mol Cl}} = 0.11\underline{4}8 \text{ g Cl (114.8 mg Cl)}$$

$$\text{Mass \% Cl} = \frac{\text{mass Cl}}{\text{mass comp}} \times 100\% = \frac{114.8 \text{ mg}}{328 \text{ mg}} \times 100\% = 35.0\% \text{ Cl}$$

To find the empirical formula, convert each mass to moles:

Mass Au = 328 mg - 114.8 mg Cl = 213.2 mg Au

$$0.1148 \text{ g Cl} \times \frac{1 \text{ mol Cl}}{35.45 \text{ g Cl}} = 0.0032\underline{3}8 \text{ mol Cl}$$

$$0.3132 \text{ g Au} \times \frac{1 \text{ mol Au}}{196.97 \text{ g Au}} = 0.0010\underline{8}2 \text{ mol Au}$$

Divide both numbers of moles by the smaller number (.001082) to find the integers:

Integer for Cl: 0.003238 mol $\div$ 0.001082 mol = 2.99, or 3

Integer for Au: 0.001082 mol $\div$ 0.001082 mol = 1.00, or 1

The empirical formula is thus $AuCl_3$.

4.109 Find the % composition of C and S from the analysis:

$$0.01665 \text{ g } CO_2 \times \frac{1 \text{ mol } CO_2}{44.01 \text{ g } CO_2} \times \frac{1 \text{ mol C}}{1 \text{ mol } CO_2} \times \frac{12.01 \text{ g C}}{1 \text{ mol C}} = 0.00454\underline{4} \text{ g C}$$

$$\% \text{C} = (0.004544 \text{ g C} \div 0.00796 \text{ g comp}) \times 100\% = 57.\underline{0}9\%$$

$$0.01196 \text{ g } BaSO_4 \times \frac{1 \text{ mol } BaSO_4}{233.39 \text{ g } BaSO_4} \times \frac{1 \text{ mol S}}{1 \text{ mol } CO_2} \times \frac{12.01 \text{ g C}}{1 \text{ mol } BaSO_4} = 0.0016\underline{4}3 \text{ g S}$$

$$\% \text{S} = (0.001643 \text{ g S} \div 0.00796 \text{ g comp}) \times 100\% = 38.\underline{1}2\%$$

$$\% \text{H} = 100.00\% - (57.09 + 38.12)\% = 4.\underline{7}9\%$$

We now obtain the empirical formula by calculating moles from the grams corresponding to each mass percentage of element:

$$57.09 \text{ g C} \times \frac{1 \text{ mol C}}{12.01 \text{ g C}} = 4.75\underline{3} \text{ mol C}$$

$$38.12 \text{ g S} \times \frac{1 \text{ mol S}}{32.06 \text{ g S}} = 1.18\underline{9} \text{ mol S}$$

$$4.79 \text{ g H} \times \frac{1 \text{ mol H}}{1.008 \text{ g H}} = 4.7\underline{5}2 \text{ mol H}$$

Dividing the moles of the elements by the smallest number (1.189), we obtain for C: 3.997, or 4; for S: 1.000, or 1; and for H: 3.996, or 4. Thus the empirical formula is C_4H_4S (formula weight = 84). Since the formula weight was given as 84 amu, the molecular formula is also C_4H_4S.

4.111 If one heme molecule contains one iron atom, then the number of moles of heme in 35.2 mg heme must be the same as the number of moles of iron in 3.19 mg of iron. Start by calculating the moles of Fe (= moles heme):

$$3.19 \times 10^{-3} \text{ g Fe} \times \frac{1 \text{ mol Fe}}{55.85 \text{ g Fe}} = 5.7\underline{1}2 \times 10^{-5} \text{ mol Fe or heme}$$

$$\text{Molar mass of heme} = \frac{35.2 \times 10^{-3} \text{ g}}{5.712 \times 10^{-5} \text{ mol}} = 61\underline{6}.2 = 616 \text{ g/mol}$$

The molecular weight of heme is thus 616 amu.

4.113 For g $CaCO_3$, use this equation: $CaCO_3 + H_2C_2O_4 \rightarrow CaC_2O_4 + H_2O + CO_2(g)$

$$0.472 \text{ g } CaC_2O_4 \times \frac{1 \text{ mol } CaC_2O_4}{128.10 \text{ g } CaC_2O_4} \times \frac{1 \text{ mol } CaCO_3}{1 \text{ mol } CaC_2O_4} \times \frac{100.1 \text{ g } CaCO_3}{1 \text{ mol } CaCO_3} = 0.36\underline{8}8 \text{ g } CaCO_3$$

$$\text{Mass \% } CaCO_3 = \frac{\text{mass } CaCO_3}{\text{mass limestone}} \times 100\% = \frac{0.36\underline{8}8 \text{ g}}{0.413 \text{ g}} \times 100\% = 89.\underline{2}9 = 89.3\%$$

4.115 Calculate the theoretical yield using this equation: $2C_2H_4 + O_2 \rightarrow 2C_2H_4O$

$$10.6 \text{ g } C_2H_4 \times \frac{1 \text{ mol } C_2H_4}{28.05 \text{ g } C_2H_4} \times \frac{2 \text{ mol } C_2H_4}{2 \text{ mol } C_2H_4} \times \frac{44.05 \text{ g } C_2H_4O}{1 \text{ mol } C_2H_4O} = 16.\underline{6}5 \text{ g } C_2H_4O$$

$$\% \text{ yield} = \frac{\text{actual yield}}{\text{theo yield}} \times 100\% = \frac{9.69 \text{ g}}{16.65 \text{ g}} = 58.\underline{1}9 \text{ g} = 58.2\%$$

4.117 To find Zn, use these equations:

$$2C + O_2 \rightarrow 2CO \quad \text{and} \quad ZnO + CO \rightarrow Zn + CO_2$$

2 mol C produces 2 mol CO; since 1 mol ZnO reacts with 1 mol CO, 2 mol ZnO will react with 2 mol CO. Thus 2 mol C is equivalent to 2 mol ZnO, or 1 mol C is equivalent to 1 mol ZnO. Using this to calculate mass of C from mass of ZnO, we have:

$$75.0 \text{ g ZnO} \times \frac{1 \text{ mol ZnO}}{81.38 \text{ g ZnO}} \times \frac{1 \text{ mol C}}{1 \text{ mol ZnO}} \times \frac{12.01 \text{ g C}}{1 \text{ mol C}} = 11.\underline{0}6 \text{ g C}$$

Thus all of the ZnO is used up in reacting with just 11.06 g of C, making ZnO the limiting reagent. Use the mass of ZnO to calculate the mass of Zn formed:

$$75.0 \text{ g ZnO} \times \frac{1 \text{ mol ZnO}}{81.38 \text{ g ZnO}} \times \frac{1 \text{ mol Zn}}{1 \text{ mol ZnO}} \times \frac{65.38 \text{ g Zn}}{1 \text{ mol Zn}} = 60.\underline{2}54 = 60.3 \text{ g Zn}$$

4.119 Divide the mass of $CaCl_2$ by its molar mass and volume to find molarity:

$$2.25 \text{ g } CaCl_2 \times \frac{1 \text{ mol } CaCl_2}{111.0 \text{ g } CaCl_2} \times \frac{1}{1.000 \text{ L soln}} = 0.020\underline{2}7 = 0.0203 \text{ M } CaCl_2$$

The $CaCl_2$ dissolves to form Ca^{2+} and $2Cl^-$ ions. Therefore the molarities of the ions are 0.0203 M Ca^{2+} and 2 x 0.0203, or 0.0406, M Cl^- ions.

4.121 Divide the mass of $K_2Cr_2O_7$ by its molar mass and volume to find molarity. Then calculate the volume needed to prepare 1.00L of a 0.100 M solution.

$$8.93 \text{ g } K_2Cr_2O_7 \times \frac{1 \text{ mol } K_2Cr_2O_7}{294.2 \text{ g } K_2Cr_2O_7} = 0.30\underline{3}5 \text{ mol } K_2Cr_2O_7$$

$$\text{Molarity} = \frac{0.30\underline{3}5 \text{ mol } K_2Cr_2O_7}{1.00 \text{ L}} = 0.30\underline{3}5 \text{ M}$$

$$V_i = \frac{V_f \times M_f}{M_i} = \frac{1.00 \text{ L} \times 0.100 \text{ M}}{0.3035 \text{ M}} = 0.32\underline{9}4 \text{ L (329 mL)}$$

4.123 Assume a volume of 1.000 L (1000 cm^3) for the 6.00% NaBr solution, and convert to moles, and then to molarity.

$$1000 \text{ cm}^3 \times \frac{1.046 \text{ g soln}}{1 \text{ cm}^3} \times \frac{6.00 \text{ g NaBr}}{100 \text{ g soln}} \times \frac{1 \text{ mol NaBr}}{102.89 \text{ g NaBr}} = 0.60\underline{9}97 \text{ mol}$$

$$\text{Molarity NaBr} = \frac{0.60997 \text{ mol}}{1.000 \text{ L}} = 0.60\underline{9}97 = 0.610 \text{ M}$$

4.125 From the equations $NH_3 + HCl \rightarrow NH_4Cl$, and $NaOH + HCl \rightarrow NaCl + H_2O$, we write:

$$\text{Mol } NH_3 = \text{mol } HCl(NH_3)$$
$$\text{Mol } NaOH = \text{mol } HCl(NaOH)$$

We can calculate the mol NaOH and the sum [mol $HCl(NH_3)$ + mol HCl(NaOH)] from the titration data. Since the sum = mol NH_3 + mol NaOH, we can calculate the unknown mol of NH_3 from the difference: mol NH_3 = sum - mol NaOH

$$\text{Mol } HCl(NaOH) + \text{mol } HCl(NH_3) = 0.0463 \text{ L} \times \frac{0.213 \text{ mol HCl}}{1.000 \text{ L}} = 0.0098\underline{6}2 \text{ mol HCl}$$

$$\text{Mol NaOH} = 0.0443 \text{ L} \times \frac{0.128 \text{ mol NaOH}}{1.000 \text{ L}} = 0.0056\underline{7}0 \text{ mol NaOH}$$

$$\text{Mol } HCl(NH_3) = 0.009862 \text{ mol} - 0.005670 \text{ mol} = 0.004192 \text{ mol}$$

$$\text{Mol } NH_3 = \text{mol } HCl(NH_3) = 0.004192 \text{ mol } NH_3$$

Since all of the N in the $(NH_4)_2SO_4$ was liberated as, and titrated as, NH_3, the amount of N in the fertilizer is equal to the amount of N in the NH_3. Thus the moles of NH_3 can be used to calculate the mass percentage of N in the fertilizer:

$$0.004192 \text{ mol } NH_3 \times \frac{1 \text{ mol N}}{1 \text{ mol } NH_3} \times \frac{14.01 \text{ g N mol}}{1 \text{ mol N}} = 0.058\underline{7}3 \text{ g N}$$

$$\% \text{ N} = \frac{\text{mass N}}{\text{mass fert}} \times 100\% = \frac{0.058\underline{7}3 \text{ g N}}{0.608 \text{ g}} \times 100\% = 9.6\underline{5}9 = 9.66\%$$

4.127 For $CaO + 3C \rightarrow CaC_2 + CO$, find the limiting reactant in terms of moles of CaC_2 obtainable:

$$\text{Mol } CaC_2 = 1.15 \times 10^3 \text{ g C} \times \frac{1 \text{ mol C}}{12.01 \text{ g C}} \times \frac{1 \text{ mol } CaC_2}{3 \text{ mol C}} = 31.\underline{9}1 \text{ mol}$$

$$\text{Mol } CaC_2 = 1.15 \times 10^3 \text{ g CaO} \times \frac{1 \text{ mol CaO}}{56.08 \text{ g CaO}} \times \frac{1 \text{ mol } CaC_2}{1 \text{ mol CaO}} = 17.\underline{8}3 \text{ mol}$$

Since CaO is the limiting reactant, calculate the mass of CaC_2 from it:

$$\text{Mass } CaC_2 = 20.506 \text{ mol } CaC_2 \times \frac{64.10 \text{ g } CaC_2}{1 \text{ mol } CaC_2} = 1.3\underline{1}4 \times 10^3 = 1.31 \times 10^3 \text{ g } CaC_2$$

4.129 From the equation $2Na + H_2O \rightarrow 2NaOH + H_2$, convert the mass of H_2 to mass of Na, and then use the mass to calculate the percentage:

$$0.108 \text{ g } H_2 \times \frac{1 \text{ mol } H_2}{2.016 \text{ g } H_2} \times \frac{2 \text{ mol Na}}{1 \text{ mol } H_2} \times \frac{22.99 \text{ g Na}}{1 \text{ mol Na}} = 2.4\underline{6}3 \text{ g Na}$$

$$\% \text{ Na} = \frac{\text{mass Na}}{\text{mass amalgam}} \times 100\% = \frac{2.463 \text{ g}}{15.23 \text{ g}} \times 100\% = 16.\underline{1}7 = 16.2\%$$

Cumulative-Skills Problems (require skills from Chs 1, 2, and 3 including problems)

4.131 After finding the volume of the alloy, convert it to mass Fe using density and % Fe. Then use Avogadro's number and the atomic weight for the number of atoms.

$$\text{Vol} = 10.0 \text{ cm} \times 20.0 \text{ cm} \times 15.0 \text{ cm} = 3.00 \times 10^3 \text{ cm}^3$$

$$\text{Mass Fe} = 3.00 \times 10^3 \text{ cm}^3 \times \frac{8.17 \text{ g alloy}}{1 \text{ cm}^3} \times \frac{54.7 \text{ g Fe}}{100.0 \text{ g alloy}} = 1.3\underline{4}07 \times 10^4 \text{ g}$$

$$\text{No. of Fe atoms} = 1.3\underline{4}07 \times 10^4 \text{ g Fe} \times \frac{1 \text{ mol Fe}}{55.85 \text{ g Fe}} \times \frac{6.02 \times 10^{23} \text{ Fe atoms}}{1 \text{ mol Fe}}$$

$$\text{No. of Fe atoms} = 1.4451 \times 10^{26} = 1.45 \times 10^{26} \text{ Fe atoms}$$

4.133 Use the density, formula weight, and percentage to convert to molarity. Then combine the 0.200 mol with mol/L to obtain the volume in liters.

$$\frac{0.807 \text{ g soln}}{1 \text{ mL}} \times \frac{0.94 \text{ g ethanol}}{1.00 \text{ g soln}} \times \frac{1 \text{ mol ethanol}}{46.07 \text{ g ethanol}} \times \frac{1000 \text{ mL}}{1 \text{ L}} = \frac{16.46 \text{ mol ethanol}}{\text{L ethanol}}$$

$$\text{L ethanol} = 0.200 \text{ mol ethanol} \times \frac{\text{L ethanol}}{16.46 \text{ mol ethanol}} = 0.012\underline{1}506 = 0.0122 \text{ L}$$

4.135 Convert the 2.290 g of Ag to mol AgI, which is chemically equivalent to moles of KI. Use that to calculate the molarity of the KI.

$$2.290 \text{ g AgI} \times \frac{1 \text{ mol AgI}}{234.77 \text{ g AgI}} = 9.75\underline{4}2 \times 10^{-3} \text{ mol AgI (eq. to } 9.75\underline{4}2 \times 10^{-3} \text{ mol KI)}$$

$$\text{Molarity} = \frac{9.7542 \times 10^{-3} \text{mol KI}}{0.0100 \text{ L}} = 0.975\underline{4}2 = 0.975 \text{ M}$$

4.137 Convert the 6.026 g of $BaSO_4$ to mol $BaSO_4$; then from the equation deduce that 3 mol $BaSO_4$ is equivalent to 1 mol $M_2(SO_4)_3$ and is equivalent to 2 mol of M. Use that with 1.200 g of the metal M to calculate the atomic weight of M.

$$6.026 \text{ g BaSO}_4 \times \frac{1 \text{ mol BaSO}_4}{233.39 \text{ g BaSO}_4} \times \frac{2 \text{ mol M}}{3 \text{ mol BaSO}_4} = 0.0172\underline{1}3 \text{ mol M}$$

$$\text{Atomic wt of M in g/mol} = \frac{1.200 \text{ g M}}{0.017213 \text{ mol M}} = 69.7\underline{1}5 \text{ g/mol (= gallium)}$$

4.139 Use the density, formula weight, percentage, and volume to convert to mol H_3PO_4. Then from the equation $P_4O_{10} + 6H_2O \rightarrow 4H_3P_4O_{10}$, deduce that 4 mol H_3PO_4 is equivalent to 1 mol of P_4O_{10}, and use that to convert to mol P_4O_{10} .

$$15\underline{0}0 \text{ mL} \times \frac{1.025 \text{ g soln}}{1 \text{ mL}} \times \frac{0.0500 \text{ g H}_3\text{PO}_4}{1 \text{ g soln}} \times \frac{1 \text{ mol H}_3\text{PO}_4}{97.99 \text{ g H}_3\text{PO}_4} = 0.78\underline{4}5 \text{ mol H}_3\text{PO}_4$$

$$0.7845 \text{ mol } H_3PO_4 \times \frac{1 \text{ mol } P_4O_{10}}{4 \text{ mol } H_3PO_4} = 0.19\underline{6}1 \text{ mol } P_4O_{10}$$

$$\text{Mass } P_4O_{10} = 0.1961 \text{ mol } P_4O_{10} \times \frac{283.88 \text{ g } P_4O_{10}}{\text{mol } P_4O_{10}} = 55.\underline{6}68 = 55.7 \text{ g } P_4O_{10}$$

4.141 Convert the 0.1068 g of hydrogen to mol H_2; then deduce from the equation that 3 mol H_2 is equivalent to 2 mol Al. Use the moles of Al to calculate mass Al and the percentage Al.

$$0.1068 \text{ g } H_2 \times \frac{1 \text{ mol } H_2}{2.016 \text{ g } H_2} \times \frac{2 \text{ mol Al}}{3 \text{ mol } H_2} = 0.0353\underline{1}75 \text{ mol Al}$$

$$\% \text{ Al} = \frac{0.0353\underline{1}75 \text{ mol Al} \times \dfrac{26.98 \text{ g Al}}{\text{mol Al}}}{1.118 \text{ g a alloy}} \times 100\% = 85.2\underline{2}9 = 85.23\%$$

4.143 Use the formula weight of $Al_2(SO_4)_3$ to convert to mol $Al_2(SO_4)_3$. Then deduce from the equation that 1 mol $Al_2(SO_4)_3$ is equivalent to 3 mol H_2SO_4, and calculate the moles of H_2SO_4 needed. Combine density, percentage, and formula weight to obtain molarity of H_2SO_4. Then combine molarity and moles to obtain volume.

$$18.7 \text{ g } Al_2(SO_4)_3 \times \frac{1 \text{ mol } Al_2(SO_4)_3}{342.19 \text{ g } Al_2(SO_4)_3} \times \frac{3 \text{ mol } H_2SO_4}{1 \text{ mol } Al_2(SO_4)_3} = 0.16\underline{3}9 \text{ mol } H_2SO_4$$

$$\frac{1.104 \text{ g soln}}{1 \text{ mL}} \times \frac{0.150 \text{ g } H_2SO_4}{1 \text{ g soln}} \times \frac{1 \text{ mol } H_2SO_4}{98.08 \text{ g } H_2SO_4} \times \frac{1000 \text{ mL}}{\text{L}} = 1.6\underline{8}8 \text{ mol } H_2SO_4\text{/L}$$

$$0.1639 \text{ mol } H_2SO_4 \times \frac{\text{L } H_2SO_4}{1.688 \text{ mol } H_2SO_4} = 0.097\underline{0}9 \text{ L (97.1 mL)}$$

4.145 The equations for the neutralization are:

$$2HCl + Mg(OH)_2 \rightarrow MgCl_2 + 2H_2O$$
$$3HCl + Al(OH)_3 \rightarrow AlCl_3 + 3H_2O$$

Calculate the moles of HCl and write two equations in two unknowns using the relations that mol $Mg(OH)_2$ = mol $MgCl_2$, and mol $Al(OH)_3$ = mol $AlCl_3$:

$$0.0485 \text{ L HCl} \times \frac{0.187 \text{ mol HCl}}{\text{L HCl}} = 0.0090\underline{6}95 \text{ mol HCl}$$

Rearrange the equation 0.0090695 mol HCl = 2 mol $Mg(OH)_2$ + 3 mol $Al(OH)_3$ to:

1) 0.0090695 - 2 [mol $Mg(OH)_2$] = 3 [mol $Al(OH)_3$], or
1a) 0.0030231 - 2/3 [mol $Mg(OH)_2$] = mol $Al(OH)_3$

Substitute mol $Mg(OH)_2$ for mol $MgCl_2$ (molar mass = 95.21), and mol $Al(OH)_3$ for mol $AlCl_3$ (molar mass = 133.34) in the second equation for the sum of the weights of the two chlorides:

2) [95.21 g/mol x mol $Mg(OH)_2$] + [133.34 g/mol x mol $Al(OH)_3$] = 0.4200 g

Substitute equation **1a** into equation **2** for the mol of $Al(OH)_3$:

$[95.21 \times \text{mol } Mg(OH)_2] + [133.34 \times (0.0030231 - 2/3 \text{ mol } Mg(OH)_2)] = 0.4200$

$6.317 \text{ mol } Mg(OH)_2 + 0.4031 = 0.4200$

$\text{Mol } Mg(OH)_2 = 0.0169 \div 6.317 = 0.0026\underline{7}53 \text{ mol } Mg(OH)_2$

$0.0026753 \text{ mol } Mg(OH)_2 \times 58.32 \text{ g } Mg(OH)_2/\text{mol } Mg(OH)_2 = 0.15\underline{6}02 \text{ g } Mg(OH)_2$

$\text{Mol } Al(OH)_3 = 0.0030231 - 2/3(0.0026753 \text{ mol } Mg(OH) = 0.0012395 \text{ mol } Al(OH)_3$

$0.0012395 \text{ mol } Al(OH)_3 \times 77.99 \text{ g } Al(OH)_3/\text{mol } Al(OH)_3 = 0.096668 \text{ g } Al(OH)_3$

$\% \; Mg(OH)_2 = [0.15602 \text{ g} \div (0.15602 + 0.096668) \text{ g}] \times 100\% = 61.\underline{7}4 = 61.7\%$

CHAPTER 5
THE GASEOUS STATE

SOLUTIONS TO EXERCISES

Note on significant figures: The final answer to all mathematical solutions is given first with one nonsignificant figure (last significant figure underlined) and is then rounded to the correct number of figures. Intermediate answers usually also have at least one nonsignificant figure.

5.1 Solve by rearrange the equation $gd_{Hg}h_{Hg} = gd_oh_o$ to solve for h_{Hg}:

$$h_{Hg} = \frac{d_oh_o}{d_{Hg}} = \frac{0.775 \text{ g/cm}^3 \times 7.68 \text{ cm oil}}{13.596 \text{ g/cm}^3} = 0.43\underline{8}9 \text{ cm Hg} = 4.39 \text{ mmHg}$$

5.2 Rearrange $P_fV_f = P_iV_i$ to solve for V_f (at constant T and n):

$$V_f = V_i \times \frac{P_i}{P_f} = 20.0 \text{ L} \times \frac{1.00 \text{ atm}}{0.830 \text{ atm}} = 24.\underline{0}96 = 24.1 \text{ L}$$

5.3 Rearrange $V_f/T_f = V_i/T_i$ to solve for V_f (at constant P and n). Use T_i as the sum of 19 + 273 (= 292 K), and T_f as the sum of 25 + 273 (= 298 K) in the equation:

$$V_f = V_i \times \frac{T_f}{T_i} = 4.38 \text{ dm}^3 \times \frac{298 \text{ K}}{292 \text{ K}} = 4.4\underline{7}0 = 4.47 \text{ dm}^3$$

5.4 Use the combined gas law to solve for V_f with the usual form of the equation. Use T_i as the sum of 24 + 273 (= 297), and T_f equal to the sum of 35 + 273 (= 308 K) in the equation:

$$V_f = V_i \times \frac{P_i}{P_f} \times \frac{T_f}{T_i} = 5.41 \text{ dm}^3 \times \frac{101.5 \text{ kPa}}{102.8 \text{ kPa}} \times \frac{308 \text{ K}}{297 \text{ K}} = 5.5\underline{3}9 = 5.54 \text{ dm}^3$$

5.5 Use the ideal gas law, PV = nRT, and solve for n:

$$n = \frac{PV}{RT} = P \times \frac{V}{RT}$$

If V and T are constant, n is proportional to P (R is a constant).

5.6 Convert kg of O_2 to moles O_2 and convert temperature to Kelvin. Then use the ideal gas law.

$$3.30 \text{ kg } O_2 \times \frac{1000 \text{ g}}{1 \text{ kg}} \times \frac{1 \text{ mol } O_2}{32.00 \text{ g } O_2} = 94.\underline{6}88 \text{ mol } O_2$$

$$23 + 273 = 296 \text{ K}$$

$$P = \frac{nRT}{V} = \frac{(94.688)(0.0821 \text{ L}\cdot\text{atm/K}\cdot\text{mol})(296 \text{ K})}{50.0 \text{ L}} = 46.\underline{0}2 = 46.0 \text{ atm}$$

5.7 Because density equals mass per unit volume, calculating the mass of 1 L (an exact number) of helium will yield the density of helium. Tablulate the values of the variables:

Variable	Value
P	752 mmHg x (1 atm/760 mmHg) = 0.98947 atm
V	1 L (exact number)
T	21 + 273 = 294 K
n	unknown

Using the ideal gas law, solve for n, and convert moles to grams of helium:

$$n = \frac{PV}{RT} = \frac{(0.98947 \text{ atm})(1 \text{ L})}{(0.0821 \text{ L}\cdot\text{atm/K}\cdot\text{mol})(294 \text{ K})} = 0.040\underline{9}9 \text{ mol}$$

$$0.040\underline{9}9 \text{ mol He} \times \frac{4.00 \text{ g He}}{1 \text{ mol He}} = 0.16\underline{3}96 \text{ g He}$$

$$\text{Mass air} - \text{mass He} = 1.188 \text{ g} - 0.16396 \text{ g} = 1.02\underline{4}04 = 1.024 \text{ g difference}$$

5.8 We will calculate the moles of the vapor from the ideal gas law, and then calculate the molar mass of the vapor. Tabulate the values of the variables:

Variable	Value
P	0.862 atm
V	1 L (exact number)
T	298 K
n	unknown

$$n = \frac{PV}{RT} = \frac{(0.862 \text{ atm})(1 \text{ L})}{(0.0821 \text{ L}\cdot\text{atm/K}\cdot\text{mol})(298 \text{ K})} = 0.035\underline{2}3 \text{ mol}$$

$$\text{Molar mass} = \frac{\text{grams vapor}}{\text{moles vapor}} = \frac{2.26 \text{ g}}{0.03523 \text{ mol}} = 64.\underline{1}49 \text{ g/mol}$$

Therefore the molecular weight is 64.1 amu.

5.9 Convert grams of LiOH to moles; then determine the moles of CO_2 from the balanced equation [Equation: $2LiOH(s) + CO_2(g) \rightarrow Li_2CO_3(s) + H_2O(l)$]. Finally, obtain the volume of CO_2 using the ideal gas law.

$$1.00 \text{ g LiOH} \times \frac{1 \text{ mol LiOH}}{23.9 \text{ g LiOH}} \times \frac{1 \text{ mol } CO_2}{2 \text{ mol LiOH}} = 0.020\underline{9}2 \text{ mol } CO_2$$

$$T = 22 + 273 = 295 \text{ K}$$

$$V = \frac{nRT}{P} = \frac{(0.02092 \text{ mol})(0.0821 \text{ L•atm/K•mol})(295 \text{ K})}{0.9842 \text{ atm}} = 0.51\underline{4}8 = 0.515 \text{ L}$$

5.10 Each gas obeys the ideal gas law. In each case, convert grams to moles and substitute into the ideal gas law to determine the partial pressure of each.

$$1.031 \text{ g } O_2 = \frac{1 \text{ mol } O_2}{32.00 \text{ g } O_2} = 0.0322\underline{1}88 \text{ mol } O_2$$

$$P = \frac{nRT}{V} = \frac{(0.0322188)(0.0821 \text{ L•atm/K•mol})(291 \text{ K})}{10.0 \text{ L}} = 0.076\underline{9}7 \text{ atm}$$

$$0.572 \text{ g } CO_2 \times \frac{1 \text{ mol } CO_2}{44.0 \text{ g } CO_2} = 0.012\underline{9}97 \text{ mol } CO_2$$

$$P = \frac{nRT}{V} = \frac{(0.012997)(0.0821 \text{ L•atm/K•mol})(291 \text{ K})}{10.0 \text{ L}} = 0.031\underline{0}51 \text{ atm}$$

The total pressure is equal to the sum of the partial pressures:

$$P = P_{O_2} + P_{CO_2} = 0.07697 + 0.031051 = 0.10\underline{8}02 = 0.108 \text{ atm}$$

$$\text{Mole frac } O_2 = 0.0322188 \div (0.0322188 + 0.012997) = 0.71\underline{2}55 = 0.713$$

5.11 Determine the number of moles of O_2 from the mass of $KClO_3$:

$$1.300 \text{ g } KClO_3 \times \frac{1 \text{ mol } KClO_3}{122.5 \text{ } KClO_3} \times \frac{3 \text{ mol } O_2}{2 \text{ mol } KClO_3} = 0.0159\underline{1}84 \text{ mol } O_2$$

Find the partial pressure of O_2 using Dalton's law:

$$P = P_{O_2} + P_{H_2O}$$

$$P_{O_2} = P - P_{H_2O} = (745 - 21.1) \text{ mmHg} = 72\underline{3}.9 \text{ mmHg}$$

Solve for the volume using the ideal gas law.

Variable	Value
P	723.9 mmHg x 1 atm/760.0 mmHg = 0.95$\underline{2}$5 atm
V	unknown
T	296 K
n	0.0159184

$$V = \frac{nRT}{P} = \frac{(0.0159184 \text{ mol})(0.0821 \text{ L•atm/K•mol})(298 \text{ K})}{0.9525 \text{ atm}} = 0.40\underline{6}1 = 0.406 \text{ L}$$

5.12 $$u = \sqrt{\frac{3RT}{M_m}} = \sqrt{\frac{3 \times 8.31\ \text{kg}\cdot\text{m}^2/(\text{s}^2\cdot\text{K}\cdot\text{mol}) \times 295\ \text{K}}{153.8 \times 10^{-3}\ \text{kg/mol}}} = 21\underline{8}.7 = 219\ \text{m/s}$$

5.13 Determine the average molecular speed for N_2 at 455°C (728 K):

$$u = \sqrt{\frac{3RT}{M_m}} = \sqrt{\frac{3 \times 8.31\ \text{kg}\cdot\text{m}^2/(\text{s}^2\cdot\text{K}\cdot\text{mol}) \times 728\ \text{K}}{28.01 \times 10^{3}\ \text{kg/mol}}} = 80\underline{4}.95\ \text{m/s}$$

After writing this equation with the same speed for H_2, square both sides and solve for T:

$$u^2 = \frac{3RT}{M_m}$$

$$T = \frac{u^2 M_m}{3R} = \frac{(804.95)^2(2.016 \times 10^{-3}\ \text{kg/mol})}{(3)(8.31\ \text{kg}\cdot\text{m}^2/\text{s}^2\cdot\text{K}\cdot\text{mol})} = 52.\underline{3}97 = 52.4\ \text{K}$$

Since the average kinetic energy of a molecule is proportional to T, the absolute temperature at which an H_2 molecule has the same average kinetic energy as an N_2 molecule at 728 K is 728 K.

5.14 The rate of effusion for He = 10.0 mL/3.52 s; that of O_2, rate(O_2), is unknown.

$$\frac{\text{rate}\ (O_2)}{\text{rate}\ (\text{He})} = \sqrt{\frac{M\ (\text{He})}{M\ (O_2)}}$$

$$\frac{\text{rate}\ (O_2)}{10.0\ \text{mL}/3.52\ \text{s}} = \sqrt{\frac{4.00}{32.00}} = 0.35\underline{3}55$$

$$\text{Rate}\ (O_2) = 0.35\underline{3}55 \times \frac{10.0\ \text{m}}{3.52\ \text{s}} = 1.0\underline{0}44\ \text{mL/s}$$

$$\text{Time for } O_2 \text{ to diffuse} = 1.0044\ \text{mL/s}/10.0\ \text{mL} = 9.9\underline{5}6 = 9.96\ \text{s}$$

5.15 The rate of effusion is inversely proportional to the molar mass. To solve for the molar mass of the unknown, square both sides of the equation:

$$\frac{\text{rate}\ (H_2)}{4.67 \times \text{rate}(H_2)} = \sqrt{\frac{2.016\ \text{g/mol}}{M\ (\text{gas})}}$$

$$\frac{1}{(4.67)^2} = \frac{2.016\ \text{g/mol}}{M\ (\text{gas})}$$

$$M\ (\text{gas}) = (4.67)^2(2.016\ \text{g/mol}) = 43.\underline{9}6 = 44.0\ \text{g/mol} \quad (\text{molec wt} = 44.0\ \text{amu})$$

5.16 From Table 5.7, a = 5.489 $L^2\cdot$atm/mol^2, and b = 0.06380 L/mol. Into the van der Waals equation, substitute R = 0.08206 L•atm/K•mol, T = 273.2 K, and V = 22.41 L.

$$P = \frac{nRT}{(V - nB)} - \frac{n^2a}{V^2}$$

$$P = \frac{1.000 \text{ mol} \times 0.08206 \text{ L•atm/K•mol} \times 273.2 \text{ K}}{22.41 \text{ L} - (1.000 \text{ mol} \times 0.06380 \text{ L/mol})} - \frac{(1.000 \text{ mol})^2 \times 5.489 \text{ L}^2\text{•atm/mol}^2}{(22.41 \text{ L})^2}$$

$$P = 1.00\underline{3}2 - 0.0109\underline{2}9 = 0.99\underline{2}27 = 0.992 \text{ atm}$$

Using the ideal gas law, P = $1.00\underline{0}4$ atm (larger).

ANSWERS TO REVIEW QUESTIONS

5.1 Pressure is the force exerted per unit area of surface. The SI unit of pressure is obtained:

$$P = \frac{\text{force}}{\text{area}} = \frac{\text{kg•m/s}^2}{\text{m}^2} = \text{kg/(m•s}^2\text{)} = \text{pascals}$$

5.2 A manometer measures the pressure of a gas within a vessel. A mercury manometer balances the pressure from a column of mercury against the gas pressure; the mercury column height is proportional to the gas pressure. Liquids other than mercury can be used.

5.3 The height of liquid in a manometer depends on the density of the liquid and the acceleration of gravity, g, and the pressure of the gas being measured.

5.4 From Boyle's law, PV = constant, we derive that $P_iV_i = P_fV_f$. Therefore,

$$V_f = V_i \times (P_i/P_f)$$

5.5 A linear relationship between variables such as x and y is given by the mathematical relation:

$$y = a + bx$$

The variable y is directly proportional to x only if a = 0.

5.6 When we convert -273.15°C to degrees Fahrenheit, we obtain -459.67°F. The relationship between volume, V, and temperature, T_F, in °F, is $V = a + bT_F$. At absolute zero, V = 0, so that

$$0 = a + b(-459.67), \text{ or } a = 459.67b$$

Thus we obtain:

$$V = 459.67b + bT_f, \text{ or } V = b(T_f + 459.67)$$

Thus an absolute temperature scale with °F degrees uses absolute temperature = T_F + 459.67.

5.7 From Charles's law, V/T = constant, we derive that $V_i/T_i = V_f/T_f$. Therefore

$$V_f = V_i \times (T_f/T_i)$$

5.8 Avogadro's law says that equal volumes of any two gases at the same temperature and pressure contain the same number of molecules. The reaction $N_2 + 3H_2 \rightarrow 2NH_3$ implies that Avogadro's number of N_2 molecules reacts with 3 times Avogadro's number of H_2 to form 2 times Avogadro's number of NH_3 molecules. From Avogadro's law, it follows that 1 volume of N_2 reacts with 3 volumes of H_2 to form 2 volumes of NH_3. This result is general; that is, the coefficients in a gas reaction can be interpreted in terms of volumes, in addition to numbers of molecules or moles.

5.9 The standard conditions are 0°C and 1 atm pressure.

5.10 The molar gas volume is the volume of one mole of gas. At standard conditions, it equals 22.4 L.

5.11 Boyle's law states that V is proportional to 1/P; Charles's law states that V is proportional to T. These two laws can be combined into one law: V is proportional to T/P, which can be written as an equation. For one mole of gas, $V_m = RT/P$, where V_m is the molar volume of gas. According to Avogadro's law, V_m has the same value for all gases, and $V = nV_m$, where n is the number of moles of gas. Therefore, multiplying the previous equation by n yields $nV_m = nRT/P$, $V = nRT/P$, or $PV = nRT$.

5.12 The variables in the ideal gas law are P, V, and T. The SI units of these variables are pascals(P), cubic meters(V), and kelvins(T).

5.13 Convert the value of R in units of liter-atmospheres per kelvin-mole to liter-mmHg per kelvin-mole:

$$0.0821 \frac{\text{L}\bullet\text{atm}}{\text{K}\bullet\text{mol}} \times \frac{760 \text{ mmHg}}{1 \text{ atm}} = 62.4 \frac{\text{L}\bullet\text{mmHg}}{\text{K}\bullet\text{mol}}$$

5.14 Six empirical gas laws can be obtained, and can be stated as follows:

$P \times V$ = constant (T and n constant)
P/T = constant (V and n constant)
P/n = constant (T and V constant)
V/T = constant (P and n constant)
V/n = constant (P and T constant)
$n \times T$ = constant (P and V constant)

5.15 The postulates are: (1) Gases are composed of molecules, whose sizes are negligible compared with the distance between the molecules. The fact that gases are compressible is in agreement with this postulate. (2) Molecules move randomly in linear motion. The random motion of molecules explains Brownian motion. (3) The forces of attraction or repulsion between molecules in a gas are very weak. This explains why a gas fills any container. (4) The collisions of gas molecules are elastic. If this were not true, we might expect the average speed of the molecules to continue to decrease. In that case, we would see a steady drop in pressure. No such thing is observed. (5) The average kinetic energy of a molecule in a gas is proportional to the absolute temperature. This explains why Brownian motion increases with temperature.

5.16 At a constant temperature (for Boyle's law), the average molecular force from collision is constant. Kinetic theory predicts that increasing the volume of a gas decreases the number of molecules per unit volume. This decreases the frequency of collisions per unit wall area.

5.17 Kinetic theory says that gas pressure on a container wall results from the bombardment of the wall by the gas molecules.

5.18 The rms speed of a molecule equals $(3RT/M_m)^{1/2}$, where M_m is the molar mass of the gas. The rms speed does not depend on the molar volume.

5.19 For a molecule to diffuse over a given distance, it must travel a very long, crooked path as the result of collisions with other molecules.

5.20 Effusion is the passage of a gas through a very small hole in a vessel. It results from the chance encounters of gas molecules with the hole. The rate of effusion therefore depends on the average molecular speed, which depends inversely on molecular mass.

5.21 The behavior of a gas begins to deviate significantly from that predicted by the ideal gas law at high pressures and relatively low temperatures.

5.22 The "a" constant in the van der Waals equation is related to intermolecular forces. The "b" constant is related to the molecular volume.

SOLUTIONS TO PRACTICE PROBLEMS

Note on significant figures: The final answer is given first with one nonsignificant figure (rightmost significant figure underlined), and is then rounded to the correct number of significant figures. Intermediate answers usually also have at least one nonsignificant figure. Atomic weights are rounded to two decimal places, except for that of hydrogen.

5.23 Rearrange the equation $gd_{Hg}H_{Hg} = gd_oh_o$, and solve for h_{Hg}:

$$h_{Hg} = \frac{d_{oil}}{d_{Hg}} \times h_{oil} = 75.7 \text{ mm} \times \frac{0.786 \text{ g oil/mL oil}}{13.596 \text{ g Hg/mL Hg}} = 4.3\underline{7}63 = 4.38 \text{ mmHg}$$

5.25 Rearrange the equation $gd_{Hg}H_{Hg} = gd_{liq}H_{liq}$, and solve for d_{liq}.

$$d_{liq} = \frac{h_{Hg}}{h_{liq}} \times d_{Hg} = 13.596 \text{ g Hg/mL Hg} \times \frac{8.56 \text{ mmHg}}{95.6 \text{ mm liq}} = 1.2\underline{1}77 = 1.22 \text{ g liq/mL liq}$$

5.27 Using Boyle's law, solve for V_f:

$$V_f = V_i \times \frac{P_i}{P_f} = 6.50 \text{ L} \times \frac{1.50 \text{ atm}}{2.50 \text{ atm}} = 3.9\underline{0}0 = 3.90 \text{ L}$$

5.29 Using Boyle's law, let V_f = volume at 1.25 atm, V_i = 849 L, and P_i = 1.00 atm.

$$V_f = V_i \times \frac{P_i}{P_f} = 849 \text{ L} \times \frac{1 \text{ atm}}{1.25 \text{ atm}} = 67\underline{9}.2 = 679 \text{ L}$$

5.31 Using Boyle's law, let P_i = pressure of 345 cm^2 of gas and solve for it:

$$P_i = P_f \times \frac{V_f}{V_i} = 2.51 \text{ kPa} \times \frac{0.0457 \text{ cm}^3}{345 \text{ cm}^3} = 3.3\underline{2}4 \times 10^{-4} = 3.32 \times 10^{-4} \text{ kPa}$$

5.33 Use Charles's law: T_i = 18°C + 273 = 291 K, and T_f = 0°C + 273 = 273 K.

$$V_f = V_i \times \frac{T_f}{T_i} = 2.67 \text{ mL} \times \frac{273 \text{ K}}{291 \text{ K}} = 2.5\underline{0}48 = 2.50 \text{ mL}$$

5.35 Use Charles's law: $T_i = 22^oC + 273 = 295$ K, and $T_f = -197^oC + 273 = 76$ K.

$$V_f = V_i \times \frac{T_f}{T_i} = 2.54 \text{ L} \times \frac{76 \text{ K}}{295 \text{ K}} = 0.6\underline{5}4 = 0.65 \text{ L}$$

5.37 Use Charles's law: $T_i = 25^oC + 273 = 298$ K, and V_f is the difference between the vessel's volume of 39.5 cm^3 and the 18.8 cm^3 of ethanol that is forced into the vessel.

$$T_f = T_i \times \frac{V_f}{V_i} = 298 \text{ K} \times \frac{(39.5 - 18.8) \text{ cm}^3}{39.5 \text{ cm}^3} = 15\underline{6}.1 \text{ K } (-11\underline{6}.9 \text{ or } -117°\text{C})$$

5.39 Use the combined law: $T_i = 31^oC + 273 = 304$ K, and $T_f = 0^oC + 273 = 273$ K.

$$V_f = V_i \times \frac{P_i}{P_f} \times \frac{T_f}{T_i} = 41.3 \text{ mL} \times \frac{753 \text{ mmHg}}{760 \text{ mmHg}} \times \frac{273 \text{ K}}{304 \text{ K}} = 36.\underline{7}4 = 36.7 \text{ mL}$$

5.41 The balanced equation is:

$$4NH_3 + 5O_2 \rightarrow 4NO + 6H_2O$$

The ratio of moles of NH_3 to moles of NO = 4 to 4, or 1 to 1, so 1 volume of NH_3 will produce 1 volume of NO at the same temperature and pressure.

5.43 Solve the ideal gas law for V:

$$V = \frac{nRT}{P} = nRT\left(\frac{1}{P}\right)$$

If the temperature and number of moles are held constant, then the product nRT is constant, and volume is inversely proportional to pressure:

$$V = \text{constant} \times \frac{1}{P}$$

5.45 Calculate the moles of neon, and then solve the ideal gas law for P:

$$n = 61.2 \text{ g} \times \frac{1 \text{ mol}}{20.18 \text{ g}} = 3.0\underline{3}2 \text{ mol}$$

$$P = \frac{nRT}{V} = \frac{(3.032 \text{ mol})(0.08206 \text{ L•atm/K•mol})(296 \text{ K})}{9.76 \text{ L}} = 7.5\underline{4}57 = 7.55 \text{ atm}$$

5.47 Using $T = 55^oC + 273 = 328$ K, solve the ideal gas law for V:

$$V = \frac{nRT}{P} = \frac{(2.50 \text{ mol})(0.08206 \text{ L•atm/K•mol})(328 \text{ K})}{1.50 \text{ atm}} = 44.\underline{8}59 = 44.9 \text{ L}$$

5.49 Solve the ideal gas law for temperature in K, and convert to °C:

$$T = \frac{(3.50 \text{ atm})(4.00 \text{ L})}{(0.410 \text{ mol})(0.08206 \text{ L•atm/K•mol})} = 41\underline{6}.1 = 416 \text{ K}$$

$$^{o}C = 416 - 273 = 143^{o}C$$

5.51 Because density equals mass per unit volume, calculating the mass of 1 L (exact number) of gas will give the density of the gas. Start from the ideal gas law and calculate n; then convert the moles of gas to grams using the molar mass.

$$n = \frac{(3.50 \text{ atm})(1 \text{ L})}{(398 \text{ K})(0.08206 \text{ L•atm/K•mol})} = 0.10\underline{7}1 \text{ mol}$$

$$0.10\underline{7}1 \text{ mol} \times \frac{16.04 \text{ g}}{1 \text{ mol}} = 1.7\underline{1}7 \text{ g}$$

Therefore, the density of CH_4 at 125°C is 1.72 g/L.

5.53 As above, calculate the mass of 1 L (exact number) using the ideal gas law; convert to mass.

$$n = \frac{(1.00 \text{ atm})(1 \text{ L})}{(298 \text{ K})(0.08206 \text{ L•atm/K•mol})} = 0.040\underline{8}9 \text{ mol}$$

$$0.040\underline{8}9 \text{ mol} \times \frac{58.12 \text{ g}}{1 \text{ mol}} = 2.3\underline{7}6 = 2.38 \text{ g}$$

Therefore the density of C_4H_{10} is 2.38 g/L.

5.55 The moles in 1 L (exact number) of the compound is obtained from the ideal gas law. The mass of 1.143 g is then divided by the moles to obtain the molar mass and molecular weight.

$$n = \frac{(720/760 \text{ atm})(1 \text{ L})}{(363 \text{ K})(0.08206 \text{ L•atm/K•mol})} = 0.31\underline{8}03 \text{ mol}$$

$$\text{Molar mass} = \frac{1.434 \text{ g}}{0.31803 \text{ mol}} = 45.\underline{09} \text{ g/mol}$$

The molecular weight is 45.1 amu.

5.57 The moles in 250 mL (0.250 L) of the compound is obtained from the ideal gas law. The 1.28 g mass of the gas is then divided by the moles to obtain molar mass and molecular weight.

$$n = \frac{(786/760 \text{ atm})(0.250 \text{ L})}{(394 \text{ K})(0.08206 \text{ L•atm/K•mol})} = 0.007\underline{9}96 \text{ mol}$$

$$\text{Molar mass} = \frac{1.28 \text{ g}}{0.007996 \text{ mol}} = 16\underline{0}.08 \text{ g/mol (molecular wt} = 160 \text{ amu)}$$

5.59 For a gas at a given temperature and pressure, the density depends on molecular weight (or for a mixture, the average molecular weight). Thus at the same temperature and pressure, the density of NH_4Cl gas would be greater than that of a mixture of NH_3 and HCl, since the average molecular weight of NH_3 and HCl would be lower than that of NH_4Cl.

5.61 Using the ideal gas law, find the moles of H_2 formed in the equation below; then calculate the moles of Zn from the moles of H_2.

$$Zn(s) + 2HCl(aq) \rightarrow ZnCl_2(aq) + H_2(g)$$

$$n = \frac{(765/760 \text{ atm})(2.50 \text{ L})}{(295 \text{ K})(0.08206 \text{ L}\bullet\text{atm/K}\bullet\text{mol})} = 0.10\underline{3}9 \text{ mol}$$

$$0.10\underline{3}95 \text{ mol}H_2 \times \frac{1 \text{ mol Zn}}{1 \text{ mol } H_2} \times \frac{65.38 \text{ g Zn}}{1 \text{ mol Zn}} = 6.7\underline{9}6 = 6.80 \text{ g Zn}$$

5.63 Find the moles of $CaCO_3$ and CO_2 formed in the equation below. Then, using the ideal gas law, find the volume of CO_2 formed in the equation below.

$$Ca(OH)_2(aq) + CO_2(g) \rightarrow CaCO_3(s) + H_2O(l)$$

$$2.35 \text{ g } CaCO_3 \times \frac{1 \text{ mol } CaCO_3}{100.1 \text{ g } CaCO_3} \times \frac{1 \text{ mol } CO_2}{1 \text{ mol } CaCO_3} = 0.023\underline{4}8 \text{ mol } CO_2$$

$$V = \frac{nRT}{P} = \frac{(0.023\underline{4}8 \text{ mol})(0.08206 \text{ L}\bullet\text{atm/K}\bullet\text{mol})(288 \text{ K})}{(775/760) \text{ atm}} = 0.54\underline{4}1 = 0.544 \text{ L } CO_2$$

5.65 Convert mass to moles of $KClO_3$ and then use the equation below to convert to moles of O_2. Use the ideal gas law to convert moles of O_2 to pressure at 21°C (294 K).

$$2KClO_3(s) \rightarrow 2KCl(s) + 3O_2(g)$$

$$85.0 \text{ g } KClO_3 \times \frac{1 \text{ mol } KClO_3}{122.55 \text{ g } KClO_3} = 0.69\underline{3}5 \text{ mol } KClO_3$$

$$0.69\underline{3}5 \text{ mol } KClO_3 \times \frac{3 \text{ mol } O_2}{2 \text{ mol } KClO_3} = 1.0\underline{4}02 \text{ mol } O_2$$

$$P = \frac{nRT}{V} = \frac{(1.0\underline{4}02 \text{ mol } O_2)(0.08206 \text{ L}\bullet\text{atm/K}\bullet\text{mol})(294 \text{ K})}{2.50 \text{ L}} = 10.\underline{0}3 = 10.0 \text{ atm } O_2$$

5.67 Convert mass of O_2, and mass of He, to moles. Use the ideal gas law to calculate the partial pressures, and then add to obtain the total pressures.

$$0.00103 \text{ g } O_2 \times \frac{1 \text{ mol } O_2}{32.0 \text{ g } O_2} = 3.2\underline{1}9 \times 10^{-5} \text{ mol } O_2$$

$$0.00041 \text{ g He} \times \frac{1 \text{ mol He}}{4.00 \text{ g He}} = 1.\underline{0}25 \times 10^{-4} \text{ mol He}$$

$$P_{O_2} = \frac{nRT}{V} = \frac{(3.219 \times 10^{-5}\ \text{mol})(0.08206\ \text{L•atm/K•mol})(288\ \text{K})}{0.2000\ \text{L}} = 0.0038\underline{0}4\ \text{atm}\ O_2$$

$$P_{He} = \frac{nRT}{V} = \frac{(1.025 \times 10^{-4}\ \text{mol})(0.08206\ \text{L•atm/K•mol})(288\ \text{K})}{0.2000\ \text{L}} = 0.01\underline{2}11\ \text{atm He}$$

$$P = P_{O_2} + P_{He} = (0.003804 + 0.01211)\ \text{atm} = 0.01\underline{5}91 = 0.016\ \text{atm}$$

5.69 For each gas, P(gas) = P x (mole fraction of gas)

$$P(H_2) = 760\ \text{mmHg} \times 0.250 = 19\underline{0}.0 = 190\ \text{mmHg}$$

$$P(CO_2) = 760\ \text{mmHg} \times 0.650 = 49\underline{4}.0 = 494\ \text{mmHg}$$

$$P(HCl) = 760\ \text{mmHg} \times 0.054 = 4\underline{1}.04 = 41\ \text{mmHg}$$

$$P(HF) = 760\ \text{mmHg} \times 0.028 = 2\underline{1}.28 = 21\ \text{mmHg}$$

$$P(SO_2) = 760\ \text{mmHg} \times 0.017 = 1\underline{2}.92 = 13\ \text{mmHg}$$

$$P(H_2S) = 760\ \text{mmHg} \times 0.001 = 0.\underline{7}6 = 0.8\ \text{mmHg}$$

5.71 The total pressure is the sum of the partial pressures of CO and H_2O, so:

$$P_{CO} = P - P_{water} = 689\ \text{mmHg} - 23.8\ \text{mmHg} = 66\underline{5}.2\ \text{mmHg}$$

$$n_{CO} = \frac{P_{CO}V}{RT} = \frac{(665/760\ \text{atm})(3.85\text{L})}{(0.08206\ \text{L•atm/K•mol})(298\ \text{K})} = 0.13\underline{7}8\ \text{mol CO}$$

$$0.13\underline{7}8\ \text{mol CO} \times \frac{1\ \text{mol HCOOH}}{1\ \text{mol CO}} \times \frac{46.03\ \text{g HCOOH}}{1\ \text{mol HCOOH}} = 6.3\underline{4}2 = 6.34\ \text{g HCOOH}$$

5.73 Substitute 298 K (25°C) and 398 K (125°C) into Maxwell's distribution:

$$u_{25} = \sqrt{\frac{3RT}{M_m}} = \sqrt{\frac{3 \times 8.31\ \text{kg•m}^2/(\text{s}^2\text{•K•mol}) \times 298\ \text{K}}{28.02 \times 10^{-3}\ \text{kg/mol}}} = 51\underline{4}.9 = 515\ \text{m/s}$$

$$u_{125} = \sqrt{\frac{3RT}{M_m}} = \sqrt{\frac{3 \times 8.31\ \text{kg•m}^2/(\text{s}^2\text{•K•mol}) \times 398\ \text{K}}{28.02 \times 10^{-3}\ \text{kg/mol}}} = 59\underline{5}.07 = 595\ \text{m/s}$$

Graph as in Figure 5.14.

5.75 Substitute 330 K(57°C) into Maxwell's distribution:

$$u_{330\ K} = \sqrt{\frac{3RT}{M_m}} = \sqrt{\frac{3 \times 8.31\ \text{kg•m}^2/(\text{s}^2\text{•K•mol}) \times 330\ \text{K}}{352 \times 10^{-3}\ \text{kg/mol}}} = 15\underline{2}.8 = 153\ \text{m/s}$$

5.77 Since $u(CO_2) = u(H_2)$, we can equate the two right-hand sides of the Maxwell distributions:

$$\sqrt{\frac{3RT(CO_2)}{M_m(CO_2)}} = \sqrt{\frac{3RT(H_2)}{M_m(H_2)}}$$

Squaring both sides, rearranging to solve for $T(CO_2)$, and substituting numerical values:

$$T(CO_2) = T(H_2) \times \frac{M_m(CO_2)}{M_m(H_2)} = 293 \text{ K} \times \frac{44.01 \text{ g/mol}}{2.016 \text{ g/mol}} = 63\underline{9}6 = 6400 \text{ K}$$

5.79 Since the ratio is the same at any temperature, $T(N_2) = T(O_2)$. Write a ratio of two Maxwell distributions after omitting $T(O_2)$ and $T(H_2)$ in each distribution:

$$\frac{u_{N_2} = \sqrt{\frac{3RT(N_2)}{M_m}} = \sqrt{\frac{3 \times 8.31 \text{ kg}\cdot\text{m}^2/(\text{s}^2\cdot\text{K}\cdot\text{mol})}{28.02 \times 10^{-3} \text{ kg/mol}}}}{u_{O_2} = \sqrt{\frac{3RT(O_2)}{M_m}} = \sqrt{\frac{3 \times 8.31 \text{ kg}\cdot\text{m}^2/(\text{s}^2\cdot\text{K}\cdot\text{mol})}{32.00 \times 10^{-3} \text{ kg/mol}}}}$$

$$\frac{u_{N_2} = \; = \sqrt{889.721}}{u_{O_2} = \; = \sqrt{779.062}} = \frac{29.828}{27.911} = \frac{1.0686}{1} = \frac{1.069}{1}$$

5.81 Since the ratio is the same at any temperature, $T(H_2) = T(I_2)$. A ratio of two Maxwell distributions can be written as in the previous two problems, but this can also be simplified by cancelling the 3 x 8.31 terms, and rearranging the denominators to give:

$$\frac{u_{H_2} = \; = \sqrt{M_m(I_2)}}{u_{I_2} = \; = \sqrt{M_m(H_2)}} = \frac{\sqrt{253.8}}{\sqrt{2.016}} = \frac{11.2205}{1} = \frac{11.22}{1}$$

Since hydrogen diffuses 11.22 times as fast as iodine, the time it would take would be 1/11.22 of the time required for iodine:

$$t(H_2) = 52 \text{ s} \times (1/11.22) = 4.\underline{6}3 = 4.6 \text{ s}$$

5.83 Since the diffusion occurs at the same temperature, T(gas) = T(Ar). A ratio of two Maxwell distributions can be written, but it can be simplified as in the previous two problems by cancelling the 3 x 8.31 terms, and rearranging the denominators. To simplify the definition of the rates, we assume the time is 1 second, and define the rate(Ar) as 9.23 mL/1 s, and the rate(gas) as 4.83 mL/1 s. Then we write a ratio of two Maxwell distributions:

$$\frac{u_{gas} = 4.83 \text{ mL/1 s} = \sqrt{M_m(Ar)} = \sqrt{39.95}}{u_{Ar} = 9.23 \text{ ml/1 s} = \sqrt{M_m(gas)} = \sqrt{M_m(gas)}}$$

$$\sqrt{M_m(gas)} = \frac{9.23 \text{ mL/1 s}}{4.83 \text{ ml/1 s}} \times \sqrt{39.95} = 12.\underline{0}7$$

$M_m(gas) = 14\underline{5}.8$ g/mol; molecular weight = 146 amu

5.85 Solving the van der Waals equation for n = 1 and T = 355.2 K for P gives:

$$P = \frac{RT}{(V - b)} - \frac{a}{V^2} = \frac{(0.08206\ \text{L•atm/K•mol}) \times 355.2\ \text{K}}{(35.00\ \text{L} - 0.08407\ \text{L})} - \frac{12.02\ \text{L}^2\text{•atm}}{(35.00\ \text{L})^2}$$

$$P = 0.834\underline{7}9 - 0.00981\underline{2}2 = 0.824\underline{9}8 = 0.8250\ \text{atm}$$

$$P(\text{ideal gas law}) = 0.83\underline{2}8\ \text{atm}$$

5.87 To calculate a/V^2 in the van der Waals equation, we obtain V from the ideal gas law at 1.00 atm:

$$V = \frac{RT}{P} = \frac{(0.08206\ \text{L•atm/K•mol}) \times 273\ \text{K}}{1.00\ \text{atm}} = 22.\underline{4}0\ \text{L}$$

$$\frac{a}{V^2} = \frac{5.489}{(22.4\ \text{L})^2} = 1.0\underline{9}4 \times 10^{-2}$$

At 1.00 atm, V = 22.4 L, and $a/V^2 = 1.094 \times 10^{-2}$. Substituting into the van der Waals equation:

$$V = \frac{RT}{P + \frac{a}{V^2}} + b = \frac{(0.08206\ \text{L•atm/K•mol})(273\ \text{K})}{(1.00 + 1.094 \times 10^{-2})\text{atm}} + 0.06380 = 22.\underline{2}2 = 22.2\ \text{L}$$

At 10.0 atm, the van der Waals equation gives 2.08 L. The ideal gas law and Table 5.3 give 22.4 L for 1.00 atm and 2.24 L for 10.0 atm.

5.89 Calculate the mass of 1 cm^2 of the 20.5 m of water above the air in the glass. The volume is the product of the area of 1 cm^2 and the height of 20.5 x 10^2 cm (20.5 m) of water. The density of 1.00 g/cm^3 must be used to convert volume to mass:

$$m = d \times V$$

$$m = 1.00\ \text{g/cm}^2 \times (1.00\ \text{cm}^2 \times 20.5 \times 10^2\ \text{cm}) = 2.05 \times 10^3\ \text{g, or } 2.05\ \text{kg}$$

The pressure exerted on an object at the bottom of the column of water is:

$$P = \frac{\text{force}}{\text{area}} = \frac{(m)(g)}{\text{area}} = \frac{(2.05\ \text{kg})(9.807\ \text{m/s2})}{1.00\ \text{cm2} \times \left[\frac{10^{-2}\text{m}}{1\ \text{cm}}\right]^2} = 2.01 \times 10^5\ \text{kg/ms2} = 2.01 \times 10^5\ \text{Pa}$$

The total pressure on the air in the tumbler = the barometric pressure and the water pressure:

$$P = 1.00 \times 10^2\ \text{kPa} + 2.01 \times 10^2\ \text{kPa} = 3.01 \times 10^2\ \text{kPa}$$

Multiply the initial volume by a factor accounting for the change in pressure to find V_f:

$$V_f = V_i\left[\frac{P_i}{P_f}\right] = 243\ \text{cm}^3\left[\frac{1.00 \times 10^2\ \text{kPa}}{3.01 \times 10^2\ \text{kPa}}\right] = 80.\underline{7}3 = 80.7\ \text{cm}^3$$

5.91 Use the combined gas law, and solve for V_f :

$$V_f = V_i \times \frac{P_i}{P_f} \times \frac{T_f}{T_i} = 183 \text{ mL} \times \frac{738 \text{ mmHg}}{760 \text{ mmHg}} \times \frac{273 \text{ K}}{294 \text{ K}} = 16\underline{5}.009 = 165 \text{ mL}$$

5.93 Use the combined gas law, and solve for V_f :

$$V_f = V_i \times \frac{P_i}{P_f} \times \frac{T_f}{T_i} = 5.0 \text{ dm}^3 \times \frac{100.0 \text{ kPa}}{79.0 \text{ kPa}} \times \frac{293 \text{ K}}{287 \text{ K}} = 6.\underline{4}6 = 6.5 \text{ dm}^3$$

5.95 Use the ideal gas law to calculate the moles of helium and combine this with Avogadro's number to obtain the number of helium atoms:

$$n = \frac{PV}{RT} = \frac{(765/760 \text{ atm})(0.01205 \text{ L})}{(0.08206 \text{ L·atm/K·mol})(296 \text{ K})} = 4.9\underline{9}3 \times 10^{-4} \text{ mol}$$

$$4.993 \times 10^{-4} \text{ mol He} \times \frac{6.02 \times 10^{23} \text{ He}^{2+} \text{ ions}}{1 \text{ mol He}} \times \frac{1 \text{ atom}}{1 \text{ He}^{2+} \text{ ion}} = 3.0\underline{0}57 \times 10^{20}$$

$$= 3.01 \times 10^{20} \text{ atoms}$$

5.97 Calculate the molar mass, M_m, by dividing the mass of 1 L of air by the moles of the gas from the ideal gas equation:

$$M_m = \frac{\text{mass}}{n} = 1.2929 \text{ g air} \times \frac{(0.082057 \text{ L·atm/K·mol})(273.15 \text{ K})}{(1 \text{ atm})(1 \text{ L})} = 28.97\underline{8}9 \text{ g/mol}$$

$$= 28.979 \text{ g/mol (amu)}$$

5.99 Use the ideal gas law to calculate the moles of CO_2. Then convert to mass of LiOH.

$$n = \frac{PV}{RT} = \frac{(1.00 \text{ atm})(5.8 \times 10^2 \text{ L})}{(0.08206 \text{ L·atm/K·mol})(273 \text{ K})} = 2\underline{5}.89 \text{ mol CO}_2$$

$$25.89 \text{ mol CO}_2 \times \frac{2 \text{ mol LiOH}}{1 \text{ mol CO}_2} \times \frac{23.95 \text{ g LiOH}}{1 \text{ mol LiOH}} = 1.\underline{2}4 \times 10^3 = 1.2 \times 10^3 \text{ g LiOH}$$

5.101 Use Maxwell's distribution to calculate the temperature in kelvins; then convert to °C.

$$T = \frac{u^2 M_m}{3R} = \frac{(0.510 \times 10^3 \text{ m/s})^2(17.03 \times 10^{-3} \text{ kg/mol})}{3(8.31 \text{ kg·m}^2/\text{s}^2\text{·K·mol})} = 17\underline{7}.6 \times 10^3 = 178 \text{ K } (-95°\text{C})$$

5.103 Calculate the ratio of the root-mean-square molecular speeds, which is the same as the ratio of the rates of effusion:

$$\frac{u \text{ of U(235)F}_6}{u \text{ of U(238)F}_6} = \frac{\sqrt{M_m[\text{U(238)F}_6]}}{\sqrt{M_m[\text{U(235)F}_6]}} = \frac{\sqrt{352.04 \text{ g/mol}}}{\sqrt{349.03 \text{ g/mol}}} = \frac{1.004302}{1} = \frac{1.0043}{1}$$

5.105 First calculate the apparent molar masses at each pressure, using the ideal gas law. Only the calculation of the apparent molar mass for 0.2500 atm will be shown; the other values will be summarized in a table.

$$n = \frac{PV}{RT} = \frac{(0.2500\ \text{atm})(3.1908\ \text{L})}{(0.082057\ \text{L}\bullet\text{atm/K}\bullet\text{mol})(273.15\ \text{K})} = 3.55\underline{8}96 \times 10^{-2}\ \text{mol}$$

$$\text{Apparent molar mass} = \frac{1.000\ \text{g}}{3.55896 \times 10^{-2}\ \text{mol}} = 28.0\underline{9}8 = 28.10\ \text{g/mol}$$

The following table summarizes the apparent molar masses calculated as above for all P's; these data are plotted in the graph to the right of the table.

P (atm)	App. Molar Mass (g/mol)
0.2500	28.10
0.5000	28.14
0.7500	28.19
1.0000	28.26

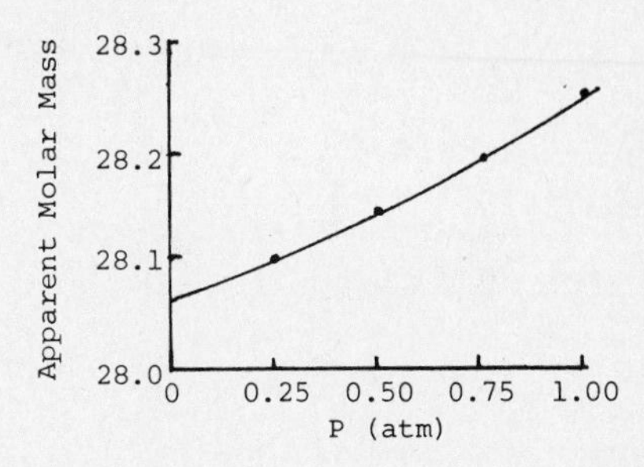

Extrapolation back to P = 0 gives 28.07 g/mol for the molar mass of the unknown gas (CO).

5.107 Use $CO + 1/2\ O_2 \rightarrow CO_2$, instead of 2CO. First find the moles of CO and O_2 by using the ideal gas law.

$$n_{CO} = \frac{PV}{RT} = \frac{(0.500\ \text{atm})(2.00\ \text{L})}{(0.08206\ \text{L}\bullet\text{atm/K}\bullet\text{mol})(300\ \text{K})} = 0.040\underline{6}2\ \text{mol}$$

$$n_{O_2} = \frac{PV}{RT} = \frac{(1.00\ \text{atm})(1.00\ \text{L})}{(0.08206\ \text{L}\bullet\text{atm/K}\bullet\text{mol})(300\ \text{K})} = 0.040\underline{6}2\ \text{mol}$$

There are equal amounts of CO and O_2, but (from the equation) only half as many moles of O_2 as CO are required for the reaction. Therefore, when 0.04062 moles of CO have been consumed, only 0.04062/2 moles of O_2 will have been used up. Then 0.04062/2 mol O_2 will remain and 0.04062 mol of CO_2 will have been produced. At the end:

$$n_{CO} = 0\ \text{mol};\ n_{O_2} = 0.0203\ \text{mol; and}\ n_C = 0.04062\ \text{mol}$$

However, the total volume with the value open is 3.00 L, so the partial pressures of O_2 and CO_2 must be calculated from the ideal gas law for each:

$$\frac{nRT}{V} = \frac{(0.0203\ \text{mol}\ O_2)(0.08206\ \text{L}\bullet\text{atm/K}\bullet\text{mol})(300\ \text{K})}{(3.00\ \text{L})} = 0.16\underline{6}58 = 0.167\ \text{atm}\ O_2$$

$$\frac{nRT}{V} = \frac{(0.04062\ \text{mol}\ CO_2)(0.08206\ \text{L}\bullet\text{atm/K}\bullet\text{mol})(300\ \text{K})}{(3.00\ \text{L})} = 0.33\underline{3}32 = 0.333\ \text{atm}\ CO_2$$

Cumulative-Skills Problems (require skills from previous chapters)

5.109 Assume a 100.0 g sample, giving 85.2 g CH_4 and 14.8 g C_2H_6. Convert each to moles:

$$85.2 \text{ g } CH_4 \times \frac{1 \text{ mol } CH_4}{16.04 \text{ g}} = 5.3\underline{1}1 \text{ mol } CH_4;\ 14.8 \text{ g } C_2H_6 \times \frac{1 \text{ mol } C_2H_6}{30.07 \text{ g}} = 0.49\underline{2}2 \text{ mol } C_2H_6$$

$$V_{CH_4} = \frac{(5.311 \text{ mol})(0.08206 \text{ L•atm/K•mol})(291 \text{ K})}{(748/760) \text{ atm}} = 12\underline{8}.9 \text{ L}$$

$$V_{C_2H_6} = \frac{(0.4922 \text{ mol})(0.08206 \text{ L•atm/K•mol})(291 \text{ K})}{(748/760) \text{ atm}} = 11.\underline{9}4 \text{ L}$$

The density is calculated as follows:

$$d = \frac{85.2 \text{ g } CH_4 + 14.8 \text{ g } C_2H_6}{(128.9 + 11.94) \text{ L}} = 0.71\underline{0}02 = 0.710 \text{ g/L}$$

5.111 First subtract the height of mercury equivalent to the 25.00 cm (250 mm) of water inside the tube from 771 mmHg to get P_{gas}. Then subtract the vapor pressure of water, 18.7 mmHg, from P_{gas} to get P_{O2}.

$$h_{Hg} = \frac{(h_w)(d_w)}{d_{Hg}} = \frac{250 \text{ mm} \times 0.99987 \text{ g/cm}^3}{13.596 \text{ g/cm}^3} = 18.\underline{3}8 \text{ mmHg}$$

$$P_{gas} = P - P_{25 \text{ cm water}} = 771 \text{ mmHg} - 18.38 \text{ mgHg} = 75\underline{2}.62$$

$$P_{O2} = 75\underline{2}.62 \text{ mmHg} - 18.7 \text{ mmHg} = 73\underline{3}.92$$

$$n = \frac{PV}{RT} = \frac{(733.92/760 \text{ atm})(0.310 \text{ L})}{(0.08206 \text{ L•atm/K•mol})(294 \text{ K})} = 0.0012\underline{4}08 \text{ mol } O_2$$

$$\text{Mass} = (2 \times 0.0012408) \text{ mol } Na_2O_2 \times \frac{77.98 \text{ g } Na_2O_2}{1 \text{ mol } Na_2O_2} = 0.19\underline{3}51 = 0.194 \text{ g } Na_2O_2$$

5.113 First find the moles of CO_2:

$$n = \frac{PV}{RT} = \frac{(785/760 \text{ atm})(1.94 \text{ L})}{(0.08206 \text{ L•atm/K•mol})(298 \text{ K})} = 0.081\underline{9}4 \text{ mol } CO_2$$

Set up one equation in one unknown: x = mol $CaCO_3$, (0.08194 - x) = mol $MgCO_3$:

$$7.85 \text{ g} = (100.1 \text{ g/mol})x + (84.32 \text{ g/mol})(0.08194 - x)$$

$$x = \frac{(7.85 - 6.9092)}{(100.1 - 84.32)} = 0.05\underline{9}62 \text{ mol } CaCO_3$$

$$(0.08194 - x) = 0.02\underline{2}32 \text{ mol } MgCO_3$$

$$\% \; CaCO_3 = \frac{0.05962 \text{ mol } CaCO_3 \times 100.1 \text{ g/mol}}{7.85 \text{ g}} \times 100\% = 7\underline{6}.02 = 76\%$$

$$\% \; MgCO_3 = 100.00\% - 76.02\% = 2\underline{3}.98 = 24\%$$

CHAPTER 6
THERMOCHEMISTRY

SOLUTIONS TO EXERCISES

Note on significant figures: The final answer to all mathematical solutions is given first with one nonsignificant figure (last significant figure underlined) and is then rounded to the correct number of figures. Intermediate answers usually also have at least one nonsignificant figure.

6.1 We substitute into the formula $E_k = 1/2\ mv^2$ using SI units:

$$E_k = 1/2 \times 9.11 \times 10^{-31}\ \text{kg} \times (5.0 \times 10^{6}\ \text{m/s})^2 = 1.\underline{1}3 \times 10^{-17} = 1.1 \times 10^{-17}\ \text{J}$$

$$1.13 \times 10^{-17}\ \text{J} \times \frac{1\ \text{cal}}{4.184\ \text{J}} = 2.\underline{7}2 \times 10^{-18} = 2.7 \times 10^{-18}\ \text{cal}$$

6.2 Heat is evolved; therefore the reaction is exothermic. The value of q is -1170 kJ.

6.3 The thermochemical equation is:

$$2N_2H_4(l) + N_2O_4(l) \rightarrow 3N_2(g) + 4H_2O(g); \quad \Delta H = -1.049 \times 10^3\ \text{kJ}$$

6.4 a. $N_2H_4(l) + 1/2N_2O_4(l) \rightarrow 3/2N_2(g) + 2H_2O(g); \quad \Delta H = -5.245 \times 10^2\ \text{kJ}$

b. $4H_2O(g) + 3N_2(g) \rightarrow 2N_2H_4(l) + N_2O_4(l); \quad \Delta H = 1.049 \times 10^3\ \text{kJ}$

6.5 The reaction is $2N_2H_4(l) + N_2O_4(l) \rightarrow 3N_2(g) + 4H_2O(g); \quad \Delta H = -1.15 \times 10^3\ \text{kJ}$

$$10.0\ \text{g}\ N_2H_4 \times \frac{1\ \text{mol}\ N_2H_4}{32.02\ \text{g}} \times \frac{1\ \text{mol}\ N_2O_4}{2\ \text{mol}\ N_2H_4} \times \frac{1.049 \times 10^3\ \text{kJ}}{1\ \text{mol}\ N_2H_4} = 16\underline{3}.80\ \text{kJ}\ (1.64 \times 10^2\ \text{J})$$

6.6 Substitute into the equation $q = s \times m \times \Delta T$ to obtain the transferred heat per gram:

$$q = s \times m \times \Delta t = \frac{0.449\ \text{J}}{\text{g}\cdot{}^\circ\text{C}} \times (100.0^\circ\text{C} - 20.0^\circ\text{C}) \times 5.00\ \text{g} = 17\underline{9}.6 = 180\ \text{J/g}$$

6.7 The heat released by the reaction, q_{rxn}, equals the negative value of the heat absorbed by the solution. Divide this by the moles of HCl to find ΔH for the reaction:

$$q = s \times m \times \Delta t = \frac{-4.184\ J}{(g)(°C)} \times (31.8°C - 25.0°C) \times (33 + 42)\ g = -2\underline{1}33.8\ J$$

$$\text{Mol HCl} = 1.20\ mol/L \times 0.033\ L = 0.03\underline{9}6\ mol$$

$$\Delta H = \frac{-2133.8\ J}{0.0396\ mol} = -5\underline{3}884\ J/mol = -54\ kJ/mol$$

$$HCl(aq) + NaOH(aq) \rightarrow NaCl(aq) + H_2O(l);\ \Delta H = -54\ kJ$$

6.8 Use Hess's law to find ΔH for $4Al(s) + 3MnO_2(s) \rightarrow 2Al_2O_3 + 3Mn(s)$ from the following data for reactions 2 and 3:

$$2Al(s) + 3/2O_2(g) \rightarrow Al_2O_3(s);\ \Delta H = -1676\ kJ$$
$$Mn(s) + O_2(g) \rightarrow MnO_2(s);\ \Delta H = -521\ kJ$$

If we take reaction 2 and multiply it by 2 and add the reverse of three times reaction 3, we obtain reaction 1 (you cancel as you would in algebraic equations). If we add the corresponding ethalpy changes, we obtain the enthalpy change of reaction 1. The layout follows:

$4Al(s) + 3O_2$	$\rightarrow 2Al_2O_3(s)$	(-1676 kJ) x (2)
$3MnO_2(s)$	$\rightarrow 3Mn(s) + 3O_2$	(- 521 kJ) x (-3)
$4Al(s) + 3MnO_2$	$\rightarrow 3Al_2O_{3(s)} + 3Mn(s)$	-1789 kJ

6.9 We will refer to each reaction below the bond-breaking reaction by numbers 1 through 5:

C(graphite)	$\rightarrow$ C(g)	715.0 kJ
$2H_2(g)$	$\rightarrow 4H(g)$	2 x 436.0 kJ
$CO_2(g)$	$\rightarrow O_2(g)$ + C(graphite)	393.5 kJ
$2H_2O(l)$	$\rightarrow 2H_2(g) + O_2(g)$	571.7 kJ
$CH_4(g) + 2O_2(g)$	$\rightarrow CO_2(g) + 2H_2O(l)$	-890.3 kJ

Multiply each reaction by a factor and add to obtain the desired equation. Since we desire that C(g) be on the right, reaction 1 is left as is. To obtain 4H(g) on the right, reaction 2 is multiplied by 2. To cancel C(graphite) and $H_2(g)$, reactions 3 and 4 are reversed. To have $CH_4(g)$ on the left, reaction 5 is added as is:

C(graphite)	$\rightarrow$ C(g)	715.0 kJ
$2H_2(g)$	$\rightarrow 4H(g)$	2 x 436.0 kJ
$CO_2(g)$	$\rightarrow O_2(g)$ + C(graphite)	393.5 kJ
$2H_2O(l)$	$\rightarrow 2H_2(g) + O_2(g)$	571.7 kJ
$CH_4(g) + 2O_2(g)$	$\rightarrow CO_2(g) + 2H_2O(l)$	-890.3 kJ
$CH_4(g)$	$\rightarrow C(g) + 4H(g)$	ΔH = 1661.9 kJ

6.10 The reaction is:

$H_2O(l) \rightarrow H_2O(g)$

-285.8 -241.8

ΔH°_{vap} = [-241.8 - (-285.8)] kJ = 44.0 kJ

6.11 The reaction is:

$3NO_2 + H_2O(l) \rightarrow 2HNO_3(aq) + NO(g)$

3 x 33.2 -285.8 2 x -206.6 90.3

ΔH° = [(2 x -206.6) + 90.3] - [(3 x 33.2) + (-285.8)] = -136.$\underline{7}$0 = -136.7 kJ

6.12 The net chemical equation is:

$2NH_4^+(aq) + 2OH^-(aq) \rightarrow 2NH_3(g) + 2H_2O(l)$

2(-132.8) 2(-229.9) 2(-45.9) 2(-285.8)

ΔH° = [2(-45.9) + 2(-285.8)] - [2(-132.8) + 2(-229.9)] = 62.$\underline{0}$0 = 62.00 kJ

6.13 The reaction is:

$4FeS_2(s) + 11O_2(g) \rightarrow 2Fe_2O_3(s) + 8SO_2(g)$

The enthalpy of reaction in terms of standard enthalpies of formation is

ΔH° = $[2\Delta H_f^{\circ}(Fe_2O_3) + 8\Delta H_f^{\circ}(SO_2)] - [4\Delta H_f^{\circ}(FeS_2) + 11\Delta H_f^{\circ}(O_2)]$

4(-828) kJ = [2(-825.5) kJ + 8(-296.8)] - [$4\Delta H_f^{\circ}(FeS_2)$ + 11(0) kJ]

Now we solve for $\Delta H_f^{\circ}(FeS_2)$:

$\Delta H_f^{\circ}(FeS_2)$ = 1/4 [4(828) kJ + 2(-825.5) kJ + 8(-296.8)] = -178.$\underline{3}$5 = -178.4 kJ

ANSWERS TO REVIEW QUESTIONS

6.1 Energy is the capacity or potential to move matter. Kinetic energy is the energy associated with an object by virtue of its motion. Potential energy is the energy an object has by virtue of its position in a field of force. Internal energy is the sum of the kinetic and potential energies of the particles making up a substance.

6.2 A joule is the SI unit of energy. One joule is the product of a force of one newton (kg•m/s^2) times a distance of 1 m. One joule is also the product of a mass of 1 kg times an acceleration of 1 m per s^2 times a distance of 1 m.

6.3 Originally a calorie was defined as the amount of energy required to raise the temperature of one gram of water by one degree Celsius. At present, the calorie is defined as 4.184 J.

6.4 At either of the two highest points above the earth in its cycle, the energy of a pendulum is all potential energy and is equal to the product mgh (m = mass of pendulum, g = constant acceleration of gravity, and h = height of pendulum). As the pendulum moves downward, its potential energy decreases from mgh to near zero, depending on how close it comes to the earth's surface. During the downward motion, its potential energy is converted to kinetic energy. When it reaches the lowest point (middle) of its cycle, the pendulum has its maximum kinetic energy and minimum potential energy. As it rises above the lowest point, its kinetic energy begins to be converted to potential energy. When it reaches the other high point in its cycle, the energy of the pendulum is again all potential energy. At rest, the energy of pendulum has been transferred to the surroundings.

6.5 As the heat flows into the gas, the gas molecules gain energy and move at a faster average speed. The internal energy of the gas increases.

6.6 An exothermic reaction evolves heat. For example, the burning of one mol of methane, $CH_4(g)$, yields carbon dioxide, water, and 890 kJ of heat. An endothermic reaction absorbs heat. For example, the reaction of one mol of barium hydroxide with ammonium thiocyanate absorbs 170.4 kJ of heat in order to form ammonia, water, and barium thiocyanate.

6.7 It means that changes in internal energy depend only on the initial and final states of the system, which are determined by variables such as temperature and pressure. Such changes do not depend on how the changes were made.

6.8 The enthalpy change equals the heat of reaction only at constant pressure (changes in temperature do not alter this equality).

6.9 The enthalpy change is positive (the enthalpy increases) for an endothermic reaction at constant pressure.

6.10 It is important to give the states when writing an equation for ΔH because ΔH depends on the states of all reactants and products. If any state changes, ΔH changes.

6.11 The value of ΔH doubles and its sign changes when the equation is doubled and reversed.

6.12 First convert the 10.0 g mass of water to moles of water, using its molar mass. Then, using the equation, multiply the moles of water by the ratio (1 mol CH_4/2 mol H_2O). Finally multiply the moles of CH_4 by -890.2 kJ/mol CH_4 (as done in Section 6.5).

6.13 The heat capacity of a substance is the quantity of heat needed to raise the temperature of the sample of substance one degree Celsius (or one kelvin). The specific heat of a substance is the quantity of heat required to raise the temperature of one gram of a substance by one degree Celsium (or one kelvin) at constant pressure.

6.14 A simple calorimeter consists of an insulated container, such as a polystyrene cup, with a thermometer positioned in the liquid held by the container. The heat of reaction is obtained by conducting the reaction in such a calorimeter. The temperature of the mixture is measured before and after the reaction. The heat capacity of the calorimeter and its contents must also be measured.

6.15 Hess's law is essentially based on the fact that if a chemical equation is the sum of multiples of other reactions, the ΔH of this equation equals a similar sum of multiples of the ΔH's of the other reactions. In other words, Hess's law is based on the additivity property of ΔH.

6.16 The heat of sublimation, $\Delta H_{sub} = \Delta H_{fus} + \Delta H_{vap}$.

6.17 The thermodynamic standard state refers to the standard thermodynamic conditions chosen for substances when listing or comparing thermodynamic data: 1 atm pressure and the specified temperature (usually 25°C).

6.18 The reference form of an element is the stablest form (physical state and allotrope) of the element under standard thermodynamic conditions. The standard enthalpy of formation of an element in its reference form is zero.

6.19 The standard enthalpy of formation of a substance is the enthalpy change for the formation of one mole of the substance from its elements at standard pressure (1 atm) and a specified temperature (25°C unless noted otherwise).

6.20 The equation for the formation of $H_2S(g)$ is:

$$H_2(g) + 1/8S_8(s) \rightarrow H_2S(g)$$

6.21 The reaction of $C(g) + 4H(g) \rightarrow CH_4(g)$ is not an appropriate equation for calculating the ΔH_f of methane because the most stable form of each element is not used. Both $H_2(g)$ and C(s), as graphite, should be used instead of H(g) and C(g).

6.22 A fuel is any substance that is burned or similarly reacted to provide heat and other forms of energy. The fossil fuels are petroleum, gas, and coal. They were formed long before our present civilization by the decomposition and compression of aquatic plants and animals beneath the sediment of swamps and seas.

6.23 One two-step set of reactions for the formation of methane from coal is:

$$C(s) + H_2O(g) \rightarrow CO(g) + H_2(g)$$

$$CO(g) + 3H_2(g) \rightarrow CH_4(g) + H_2O(g)$$

6.24 Some possible rocket fuel/oxidizer combinations are H_2/O_2 and hydrazine/dinitrogen tetraoxide. The chemical equations for their reactions are:

$$H_2(g) + 1/2O_2(g) \rightarrow H_2O(g);\ \Delta H^\circ = -242\ kJ$$

$$2N_2H_4(l) + N_2O_4(l) \rightarrow 3N_2(g) + 4H_2O(g);\ \Delta H^\circ = -1049\ kJ$$

SOLUTIONS TO PRACTICE PROBLEMS

Note on significant figures: The final answer is given first with one nonsignificant figure (rightmost significant figure underlined), and is then rounded to the correct number of significant figures. Intermediate answers usually also have at least one nonsignificant figure. Atomic weights are rounded to two decimal places, except for that of hydrogen.

6.25 The SI units for force must be kg•m/s^2 (= newton, N) to be consistent with the joule, the SI unit for energy:

$$\frac{kg \cdot m}{s^2} \times m = \frac{kg \cdot m^2}{s^2} = \text{joule, J}$$

6.27 The heat in kcal released is:

$$286\ kJ \times \frac{1000\ J}{1\ kJ} \times \frac{1\ cal}{4.184\ J} \times \frac{1\ kcal}{1000\ cal} = 68.\underline{3}55 = 68.4\ kcal$$

6.29 The kinetic energy in J, and in cal:

$$E_k = \frac{1}{2} \times 4.53 \times 10^3 \text{ lb} \times \frac{0.4536 \text{ kg}}{1 \text{ lb}} \times \left[\frac{45 \text{ mi}}{1 \text{ h}}\right]^2 \times \left[\frac{1{,}609 \text{ m}}{1 \text{ mi}}\right]^2 \times \left[\frac{1 \text{ h}}{3600 \text{ s}}\right]^2$$

$$= 4.\underline{1}807 \times 10^5 = 4.2 \times 10^5 \text{ J}$$

$$E_k = 4.\underline{1}807 \times 10^5 \text{ J} \times \frac{1 \text{ cal}}{4.184 \text{ J}} = 9.\underline{9}92 \times 10^4 = 1.0 \times 10^5 \text{ cal}$$

6.31 To insert the weight of 1 molecule of water in the formula, multiply the molar mass by the reciprocal of Avogadro's number. The kinetic energy in J:

$$E_k = \frac{1}{2} \times \frac{18.02 \text{ g}}{1 \text{ mol}} \times \frac{1 \text{ mol}}{6.02 \times 10^{23} \text{ molec}} \times \frac{1 \text{ kg}}{1000 \text{ g}} \times \left[\frac{648 \text{ m}}{1 \text{ s}}\right]^2 = 6.2\underline{8}4 \times 10^{-21}$$

$$= 6.28 \times 10^{-21} \text{ J/molec}$$

6.33 The reaction is exothermic. Since heat is evolved from the reaction, heat is subtracted from the system, and q = -939 kJ.

6.35 For 1 mole of mercury(II) oxide absorbing 90.8 kJ of heat:

$$HgO(s) \rightarrow Hg(l) + 1/2O_2(g);\ \Delta H = 90.8 \text{ kJ}$$

6.37 When the equation is reversed, the sign of ΔH is also reversed.

$$C_2H_2(g) + Ca(OH)_2(s) \rightarrow CaC_2(s) + 2H_2O(l);\ \Delta H = +128 \text{ kJ}$$

6.39 The first equation is: $1/4P_4O_{10}(s) + 3/2H_2O(l) \rightarrow H_3PO_4(aq);\ \Delta H = -107.6$ kJ

The second equation is: $P_4O_{10}(s) + H_2O(l) \rightarrow 4H_3PO_4(aq);\ \Delta H = ?$

The second equation has been obtained from the first by multiplying each coefficient by 4. Therefore, to obtain the ΔH for the second equation, the ΔH for the first equation must be multiplied by 4: -107.6 x 4 = -430.4 kJ.

6.41 Since phosphorus is written as P_4 in the equation, its molar mass = 123.88 g/mol. From the equation, 1 mole of P_4 evolves 2942 kJ of heat. Divide this by the molar mass to obtain the amount of heat evolved per gram of P_4:

$$\frac{-2942 \text{ kJ}}{1 \text{ mol } P_4} \times \frac{1 \text{ mol } P_4}{123.88 \text{ g } P_4} = -23.7\underline{4}8 = \frac{-23.75 \text{ kJ}}{1 \text{ g } P_4}$$

6.43 The molar mass of ammonia is 17.03 g/mol. From the equation, 4 moles of NH_3 evolve 1267 kJ of heat. Divide 25.6 g of NH_3 by its molar mass and the 4 moles of NH_3 in the equation to obtain the amount of heat evolved:

$$25.6 \text{ g } NH_3 \times \frac{1 \text{ mol } NH_3}{17.03 \text{ g } NH_3} \times \frac{-1267 \text{ kJ}}{4 \text{ mol } NH_3} = -47\underline{6}.1 = -476 \text{ kJ}$$

6.45 Multiply the 180 g (0.180 kg) of water by the specific heat of 4.18 J/(g•°C) and by Δt, to obtain heat in joules:

$$180\ \text{g} \times (96°\text{C} - 15°\text{C}) \times \frac{4.18\ \text{J}}{\text{g}\cdot°\text{C}} = 6.\underline{0}94 \times 10^4 = 6.1 \times 10^4\ \text{J}$$

6.47 Use the 2.26×10^3 J/g (2.26 kJ/g) heat of vaporization to calculate the heat of condensation. Then use it to calculate Δt, the temperature change.

$$\text{Heat of condens} = \frac{2.26 \times 10^3\ \text{J}}{1\ \text{g}} \times 124\ \text{g} = 2.8\underline{0}24 \times 10^5\ \text{J}$$

$$\text{Temp change} = \Delta t = \frac{2.8024 \times 10^5\ \text{J}}{6.44 \times 10^4\ \text{g}} \times \frac{\text{g}\cdot°\text{C}}{1.015\ \text{J}} = +4.2\underline{8}7 = +4.29°\text{C}$$

6.49 The enthalpy change for the reaction is equal in magnitude, and opposite in sign, to the heat energy change occurring from the cooling of the solution and calorimeter.

$$q_{calorimeter} = (1071\ \text{J/°C})(21.56°\text{C} - 25.00°\text{C}) = -36\underline{8}4.2\ \text{J}$$

Thus, 15.3 g $NaNO_3$ is equivalent to 3684.2 J of heat energy. The amount of heat absorbed by 1.000 mol of $NaNO_3$ is calculated from +3684.2 J (opposite sign):

$$1.000\ \text{mol NaNO}_3 \times \frac{85.00\ \text{g NaNO}_3}{1\ \text{mol NaNO}_3} \times \frac{3684.2\ \text{J}}{15.3\ \text{g NaNO}_3} = 2.0\underline{4}6 \times 10^4\ \text{J}$$

Thus the enthalpy change, ΔH, for the reaction is 2.05×10^4 J, or 20.5 kJ, per mol of $NaNO_3$.

6.51 The energy change for the reaction is equal in magnitude, and opposite in sign, to the heat energy produced from the warming of the solution and the calorimeter.

$$q_{calorimeter} = (9.63\ \text{kJ/°C})(33.73 - 25.00°\text{C}) = +84.\underline{0}6\ \text{kJ}$$

Thus, 2.84 g of C_2H_5OH is equivalent to 84.06 kJ of heat energy. The amount of heat released by 1.000 mol of C_2H_5OH is calculated from -84.06 kJ (opposite sign):

$$1.000\ \text{mol C}_2\text{H}_5\text{OH} \times \frac{46.07\ \text{g C}_2\text{H}_5\text{OH}}{1\ \text{mol C}_2\text{H}_5\text{OH}} \times \frac{-84.06\ \text{kJ}}{2.84\ \text{g C}_2\text{H}_5\text{OH}} = -13\underline{6}3.6\ \text{kJ}$$

Thus the enthalpy change, ΔH, for the reaction is -1.36×10^3 kJ/mol of ethanol.

6.53 Using the equations in the data, reverse the direction of the first reaction and reverse the sign of its ΔH. Then divide the second equation by 2, divide its ΔH by 2, and add. Setup:

$$H_2O(l) + 1/2O_2(g) \rightarrow H_2O_2(l);\quad \Delta H = (-98.0\ \text{kJ}) \times (-1)$$

$$H_2(g) + 1/2O_2(g) \rightarrow H_2O(l);\quad \Delta H = (-571.6\ \text{kJ}) \div (2)$$

$$H_2(g) + O_2(g) \rightarrow H_2O_2(l);\quad \Delta H = -187.8\ \text{kJ}$$

6.55 Using the equations in the data, multiply the second equation by 2 and reverse its direction; do the same to its ΔH. Then multiply the first equation by 2 and its ΔH by 2. Finally, multiply the third equation by 3 and its ΔH by 3. Then add. Setup:

$$4NH_3(g) \rightarrow 2N_2(g) + 6H_2(g); \quad \Delta H = (-91.8\ kJ) \times (-2)$$

$$2N_2(g) + 2O_2(g) \rightarrow 4NO(g); \quad \Delta H = (180.6\ kJ) \times (2)$$

$$6H_2(g) + 3O_2(g) \rightarrow 6H_2O(g); \quad \Delta H = (-483.7\ kJ) \times (3)$$

$$4NH_3(g) + 5O_2(g) \rightarrow 4NO(g) + 6H_2O(g); \quad \Delta H = -906.3\ kJ$$

6.57 After reversing the first equation in the data, add all the equations. Setup:

$$C_2H_4(g) + 3O_2(g) \rightarrow 2CO_2(g) + 2H_2O(l); \quad \Delta H = (-1401\ kJ)$$

$$2CO_2(g) + 3H_2O(l) \rightarrow C_2H_6(g) + 7/2O_2(g); \quad \Delta H = (-1550) \times (-1)$$

$$H_2(g) + 1/2O_2(g) \rightarrow H_2O(l); \quad \Delta H = (-286\ kJ)$$

$$C_2H_4(g) + H_2(g) \rightarrow C_2H_6(g); \quad \Delta H = -137\ kJ$$

6.59 Write the ΔH° values (Table 6.2) underneath each compound in the balanced equation:

$$CCl_4(l) \rightarrow CCl_4(g)$$

-139 -96.0 (kJ)

$$\Delta H^o_{vap} = [\Delta H^o_f(CCl_4(g))] - [\Delta H^o_f(CCl_4(l))] = [-96] - [-139]\ kJ = +43\ kJ$$

6.61 Write the ΔH° values (Table 6.2) underneath each compound in the balanced equation.

$$2H_2S(g) + 3O_2(g) \rightarrow 2H_2O(l) + 2SO_2(g)$$

2(-20) 3(0) 2(-285.8) 2(-296.8) (kJ)

$$\Delta H^o = \Sigma n\Delta H^o(\text{products}) - \Sigma m\Delta H^o(\text{reactants})$$

$$= [2(-285.8) + 2(-296.8)] - [2(-20) + 3(0)]\ kJ = -112\underline{5}.2 = -1125\ kJ$$

6.63 Write the ΔH° values (Appendix C) underneath each compound in the balanced equation.

$$2PbS(s) + 3O_2(g) \rightarrow 2SO_2(g) + 2PbO(s)$$

2(-98.3) 3(0) 2(-296.8) 2(-219) (kJ)

$$\Delta H^o = \Sigma n\Delta H^o(\text{products}) - \Sigma m\Delta H^o(\text{reactants})$$

$$= [2(-296.8) + 2(-219)] - [2(-98.3) + 3(0)] = -835\ kJ$$

6.65 Write the ΔH^{o} values (Table 6.2) underneath each compound in the balanced equation.

$$HCl(g) \rightarrow H^{+}(aq) + Cl^{-}(aq)$$
$$-92.3 \quad 0 \quad -167.5 \quad (kJ)$$

$$\Delta H^{o} = \Sigma n\Delta H^{o}(\text{products}) - \Delta H^{o}(\text{reactant})$$
$$= [(0) + (-167.5)] - [-92.3] = -75.2 \text{ kJ}$$

6.67 Write the ΔH^{o} values (Table 6.2) underneath each compound in the balanced equation; rearrange the ΔH^{o} equation and solve for ΔH_f^{o} for C_4H_{10}.

$$C_4H_{10}(g) + 13/2O_2(g) \rightarrow 4CO_2(g) + 5H_2O(l); \quad \Delta H_f^{o} = -2855 \text{ kJ}$$
$$\Delta H_f^{o} \quad 13/2(0) \quad 4(-393.5) \quad 5(-285.8) \quad (kJ)$$

$$\Delta H^{o} = \Sigma n\Delta H^{o}(\text{products}) - \Sigma m\Delta H^{o}(\text{reactants}) = -2855 \text{ kJ}$$

$$-2855 \text{ kJ} = [4(-393.5) + 5(-285.8)] - [\Delta H_f^{o} + 13/2(0)]$$

$$\Delta H_f^{o} = 2855 + [4(-393.5) + 5(-285.8)] - 13/2(0) = -148 \text{ kJ}$$

6.69 Using Table 1.3 and 4.184 J/cal, convert the 686 Btu/lb to J/g:

$$\frac{686 \text{ Btu}}{1 \text{ lb}} \times \frac{252 \text{ cal}}{1 \text{ Btu}} \times \frac{4.184 \text{ J}}{1 \text{ cal}} \times \frac{1 \text{ lb}}{0.4536 \text{ kg}} \times \frac{1 \text{ kg}}{10^3 \text{ g}} = 1.5\underline{9}4 \times 10^3 = 1.59 \times 10^3 \text{ J/g}$$

6.71 Substitute into the equation E_p = mgh, and convert to SI units.

$$E_p = 1.00 \text{ lb} \times \frac{9.78 \text{ m}}{s^2} \times 167 \text{ ft} \times \frac{0.4536 \text{ kg}}{1 \text{ lb}} \times \frac{0.9144 \text{ m}}{3 \text{ ft}} = \frac{22\underline{5}.8 \text{ kg}\cdot\text{m}^2}{s^2} = 226 \text{ J}$$

At the bottom, all the potential energy is converted to kinetic energy, so E_k = 225.8 kg•m^2/s^2. Since $E_k = 1/2mv^2$, solve for v, the speed (velocity):

$$\text{Speed} = \sqrt{\frac{E_k}{1/2 \times m}} = \sqrt{\frac{225.8 \text{ kg}\cdot\text{m}^2/\text{s}^2}{1/2 \times 1.00 \text{ lb} \times 0.4536 \text{ kg/lb}}} = \frac{31.552 \text{ m}}{s} = \frac{31.6 \text{ m}}{s}$$

6.73 The equation is $CaCO_3(s) \rightarrow CaO(s) + CO_2(g)$; ΔH = 178.3 kJ. Use the molar mass of 100.08 g/mol to convert the heat per mol to heat per 12.0 g.

$$12.0 \text{ g } CaCO_3 \times \frac{1 \text{ mol } CaCO_3}{100.08 \text{ g } CaCO_3} \times \frac{178.3 \text{ kJ}}{1 \text{ mol } CaCO_3} = 21.\underline{3}8 \text{ kJ (21.4 kJ abs'd by12.0 g)}$$

6.75 The equation is $2HCHO_2(l) + O_2(g) \rightarrow 2CO_2(g) + 2H_2O(l)$. Use the molar mass of 46.03 g/mol to convert -30.3 kJ/5.48 g to ΔH per mol of acid.

$$\frac{-30.3 \text{ kJ}}{5.48 \text{ g } HCHO_2} \times \frac{46.03 \text{ g } HCHO_2}{1 \text{ mol } HCHO_2} = -25\underline{4}.508 = -255 \text{ kJ/mol}$$

6.77 Divide the 235 J heat by the mass of lead and the Δt to obtain the specific heat.

$$\text{Specific heat} = \frac{235\text{ J}}{121.6\text{ g }(35.5°\text{C} - 20.4°\text{C})} = 0.12\underline{7}98 = \frac{0.128\text{ J}}{(\text{g}\cdot°\text{C})}$$

6.79 The energy used to heat the Zn comes from cooling the water. Calculate q for water:

$$q_{wat} = \text{specific heat x mass x } \Delta t$$

$$q_{wat} = \frac{4.18\text{ J}}{\text{g}\cdot°\text{C}} \text{ x } 50.0\text{ g x } (96.68°\text{C} - 100.00°\text{C}) = -69\underline{3}.88\text{ J}$$

The sign of q for the Zn is the reverse of the sign of q for water because the Zn is absorbing heat:

$$q_{met} = -(q_{wat}) = -(-693.88) = 69\underline{3}.88\text{ J}$$

$$\text{Sp heat} = \frac{693.88\text{ J}}{(25.3\text{ g})(96.68 - 25.00)\ °\text{C}} = 0.38\underline{2}6 = \frac{0.383\text{ J}}{(\text{g}\cdot°\text{C})}$$

6.81 Use Δt and the heat capacity of 547 J/°C to calculate q:

$$q = C\Delta t = (547\text{ J/°C})(36.66 - 25.00)°\text{C} = 6.3\underline{7}8 \text{ x } 10^3\text{ J } (6.3\underline{7}8\text{ kJ})$$

Energy is released in the solution process in raising the temperature, so ΔH is negative:

$$\Delta H = \frac{-6.378\text{ kJ}}{6.48\text{ g LiOH}} \text{ x } \frac{23.95\text{ g LiOH}}{1\text{ mol LiOH}} = -23.\underline{5}7 = -23.6\text{ kJ/mol}$$

6.83 Use Δt and the heat capacity of 13.43 kJ/°C to calculate q:

$$q = C\Delta t = (13.43\text{ kJ/°C})(35.81 - 25.00)°\text{C} = 145.\underline{1}7\text{ kJ}$$

As in the previous 2 problems, the sign of ΔH must be reversed, making the heat negative:

$$\Delta H = \frac{-145.17\text{ kJ}}{10.00\text{ g } HC_2H_3O_2} \text{ x } \frac{60.05\text{ g } HC_2H_3O_2}{1\text{ mol } HC_2H_3O_2} = -871.\underline{7}4 = -871.7\text{ kJ/mol}$$

6.85 Using the equations in the data, reverse the direction of the first reaction and reverse the sign of its ΔH. Then add the second and third equations and their ΔH's.

$H_2O(g) + SO_2(g) \rightarrow H_2S(g) + 3/2O_2(g)$; $\Delta H = (-519\text{ kJ}) \text{ x } (-1)$

$H_2(g) + 1/2O_2(g) \rightarrow H_2O(g)$; $\Delta H = (-242\text{ kJ})$

$1/8S_8(rh) + O_2(g) \rightarrow SO_2(g)$; $\Delta H = (-297\text{ kJ})$

$H_2(g) + 1/8S_8(rh) \rightarrow H_2S(g)$; $\Delta H = -20\text{ kJ}$

6.87 Write the ΔH^o values (Table 6.2) underneath each compound in the balanced equation.

$$CH_4(g) + H_2O(g) \rightarrow CO(g) + 3H_2(g)$$

$$-74.9 \quad -241.8 \quad -110.5 \quad 3(0) \quad (kJ)$$

$$\Delta H^o = [-110.5 + 3(0)] - [(-74.9) + (-241.8)] = 206.2 \text{ kJ}$$

6.89 Write the ΔH^o values (Table 6.2) underneath each compund in the balanced equation. The ΔH_f^o of -635 kJ/mol CaO is given in the problem.

$$CaCO_3(s) \rightarrow CaO(s) + CO_2(g)$$

$$-1206.9 \quad -635 \quad -393.5 \quad (kJ)$$

$$\Delta H^o = [(-635) + (-393.5)] - [-1206.9] = 17\underline{8}.4 = 178 \text{ kJ}$$

6.91 Let ΔH_f^o = the unknown standard enthalpy of formation for sucrose. Then write this symbol and the other ΔH_f^o values underneath each compound in the balanced equation, and solve for ΔH_f^o, using the ΔH^o of -5641 kJ for the reaction.

$$C_{12}H_{22}O_{11}(s) + 12O_2(g) \rightarrow 12CO_2(g) + 11H_2O(g)$$

$$\Delta H_f^o \quad 12(0) \quad 12(-393.5) \quad 11(-241.8) \quad (kJ)$$

$$\Delta H^o = -5641 = [12(-393.5) + 11(-241.8)] - [\Delta H_f^o + 12(0)]$$

$$\Delta H_f^o = -2225 \text{ kJ/mol sucrose}$$

Cumulative-Skills Problems (require skills from Chapters 4, 5, and 6)

6.93 First calculate the mole fraction of each gas in the product, assuming 100 g product:

$$\text{Mol CO} = 0.33 \text{ g CO} \times 1 \text{ mol CO}/28.01 \text{ g CO} = 1.\underline{1}78 \text{ mol CO}$$

$$\text{Mol } CO_2 = 0.67 \text{ g } CO_2 \times 1 \text{ mol } CO_2/44.01 \text{ g } CO_2 = 1.\underline{5}22 \text{ mol } CO_2$$

$$\text{Mol frac CO} = \frac{1.178 \text{ mol CO}}{(1.178 + 1.522) \text{ mol}} = 0.4\underline{3}63$$

$$\text{Mol frac } CO_2 = \frac{1.522 \text{ mol } CO_2}{(1.178 + 1.522) \text{ mol}} = 0.5\underline{6}37$$

Now calculate the starting moles of C, which equals the total moles of CO and CO_2:

$$\text{Starting mol C(s)} = 1.00 \text{ g C} \times 1 \text{ mol C}/12.01 \text{ g C} = 0.083\underline{2}6 \text{ mol C} = \text{mol CO} + CO_2$$

Use the mol fractions to convert mol CO + CO_2 to mol CO and mol CO_2:

$$\text{Mol CO} = \frac{0.4363 \text{ mol CO}}{1 \text{ mol total}} \times 0.08326 \text{ mol total} = 0.03\underline{6}33 \text{ mol CO}$$

$$\text{Mol } CO_2 = \frac{0.5637 \text{ mol } CO_2}{1 \text{ mol total}} \times 0.08326 \text{ mol total} = 0.04\underline{6}93 \text{ mol } CO_2$$

Now use enthalpies of formation (Table 6.2) to calculate the heat of combustion for both:

$C(s)$ +	$1/2O_2(g)$ →	$CO(g)$	
0.03633	excess	0.03633	(mol)
0	0	-110.5	(kJ/mol)
0	0	-4.014	(kJ/0.03633 mol)

$C(s)$ +	$O_2(g)$ →	$CO_2(g)$	
0.04693	excess	0.04693	(mol)
0	0	-393.5	(kJ/mol)
0	0	-18.47	(kJ/0.04693 mol)

Total ΔH = -4.014 + (-18.47) = -2$\underline{2}$.48 = -22 kJ
Heat released = 22 kJ

6.95 The equation is: $4NH_3(g) + 5O_2(g) \rightarrow 4NO(g) + 6H_2O(g)$; $\Delta H^\circ = -906$ kJ

First determine the limiting reactant by calculating the moles of NH_3 and of O_2; then, assuming one of the reactants is totally consumed, calculate the moles of the other reactant needed for the reaction.

$$10.0 \text{ g } NH_3 \times \frac{1 \text{ mol } NH_3}{17.03 \text{ g } NH_3} = 0.58\underline{7}2 \text{ mol } NH_3$$

$$20.0 \text{ g } O_2 \times \frac{1 \text{ mol } O_2}{32.00 \text{ g } O_2} = 0.62\underline{5}0 \text{ mol } O_2$$

$$0.62\underline{5}0 \text{ mol } O_2 \times \frac{4 \text{mol } NH_3}{5 \text{ mol } O_2} = 0.500 \text{ mol } NH_3 \text{ needed if } O_2 \text{ is totally consumed}$$

Since NH_3 is present in excess of what is needed, O_2 must be the limiting reactant.

Now calculate the heat released on the basis of complete reaction of 0.622250 mol O_2:

$$\Delta H = \frac{-906 \text{ kJ}}{5 \text{ mol } O_2} \times 0.6250 \text{ mol } O_2 = -11\underline{3}.25 = -113 \text{ kJ}$$

The heat released by the complete reaction of the 20.0 g (0.6250 mol) of $O_2(g)$ is 113 kJ.

6.97 The equation is: $N_2(g) + 3H_2(g) \rightarrow 2NH_3(g)$; $\Delta H^\circ = -91.8$ kJ

a. To find the heat evolved from the production of 1.00 L of NH_3, convert the 1.00 L to mol NH_3, using the molar gas volume of 24.46 L/mol (= V_m) at 25°C (from 22.41 x 298/273), and use $V = nV_m$; then convert the moles to heat (ΔH) using ΔH°:

$$n = \frac{V}{V_m} = \frac{1.00 \text{ L } NH_3}{24.46 \text{ L } NH_3/\text{mol } NH_3} = 0.040\underline{8}8 \text{ mol } NH_3$$

$$\Delta H = \frac{-91.8 \text{ kJ}}{2 \text{ mol } NH_3} \times 0.040\underline{8}8 \text{ mol } NH_3 = -1.8\underline{7}6 \text{ (1.88 kJ heat evolved)}$$

b. First find the moles of N_2 using the molar gas volume of 24.46 L/mol (part a). Then convert the heat needed to raise the N_2 from 25° to 400°C:

$$\frac{0.500 \text{ L } N_2}{24.46 \text{ L } N_2/\text{mol } N_2} = 0.020\underline{4}4 \text{ mol } N_2$$

$$0.020\underline{4}4 \text{ mol } N_2 \times \frac{29.12 \text{ J}}{(\text{mol}\cdot°\text{C})} \times (400 - 25)°\text{C} = 22\underline{3}.2 \text{ J} \quad (0.22\underline{3}2 \text{ kJ})$$

$$\% \text{ heat for } N_2 = \frac{0.2232 \text{ kJ}}{187.6 \text{ kJ}} \times 100\% = 11.\underline{8}9 = 11.9\%$$

6.99 The glucose equation is: $C_6H_{12}O_6 + 6O_2 \rightarrow 6CO_2 + 6H_2O$; $\Delta H° = -2802.8$ kJ

Convert the 2.50×10^3 kCal to mol of glucose, using the $\Delta H°$ of -2802.8 kJ for the reaction, and the conversion factor of 4.184 kJ/kCal:

$$2.50 \times 10^3 \text{ kCal} \times \frac{4.184 \text{ kJ}}{1.000 \text{ kCal}} \times \frac{1 \text{ mol glucose}}{2802.8 \text{ kJ}} = 3.7\underline{3}1 \text{ mol glucose}$$

Next convert mol glucose to mol LiOH using the above equation for glucose, and the equation for LiOH: $2LiOH(s) + CO_2(g) \rightarrow Li_2CO_3(s) + H_2O(l)$:

$$3.7\underline{3}1 \text{ mol glucose} \times \frac{6 \text{ mol } CO_2}{1 \text{ mol glucose}} \times \frac{2 \text{ mol LiOH}}{1 \text{ mol } CO_2} = 44.\underline{7}7 \text{ mol LiOH}$$

Finally use the molar mass of LiOH to convert moles to mass:

$$44.77 \text{ mol LiOH} \times \frac{23.95 \text{ g LiOH}}{1 \text{ mol LiOH}} = 1.0\underline{7}2 \times 10^3 \text{ g (1.07 kg) LiOH}$$

CHAPTER 7

ATOMIC STRUCTURE

SOLUTIONS TO EXERCISES

Note on significant figures: The final answer to all mathematical solutions is given first with one nonsignificant figure (last significant figure underlined) and is then rounded to the correct number of figures. Intermediate answers usually also have at least one nonsignificant figure.

7.1 Uranium has an atomic number of 92. Therefore, the neutral atom contains 92 protons and 92 electrons. The U^{2+} ion has lost 2 electrons, producing a charge of +2, and leaving 90 electrons.

7.2 The element whose nucleus has 17 protons has an atomic number of 17 and is therefore chlorine (symbol = Cl). The mass number is 17 + 18 = 35. The symbol is ${}^{35}_{17}Cl$.

7.3 Carbon-14 is an isotope of carbon. A carbon-14 atom contains 6 protons, 6 electrons, and (14 - 6), or 8 neutrons.

7.4 Multiply each isotopic mass by its fractional abundance, and then sum:

$$34.96885 \text{ amu} \times 0.75771 = 26.49\underline{6}247$$
$$36.96590 \text{ amu} \times 0.24229 = \underline{8.956467}$$
$$\text{The atomic weight of chlorine} = 35.45\underline{2}714 = 35.453 \text{ amu}$$

7.5 Rearrange the equation $c = \nu\lambda$, which relates wavelength to frequency and the speed of light (3.00×10^8 m/s):

$$\lambda = \frac{c}{\nu} = \frac{3.00 \times 10^8 \text{ m/s}}{3.91 \times 10^{14} \text{ /s}} = 7.6\underline{7}2 \times 10^{-7} = 7.67 \times 10^{-7} \text{ m, or 767 nm}$$

7.6 Rearrange the equation $c = \nu\lambda$, which relates frequency to wavelength and the speed of light (3.00×10^8 m/s). Recognize that 456 nm = 4.56×10^{-7} m.

$$\nu = \frac{c}{\lambda} = \frac{3.00 \times 10^8 \text{ m/s}}{4.56 \times 10^{-7} \text{ m}} = 6.5\underline{7}8 \times 10^{14} = 6.58 \times 10^{14}\text{/s}$$

7.7 First use the wavelength to calculate the frequency from $c = \nu\lambda$. Then calculate the energy using $E = h\nu$.

$$\nu = \frac{c}{\lambda} = \frac{3.00 \times 10^{8}\ \text{m/s}}{1.0 \times 10^{-6}\ \text{m}} = 3.\underline{0}0 \times 10^{14}/\text{s}$$

$$\nu = \frac{c}{\lambda} = \frac{3.00 \times 10^{8}\ \text{m/s}}{1.0 \times 10^{-8}\ \text{m}} = 3.\underline{0}0 \times 10^{16}/\text{s}$$

$$\nu = \frac{c}{\lambda} = \frac{3.00 \times 10^{8}\ \text{m/s}}{1.0 \times 10^{-10}\ \text{m}} = 3.\underline{0}0 \times 10^{18}/\text{s}$$

$$E = h\nu = 6.63 \times 10^{-34}\ \text{J}\cdot\text{s} \times 3.00 \times 10^{14}/\text{s} = 1.\underline{9}8 \times 10^{-19} = 2.0 \times 10^{-19}\ \text{J (IR)}$$

$$E = h\nu = 6.63 \times 10^{-34}\ \text{J}\cdot\text{s} \times 3.00 \times 10^{16}/\text{s} = 1.\underline{9}8 \times 10^{-17} = 2.0 \times 10^{-17}\ \text{J (UV)}$$

$$E = h\nu = 6.63 \times 10^{-34}\ \text{J}\cdot\text{s} \times 3.00 \times 10^{18}/\text{s} = 1.\underline{9}8 \times 10^{-15} = 2.0 \times 10^{-15}\ \text{J (X-ray)}$$

The X-ray photon (shortest wavelength) has the greatest amount of energy; the IR photon (longest wavelength) has the least amount of energy.

7.8 Solve the equation $E = -R_H/n^2$ for both E_i and E_f, and calculate the energy change for the transition from n = 3 to n = 1. Convert E to wavelength using $E = h\nu$ and $c = \nu\lambda$.

$$E_i = \frac{-R_H}{3^2} = \frac{-R_H}{9}; \quad E_f = \frac{-R_H}{1^2} = \frac{-R_H}{1}$$

$$\left[\frac{-R_H}{9}\right] - \left[\frac{-R_H}{1}\right] = \frac{8R_H}{9} = h\nu$$

The frequency of the emitted radiation is:

$$\nu = \frac{8R_H}{9h} = \frac{8}{9} \times \frac{2.180 \times 10^{-18}\ \text{J}}{6.63 \times 10^{-34}\ \text{J}\cdot\text{s}} = 2.9\underline{2}2 \times 10^{15} = 2.92 \times 10^{15}/\text{s}$$

$$\lambda = \frac{c}{\nu} = \frac{3.00 \times 10^{8}\ \text{m/s}}{2.92 \times 10^{15}/\text{s}} = 1.0\underline{2}7 \times 10^{-7} = 1.03 \times 10^{-7}\ \text{m, or 103 nm}$$

7.9 Calculate the frequency from $c = \nu\lambda$, recognizing that 589 nm is 5.89×10^{-7} m.

$$\lambda = \frac{c}{\lambda} = \frac{3.00 \times 10^{8}\ \text{m/s}}{5.89 \times 10^{-7}\ \text{m}} = 5.0\underline{9}3 \times 10^{14} = 5.09 \times 10^{14}/\text{s}$$

Finally calculate the energy difference.

$$E = h\nu = 6.63 \times 10^{-34}\ \text{J}\cdot\text{s} \times 5.093 \times 10^{14}/\text{s} = 3.376 \times 10^{-19} = 3.38 \times 10^{-19}\ \text{J}$$

7.10 To calculate wavelength, use the mass of an electron, m = 9.11×10^{-31} kg, and Planck's constant (h = 6.63×10^{-34} J•s, or 6.63×10^{-34} kg•m²/s²).

$$\lambda = \frac{h}{mv} = \frac{6.63 \times 10^{-34}\ \text{kgm}^2/\text{s}}{9.11 \times 10^{-31}\ \text{kg} \times 2.19 \times 10^{6}\ \text{m/s}} = 3.3\underline{2}3 \times 10^{-10} = 3.32 \times 10^{-10}\ \text{m}\ (3.32\ \text{Å})$$

7.11
a. The value of n must be a positive whole number.
b. The values for l can only range from 0 to (n - 1). Here, l has a value greater than n.
c. The values for m_l range from $-l$ to $+l$. Here, m_l has a value greater than 1.
d. The values for m_s are either +1/2 or -1/2, not 0.

ANSWERS TO REVIEW QUESTIONS

7.1 A cathode ray tube consists of an evacuated tube with electrodes. When a high voltage is placed across the electrodes, cathode rays are emitted by the negative electrode. Two electrically charged plates placed inside the tube will bend the rays towards the positive plate, showing that the cathode rays are negatively charged.

7.2 The evidence is that cathode rays have the same characteristics regardless of which gas occupies the tube or which material is used to make the electrodes. This suggests that the particles in cathode rays (electrons) are part of all matter, regardless of composition.

7.3 When oil drops are formed, they may acquire an electrical charge. By observing how a particular drop behaves during free fall and what voltage is required to balance the force of gravity, we can obtain the charge on the oil drop. After observing several oil drops, we can obtain the smallest difference in charge, which is assumed to be the charge on one electron.

7.4 Three kinds of radiation are observed in natural radioactivity. They can be distinguished by using electric or magnetic fields, which split the radiation into three rays, called alpha, beta, and gamma rays. Alpha rays, being positively charged particles, are repelled by a positively charged plate. Beta rays, being negatively charged particles, are repelled by a negatively charged plate. Gamma rays are unaffected by these fields and pass through them.

7.5 The nuclear model of the atom states that most of the mass of the atom is concentrated in a positively charged center called the nucleus, around which exist the negatively charged electrons. Thus when alpha particles are directed toward a metal foil, most pass through the empty space of the atoms more or less undeflected. However, if an alpha particle happens to strike a nucleus, it is scattered backward by the repulsion of positive charges.

7.6 The plum pudding model predicts that alpha particles will be deflected through small angles. What is actually observed, however, is that occasionally alpha particles are deflected by large angles, perhaps even backward. These results are correctly explained by the nuclear model.

7.7 The protons and neutrons are the particles in the nucleus. They have about the same mass: the neutron is electrically neutral but the proton is positively charged. An electron has a much smaller mass then either a proton or a neutron and is negatively charged.

7.8 Protons (hydrogen nuclei) were discovered as products during experiments involving the collision of alpha particles (helium nuclei) with nitrogen atoms. Experiments with other nuclei also produced protons. Neutrons were discovered as the radiation product of collisions of alpha particles with beryllium atoms. The resulting radiation was discovered to consist of particles having a mass approximately equal to that of a proton and having no charge (neutral particles, or neutrons). Experiments on other elements also showed they contained neutrons.

7.9 In a mass spectrometer, atoms or molecules of a substance are bombarded with a beam of electrons (cathode rays). The beam of electrons displace electrons from the atoms or molecules, producing positively charged ions. These ions, as a beam, are accelerated down a tube by the attraction of a negatively charged plate, or grid. The beam is separated by a magnetic field into many beams of ions with different mass-to-charge ratios. From the measurement of these beams, one can obtain the masses of the ions and their relative numbers or abundances.

7.10 Light consists of oscillating electric and magnetic fields that travel through space rapidly. Light waves are characterized by wavelength and frequency. Visible light, or "light," covers the wavelengths from 400 to 800 nm.

7.11 Gamma rays and X rays are the types of radiation with the shortest wavelengths. As wavelength is increased, we encounter ultraviolet (UV) radiation, visible light ("light"), infrared radiation, microwave radiation, and radio waves. Radio waves have the longest wavelengths.

7.12 The photoelectric effect is the ejection of electrons from a surface of a metal when light shines on it. Electrons are ejected only if the frequency of light is high enough. The photon concept explains this effect in terms of energy. If a single photon has enough energy, it can eject one electron from the surface. If the photon does not have enough energy, it cannot eject one electron. Since the energy of a photon is directly proportional to frequency, we can also say that electrons are ejected only if the frequency is high enough.

7.13 The wave theory of light pictures light as consisting of oscillating electric and magnetic fields. A wave is characterized by both wavelength and frequency. The particle theory of light pictures light as consisting of photons, each photon having an energy given by $E = h\nu$. The wave and particle pictures of light are complementary views of the same physical entity; neither is a complete description of all of the properties of light.

7.14 According to existing theory at Rutherford's time, an electrically charged particle revolving around a center will continuously radiate energy. This would imply that the electron in the atom would continuously lose energy, spiraling into smaller and smaller orbits in the process, ultimately spiraling into the nucleus. Thus, the theory apparently could not explain the stability of the electrons around the atom and the atom itself.

7.15 According to Bohr, an electron in an atom can have only specific energy values. An electron in an atom can change energy only by going from one energy level (of allowed energy) to another energy level (of allowed energy). An electron in a higher energy level can go to a lower energy level by emitting a photon of an energy equal to the difference in energy. However, when an electron is in its lowest energy level, no further changes in energy can occur. Thus the electron does not continuously radiate energy, as thought at Rutherford's time.

7.16 Emission of a photon occurs when an electron in a higher energy level undergoes a transition to a lower energy level. The energy lost is emitted as a photon.

7.17 Absorption of a photon occurs when a photon of a certain required energy is absorbed by a certain electron in an atom. The energy of the photon must be equal to the energy necessary to excite the electron of the atom from a lower energy level, usually the lowest, to a higher energy level. The photon's energy is converted into electronic energy.

7.18 The diffraction of an electron beam is evidence for electron waves. A practical example of diffraction is the operation of the electron microscope.

7.19 The square of a wave function, rather than the value of the function itself, equals the probability for finding an electron at a given point in space. More complex mathematical manipulations of the wave function yield values for other parameters.

7.20 a. The principal quantum number can have an integer value between 1 and infinity.
b. The angular momentum quantum number can have any integer value between 0 and (n - 1).
c. The magnetic quantum number can have any integer value between -l and +l.
d. The spin quantum number can be either +1/2 or -1/2.

7.21 The notation is 4f. This subshell contains 7 orbitals.

7.22 An s orbital has a spherical shape. A p orbital has two lobes positioned along a straight line through the nucleus at the center of the line (= a dumbbell shape).

SOLUTIONS TO PRACTICE PROBLEMS

Note on significant figures: **The final answer is given first with one nonsignificant figure (rightmost significant figure underlined), and is then rounded to the correct number of significant figures. Intermediate answers usually also have at least one nonsignificant figure. Atomic weights are rounded to two decimal places, except for that of hydrogen.**

7.23 Multiply the two quantities to obtain mass:

$$1.605 \times 10^{-19}\ \text{C} \times \frac{5.64 \times 10^{-12}\ \text{kg}}{\text{C}} = 9.0\underline{5}2 \times 10^{-31} = 9.05 \times 10^{-31}\ \text{kg}$$

7.25 For the Eu atom to be neutral, the number of electrons must = the number of protons, so a neutral europium atom has 63 electrons. The +3 charge on the Eu^{3+} indicates there are 3 more protons than electrons, so the number of electrons = (63 - 3) = 60.

7.27 A nucleus with 34 protons is the element selenium (Se). The mass number = 34 + 36 = 70. The notation for the nucleus is ${}^{70}_{34}Se$.

7.29 The number of neutrons = mass number - atomic number:

Cl-35: no. of neutrons = 35 - 17 = 18; no. of protons = no. of electrons = 17
Cl-37: no. of neutrons = 37 - 17 = 20; no. of protons = no. of electrons = 17

7.31 The number of protons = mass number - number of neutrons = 69 - 38 = 31. The element with Z = 31 is gallium (Ga).

The ionic charge = number of protons - number of electrons = 31 - 28 = +3. Symbol: ${}^{69}_{31}Ga^{3+}$

7.33 B-10: 10.013 x 0.1978 = $1.98\underline{0}5$
B-11: 11.009 x 0.8022 = 8.8314
average = $10.81\underline{1}9$ = 10.812 amu (= atomic weight)

7.35 Mg-24: 23.985 x 0.7870 = $18.8\underline{7}6$
Mg-25: 24.986 x 0.1013 = $2.53\underline{1}1$
Mg-26: 25.983 x 0.1117 = 2.9023
average = $24.3\underline{0}9$ = 24.31 amu (= atomic weight).

7.37 The sum of the fractional abundances must equal 1. Thus, the abundance of one isotope can be expressed in terms of the other. Let y = the fractional abundance of Ag-107. Then the fractional abundance of Ag-109 = (1 - y). We can write one equation in one unknown:

$$\text{Atomic weight} = 107.868 = 106.91y + 108.90(1 - y)$$
$$107.868 = 108.90 - 1.99y$$
$$y = \frac{108.90 - 107.868}{1.99} = 0.51\underline{8}59 = 0.519$$

The fractional abundance of Ag-107 = 0.519.
The fractional abundance of Ag-109 = 1 - 0.51859 = 0.48$\underline{1}$41 = 0.481.

7.39 Radio waves travel at the speed of light, so divide the distance by c:

$$56 \times 10^9 \text{ m} \times \frac{1 \text{ s}}{3.00 \times 10^8 \text{ m}} = 1.\underline{8}6 \times 10^2 = 1.9 \times 10^2 \text{ s}$$

7.41 Using c = 2.998 x 10^{10} for 4 significant figures, solve c = $\lambda\nu$ for λ:

$$\lambda = \frac{c}{\nu} = \frac{2.988 \times 10^8 \text{ m/s}}{1.255 \times 10^6\text{/s}} = 2.38\underline{8}8 \times 10^2 = 238.9 \text{ m}$$

7.43 Solve c = $\lambda\nu$ for ν. Recognize that 465 nm = 465 x 10^{-9} m, or 4.65 x 10^{-7} m.

$$\nu = \frac{c}{\lambda} = \frac{3.00 \times 10^8 \text{ m/s}}{4.65 \times 10^{-7} \text{ m}} = 6.4\underline{5}1 \times 10^{14} = 6.45 \times 10^{14}\text{/s}$$

7.45 Solve for E, using E = hν, and 4 significant figures for h:

$$E = h\nu = (6.626 \times 10^{-34} \text{ J}\cdot\text{s}) \times (1.255 \times 10^6\text{/s}) = 8.31\underline{5}6 \times 10^{-28} = 8.316 \times 10^{-28} \text{ J}$$

7.47 Recognize that 535 nm = 535 x 10^{-9} m = 5.35 x 10^{-7} m. Then calculate ν and E.

$$\nu = \frac{c}{\lambda} = \frac{3.00 \times 10^8 \text{ m/s}}{5.35 \times 10^{-7} \text{ m}} = 5.6\underline{0}7 \times 10^{14}\text{/s}$$

$$E = h\nu = (6.63 \times 10^{-34} \text{ J}\cdot\text{s}) \times (5.607 \times 10^{-14}\text{/s}) = 3.7\underline{1}7 \times 10^{-19} = 3.72 \times 10^{-19} \text{ J}$$

7.49 Solve the equation E = $-R_H/n^2$ for both E_4 and E_3; equate to hν and solve for ν.

$$E_4 = \frac{-R_H}{4^2} = \frac{-R_H}{16}; \quad E_3 = \frac{-R_H}{3^2} = \frac{-R_H}{9}$$

$$\left[\frac{-R_H}{16}\right] - \left[\frac{-R_H}{9}\right] = \frac{7R_H}{144} = h\nu$$

The frequency of the emitted radiation is:

$$\nu = \frac{7R_H}{144h} = \frac{7}{144} \times \frac{2.180 \times 10^{-18}\ J}{6.63 \times 10^{-34}\ J\cdot s} = 1.5\underline{9}8 \times 10^{14} = 1.60 \times 10^{14}/s$$

7.51 Solve the equation $E = -R_H/n^2$ for both E_2 and E_1; solve for ν and convert to λ.

$$E_2 = \frac{-R_H}{2^2} = \frac{-R_H}{4}; \quad E_1 = \frac{-R_H}{1^2} = \frac{-R_H}{1}$$

$$\left[\frac{-R_H}{4}\right] - \left[\frac{-R_H}{1}\right] = \frac{3R_H}{4} = h\nu$$

$$\nu = \frac{3R_H}{4h} = \frac{3}{4} \times \frac{2.180 \times 10^{-18}\ J}{6.63 \times 10^{-34}\ J\cdot s} = 2.4\underline{6}6 \times 10^{15}/s$$

$$\lambda = \frac{c}{\nu} \times \frac{3.00 \times 10^{8}\ m/s}{2.466 \times 10^{15}/s} = 1.2\underline{1}6 \times 10^{-7} = 1.22 \times 10^{-7}\ m\ \text{(near UV)}$$

7.53 Use 397 nm = 3.97×10^{-7} m, and convert to frequency and then to energy.

$$\nu = \frac{c}{\lambda} \times \frac{3.00 \times 10^{8}\ m/s}{3.97 \times 10^{-7}\ m} = 7.5\underline{5}6 \times 10^{14}/s$$

$$E = h\nu = (6.63 \times 10^{-34}\ J\cdot s) \times (7.556 \times 10^{14}) = 5.0\underline{0}9 \times 10^{-19}\ J$$

Substitute this energy into the Balmer formula, recalling that the Balmer series is an emission spectrum, so ΔE is negative:

$$E_2 = \frac{-R_H}{2^2} = \frac{-R_H}{4}; \quad E_i = \frac{-R_H}{n_i^2}$$

$$\left[\frac{-R_H}{4}\right] - \left[\frac{-R_H}{n_i^2}\right] = -R_H\left[\frac{1}{4} - \frac{1}{n_i^2}\right] = E\ \text{(of line)}$$

$$\left[\frac{1}{4} - \frac{1}{n_i^2}\right] = \frac{5.009 \times 10^{-19}\ J}{2.180 \times 10^{-18}\ J} = 0.22\underline{9}8$$

$$\frac{1}{n_i^2} = \frac{1}{4} - 0.22\underline{9}8 = 0.02\underline{0}2$$

$$n_i = \left[\frac{1}{0.0202}\right]^{1/2} = 7.\underline{0}35 = 7.0\ (= n)$$

7.55 Convert to λ and then to ΔE, using $\Delta E = h\nu$. Recognize that 795 nm = 7.95×10^{-7} m.

$$n = \frac{c}{\lambda} \times \frac{3.00 \times 10^{8}\ m/s}{7.95 \times 10^{-7}\ m} = 3.7\underline{7}4 \times 10^{14}/s$$

$$\Delta E = h\nu = (6.63 \times 10^{-34}\ J\cdot s) \times (3.774 \times 10^{14}/s) = 2.5\underline{0}2 \times 10^{-19} = 2.50 \times 10^{-19}\ J$$

7.57 The mass of a neutron = 1.67495 x 10^{-27} kg. Its speed or velocity, v, of 3.65 km/s = 3.65 x 10^{3} m/s. Substitute these parameters into the de Broglie relation and solve for λ:

$$\lambda = \frac{h}{mv} = \frac{6.63 \times 10^{-34}\ \text{kgm}^2/\text{s}}{1.67495 \times 10^{-27}\ \text{kg} \times 3.65 \times 10^{3}\ \text{m/s}} = 1.0\underline{8}4 \times 10^{-10}\ \text{m, or } 1.08\ \text{Å}$$

7.59 The mass of an electron = 9.10953 x 10^{-31} kg. The wavelength, λ, given as 0.125 Å, is equivalent to 1.25 x 10^{-11} m. Substitute these parameters into the de Broglie relation and solve for the frequency, ν:

$$\nu = \frac{h}{m\lambda} = \frac{6.63 \times 10^{-34}\ \text{kgm}^2/\text{s}}{9.10953 \times 10^{-31}\ \text{kg} \times 1.25 \times 10^{-11}\ \text{m}} = 5.8\underline{2}2 \times 10^{7} = 5.82 \times 10^{7}\ \text{m/s}$$

7.61 The possible values of l range from 0 to (n - 1), so l may be 0, 1, 2, or 3. The possible values of m_l range from $-l$ to $+l$, so m_l may be -3, -2, -1, 0, +1, +2, or +3.

7.63 For the M shell, n = 3; there are 3 subshells in this shell (l = 0, 1, and 2). An f subshell has l = 3; the number of orbitals in this subshell is 2(3) + 1 = 7 (m_l = -3, -2, -1, 0, 1, 2, and 3).

7.65 a. 3p b. 4d c. 4s d. 5f

7.67
a. Impossible: n starts at 1, not at zero.
b. Impossible: l may only be as large as (n - 1).
c. Possible
d. Impossible: m_l may not exceed l in magnitude.
e. Possible

7.69 The difference between -1.12 x 10^{-18} C and 9.60 x 10^{-19} C is -1.6 x 10^{-19} C. If this charge is equivalent to 1 electron, the number of excess electrons on a drop may be found by dividing the negative charge by the charge of 1 electron.

Drop 1: $\frac{-3.20 \times 10^{-19}\ \text{C}}{-1.6 \times 10^{-19}\ \text{C}} = 2.0 \cong 2$ electrons

Drop 2: $\frac{-6.40 \times 10^{-19}\ \text{C}}{-1.6 \times 10^{-19}\ \text{C}} = 4.0 \cong 4$ electrons

Drop 3: $\frac{-9.60 \times 10^{-19}\ \text{C}}{-1.6 \times 10^{-19}\ \text{C}} = 6.0 \cong 6$ electrons

Drop 4: $\frac{-1.12 \times 10^{-18}\ \text{C}}{-1.6 \times 10^{-19}\ \text{C}} = 7.0 \cong 7$ electrons

7.71 Use c = νλ to calculate frequency; then use E = hν to calculate energy.

$$\nu = \frac{c}{\lambda} \times \frac{3.00 \times 10^{8}\ \text{m/s}}{4.61 \times 10^{-7}\ \text{m}} = 6.5\underline{0}7 \times 10^{14} = 6.51 \times 10^{14}/\text{s}$$

$$E = h\nu = (6.63 \times 10^{-34}\ \text{J}\cdot\text{s}) \times (6.507 \times 10^{14}/\text{s}) = 4.3\underline{1}4 \times 10^{-19} = 4.31 \times 10^{-19}\ \text{J}$$

7.73 $^{28}_{13}Al$: 13 protons; no. of neutrons = mass no. - atomic no. = 28 - 13 = 15

$^{41}_{20}Ca$: 20 protons; no. of neutrons = mass no. - atomic no. = 41 - 20 = 21

$^{59}_{28}Ni$: 28 protons; no. of neutrons = mass no. - atomic no. = 59 - 28 = 31

7.75 The element with Z = 78 is platinum (Pt): $^{196}_{78}Pt$

7.77 The sum of the fractional abundances must equal 1. Let y = the fractional abundance of ^{121}Sb. Then the fractional abundance of ^{123}Sb = (1 - y). We write one equation in one unknown:

$$\begin{aligned}
\text{Atomic weight} = 121.75 &= 120.90y + 122.90(1 - y) \\
121.75 &= 122.90 - 2.00y \\
y &= \frac{122.90 - 121.75}{2.00} = 0.57\underline{5}0
\end{aligned}$$

The fractional abundance of ^{121}Sb is 0.575.

7.79 The charge on an electron is -1.60219×10^{-19} C. Since all the isotopic charge-to-mass ratios are positive, all of the ions must be negatively charged. Use the magnitude of the electron charge to convert each charge-to-mass ratio to mass in amu.

$$\text{Ion 1: } \frac{3.3137 \times 10^{-7}\ \text{kg}}{\text{C}} \times \frac{1.60219 \times 10^{-19}\ \text{C}}{e^-} \times \frac{1\ \text{amu}}{1.66056 \times 10^{-27}\ \text{kg}} = 31.97\underline{2}2\ \text{amu}/e^-$$

$$\text{Ion 2: } \frac{3.5205 \times 10^{-7}\ \text{kg}}{\text{C}} \times \frac{1.60219 \times 10^{-19}\ \text{C}}{e^-} \times \frac{1\ \text{amu}}{1.66056 \times 10^{-27}\ \text{kg}} = 33.96\underline{7}51\ \text{amu}/e^-$$

$$\text{Ion 3: } \frac{1.6568 \times 10^{-7}\ \text{kg}}{\text{C}} \times \frac{1.60219 \times 10^{-19}\ \text{C}}{e^-} \times \frac{1\ \text{amu}}{1.66056 \times 10^{-27}\ \text{kg}} = 15.98\underline{5}6\ \text{amu}/e^-$$

$$\text{Ion 4: } \frac{1.7603 \times 10^{-7}\ \text{kg}}{\text{C}} \times \frac{1.60219 \times 10^{-19}\ \text{C}}{e^-} \times \frac{1\ \text{amu}}{1.66056 \times 10^{-27}\ \text{kg}} = 16.98\underline{4}2\ \text{amu}/e^-$$

S has an atomic weight of 32.06. Thus a -1 sulfur ion will have a mass-to-charge ratio of about 32 amu/e^-, and a sulfur ion with a charge of -2 will have a mass-to-charge ratio of about 32 amu/2e^- or 16 amu/e^-. Therefore the four isotopic ions may be described as follows:

Ion 1:	-1 charge	31.972 amu/e^-	31.972 amu mass
Ion 2:	-1 charge	33.968 amu/e^-	33.968 amu mass
Ion 3:	-2 charge	15.986 amu/e^-	31.971 amu mass
Ion 4:	-2 charge	16.984 amu/e^-	33.968 amu mass

7.81 Solve for frequency, using $E = h\nu$.

$$\nu = \frac{E}{h} = \frac{4.34 \text{ x } \text{ x } 10^{-19} \text{ J}}{6.63 \text{ x } 10^{-34} \text{ J•s}} = 6.5\underline{4}6 \text{ x } 10^{14} = 6.55 \text{ x } 10^{14}\text{/s}$$

7.83 Employ the Balmer formula, using Z = 2 for the He^{+} ion.

$$E_3 = (2)^2 \frac{-R_H}{3^2} = \frac{-R_H}{9}; \quad E_2 = (2)^2 \frac{-R_H}{2^2} = \frac{-R_H}{4}$$

$$4\left[\frac{-R_H}{9}\right] - 4\left[\frac{-R_H}{4}\right] = 4 \text{ x } \frac{5R_H}{36} = h\nu$$

The frequency of the radiation is:

$$\nu = \frac{4 \text{ x } 5R_H}{36h} = \frac{20}{36} \text{ x } \frac{2.180 \text{ x } 10^{-18} \text{ J}}{6.63 \text{ x } 10^{-34} \text{ J•s}} = 1.8\underline{2}67 \text{ x } 10^{15}\text{/s}$$

$$\lambda = \frac{c}{\nu} = \frac{3.00 \text{ x } 10^{8} \text{ m/s}}{1.8267 \text{ x } 10^{15}\text{/s}} = 1.6\underline{4}2 \text{ x } 10^{-7} \text{ m (164 nm = near UV)}$$

Cumulative-Skills Problems (require skill from Chapters 1, 4, 5, and 6)

7.85 First use Avogadro's number to calculate the energy for one Cl_2 molecule.

$$\frac{239 \text{ kJ}}{1 \text{ mol}} \text{ x } \frac{1000 \text{ J}}{1 \text{ kJ}} \text{ x } \frac{1 \text{ mol}}{6.02 \text{ x } 10^{23} \text{ molec}} = 3.9\underline{7}01 \text{ x } 10^{-19} \text{ J/molec}$$

Then convert energy to frequency and finally to wavelength.

$$\nu = \frac{E}{h} = \frac{3.9701 \text{ x } 10^{-19} \text{ J}}{6.63 \text{ x } 10^{-34} \text{ J•s}} = 5.9\underline{8}8 \text{ x } 10^{14}\text{/s}$$

$$\lambda = \frac{c}{\nu} = \frac{3.00 \text{ x } 10^{8} \text{ m/s}}{5.988 \text{ x } 10^{14}\text{/s}} = 5.0\underline{1}0 \text{ x } 10^{-7} \text{ m (= 501 nm, in the visible)}$$

7.87 First calculate the energy needed to heat the 0.250 L of water from 20.0 to 100.0°C:

$$0.250 \text{ L x } \frac{1000 \text{ g}}{1 \text{ L}} \text{ x } \frac{4.184 \text{ J}}{(\text{g•°C})} \text{ x } (100.0 - 20.0)\text{°C} = 8.3\underline{6}8 \text{ x } 10^{4} \text{ J}$$

Then calculate the frequency, the energy of one photon, and the number of photons:

$$\nu = \frac{c}{\lambda} = \frac{3.00 \text{ x } 10^{8} \text{ m/s}}{0.125 \text{ m}} = 2.4\underline{0}0 \text{ x } 10^{9}\text{/s}$$

$$\text{E of 1 photon} = h\nu = (6.63 \text{ x } 10^{-34} \text{ J•s}) \text{ x } (2.40 \text{ x } 10^{9}\text{/s}) = 1.5\underline{9}1 \text{ x } 10^{-24} \text{ J}$$

$$\text{No. photons} = h\nu = 8.368 \times 10^{4} \text{ J} \times \frac{1 \text{ photon}}{1.591 \times 10^{-24} \text{ J}} = 5.2\underline{5}9 \times 10^{28}$$

$$= 5.26 \times 10^{28} \text{ photons}$$

7.89 First write the following equality for the energy to remove one electron, $E_{removal}$:

$$E_{removal} = E_{425 \text{ nm}} - E_k \text{ of ejected photon}$$

Use E = hv to calculate the energy of the photon. Then recall from chapter 1 that E_k, kinetic energy, = 1/2 mv^2. Use this to calculate E_k.

$$E_{405rm} = \frac{hc}{\lambda} = \frac{(6.63 \times 10^{-34} \text{ J•s}) \times (3.00 \times 10^{8} \text{ m/s})}{4.25 \times 10^{-7} \text{ m}} = 4.6\underline{8}0 \times 10^{-19} \text{ J}$$

$$E_k = 1/2mv^2 = 1/2 \times (9.10953 \times 10^{-31} \text{ kg}) \times (4.88 \times 10^{5} \text{ m/s})^2 = 1.0\underline{8}47 \times 10^{-19} \text{ J}$$

Subtract to find $E_{removal}$ and convert it to kJ/mol:

$$E_{removal} = 4.680 \times 10^{-19} \text{ J} - (1.0847 \times 10^{-19} \text{ J}) = 3.5\underline{9}53 \times 10^{-19}$$

$$= 3.60 \times 10^{-19} \text{ J/electron}$$

$$E_{removal} = \frac{3.5953 \times 10^{-19} \text{ J}}{e^-} \times \frac{6.02 \times 10^{23} \text{ e}^-}{1 \text{ mol}} \times \frac{1 \text{ kJ}}{1000 \text{ J}} = 21\underline{6}.4 = 216 \text{ kJ/mol}$$

7.91 First calculate the energy, E, in joules using the product of voltage and charge:

$$E = (4.00 \times 10^{3} \text{ V}) \times (1.602 \times 10^{-19} \text{ C}) = 6.4\underline{0}8 \times 10^{-16} \text{ J}$$

Now use the kinetic energy equation $E_k = 1/2\ mv^2$ and solve for velocity:

$$v = \sqrt{\frac{2E_k}{m}} = \sqrt{\frac{2 \times 6.408 \times 10^{-16} \text{ J}}{9.11 \times 10^{-31} \text{ kg}}} = 3.7\underline{5}07 \times 10^{7} \text{ m/s}$$

$$\lambda = \frac{h}{mv} = \frac{6.626 \times 10^{-34} \text{ J•s}}{(9.11 \times 10^{-31} \text{ kg}) \times (3.7507 \times 10^{7} \text{ m/s})} = 1.9\underline{3}9 \times 10^{-11} \text{ m (19.4 pm)}$$

CHAPTER 8

ELECTRON CONFIGURATIONS AND PERIODICITY

SOLUTIONS TO EXERCISES

Note on significant figures: The final answer to all mathematical solutions is given first with one nonsignificant figure (last significant figure underlined) and is then rounded to the correct number of figures. Intermediate answers usually also have at least one nonsignificant figure.

8.1
a. Possible orbital diagram.
b. Possible orbital diagram.
c. Impossible orbital diagram: there are two electrons in a 2p orbital with the same spin.
d. Possible electronic configuration
e. Impossible electronic configuration: only two electrons are allowed in an s subshell.
f. Impossible electronic configuration: only six electrons are allowed in a p subshell.

8.2 Look at the periodic table. Start with hydrogen and go through the periods, writing down the subshells being filled, stopping with manganese (Z = 25). We obtain the following order:

Order:	1s	2s2p	3s3p	4s3d
Period:	first	second	third	fourth

Now we fill the subshells with electrons, remembering that we have a total of 25 electrons to distribute. We obtain:

$1s^22s^22p^63s^23p^64s^23d^5$ or by shells: $1s^22s^22p^63s^23p^63d^54s^2$

8.3 Arsenic is in Group VA and in period 4, so that the five outer electrons should occupy the 4s and 4p subshells; the five valence electrons have the configuration $4s^24p^3$.

8.4 Since the sum of the $6s^2$ and $6p^2$ electrons gives four outer (valence) electrons, lead should be in Group IVA, and it is. Looking at the table, it is in period 6. From its position, it would be classified as a main-group element.

8.5 The electronic configuration of phosphorus is $1s^22s^22p^63s^23p^3$. The orbital diagram is:

(↑↓)	(↑↓)	(↑↓)(↑↓)(↑↓)	(↑↓)	(↑)(↑)(↑)
1s	2s	2p	3s	3p

8.6 The radius tends to decrease across a row of the periodic table from left to right, and it tends to increase from the top of a column to the bottom. Therefore in order of increasing radius:

Be < Mg < Na

8.7 It is more likely that 1000 kJ/mol is the ionization energy for iodine because ionization energies tend to decrease with atomic number in a group (I is below Cl in group VIIA).

8.8 Fluorine should have a more negative electron affinity because: (1) carbon has a stable half-filled p subshell, (2) the -1 fluoride ion has a stable noble-gas configuration, and (3) the electron can approach the fluorine nucleus more closely than the carbon nucleus.

8.9 a. The formula is H_2Se. (Group VIA elements form hydrogen compounds of the form H_2X.)
b. The formula is $CaSeO_4$.

ANSWERS TO REVIEW QUESTIONS

8.1 In the Stern-Gerlach experiment, a beam of atoms, such as hydrogen atoms, is channeled into the field of a specially constructed magnet. If the beam is composed of hydrogen atoms, the beam is split into two, half the beam being bent into one direction, and the other half being bent in the other direction.

8.2 Visualize the electron as a ball of spinning charge. The circulating charge creates a magnetic field with a spin axis that has more than one possible direction. Quantum mechanics predicts that the spin axis is restricted to one of two directions, corresponding to the m_s quantum numbers +1/2 and -1/2.

8.3 The Pauli principle excludes the configurations in which two or more electrons of the same atom have the same four quantum numbers.

8.4 The g subshell has nine orbitals, each containing a maximum of two electrons. Thus the g subshell can hold a maximum of eighteen electrons.

8.5 Orbitals in order of increasing orbital energy: 1s, 2s, 2p, 3s, 3p.

8.6 Noble-gas core: this has an electronic configuration corresponding to one of the noble gases.
Pseudo-noble-gas core: this has a noble-gas core, outside of which are $(n-1)d^{10}$ electrons; the pseudo-noble gas core is below other ns or nsnp electrons (valence electrons).
Valence electrons: these are electrons outside the noble-gas or pseudo-noble-gas core.

8.7 The orbital diagram for the $1s^22s^22p^4$ ground state of oxygen is:

(↑↓) (↑↓) (↑↓)(↑)(↑)

1s 2s 2p

Another possible oxygen orbital diagram, but not a ground state, is:

(↑↓) (↑↓) (↑↓)(↑↓)( )

1s 2s 2p

8.8 Diamagnetic substance: an atomic substance that is not attracted to, or is slightly repelled by, a magnetic field.
Paramagnetic substance: an atomic substance that is attracted by a magnetic field.
A ground-state oxygen atom has unpaired electrons and is therefore paramagnetic.

8.9 An s subshell is being filled in the elements in Groups IA and IIA; a p subshell is being filled in Groups IIIA to VIIIA; a d subshell is being filled in the transition elements; and an f subshell is being filled in the lanthanides and actinides.

8.10 Mendeleev found that the elements could be arranged in order by their atomic weights (more correctly, atomic numbers) in rows and columns, so that the elements in the same row have similar properties. This arrangement is known as a periodic table. In first setting up his periodic table, he found it necessary to leave certain positions blank. He postulated that these blank positions corresponded to undiscovered elements. The properties of these elements could be predicted from their positions in the periodic table. Gallium was one of the elements whose properties Mendeleev predicted from the periodic table.

8.11 The major trends seen when atomic radii are plotted vs. atomic number are: (1) within each period, the atomic radius tends to decrease with atomic number, and (2) within each group, the atomic radius tends to increase with the period number. The major trends seen when ionization energies are plotted against atomic number are: (1) within a period, the ionization energy increases with atomic number, and (2) within a group, the ionization energy tends to decrease going down the group.

8.12 The alkaline-earth element with the smallest radius is beryllium (Be).

8.13 Group VIIA (halogens) is the main group with the most negative electron affinities. Configurations with filled subshells (ground states of the noble-gas elements) when adding one electron per atom would form unstable negative ions.

8.14 The Na^{+} and Mg^{2+} ions are stable because they are isoelectronic with the noble gas neon. If Na^{2+} and Mg^{3+} ions were to exist, they would be very unstable because they would not be isoelectronic with any noble gas structure.

8.15 The elements tend to increase in metallic character in going from right to left in any period. They also tend to increase in metallic character in going down any column (group) of elements.

8.16 A basic oxide is an oxide that reacts with acids, and gives basic solutions with water. An example is calcium oxide, CaO. An acidic oxide is an oxide that reacts with bases; it gives acidic solutions with water. An example is carbon dioxide, CO_2.

8.17 Rubidium is the alkali-metal atom with a $5s^1$ configuration.

8.18 Atomic number = 117 (protons in last known element + those needed to reach Group VIIA).

8.19 The following elements are in Groups IIIA to VIA:

Group IIIA	Group IVA	Group VA	Group VIA
B: metalloid	C: nonmetal	N: nonmetal	O: nonmetal
Al: metal	Si: metalloid	P: nonmetal	S: nonmetal
Ga: metal	Ge: metalloid	As: metalloid	Se: nonmetal
In: metal	Sn: metal	Sb: metalloid	Te: metalloid
Tl: metal	Pb: metal	Bi: metal	Po: metal

Yes, each column displays the expected increasing metallic character.

8.20 The oxides of the following elements are listed as either acidic, basic, amphoteric, or t.g.n.i. (text gives no information):

Group IIIA	Group IVA	Group VA	Group VIA
B: acidic	C: acidic	N: acidic	O: amphoteric (H_2O)
Al: amphoteric	Si: acidic	P: acidic	S: acidic
Ga: amphoteric	Ge: t.g.n.i.	As: t.g.n.i.	Se: t.g.n.i.
In: basic	Sn: amphoteric	Sb: t.g.n.i.	Te: t.g.n.i.
Tl: basic	Pb: amphoteric	Bi: t.g.n.i.	Po: t.g.n.i.

8.21 The reaction is: $2K(s) + 2H_2O(l) \rightarrow 2KOH(aq) + H_2(g)$

8.22 Barium should be a soft, reactive metal. Barium should form the basic oxide, BaO. Barium metal, for example, would be expected to react with water according to this equation:

$$Ba(s) + 2H_2O(l) \rightarrow Ba(OH)_2(aq) + H_2(g)$$

8.23 The two allotropes of carbon are graphite, a soft black substance, and diamond, a very hard clear crystalline substance.

8.24 (a) white phosphorus (b) sulfur (c) bromine (d) sodium

SOLUTIONS TO PRACTICE PROBLEMS

Note on significant figures: The final answer is given first with one nonsignificant figure (rightmost significant figure underlined), and is then rounded to the correct number of significant figures. Intermediate answers usually also have at least one nonsignificant figure. Atomic weights are rounded to two decimal places, except for that of hydrogen.

8.25
a. Allowed
b. Not allowed (the two electrons in the first 2p orbital must have opposite spins)
c. Allowed
d. Not allowed (the 2s orbital should have just the first two electrons, not the 3rd)

8.27 a. Impossible (the 2s orbital can only hold 2 electrons)
b. Possible
c. Impossible (the 3p subshell can only hold 6 electrons and the 4s orbital fills before the 3d subshell)
d. Possible

8.29 The six possible orbital diagrams for $1s^2 2p^1$ are:

$1s^2$	$2p^1$
(↑↓)	(↑) ( ) ( )
(↑↓)	(↓) ( ) ( )
(↑↓)	( ) (↑) ( )
(↑↓)	( ) (↓) ( )
(↑↓)	( ) ( ) (↑)
(↑↓)	( ) ( ) (↓)

8.31 Phosphorus (Z = 15): $1s^2 2s^2 2p^6 3s^2 3p^3$

8.33 Vanadium (Z = 23): $1s^2 2s^2 2p^6 3s^2 3p^6 3d^3 4s^2$

8.35 Bromine (Z = 35): $4s^2 4p^5$

8.37 Titanium (Z = 22): $3d^2 4s^2$

8.39 The highest value of n is 6, so thallium (Tl) is in the sixth period. The 5d subshell is filled, and there is a 6p electron, so it belongs in an A group. There are three valence electrons, so Tl is in Group IIIA. It is a main-group element.

8.41 Nickel (Z = 28): [Ar] (↑↓)(↑↓)(↑↓)(↑)(↑) (↑↓)
3d 4s

8.43 Potassium (Z = 19): [Ar] (↑)
4s

All the subshells are filled in the argon core; however, the 4s electron is unpaired, causing the ground state of the potassium atom to be a paramagnetic substance.

8.45 Atomic radius increases going down a column (group), from F to Cl, and increases going from right to left in a row, from S to Cl. Thus the order by increasing atomic radius is F, Cl, S.

8.47 Ionization energy increases going up a column (group), from Ca to Mg, and increases going left to right in a row, from Mg to S. Thus the order by increasing ionization energy is Ca, Mg, S.

8.49 a. In general, the electron affinity becomes more negative on going from left to right within a period. Thus Cl has a more negative electron affinity than S.

b. In general, the electron affinity of a nonmetal is more negative than that of a metal. Thus Se has a more negative electron affinity than K.

8.51 The expected positive oxidation states of tellurium are +4 and +6. The corresponding oxides have the simplest formulas of TeO_2 and TeO_3.

8.53 Strontium: $1s^2 2s^2 2p^6 3s^2 3p^6 3d^{10} 4s^2 4p^6 5s^2$

8.55 Polonium: $6s^2 6p^4$

8.57 The orbital diagram for arsenic is:

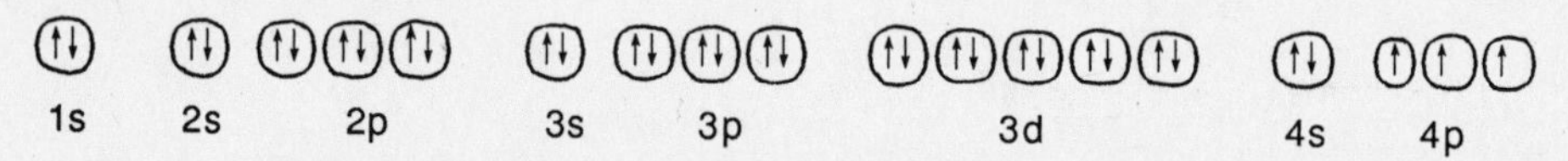

8.59 For eka-lead: [Rn] $5f^{14} 6d^{10} 7s^2 7p^2$. It is a metal, and the oxide is eka-PbO or eka-PbO_2.

8.61 The ionization energy of Fr is ~370 kJ/mol (slightly less than that of Cs).

8.63 Niobium: [Kr] (↑)(↑)(↑)(↑)(↑) (↑)

4d 5s

8.65 a. Cl_2 b. Na c. Sb d. Ar

8.67 Element with Z = 23: $1s^2 2s^2 2p^6 3s^2 3p^6 3d^3 4s^2$. The element is in Group VB (three of the five valence electrons are d electrons), and in Period 4 (largest n is 4). It is a d-block transition element.

Cumulative-Skills Problems (require skills from previous chapters)

8.69 The equation is: $Ba(s) + 2H_2O(l) \rightarrow Ba(OH)_2(aq) + H_2(g)$
Using the equation, calculate the moles of H_2; then use the ideal gas law to convert to volume.

$$\text{Mol } H_2 = 2.50 \text{ g Ba} \times \frac{1 \text{ mol Ba}}{137.33} \times \frac{1 \text{ mol } H_2}{1 \text{ mol Ba}} = 0.018\underline{2}04 \text{ mol } H_2$$

$$V = \frac{nRT}{P} = \frac{[0.018204 \text{ mol}][0.082057 \text{ L}\cdot\text{atm}/(\text{K}\cdot\text{mol})][204.2 \text{ K}]}{(748/760) \text{ atm}} = 0.44\underline{6}51 \text{ L} \quad (447 \text{ mL})$$

8.71 Radium is in Group IIA; hence the radium cation is Ra^{2+}, and its oxide is RaO. Use the atomic weights to calculate the percentage of Ra in RaO:

$$\% \text{ Ra} = \frac{226 \text{ amu Ra}}{226 \text{ amu Ra} + 16.000 \text{ amu O}} \times 100\% = 93.\underline{3}8 = 93.4\% \text{ Ra}$$

8.73 Convert 5.00 mg (0.00500 g) of Na to moles of Na; then convert to energy using the first ionization energy of 496 kJ/mol Na.

$$\text{Mol Na} = 0.00500 \text{ g Na} \times \frac{1 \text{ mol Na}}{22.99 \text{ g Na}} = 2.1\underline{7}4 \times 10^{-4} \text{ mol Na}$$

$$2.1\underline{7}4 \times 10^{-4} \text{ mol Na} \times \frac{496 \text{ kJ}}{1 \text{ mol Na}} = 0.10\underline{7}8 = 0.108 \text{ kJ}$$

8.75 Use the Bohr formula, where $n_f = \infty$ and $n_i = 1$.

$$\Delta E = -R_H \left[\frac{1}{\infty^2} - \frac{1}{1^2} \right] = -R_H[-1] = R_H = \frac{2.180 \times 10^{-18} \text{ J}}{1 \text{ H atom}}$$

$$\text{Ioniz en} = \frac{2.180 \times 10^{-18} \text{ J}}{1 \text{ H atom}} \times \frac{6.02 \times 10^{23} \text{ H atoms}}{1 \text{ mol H}} = \frac{1.312 \times 10^{6} \text{ J}}{1 \text{ mol H}} = \frac{1.31 \times 10^{3} \text{ kJ}}{1 \text{ mol H}}$$

8.77 Add the three equations after reversing the equation for the lattice energy and its ΔH:

$Na(g) \rightarrow Na^+(g) + e^-$		$\Delta H = +496$ kJ/mol
$Cl(g) + e^- \rightarrow Cl^-(g)$		$\Delta H = -349$ kJ/mol
$Na^+(g) + Cl^-(g) \rightarrow NaCl(s)$		$-1(\Delta H = 786$ kJ/mol)
$Na(g) + Cl(g) \rightarrow NaCl(s)$		$\Delta H = -639$ kJ/mol

CHAPTER 9

IONIC AND COVALENT BONDING

SOLUTIONS TO EXERCISES

9.1 The Lewis symbol for oxygen is $:\dot{\underset{..}{O}}\cdot$ and the Lewis symbol for magnesium is $\cdot Mg \cdot$. The magnesium atom loses two electrons and the oxygen atom accepts two electrons. We can represent this electron transfer as follows:

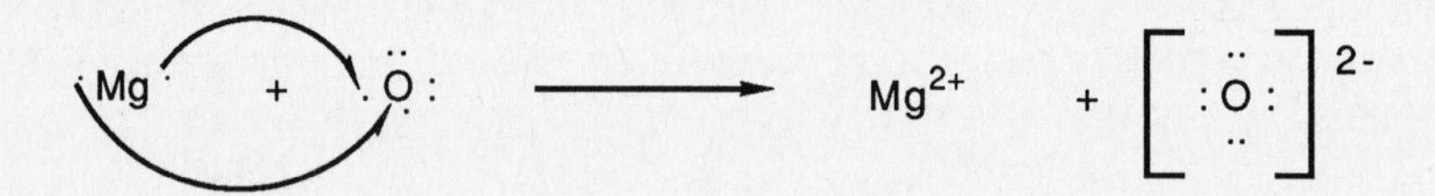

9.2 The electron configuration of the Ca atom is $[Ar]4s^2$. By losing two electrons the atom assumes a 2+ charge and the argon configuration. The Lewis symbol is Ca^{2+}. The S atom has the configuration $[Ne]3s^2 3p^4$. By gaining two electrons, the atom assumes a 2- charge and the argon configuration. The Lewis symbol is

$$\left[:\ddot{\underset{..}{S}}:\right]^{2-}$$

9.3 The electron configuration of lead is $[Xe]4f^{14}5d^{10}6s^2 6p^2$. The electron configuration of Pb^{2+} is $[Xe]4f^{14}5d^{10}6s^2$.

9.4 The electron configuration of manganese is $[Ar]3d^5 4s^2$. To find the ion configuration, first remove the 4s electrons, then the 3d electrons. In this case, only two electrons need to be removed. The electron configuration of Mn^{2+} is $[Ar]3d^5$.

9.5 S^{2-} has a larger radius than S. The anion has more electrons than the atom. The electron-electron repulsion is greater; hence, the valence orbitals expand. The anion radius is larger than the atomic radius.

9.6 The ionic radii increase down any column because of the addition of electron shells. All of these ions are from the IIA family; therefore, $Mg^{2+} < Ca^{2+} < Sr^{2+}$.

9.7 Cl^-, Ca^{2+}, and P^{3-} are isoelectronic. In an isoelectronic sequence, the ionic radius decreases with increasing atomic number. Therefore, in order of increasing ionic radius, we have Ca^{2+}, Cl^-, and P^{3-}.

9.8 The absolute value of the electronegativity differences are C-O, 1.0; C-S, 0.0; H-Br, 0.7. C-O is the most polar bond.

9.9 First, calculate the total number of valence electrons. C has 4, Cl has 7, F has 7. The total number is 4 + (2 x 7) + (2 x 7) = 32. From rule 1, the expected skeleton consists of a carbon atom surrounded by Cl and F atoms. Distribute the electron pairs to the surrounding atoms to satisfy the octet rule. All 32 electrons (16 pairs) are accounted for.

```
      ..
     : Cl :
  ..  ..  ..
 : F : C : F :
  ..  ..  ..
     : Cl :
      ..
```

9.10 The total number of electrons in CO_2 is 4 + (2 x 6) = 16. Carbon, since it is more electropositive than oxygen, is expected to be the central atom. Distribute the electrons to the surrounding atoms to satisfy the octet rule.

```
  ..     ..
 : O : C : O :
  ..     ..
```

All 16 electrons have been used, but notice there are only 4 electrons on carbon. This is 4 electrons short of a complete octet, suggesting the existence of double bonds. Move a pair of electrons from each oxygen to the carbon-oxygen bonds.

```
 ..      ..           ..        ..
 O : : C : : O   or   O ═ C ═ O
 ..      ..           ..        ..
```

9.11 a. There are (3 x 1) + 6 = 9 valence electrons in H_3O. The H_3O^+ ion has one less electron than is provided by the neutral atoms because the charge on the ion is +1. Hence, there are 8 valence electrons in H_3O^+. The electron dot formula is

```
 ┌        ┐+
 │   H    │
 │   ..   │
 │ H : O : H │
 │   ..   │
 └        ┘
```

b. Cl has 7 valence electrons and O has 6 valence electrons. The total number of valence electrons from the neutral atoms is 7 + (2 x 6) = 19. The charge on the ClO_2^- is -1, which provides one more electron than the neutral atoms. This makes a total of 20 valence electrons. The electron dot formula for ClO_2^- is

```
 ┌  ..   ..   ..  ┐-
 │ : O : Cl : O : │
 └  ..   ..   ..  ┘
```

9.12 Be has 2 valence electrons and Cl has 7 valence electrons. The total number of valence electrons is 2 + (2 x 7) = 16 in the $BeCl_2$ molecule. Be, a group IIA element, can have fewer than eight electrons around it. The electron-dot formula of $BeCl_2$ is

```
   ..         ..
 : Cl : Be : Cl :
   ..         ..
```

9.13 The number of valence electrons in SF_4 is 6 + (4 x 7) = 34. The skeleton structure is a sulfur atom surrounded by fluorine atoms. After placing the electron pairs on the F atoms to satisfy the octet rule, two electrons remain.

These additional two electrons are put on the sulfur atom since it can expand its octet.

9.14 The resonance formulas for NO_3^- are

9.15 The bond length can be predicted by adding the covalent radii of the two atoms. For O-H, we have 0.66Å + 0.37Å = 1.03Å.

9.16 As the bond order increases, the bond length decreases. Since the C═O is a double bond, one would expect it to be the shorter one, 1.23Å.

9.17

$$H_2C{=}CH_2 + 3\ O_2 \longrightarrow 2\ O{=}C{=}O + 2\ H{-}O{-}H$$

One C=C bond, four C-H bonds, and three O_2 bonds are broken. There are four C=O bonds and four O-H bonds formed.

$$\Delta H = \{[602 + (4 \times 411) + (3 \times 494)] - [(4 \times 799) + (4 \times 459)]\}\ kJ$$
$$= -1304\ kJ$$

ANSWERS TO REVIEW QUESTIONS

9.1 We may think of the formation of the NaCl crystal from atoms in the following way. As an Na atom approaches a Cl atom, the outer electron of the Na atom is transferred to the Cl atom. The result is an Na^+ and a Cl^- ion. Positively charged ions attract negatively charged ions, so that finally the NaCl crystal consists of Na^+ ions surrounded by six Cl^- ions surrounded by six Na^+ ions.

9.2 Ions tend to attract as many ions of opposite charge about them as possible. The result is that ions tend to form crystalline solids rather than molecular substances.

9.3 The energy terms involved in the formation of an ionic solid from atoms are the ionization energy of the metal atom, the electron affinity of the nonmetal atom, and the energy of the attraction of the ions forming the ionic solid. The energy for the solid will be low if the ionization energy of the metal is low, the electron affinity of the nonmetal is high, and the energy of the attraction of the ions is large.

9.4 The lattice energy for potassium bromide is the change in energy that occurs when KBr(s) is separated into isolated $K^+(g)$ and $Br^-(g)$ ions in the gas phase.

$$KBr\,(s) \longrightarrow K^+\,(g) + Br^-\,(g)$$

9.5 A monatomic cation with a charge equal to the group number corresponds to the loss of all valence electrons. This loss of electrons would give a noble gas configuration, which is especially stable. A monatomic anion with a charge equal to the group number minus eight would have a noble gas configuration.

9.6 Most of the transition elements have configurations in which the outer s subshell is doubly occupied. These electrons will be lost first, and we might expect each to be lost with equal ease, resulting in 2+ ions.

9.7 If we assume that the ions are spheres that are just touching, the distances between centers of the spheres will be related to the radii of the spheres. For example, in LiI, we assume that the I^- ions are large spheres that are touching. The distance between centers of the I^- ions equals two times the radius of the I^- ion.

9.8 In going across a period, the cations decrease in radius. When we reach the anions, there is an abrupt increase in radius, and then the radii again decrease. Ionic radii increase in going down any column of the periodic table.

9.9 As the H atoms approach one another, their 1s orbitals begin to overlap. Each electron can then occupy the space around both atoms; that is, the two electrons are shared by the atoms.

9.10

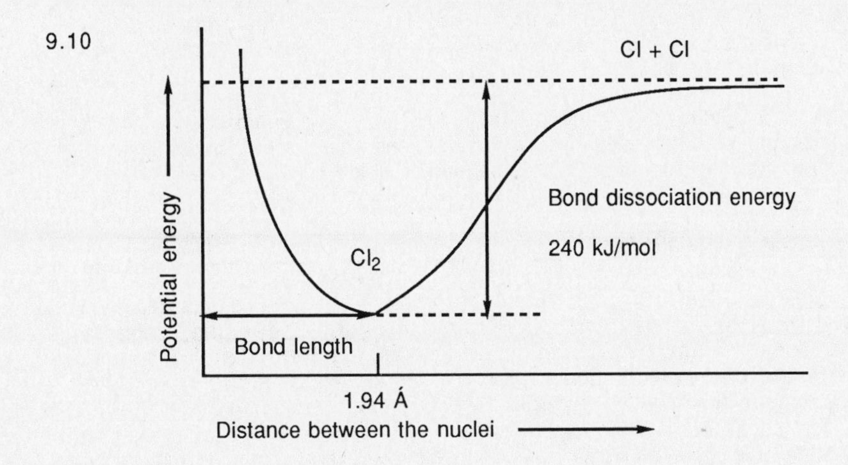

9.11 An example is thionyl chloride, $SOCl_2$ (see example 9.6 for a discussion of the electron dot formula) :

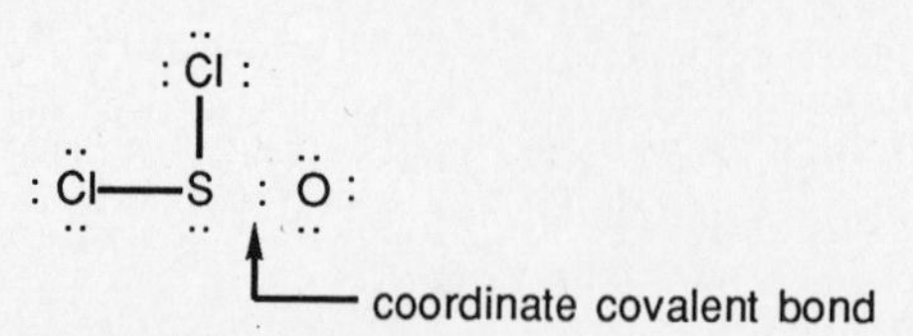

Note that the O atom has eight electrons around it; that is, it has two more electrons than the neutral atom. These two electrons must have come from the S atom. Thus, this bond is a coordinate covalent bond.

9.12 In many atoms of the main group elements, bonding uses an s orbital and the three p orbitals of the valence shell. These four orbitals are filled with eight electrons, thus accounting for the octet rule.

9.13 Electronegativity increases from left to right and decreases from top to bottom in the periodic table.

9.14 The absolute difference in the electronegativities of the two atoms in a bond gives a rough measure of the polarity of the bond.

9.15 Molecules having an odd number of electrons do not obey the octet rule. The other exceptions fall into two groups. In one group are molecules with an atom having fewer than eight valence electrons around it. In the other group are molecules with an atom having more than eight valence electrons around it.

9.16 Resonance is used to describe the electron structure of a molecule in which bonding electrons are delocalized. In a resonance description, the molecule is described in terms of two or more Lewis formulas. If we wish to retain Lewis formulas, resonance is required because each Lewis formula assumes that a bonding pair of electrons occupies the region between two atoms. We must imagine that the actual electron structure of the molecule is a composite of all resonance formulas.

9.17 As the bond order increases, the bond length decreases. For example, the average carbon-carbon single bond length is 1.54 Å, whereas the carbon-carbon double bond length is 1.34 Å, and the carbon-carbon triple bond length is 1.20 Å.

9.18 Bond energy is the average enthalpy change for the breaking of a bond in a molecule. The enthalpy of a reaction can be determined by summing the bond energies of all the bonds that are broken and subtracting the sum of the bond energies of all the bonds that are formed.

SOLUTIONS TO PRACTICE PROBLEMS

Note on significant figures: The final answer to cumulative-skills problems is given first with one nonsignificant figure (rightmost significant figure underlined), and is then rounded to the correct number of significant figures. Intermediate answers usually also have at least one nonsignificant figure. Atomic weights are rounded to two decimal places, except for that of hydrogen.

9.19 a. Ba has the valence configuration $[Xe]6s^2$. It has 2 valence electrons ($6s^2$), giving the Lewis formula

$$\cdot Ba \cdot$$

b. Ba^{2+} has 2 fewer electrons than barium. It has lost the 2 valence electrons. The Lewis formula is

Ba^{2+}

c. Iodine has the electron configuration $[Kr]4d^{10}5s^{2}5p^{5}$. It has 7 valence electrons ($5s^{2}5p^{5}$), giving the Lewis formula

:I·

d. I^{-} has 1 more electron than iodine. It has gained 1 electron in its valence shell. The Lewis formula is

$[:I:]^{-}$

9.21 a. The Lewis symbols for Na and I are Na · and :I· . If the sodium atom loses 1 electron and the iodine atom gains 1 electron, both will assume noble gas configurations. This can be represented as follows:

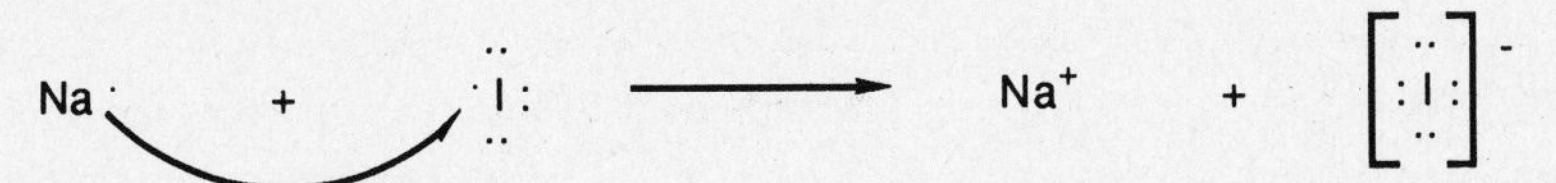

b. The Lewis symbols for Na and S are Na · and :S· . The S atom gains two electrons to assume a noble gas configuration. Since the sodium atom has only one valence electron to lose, two sodium atoms must take part in the electron transfer. This can be represented as follows:

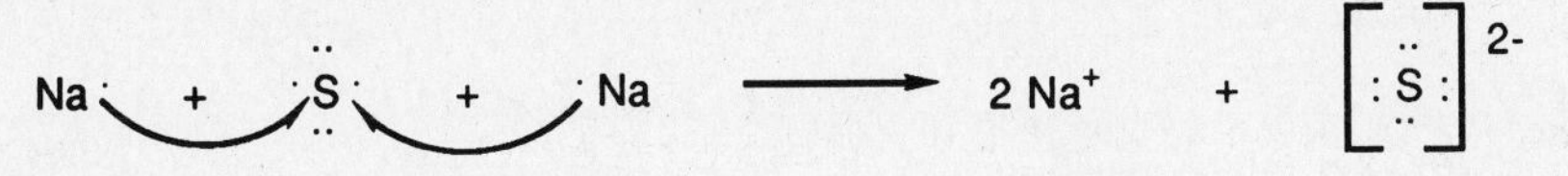

9.23 a. Mg : $[Ne]3s^{2}$ · Mg ·

b. Mg^{2+} : $[Ne] = 1s^{2}2s^{2}2p^{6}$ Mg^{2+}

c. Se^{2-} : $[Kr] = 1s^{2}2s^{2}2p^{6}3s^{2}3p^{6}3d^{10}4s^{2}4p^{6}$ $[:Se:]^{2-}$

d. Br^{-} : $[Kr] = 1s^{2}2s^{2}2p^{6}3s^{2}3p^{6}3d^{10}4s^{2}4p^{6}$ $[:Br:]^{-}$

9.25 a. Sn : $[Kr]4d^{10}5s^{2}5p^{2}$

b. Sn^{2+} : The two 5p electrons are lost from the valence shell.
$[Kr]4d^{10}5s^{2}$

9.27 The 2+ ion is formed by loss of electrons from the 4s subshell.

Ni^{2+} : $[Ar]3d^8$

The 3+ ion is formed by loss of electrons from the 4s and 3d subshells.

Ni^{3+} : $[Ar]3d^7$

9.29 a. Rb^+, Rb
The cation is smaller than the neutral atom because it has lost all its valence electrons; hence, it has one less shell of electrons. The electron-electron repulsion is reduced, so the orbitals shrink from the increased attraction of the electrons to the nucleus.

b. Se, Se^{2-}
The anion is larger than the neutral atom because it has more electrons. The electron-electron repulsion is greater, so the valence orbitals expand to give a larger radius.

9.31 S^{2-}, Se^{2-}, Te^{2-}
All have the same number of electrons in the valence shell. The radius increases with increasing number of filled shells.

9.33 Smallest Na^+ (Z=11), F^- (Z=9), N^{3-} (Z=7) Largest
These ions are isoelectronic. The atomic radius increases with decreasing nuclear charge (Z).

9.35

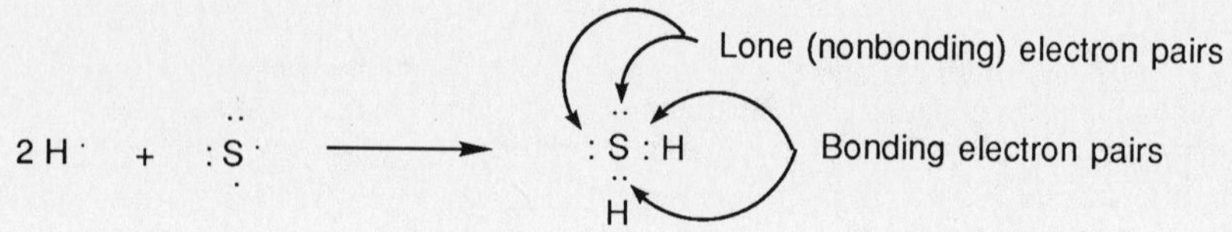

2 H· + :S̈· ⟶ :S̈:H (with H bonded below S)

Lone (nonbonding) electron pairs

Bonding electron pairs

9.37 The Lewis symbols for the atoms are :C̈l· and ·Ṡi· .
The silicon atom has four valence electrons. Each of these can be paired with the unpaired electron of the chlorine atom. The expected formula is $SiCl_4$.

9.39 a. Cs, Ba, Sr
Electronegativity increases from left to right and bottom to top in the periodic table.

b. Ca, Ga, Ge
Electronegativity increases from left to right within a period.

c. As, P, S
Electronegativity increases from left to right and bottom to top in the periodic table.

9.41 $X_{Se} - X_H = 2.4 - 2.1 = 0.3$

$X_{Cl} - X_P = 3.0 - 2.1 = 0.9$

$X_{Cl} - X_N = 3.0 - 3.0 = 0.0$

The difference in electronegativity is smallest for the N-Cl bond; hence, it is the least polar.

9.43 The atom with the greater electronegativity has the partial negative charge.

a. H—Se
$\delta+$ $\delta-$

b. P—Cl
$\delta+$ $\delta-$

c. nonpolar

9.45 a. Total number of valence electrons = 7 + 7 = 14. Br-Br is the skeleton. Distribute the 12 electrons remaining.

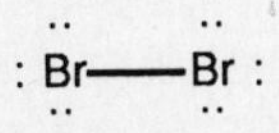

```
  ..    ..
: Br————Br :
  ..    ..
```

b. Total valence electrons = (2 x 1) + 6 = 8. By rule 1, the skeleton is H-Se-H. Distribute the remaining 4 electrons.

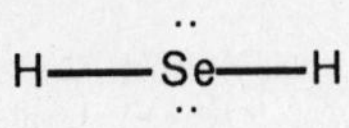

```
       ..
H————Se————H
       ..
```

c. Total valence electrons = (2 x 7) + 6 + 6 = 26. By rule 1, the skeleton is:

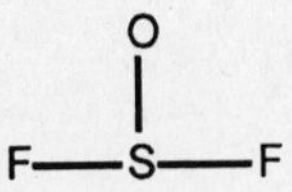

```
      O
      |
F ——— S ——— F
```

Distribute the remaining 20 electrons.

```
       ..
      :O:
       |
  ..   |    ..
: F————S————F :
  ..   ..   ..
```

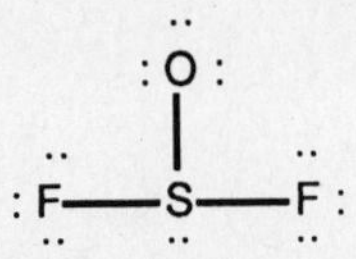

d. Total valence electrons = (2 x 1) + 6 + (4 x 6) = 32. By rule 3, the skeleton is:

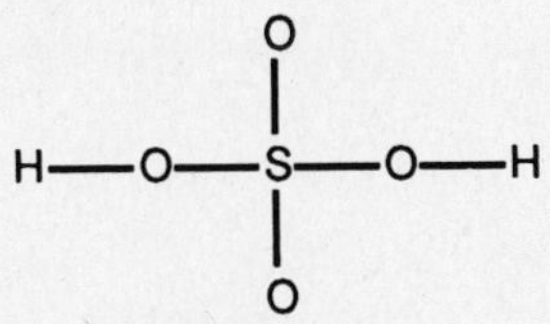

```
            O
            |
H ——— O ——— S ——— O ——— H
            |
            O
```

Distribute the remaining 20 electrons.

```
             ..
            :O:
             |
       ..    |     ..
H ——— O ——— S ——— O ——— H
       ..    |     ..
             |
            :O:
             ..
```

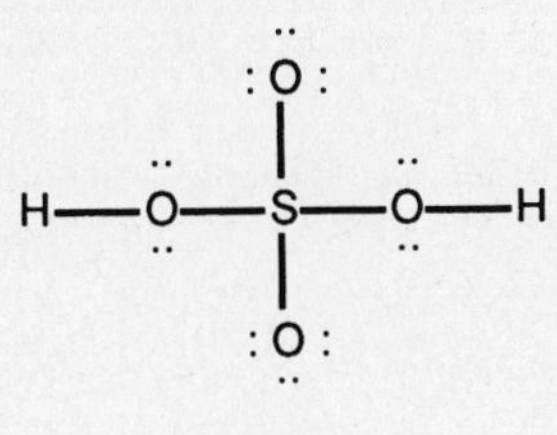

9.47 a. Total valence electrons = 2 x 5 = 10. The skeleton is P-P. Distribute the remaining electrons symmetrically:

```
   ..    ..
 : P-----P :
```

Neither P atom has an octet. There are four fewer electrons than are needed. This suggests the presence of a triple bond. Make one lone pair from each P a bonding pair.

```
 : P≡P :
```

b. Total valence electrons = 4 + 6 + 6 = 16. From rule 1, the skeleton is O-C-Se. Distribute the remaining 12 electrons on O and Se:

```
   ..          ..
 : O-----C-----Se :
   ..          ..
```

The carbon atom does not have an octet. It is four electrons short. There are probably two double bonds. By rule 4, the most likely structure is:

```
   ..          ..
 : O=====C=====Se :
```

c. Total valence electrons = 4 + 6 + (2 x 7) = 24. By rule 1, the skeleton is:

```
         Br
         |
 O-------C-------Br
```

Distribute the remaining 18 electrons.

```
          ..
        : Br :
          |
   ..     |      ..
 : O------C------Br :
   ..            ..
```

Notice that carbon is two electrons short of an octet. This suggests the presence of a double bond. By rule 4, the most likely double bond is between C and O.

```
          ..
        : Br :
          |
          |      ..
 : O======C------Br :
   ..            ..
```

d. Total valence electrons = 1 + 5 + (2 x 6) = 18. By rule 3, the skeleton is H-O-N-O. Distribute the remaining 12 electrons:

```
        ..     ..     ..
 H------O------N------O :
        ..            ..
```

Notice that the N atom does not have an octet. It is two electrons short. By rule 4, the most likely double bond is between N and O.

```
        ..     ..     ..
 H------O------N======O :
        ..
```

9.49 a. Total valence electrons = 7 + 6 + 1 = 14. The skeleton is Cl-O. Distribute the remaining 12 electrons.

```
 [  ..    ..  ] -
 [ : Cl——O :  ]
 [  ..    ..  ]
```

b. Total valence electrons = 7 + (2 x 7) - 1 = 20. By rule 2, the skeleton is F-Cl-F. Distribute the remaining 16 electrons.

```
 [  ..    ..    ..  ] +
 [ : F——Cl——F :     ]
 [  ..    ..    ..  ]
```

c. Total valence electrons = 4 + (3 x 7) + 1 = 26. By rule 1, the skeleton is:

```
       Cl
       |
 Cl——Sn——Cl
```

Distribute the remaining 20 electrons such that each atom has an octet.

```
 [          ..            ] -
 [        : Cl :          ]
 [          |             ]
 [   ..     |      ..     ]
 [ : Cl——Sn——Cl :         ]
 [   ..     ..     ..     ]
```

d. Total valence electrons = (2 x 6) + 2 = 14. The skeleton is S-S. Distribute the remaining 12 electrons.

```
 [  ..    ..  ] 2-
 [ : S——S :   ]
 [  ..    ..  ]
```

9.51 a. Total valence electrons = 3 + (3 x 7) = 24. By rule 1, the skeleton is:

```
       Cl
       |
 Cl——B——Cl
```

Distribute the remaining 18 electrons.

```
          ..
        : Cl :
          |
   ..     |     ..
 : Cl——B——Cl :
   ..           ..
```

Although boron has only 6 electrons, it has the normal number of covalent bonds.

b. Total valence electrons = 3 + (2 x 7) - 1 = 16. The skeleton is Cl-Tl-Cl. Distribute the remaining 12 electrons.

```
 [   ..          ..   ] +
 [ : Cl——Tl——Cl :     ]
 [   ..          ..   ]
```

Tl has only 4 electrons around it.

c. Total valence electrons = 2 + (2 x 7) = 16. The skeleton is Br-Be-Br. Distribute the remaining 12 electrons.

```
 ..              ..
: Br——Be——Br :
 ..              ..
```

In covalent compounds, beryllium frequently has two bonds, although it does not have an octet.

9.53 a. Total valence electrons = 8 + (2 x 7) = 22. The skeleton is F-Xe-F. Place 6 electrons around each fluorine atom to satisfy its octet.

```
 ..          ..
: F——Xe——F :
 ..          ..
```

There are three electron pairs remaining. Place them on the xenon atom.

```
 ..   . ˙ .   ..
: F——Xe——F :
 ..    ..    ..
```

b. Total valence electrons = 6 + (4 x 7) = 34. The skeleton is:

```
       F
       |
  F——Se——F
       |
       F
```

Distribute 24 of the remaining 26 electrons on the fluorine atoms. The remaining pair of electrons is placed on the selenium atom.

```
        ..
       : F :
         |
  ..    .|      ..
: F——:Se——F :
  ..     |      ..
       : F :
        ..
```

c. Total valence electrons = 6 + (6 x 7) = 48. The skeleton is:

```
     F   F
      \ /
  F——Te——F
      / \
     F   F
```

Distribute the remaining 36 electrons on the fluorine atoms.

```
      ..    ..
    : F :  : F :
        \   /
  ..                ..
: F——Te——F :
  ..                ..
        /   \
    : F :  : F :
      ..    ..
```

d. Total valence electrons = 8 + (5 x 7) - 1 = 42. The skeleton is:

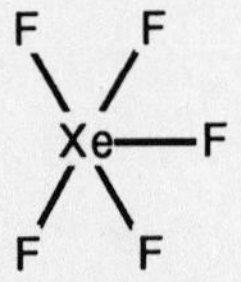

Use 30 of the remaining 32 electrons on the fluorine atoms to complete their octets. The remaining 2 electrons form a lone pair on the xenon atom.

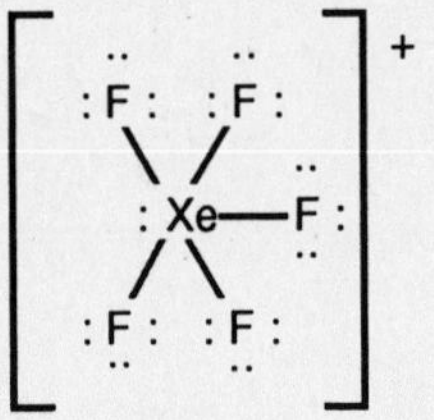

9.55 a. One possible electron dot formula for FNO_2 is

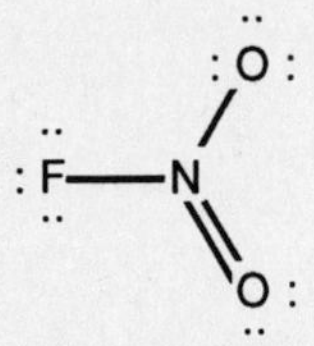

Since the nitrogen-oxygen bonds are expected to be equivalent, the structure must be described in resonance terms.

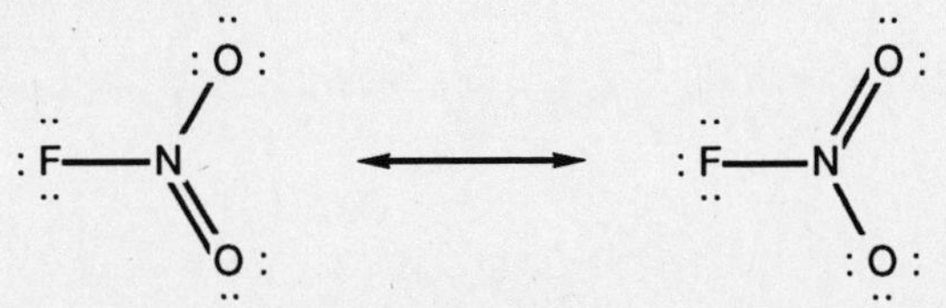

One electron pair is delocalized over the nitrogen atom and the two oxygen atoms.

b. One possible electron dot formula for SO_3 is

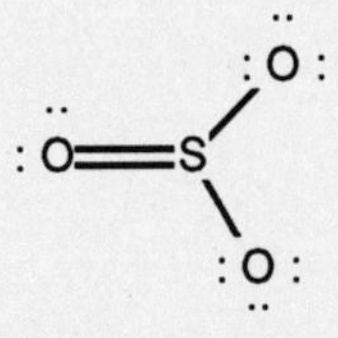

Since the sulfur-oxygen bonds are expected to be equivalent, the structure must be described in resonance terms.

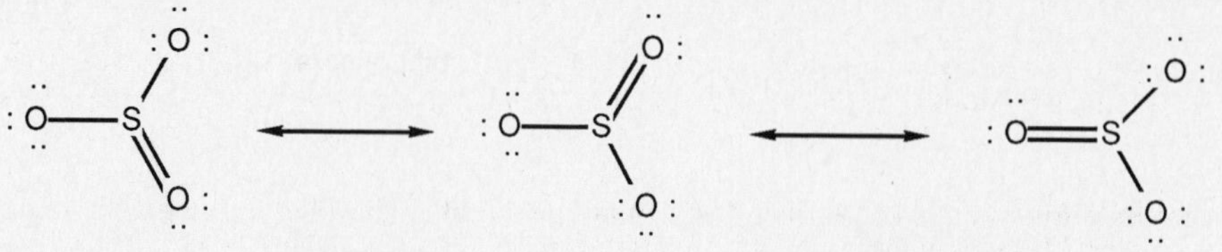

One electron pair is delocalized over the region of all three sulfur-oxygen bonds.

9.57

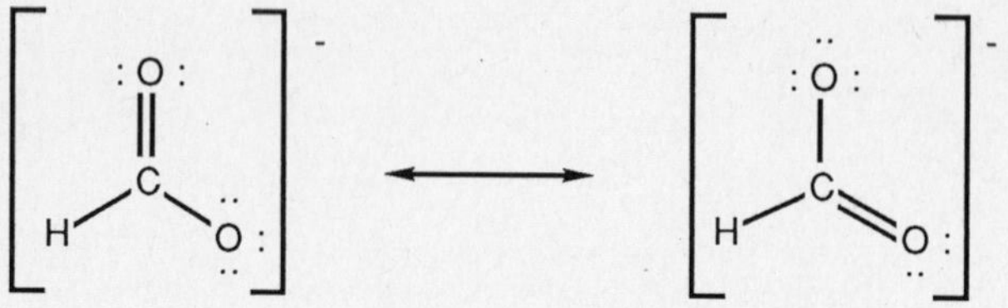

The two carbon-oxygen bonds are expected to be the same. This means that one pair of electrons is delocalized over the region of the O-C-O bonds.

One pair of electrons is delocalized over the O-N-O bonds.

9.59 $r_F = 0.64$ Å

$$d_{P\text{-}F} = r_F + r_P = 0.64\ \text{Å} + 1.10\ \text{Å} = 1.74\ \text{Å}$$

$r_P = 1.10$ Å

9.61 a. $d_{C\text{-}H} = r_C + r_H = 0.77\ \text{Å} + 0.37\ \text{Å} = 1.14\ \text{Å}$
b. $d_{S\text{-}Cl} = r_S + r_{Cl} = 1.04\ \text{Å} + 0.99\ \text{Å} = 2.03\ \text{Å}$
c. $d_{Br\text{-}Cl} = r_{Br} + r_{Cl} = 1.14\ \text{Å} + 0.99\ \text{Å} = 2.13\ \text{Å}$
d. $d_{Si\text{-}O} = r_{Si} + r_O = 1.17\ \text{Å} + 0.66\ \text{Å} = 1.83\ \text{Å}$

9.63

Methylamine	1.47Å	single C-N bond is longer
Acetonitrile	1.16Å	triple C-N bond is shorter

9.65

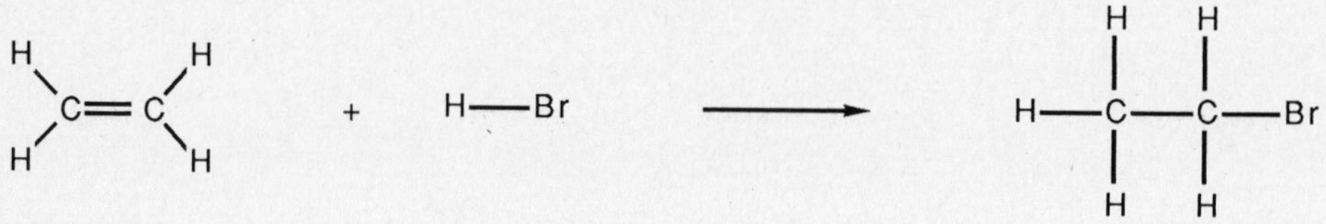

In the reaction, a C=C double bond is converted to a C-C single bond. An H-Br bond is broken and one C-H bond and one C-Br bond are formed.

$$\Delta H \cong BE(C{=}C) + BE(H\text{-}Br) - BE(C\text{-}C) - BE(C\text{-}H) - BE(C\text{-}Br)$$
$$= (602 + 362 - 346 - 411 - 285)\ \text{kJ} = -78\ \text{kJ}$$

9.67 a. Strontium is a metal and oxygen is a nonmetal. The binary compound is likely to be ionic. Strontium, in group IIA, forms Sr^{2+} ions and oxygen, from group VIA, forms O^{2-} ions. The binary compound has the formula SrO and is named strontium oxide.

b. Carbon and bromine are both nonmetals; hence, the binary compound is likely to be covalent. Carbon usually forms four bonds and bromine usually forms one bond. The formula for the binary compound is CBr_4. It is called carbon tetrabromide.

c. Gallium is a metal and fluorine is a nonmetal. The binary compound is likely to be ionic. Gallium is in group IIIA and forms Ga^{3+} ions. Fluorine is in group VIIA and forms F^- ions. The binary compound is GaF_3 and is named gallium(III) fluoride.

d. Nitrogen and bromine are both nonmetals; hence, the binary compound is likely to be covalent. Nitrogen usually forms three bonds and bromine usually forms one bond. The formula for the binary compound is NBr_3. It is called nitrogen tribromide.

9.69 Total valence electrons = 5 + (4 x 6) + 3 = 32. By rule 1, the skeleton is:

```
       O
       |
  O—— As ——O
       |
       O
```

Distribute the remaining 24 electrons to complete the octets around the oxygen atoms.

```
 ┌                  ┐ 3-
 |        ..        |
 |       :O:        |
 |        |         |
 |  ..    |    ..   |
 | :O—— As ——O:     |
 |  ..    |    ..   |
 |        |         |
 |       :O:        |
 |        ..        |
 └                  ┘
```

The formula for lead(II) arsenate is $Pb_3(AsO_4)_2$.

9.71 Total valence electrons = 1 + 7 + (3 x 6) = 26. By rule 3, the skeleton is:

```
H—— O —— I —— O
          |
          O
```

Distribute the remaining 18 electrons to satisfy the octet rule.

```
      ..   ..   ..
H—— O —— I —— O :
      ..   |    ..
           |
          :O:
           ..
```

9.73 Total valence electrons = 5 + (2 x 1) + 1 = 8. By rule 1, the skeleton is H-N-H. Distribute the remaining 4 electrons to complete the octet of the nitrogen atom.

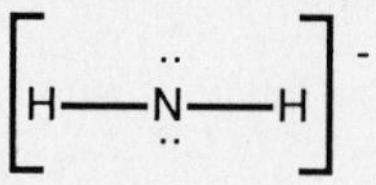

[H—N̈—H]⁻ (N with two lone pairs)

9.75 Total valence electrons = 5 + (2 x 6) - 1 = 16. The skeleton is O-N-O. Distribute the remaining electrons on the oxygen atoms.

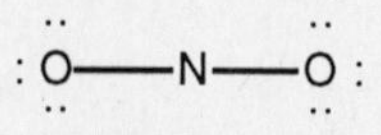

:Ö—N—Ö:

The nitrogen atom is short four electrons. Rule 2 allows the use of two double bonds, one with each oxygen.

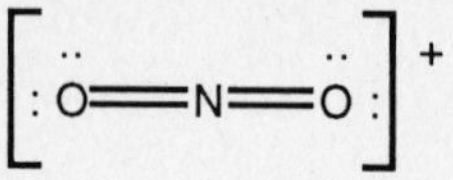

[:Ö=N=Ö:]+

9.77 a. $SeOCl_2$: :Ö: single-bonded to Se; :C̈l—S̈e—C̈l:

b. :S̈e=C=S̈e:

c. [$GaCl_4$]⁻: Ga bonded to four :C̈l: atoms

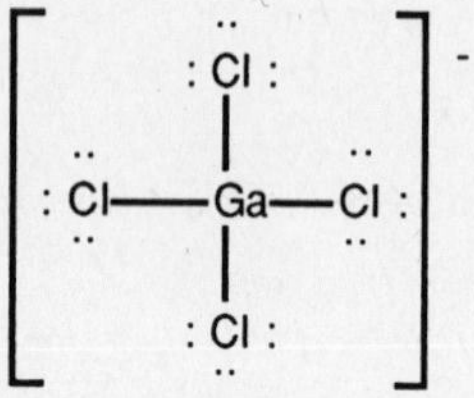

d. [:C≡C:]2-

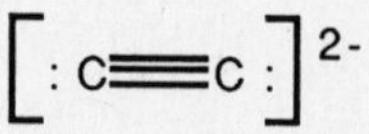

9.79 a. Total valence electrons = 5 + (3 x 7) = 26. By rule 1, the skeleton is:

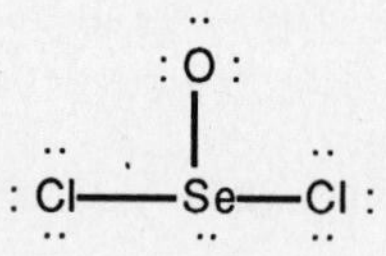

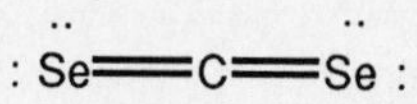

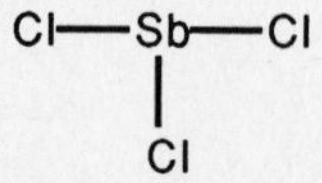

Cl—Sb—Cl
with a third Cl bonded below Sb

Distribute the remaining 20 electrons to the chlorine atoms and the antimony atom to complete their octets.

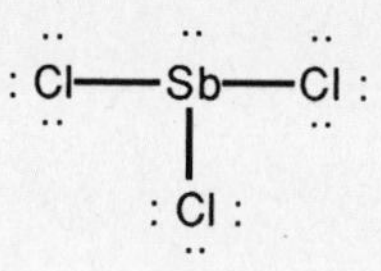

:C̈l—S̈b—C̈l:
with :C̈l: bonded below Sb

b. Total valence electrons = 7 + 4 + 5 = 16. The skeleton is I-C-N. Distribute the remaining 12 electrons.

```
 ..        ..
:I — C — N :
 ..        ..
```

Notice that the carbon atom is four electrons short of an octet. Rule 4 suggests a triple bond between C and N made from the four nonbonding electrons on the nitrogen.

```
 ..
:I — C ≡ N :
 ..
```

c. Total valence electrons = 7 + (3 x 7) = 28. By rule 1, the skeleton is:

```
Cl — I — Cl
     |
     Cl
```

Distribute 18 of the remaining 22 electrons to complete the octets of the chlorine atoms. The four remaining electrons form two sets of lone pairs on the iodine atom.

```
 ..   .·.·   ..
:Cl — I — Cl :
 ..          ..
      |
    :Cl :
     ..
```

d. Total valence electrons = 7 + (5 x 7) = 42. By rule 1, the skeleton is:

```
F   F
 \ /
  I — F
 / \
F   F
```

Use 30 of the remaining 32 electrons to complete the octets of the fluorine atoms. The two electrons remaining form a lone pair on the iodine atom.

```
 ..    ..
:F:   :F:
   \  /      ..
   :I —— F :
   /  \      ..
:F:   :F:
 ..    ..
```

9.81 a. One possible electron-dot structure is:

```
 ..   ..    ..
:O = Se — O :
            ..
```

Since the selenium-oxygen bonds are expected to be equivalent, the structure must be described in resonance terms.

:O=Se—O: ⟷ :O—Se=O:

One electron pair is delocalized over the selenium atom and the two oxygen atoms.

b. The possible electron-dot structures are:

At each end of the molecule, a pair of electrons is delocalized over the region of the nitrogen atom and the two oxygen atoms.

9.83 The possible electron-dot structures are:

Since double bonds are shorter, the terminal N-O bonds are 1.18Å and the central N-O bonds are 1.36Å.

9.85 $\Delta H = BE(H\text{-}H) + BE(O{=}O) - 2BE(H\text{-}O) - BE(O\text{-}O)$
$= (432 + 494 - 2 \times 459 - 142)\ kJ = -134\ kJ$

9.87 $\Delta H = BE(N{=}N) + BE(F\text{-}F) - BE(N\text{-}N) - 2BE(N\text{-}F)$
$= (418 + 155 - 167 - 2 \times 283)\ kJ = -160\ kJ$

Cumulative-Skills Problems (Require skills from previous chapters 3 and 4.)

9.89 After assuming a 100.0g sample, convert to moles:

$$10.9\ g\ Mg \times \frac{1\ mol\ Mg}{24.3\ g\ Mg} = 0.44\underline{8}5\ mol\ Mg$$

$$31.8 \text{ g Cl} \times \frac{1 \text{ mol Cl}}{35.453 \text{ g Cl}} = 0.89\underline{7}03 \text{ mol Cl}$$

$$57.3 \text{ g O} \times \frac{1 \text{ mol O}}{16.00 \text{ g O}} = 3.5\underline{8}1 \text{ mol O}$$

Divide by 0.4485:

$$\text{Mg: } \frac{0.4485}{0.4485} = 1; \quad \text{Cl: } \frac{0.89703}{0.4485} = 2.00; \quad \text{O: } \frac{3.581}{0.4485} = 7.98$$

The simplest formula is $Mg(ClO_4)_2$, magnesium perchlorate.

Lewis formulas: Mg^{2+} and $[ClO_4]^-$ (Cl bonded to four O atoms, each O with three lone pairs)

9.91 After assuming a 100.0g sample, convert to moles:

$$25.0 \text{ g C} \times \frac{1 \text{ mol C}}{12.01 \text{ g C}} = 2.0\underline{8}1 \text{ mol C}$$

$$2.1 \text{ g H} \times \frac{1 \text{ mol H}}{1.008 \text{ g H}} = 2.\underline{0}8 \text{ mol H}$$

$$33.3 \text{ g O} \times \frac{1 \text{ mol O}}{16.00 \text{ g O}} = 2.0\underline{8}1 \text{ mol O}$$

$$39.6 \text{ g F} \times \frac{1 \text{ mol F}}{18.99 \text{ g F}} = 2.0\underline{8}5 \text{ mol F}$$

The simplest formula is CHOF. Since the molecular mass of 48.0 divided by the formula mass of 48.0 = 1, the molecular formula is CHOF also.

Lewis formula: H—C(=O:)—F: (C bonded to H, double-bonded to O with two lone pairs, and bonded to F with three lone pairs)

9.93 First calculate the number of moles in one liter:

$$n = \frac{PV}{RT} = \frac{1.00 \text{ atm} \times 1.00 \text{ L}}{0.0821 \text{ L-atm/K-mol} \times 424 \text{ K}} = 0.028\underline{7}27 \text{ mol}$$

$$MW = \frac{g}{mol} = \frac{7.49 \text{ g}}{0.028727 \text{ mol}} = 26\underline{0}.7 \text{ g/mol}$$

MW = 260.7 amu = 118.69 amu Sn + n(35.453 amu Cl)

$$n = \frac{260.7 - 118.69}{35.453} = 4.0055$$

The formula is $SnCl_4$, which is molecular since the electronegativity difference = 1.2.

Lewis formula:

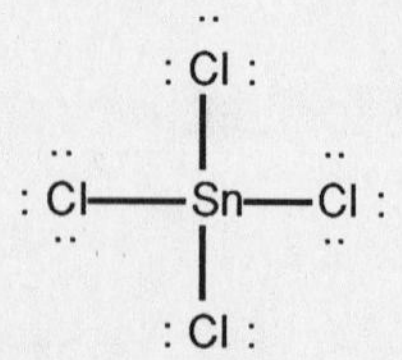

9.95

HCN (g)	⟶	H (g)	+	C (g)	+	N (g)	
[135		218.00		715.0		473]	kJ / mol

ΔH = 1271 kJ / mol

BE(C≡N) = ΔH - BE(C-H) = [1271 - 411] kJ / mol

= 860 kJ / mol Table 9.5 has 887 kJ/mol.

9.97 Use the O-H bond and its bond energy of 459 kJ / mol to calculate X_O.

$$BE(O\text{-}H) = \frac{1}{2}[BE(H\text{-}H) + BE(O\text{-}O)] + k(X_O - X_H)^2$$

$$459 \text{ kJ/mol} = \frac{1}{2}[432 \text{ kJ/mol} + 142 \text{ kJ/mol}] + 98.6 \text{ kJ } (X_o - X_H)^2$$

Collecting the terms gives:

$$\frac{459 - 287}{98.6} = (X_O - X_H)^2$$

Taking the square root of both sides gives:

$$1.3207 = 1.32 = (X_O - X_H)$$

Since X_H = 2.1, X_O = 2.1 + 1.32 = 3.42 = 3.4 Table 9.5 has 3.5

9.99 $$X = \frac{\text{I.E.} - \text{E.A.}}{2} = \frac{[1250 - (-349)] \text{ kJ / mol}}{2} = 799.5 \text{ kJ/mol}$$

$$\frac{799.5 \text{ kJ / mol}}{230 \text{ kJ / mol}} = 3.47 \quad \text{(Pauling's X = 3.0)}$$

CHAPTER 10

MOLECULAR GEOMETRY AND CHEMICAL BONDING THEORY

SOLUTIONS TO EXERCISES

10.1 a. The Lewis structure of ClO_3^- is

```
 [      ..         ] -
 [     : O :       ]
 [       |         ]
 [  ..   ..   ..   ]
 [ : O — Cl — O :  ]
 [  ..   ..   ..   ]
```

There are four electron pairs arranged tetrahedrally about the central atom. Three pairs are bonding and one pair is nonbonding. The expected geometry is trigonal pyramidal.

b. The Lewis structure of OF_2 is

```
  ..   ..   ..
: F —— O —— F :
  ..   ..   ..
```

There are four electron pairs arranged tetrahedrally about the central atom. Two pairs are bonding and two pairs are nonbonding. The expected geometry is bent.

c. The Lewis structure of SiF_4 is

```
  ..       ..
: F        F :
  ..  \  / ..
       Si
  ..  /  \ ..
: F        F :
  ..       ..
```

There are four bonding electron pairs arranged tetrahedrally around the central atom. The expected geometry is tetrahedral.

10.2 First, distribute the valence electrons to the bonds and the chlorine atoms. Then distribute the remaining electrons to iodine.

```
  ..     . .    ..
: Cl —— I —— Cl :
  ..     |      ..
       : Cl :
         ..
```

The five electron pairs around iodine should have a trigonal bipyramidal arrangement, with two lone pairs occupying equatorial positions. The molecule is T-shaped.

10.3 Both (b) trigonal pyramidal and (c) T-shaped geometries are consistent with a nonzero dipole moment. In trigonal planar geometry, the Br-F contributions to the dipole moment would cancel.

10.4 On the basis of symmetry, SiF_4 (b) would be expected to have a dipole moment of zero. The bonds are all symmetric about the central atom.

10.5 The Lewis structure for ammonia, NH_3, is

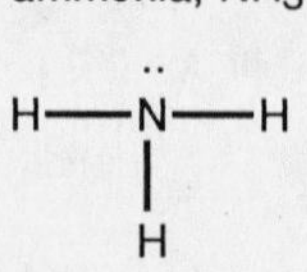

```
      ..
  H—N—H
     |
     H
```

There are four pairs of electrons arranged tetrahedrally around the nitrogen atom. sp^3 hybridization is consistent with a tetrahedral arrangement. For the nitrogen atom:

N: ground state [He] (↑↓) (↑)(↑)(↑)
2s 2p

N: hybridized [He] (↑↓)(↑)(↑)(↑)
sp^3

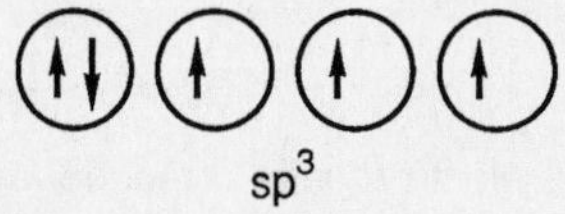

Each N-H bond is formed by the overlap of a 1s orbital of a hydrogen atom with one of the singly occupied sp^3 hybrid orbitals of the nitrogen atom.

10.6 The Lewis structure for PCl_5 is

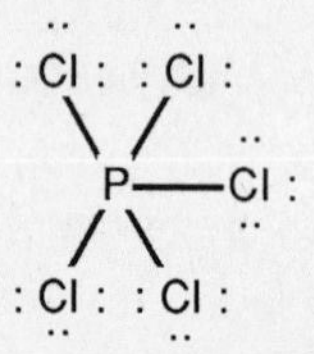

```
  ..    ..
 : Cl :  : Cl :
     \  /      ..
      P——Cl :
     /  \      ..
 : Cl :  : Cl :
  ..    ..
```

There are five σ bonds to P. This suggests a hybridization of sp^3d. Each Cl atom has one singly occupied 3p orbital (recall the valence shell configuration is $3s^2 3p^5$). Each P-Cl bond is formed by the overlap of a phosphorus sp^3d hybrid orbital with a singly occupied chlorine 3p atomic orbital.

10.7 The Lewis structure of CO_2 is

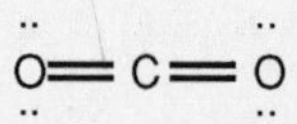

```
 ..      ..
 O═C═O
 ..      ..
```

Since there are two bonds to the carbon atom, sp hybridization is suggested. The changes on this atom are:

C atom: ground state (↑↓) (↑↓) (↑)(↑)( )
1s 2s 2p

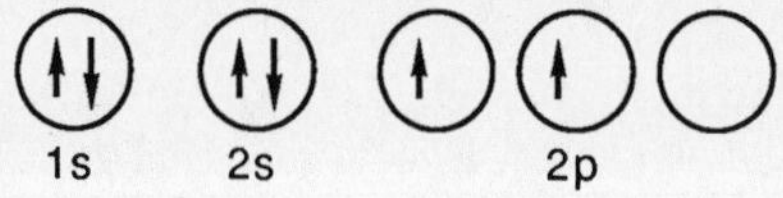

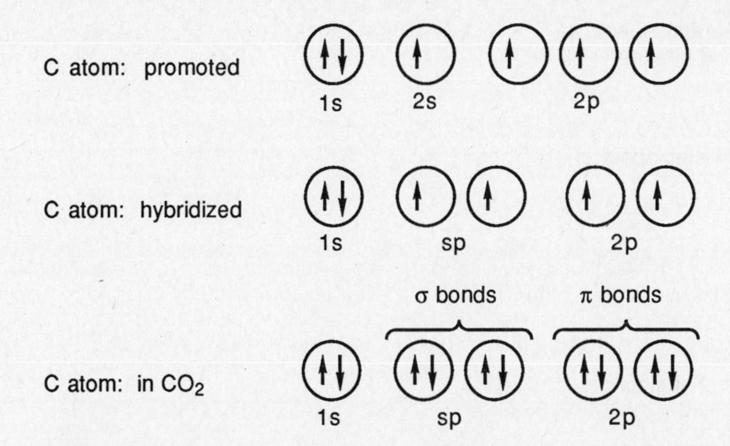

The hybrid orbitals on C are used to describe σ bonds. Each hybrid orbital on the C atom overlaps with a 2p orbital that is pointing along the bond axis on the O atoms to form two σ bonds. Each C atom 2p orbital that is perpendicular to the bond axis overlaps another 2p orbital that is parallel to it on each O atom. The result is a π bond to each O atom. Each C=O consists of one σ bond and one π bond.

10.8 The structural formulas for the isomers are as follows:

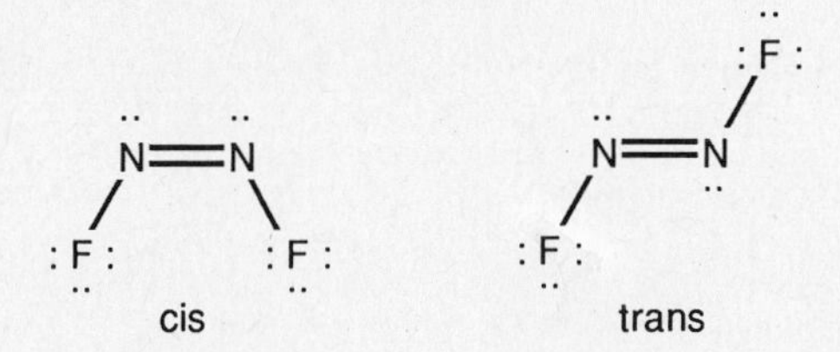

These compounds exist as separate isomers. For these to interconvert, one end of the molecule would have to rotate with respect to the other end. This would require breaking the π bond and expending considerable energy.

10.9 There are 2 x 6 = 12 electrons in C_2. They occupy the orbitals as shown below.

The electron configuration is $KK(\sigma_{2s})^2(\sigma_{2s}^*)^2(\pi_{2p})^4$. There are no unpaired electrons; therefore, C_2 is diamagnetic. There are 8 bonding and 4 antibonding electrons. The bond order is 1/2(8 - 4) = 2.

10.10 There are 6 + 8 = 14 electrons in CO. The orbital diagram is

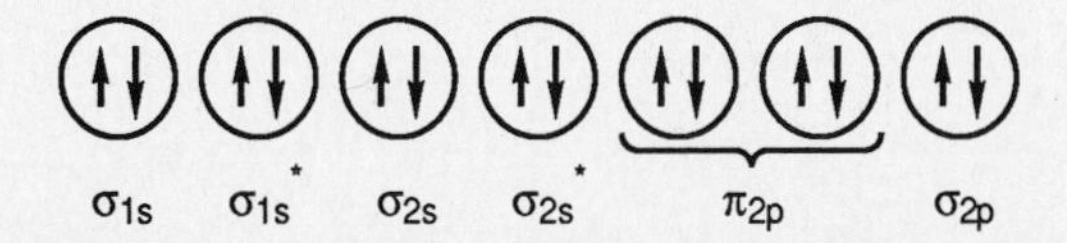

The electron configuration is KK$(\sigma_{2s})^2(\sigma_{2s}^*)^2(\pi_{2p})^4(\sigma_{2p})^2$. There are 10 bonding and 4 anti-bonding electrons. The bond order is 1/2(10 - 4) = 3. There are no unpaired electrons; hence, CO is diamagnetic.

ANSWERS TO REVIEW QUESTIONS

10.1 The VSEPR model is used to predict the geometry of molecules. The electron pairs around an atom are assumed to arrange themselves to reduce electron repulsion. The molecular geometry is determined by the positions of the bonding electron pairs.

10.2 The arrangements are linear, trigonal planar, tetrahedral, trigonal bipyramidal, and octahedral.

10.3 A lone pair is "larger" than a bonding pair; therefore, it will occupy an equatorial position, where it encounters less repulsion than if it were in an axial position.

10.4 The bonds could be polar, but if they are arranged symmetrically, the molecule will be nonpolar. The bond dipoles will cancel.

10.5 Nitrogen trifluoride has three N-F bonds arranged to form a trigonal pyramid. These bonds are polar and would give a polar molecule with partial negative charges on the fluorine atoms and a partial positive charge on the nitrogen atom. However, there is also a lone pair of electrons on nitrogen that is directed away from the bonds. The result is that the lone pair nearly cancels the polarity of the bonds and gives a molecule with a very small dipole moment.

10.6 Certain orbitals, such as p orbitals and hybrid orbitals, have lobes in given directions. Bonding to these orbitals is directional; that is, the bonding is in preferred directions. This explains why the bonding gives a particular molecular geometry.

10.7 The angle is 109.5° .

10.8 A sigma bond has a cylindrical shape about the bond axis. A pi bond has a distribution of electrons above and below the bond axis.

10.9 The changes on a given carbon atom may be described as follows:

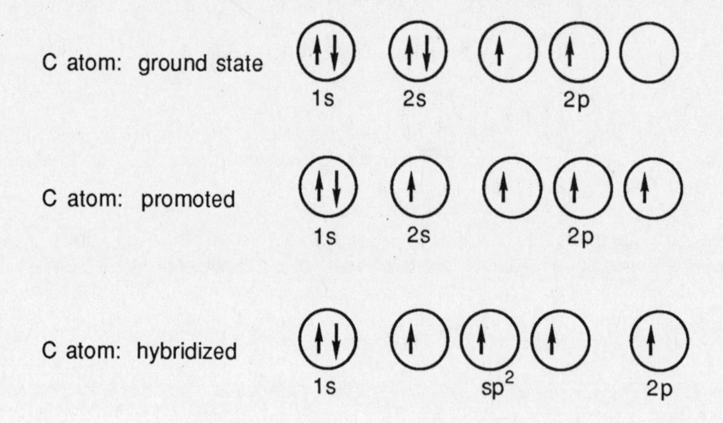

An sp^2 hybrid orbital on one carbon atom overlaps a similar hybrid orbital on the other carbon atom to form a σ bond. The remaining hybrid orbitals on the two carbon atoms overlap 1s orbitals from the hydrogen atoms to form four C-H bonds. The unhybridized 2p orbital on one carbon atom overlaps the unhybridized 2p orbital on the other carbon atom to form a π bond. The σ and π bonds together constitute a double bond.

10.10 Both of the unhybridized 2p orbitals, one from each carbon atom, are perpendicular to their CH_2 planes. When these orbitals overlap each other, they fix both planes to be in the same plane. The two ends of the molecule cannot twist around without breaking the π bond, which requires considerable energy. Therefore, it is possible to have stable molecules with the following structures:

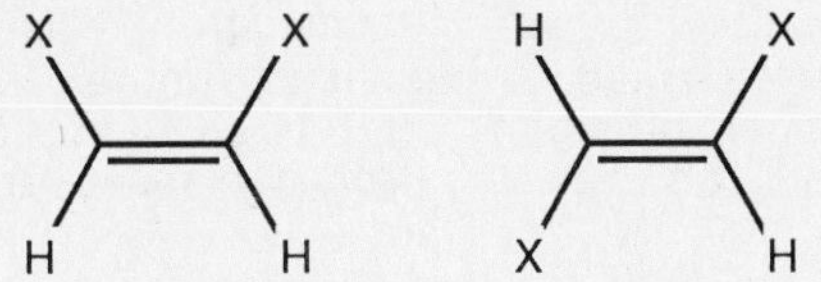

X X H X

H H X H

cis trans

Since these have the same molecular formulas, they are isomers. In this case, they are called cis-trans isomers or geometrical isomers.

10.11 In a bonding orbital, the probability of finding electrons between the two nuclei is high. For this reason, the energy of the bonding orbital is lower than that of the separate atomic orbitals. In an antibonding orbital, the probability of finding electrons between the two nuclei is low. For this reason, the energy of the antibonding orbital is higher than that of the separate atomic orbitals.

10.12 The factors determining the strength of interaction of two atomic orbitals are (1) the energy difference between the interacting orbitals and (2) the magnitude of their overlap.

10.13 When two 2s orbitals overlap, they interact to form a bonding orbital, σ_{2s}, and an antibonding orbital, σ_{2s}^*. The bonding orbital is at lower energy than the antibonding orbital.

10.14 When two 2p orbitals overlap, they interact to form a bonding sigma orbital, σ_{2p}, and an antibonding sigma orbital, σ_{2p}^*. Two bonding pi orbitals, π_{2p} (each have the same energy), and two antibonding pi orbitals, π_{2p}^* (each have the same energy), are also formed.

10.15 A σ bonding orbital is formed by the overlap of the 1s orbital on the H atom with the 2p orbital on the F atom. This H-F orbital is made up primarily of the fluorine orbital.

10.16 The SO_2 molecule consists of a framework of localized orbitals and of delocalized pi molecular orbitals. The localized framework is formed from sp^2 hybrid orbitals on each atom. Thus, an S-O bond is formed by the overlap of a hybrid orbital on the S atom with a hybrid orbital on one of the O atoms. Another S-O bond is formed by the overlap of another hybrid orbital on the S atom with a hybrid orbital on the other O atom. The remaining hybrid orbitals are occupied by lone pairs of electrons. Also, there is one unhybridized p orbital on each of the atoms. These p orbitals are perpendicular to the plane of the molecule and overlap sidewise to give three pi molecular orbitals that are delocalized. The two orbitals of lowest energy are occupied by pairs of electrons.

10.17 When a large number, N (on the order of Avogadro's number), of metal atoms have been brought together to form a crystal, the atoms will have formed N molecular orbitals encompassing the entire crystal. As the metal atoms are added, the number of energy levels grows until the molecular orbital levels merge into a *band* of continuous energies. The metal can be thought of as a regular array of positive atomic cores surrounded by a "sea" of electrons from the valence shells. These electrons are free to move throughout the crystal. When a voltage is applied to the metal, the highest energy electrons are easily excited into the unoccupied orbitals, creating an electrical conductor.

10.18 A semiconductor becomes more conducting when the temperature increases because electrons are excited to the conduction band. When a metal is heated, there is an increase in collisions between conduction electrons and atomic cores. This results in greater electrical resistance; that is, less conductance.

SOLUTIONS TO PRACTICE PROBLEMS

Note on significant figures: The final answer to cumulative skills problems is given first with one nonsignificant figure (rightmost significant figure underlined), and is then rounded to the correct number of significant figures. Intermediate answers usually have at least one nonsignificant figure. Atomic masses are rounded to two decimal places, except for hydrogen.

10.19

	Electron Dot Structure	Number of Electron Pairs	Number of Lone Pairs	Geometry
a.	(see structure a below)	4	0	tetrahedral
b.	(see structure b below)	4	2	bent (angular)
c.	(see structure c below)	4	1	trigonal pyramidal
d.	(see structure d below)	3	0	trigonal planar

a.
```
        ..
       : Cl :
         |
   ..         ..
 : Cl — C — Cl :
   ..         ..
         |
       : Cl :
        ..
```

b.
```
         ..
  H — Se — H
         ..
```

c.
```
        ..
       : F :
         |
   ..         ..
 : F — As — F :
   ..    ..   ..
```

d.
```
        ..
      : Cl :
         |
        Al
   ..  /    \  ..
 : Cl        Cl :
   ..          ..
```

10.21

	Electron Dot Structure	Number of Electron Pairs	Number of Lone Pairs	Geometry
a.	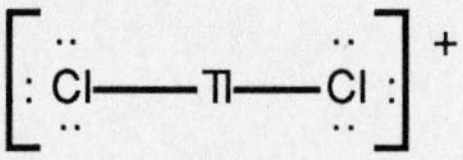 $[:\ddot{Cl}—Tl—\ddot{Cl}:]^+$	2	0	linear
b.	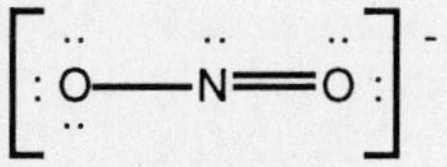 $[:\ddot{O}—\ddot{N}=\ddot{O}:]^-$	3	1	bent
c.	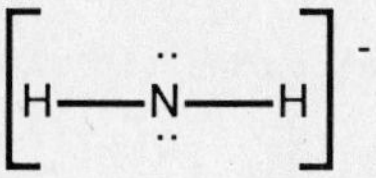 $[H—\ddot{N}—H]^-$	4	2	bent (angular)
d.	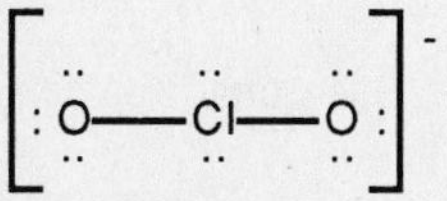 $[:\ddot{O}—\ddot{Cl}—\ddot{O}:]^-$	4	2	bent (angular)

10.23

	Electron Dot Structure	Number of Electron Pairs	Number of Lone Pairs	Geometry
a.	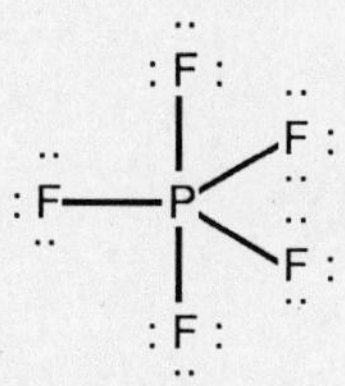 PF_5 (P bonded to five :F:)	5	0	trigonal bipyramidal
b.	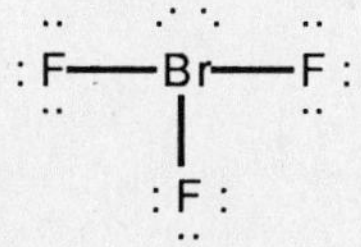 BrF_3 (:F—Br—F:, with :F: below Br)	5	2	T-shaped
c.	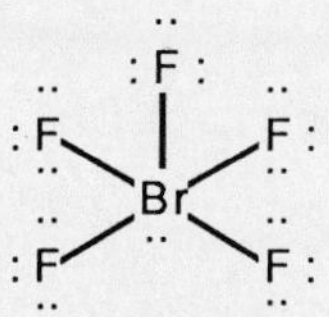 BrF_5 (Br bonded to five :F:)	6	1	square pyramidal

d.	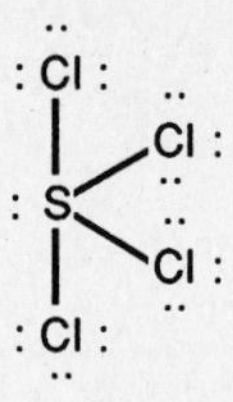 5	1	distorted tetrahedron (see-saw)

10.25

	Electron Dot Structure	Number of Electron Pairs	Number of Lone Pairs	Geometry
a.	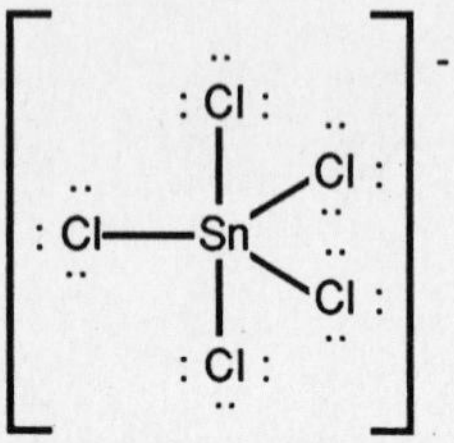	5	0	trigonal bipyramidal
b.	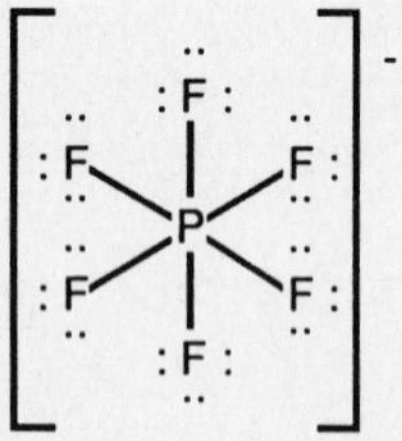	6	0	octahedral
c.	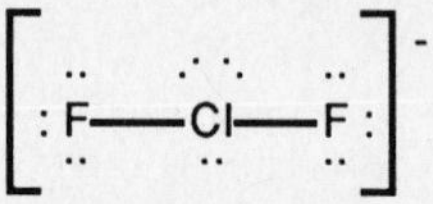	5	3	linear
d.	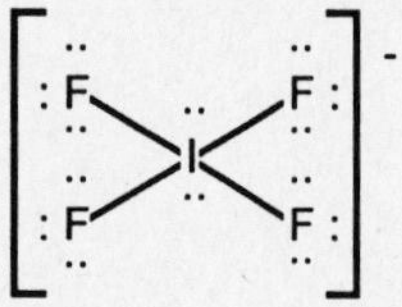	6	2	square planar

10.27 a. Trigonal pyramidal and T-shaped. Trigonal planar would have a dipole moment of zero.

b. Bent. Linear would have a dipole moment of zero.

10.29 TeF_6 (octahedral) and BeF_2 (linear) have zero dipole moments.

10.31 a. Aluminum has three valence electrons. Each bromine contributes one electron; hence, the Al has a total of six valence electrons, or three electron pairs. The hybridization of Al is sp^2.

b. Beryllium has two valence electrons. The chlorine atoms each donate one electron to their bonds. There are four electrons, or two electron pairs, around Be. The bonding can be described in terms of sp hybridization of the beryllium.

c. Silicon has four valence electrons. Each Cl atom contributes an electron to give a total of eight electrons, or four electron pairs, around the Si atom. The hybridization of Si is sp^3.

d. Beryllium has two valence electrons, to which may be added one electron from each of the three fluorine atoms and one electron to account for the charge on the ion. This gives a total of six electrons, or three electron pairs, around the Be atom. The hybridization of Be is sp^2.

10.33 a. The Lewis structure is :Cl—Hg—Cl: . The presence of two single bonds and no lone pairs suggests sp hybridization. Thus, an Hg atom with configuration $[Xe]4f^{14}5d^{10}6s^2$ is promoted to $[Xe]4f^{14}5d^{10}6s^16p^1$, then hybridized. An Hg-Cl bond is formed by overlapping an Hg hybrid orbital with a 3p orbital of Cl.

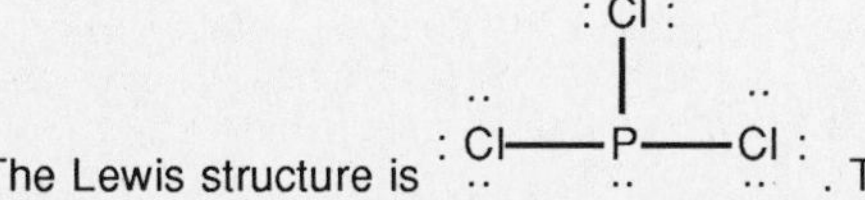

b. The Lewis structure is (shown above). The presence of three single bonds and one lone pair suggests sp^3 hybridization of the P atom. Three hybrid orbitals each overlap a 3p orbital of a Cl atom to form a P-Cl bond. The fourth hybrid orbital contains the lone pair.

10.35 a. Bromine has seven valence electrons. Each F atom donates one electron to give a total of twelve electrons, or six electron pairs, around the Br atom. The hybridization is sp^3d^2.

b. Bromine has seven valence electrons. Each F atom donates one electron to give a total of ten electrons, or five electron pairs, around the Br atom. The hybridization is sp^3d.

c. Arsenic has five valence electrons. The Cl atoms each donate one electron to give a total of ten electrons, or five electron pairs, around the As atom. The hybridization is sp^3d.

d. Chlorine has seven valence electrons, to which may be added one electron from each F atom minus one electron for the charge on the ion. This gives a total of ten electrons, or five electron pairs, around chlorine. The hybridization is sp^3d.

10.37 The P atom in PCl_6^- has six single bonds around it and no lone pairs. This suggests sp^3d^2 hybridization. Each bond in this ion is a σ bond formed by overlap of an sp^3d^2 hybrid orbital on P with a 3p orbital on Cl.

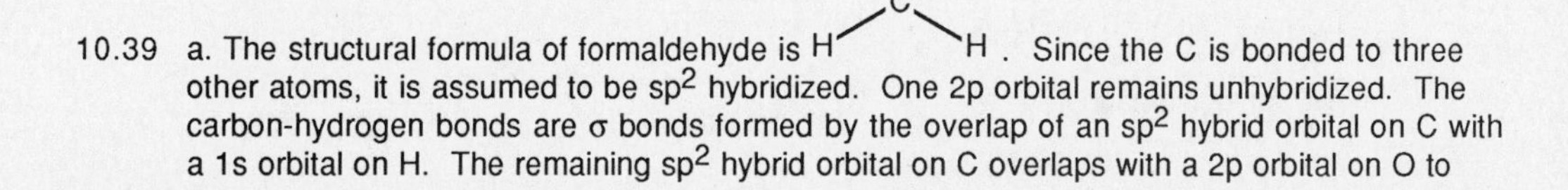

10.39 a. The structural formula of formaldehyde is (shown above). Since the C is bonded to three other atoms, it is assumed to be sp^2 hybridized. One 2p orbital remains unhybridized. The carbon-hydrogen bonds are σ bonds formed by the overlap of an sp^2 hybrid orbital on C with a 1s orbital on H. The remaining sp^2 hybrid orbital on C overlaps with a 2p orbital on O to form a σ bond. The unhybridized 2p orbital on C overlaps with a parallel 2p orbital on O to form a π bond. Together the σ and π bond constitute a double bond.

b. The nitrogen atoms are sp hybridized. A σ bond is formed by the overlap of an sp hybrid orbital from each N. The remaining sp hybrid orbitals contain lone pairs of electrons. The two unhybridized 2p orbitals on one N overlap with the parallel unhybridized 2p orbitals on the other N to form two π bonds.

10.41

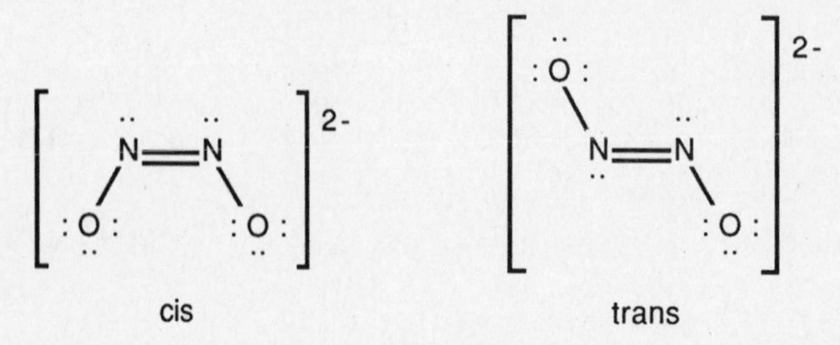

cis trans

Each of the N atoms has a lone pair of electrons and is bonded to two atoms. The N atoms are sp^2 hybridized. The two possible arrangements of the O atoms relative to one another are shown above. Since the π bond between the N atoms must be broken to interconvert these two forms, it is to be expected that the hyponitrite ion will exhibit cis-trans isomerism.

10.43 a. Total electrons = 2 x 5 = 10.
The electron configuration is $KK(\sigma_{2s})^2(\sigma_{2s}^*)^2(\pi_{2p})^2$.

$$\text{bond order} = \frac{1}{2}(n_b - n_a) = \frac{1}{2}(6 - 4) = 1$$

The B_2 molecule is stable. It is paramagnetic because the two electrons in the π_{2p} subshell occupy separate orbitals.

b. Total electrons = 2 x 5 - 1 = 9.
The electron configuration is $KK(\sigma_{2s})^2(\sigma_{2s}^*)^2(\pi_{2p})^1$.

$$\text{bond order} = \frac{1}{2}(n_b - n_a) = \frac{1}{2}(5 - 4) = \frac{1}{2}$$

The B_2^+ molecule should be stable and is paramagnetic since there is one unpaired electron in the π_{2p} subshell.

c. Total electrons = 2 x 8 + 1 = 17.
The electron configuration is $KK(\sigma_{2s})^2(\sigma_{2s}^*)^2(\pi_{2p})^4(\sigma_{2p})^2(\pi_{2p}^*)^3$.

$$\text{bond order} = \frac{1}{2}(n_b - n_a) = \frac{1}{2}(10 - 7) = \frac{3}{2}$$

The O_2^- molecule should be stable and is paramagnetic since there is one unpaired electron in the π_{2p}^* subshell.

10.45 Total electrons = 6 + 7 + 1 = 14.
The electron configuration is $KK(\sigma_{2s})^2(\sigma_{2s}^*)^2(\pi_{2p})^4(\sigma_{2p})^2$.

$$\text{bond order} = \frac{1}{2}(n_b - n_a) = \frac{1}{2}(10 - 4) = 3$$

The CN^- ion is diamagnetic.

10.47

	Electron Dot Structure	Number of Electron Pairs	Number of Lone Pairs	Geometry
a.	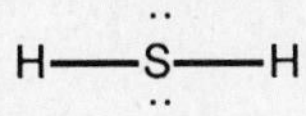 H—S—H (two lone pairs on S)	2	2	bent
b.	 $[\,:I—I—I:\,]^-$ (three lone pairs on each I)	5	3	linear
c.	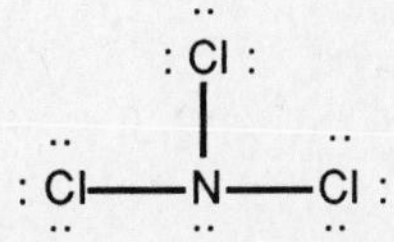 :Cl—N—Cl: with :Cl: bonded to N (one lone pair on N)	4	1	trigonal pyramidal
d.	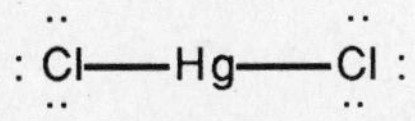 :Cl—Hg—Cl:	2	0	linear

10.49 a. 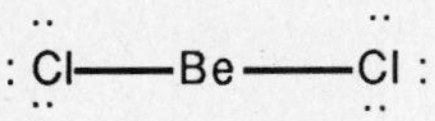 :Cl—Be—Cl: linear

c. 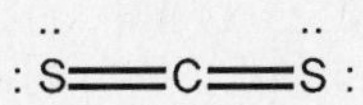 :S=C=S: linear

b. 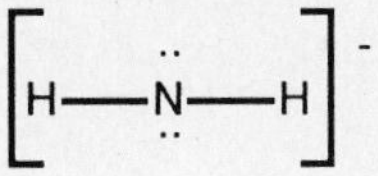 $[\,H—N—H\,]^-$ (two lone pairs on N) bent

d. 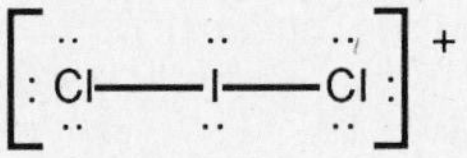 $[\,:Cl—I—Cl:\,]^+$ (two lone pairs on I) bent

10.51

$H_2C=CH—CH_2—O—H$ (carbons labelled a, b, c)

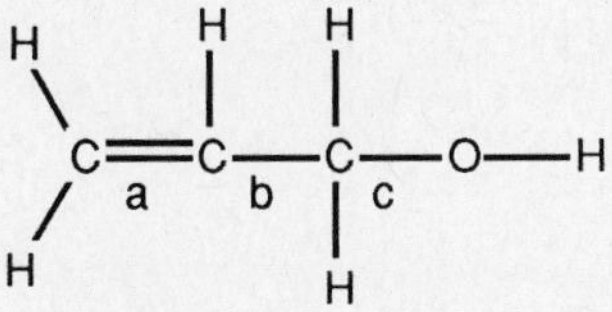

C_a and C_b : 3 electron pairs around each. They are sp^2 hybridized.

C_c : 4 electron pairs around it. It is sp^3 hybridized.

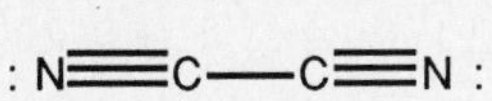

:N≡C—C≡N:

Both C atoms are bonded to two other atoms and have no lone pairs of electrons. They are sp hybridized.

10.53

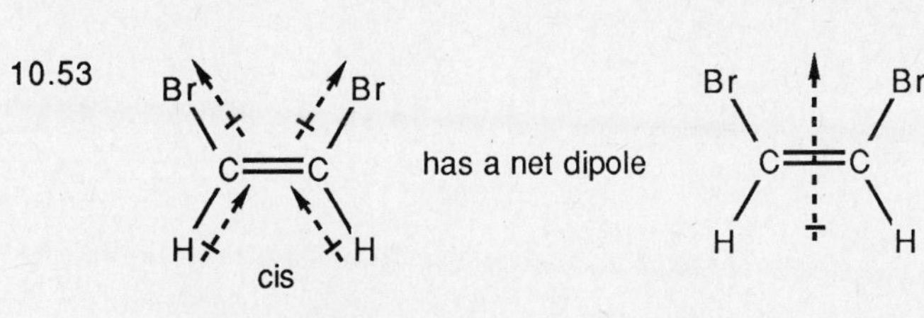

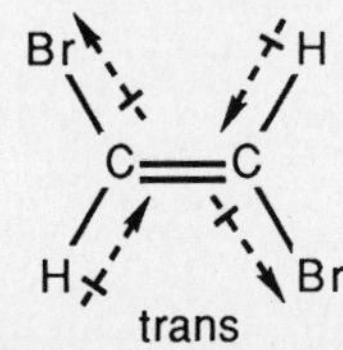

has no net dipole. The two C-Br bond dipoles cancel and the two C-H bond dipoles cancel.

10.55 Total electrons = 2 + 1 - 1 = 2.
The electron configuration is $(\sigma_{1s})^2$.
The two electrons in the HeH^+ ion reside in the σ_{1s} molecular orbital. There are no electrons n any antibonding orbitals. The ion is expected to be stable.

10.57 Total electrons = 2 x 6 + 2 = 14.
The electron configuration is $KK(\sigma_{2s})^2(\sigma_{2s}^*)^2(\pi_{2p})^4(\sigma_{2p})^2$.

$$\text{bond order} = \frac{1}{2}(n_b - n_a) = \frac{1}{2}(10 - 4) = 3$$

10.59 The molecular orbital configuration of O_2 is $KK(\sigma_{2s})^2(\sigma_{2s}^*)^2(\pi_{2p})^4(\sigma_{2p})^2(\pi_{2p}^*)^2$. O_2^+ has one electron less than O_2. The difference is in the number of electrons in the π_{2p}^* antibonding orbital. This means that the bond order is larger for O_2^+ than for O_2.

$$O_2: \text{ bond order} = \frac{1}{2}(n_b - n_a) = \frac{1}{2}(10 - 6) = 2$$

$$O_2^+: \text{ bond order} = \frac{1}{2}(n_b - n_a) = \frac{1}{2}(10 - 5) = \frac{5}{2}$$

It is expected that the species with the higher bond order, O_2^+, has the shorter bond length. In O_2^-, there is one more electron than in O_2. This additional electron occupies a π_{2p}^* orbital. Increasing the number of electrons in antibonding orbitals decreases the bond order; hence, O_2^- should have a longer bond length than O_2.

Cumulative-Skills problems: (Require skills from previous chapters)

10.61 After assuming a 100.0 g sample, convert to moles:

$$60.4 \text{ g Xe} \times \frac{1 \text{ mol Xe}}{131.29 \text{ g Xe}} = 0.46\underline{0}05 \text{ mol Xe}$$

$$22.1 \text{ g O} \times \frac{1 \text{ mol O}}{16.00 \text{ g O}} = 1.3\underline{8}1 \text{ mol O}$$

$$17.5 \text{ g F} \times \frac{1 \text{ mol F}}{18.99 \text{ g F}} = 0.92\underline{1}5 \text{ mol F}$$

Divide by 0.460 : $\text{Xe: } \frac{0.460}{0.460} = 1$ $\quad \text{O: } \frac{1.381}{0.460} = 3.00$ $\quad \text{F: } \frac{0.9215}{0.460} = 2.00$

The simplest formula is XeO_3F_2. This is also the molecular formula.

Lewis formula:

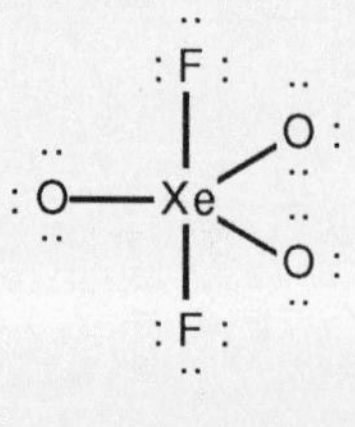

Number of electron pairs = 5, number of lone pairs = 0; hence, the geometry is trigonal bipyramidal. Since xenon has five single bonds, it will require five orbitals to describe the bonding. This suggests sp^3d hybridization.

10.63 $U\ (s) + ClF_n \rightarrow UF_6 + ClF\ (g)$

$$\#\ \text{mol } UF_6 = 3.53\ \text{g } UF_6 \times \frac{1\ \text{mol } UF_6}{352.07\ \text{g } UF_6} = 0.01\underline{0}03\ \text{mol } UF_6$$

$$\#\ \text{mol ClF} = n = \frac{PV}{RT} = \frac{2.50\ \text{atm} \times 0.343\ \text{L}}{0.082057\ \text{L-atm/K-mol} \times 348\text{K}} = 0.030\underline{0}2\ \text{mol ClF}$$

0.010 mol UF_6 = 0.060 mol F and 0.030 mol ClF = 0.030 mol F; therefore, the total moles of F from ClF_n = 0.090 mol F. Since mol ClF_n must equal mol ClF, mol ClF_n = 0.030 mol and n = 0.090 mol F / 0.030 mol ClF_n = 3.

Lewis formula:

```
  ..    .:.    ..
: F —— Cl —— F :
  ..     |    ..
         |
       : F :
         ..
```

Number of electron pairs = 5, number of lone pairs = 2; hence, the geometry is T-shaped. Since chlorine has five electron pairs, it will require five orbitals to describe the bonding. This suggests sp^3d hybridization.

10.65 N_2 : Triple bond, bond length = 1.10 Å. Geometry is linear. sp hybrid orbitals are needed for one lone pair and one σ bond.

N_2F_2 : Double bond, bond length = 1.22 Å. Geometry is trigonal planar. sp^2 hybrid orbitals are needed for one lone pair and two σ bonds.

N_2H_4 : Single bond, bond length = 1.45 Å. Geometry is tetrahedral. sp^3 hybrid orbitals are needed for one lone pair and three σ bonds.

10.67 HNO_3 resonance formulas:

The geometry around the nitrogen is trigonal planar; therefore, the hybridization is sp^2.

Formation reaction: $H_2\ (g) + 3\ O_2\ (g) + N_2\ (g) \rightarrow 2\ HNO_3\ (g)$

$$\begin{aligned} 2 \times \Delta H_f^\circ &= BE(H\text{-}H) + 3BE(O_2) + BE(N_2) - 2BE(H\text{-}O) - 4BE(N\text{-}O) - 2BE(N{=}O) \\ &= (432 + 3 \times 494 + 942 - 2 \times 459 - 4 \times 201 - 2 \times 607)\ \text{kJ/2 mol} \\ &= -80\ \text{kJ / 2 mol} = -40\ \text{kJ/mol} \end{aligned}$$

Resonance energy = -40 kJ - (-135 kJ) = 95 kJ

CHAPTER 11

STATES OF MATTER; LIQUIDS AND SOLIDS

SOLUTIONS TO EXERCISES

Note on significant figures: The final answer to all mathematical solutions is given first with one nonsignificant figure (last significant figure underlined) and is then rounded to the correct number of figures. Intermediate answers usually also have at least one nonsignificant figure.

11.1 First calculate that the heat required to vaporize 1.00 kg of ammonia is

$$1.00 \text{ kg } NH_3 \times \frac{1000 \text{ g}}{1 \text{ kg}} \times \frac{1 \text{ mol } NH_3}{17.03 \text{ g}} \times \frac{23.4 \text{ kJ}}{1 \text{ mol}} = 13\underline{7}4.04 \text{ kJ}$$

The amount of water at 0°C that can be frozen to ice at 0°C with this heat is

$$1374.04 \text{ kJ} \times \frac{1 \text{ mol } H_2O}{6.01 \text{ kJ}} \times \frac{18.01 \text{ g } H_2O}{1 \text{ mol } H_2O} = 41\underline{1}7.54 \text{ g} = 4.12 \text{ kg } H_2O$$

11.2 Use the two-point form of the Clausius-Clapeyron equation to calculate P_2:

$$\text{Log} \frac{P_2}{760 \text{ mmHg}} = \frac{26.8 \times 10^3 \text{ J/mol}}{2.303 \times 8.31 \text{ J/(K•mol)}} \left[\frac{1}{319\text{K}} - \frac{1}{308\text{K}}\right] = 1400.4 \text{ K} \left[\frac{-1.12 \times 10^{-4}}{\text{K}}\right]$$

$$= -0.15\underline{6}8$$

Converting to antilogs gives: $\frac{P_2}{760 \text{ mmHg}} = 0.69\underline{6}8$

$$P_2 = 0.6968 \times 760 \text{ mmHg} = 52\underline{9}.6 = 530 \text{ mmHg}$$

11.3 Use the two-point form of the Clausius-Clapeyron equation to solve for ΔH_{vap}:

$$\text{Log} \frac{757 \text{ mmHg}}{522 \text{ mmHg}} = \frac{\Delta H_{vap}}{2.303 \times 8.31 \text{ J/(K•mol)}} \left[\frac{1}{368 \text{ K}} - \frac{1}{378 \text{ K}}\right]$$

$$0.16\underline{1}4 = \frac{\Delta H_{vap}}{19.\underline{1}37\,\text{J/(K}\cdot\text{mol)}}\left[\frac{7.188 \times 10^{-5}}{\text{K}}\right]$$

$$\Delta H_{vap} = 4.2\underline{9}6 \times 10^{4}\ \text{J/mol}\quad (43.0\ \text{kJ/mol})$$

11.4
a. Liquefy methyl chloride by a sufficient increase in pressure below 144°C.
b. Liquefy oxygen by compressing to 50 atm below -119°C.

11.5
a. Propanol has a hydrogen atom bonded to an oxygen atom. Therefore, hydrogen bonding is expected. Since propanol is polar (from the O-H bond), we also expect dipole-dipole forces. Weak London forces exist too because such forces exist between all molecules.
b. Linear carbon dioxide (Chapter 10) is not polar so only London forces exists between CO_2 molecules.
c. Bent sulfur dioxide (Chapter 10) is polar, so we expecte dipole-dipole forces; we also expect the usual London forces.

11.6 The order of increasing vapor pressure is: butane (C_4H_{10}), propane (C_3H_8), and ethane (C_2H_6). Because London forces tend to increase with increasing molecular weight, one would expect the molecule with the highest molecular weight to have the lowest vapor pressure.

11.7 Because ethanol has an H atom bonded to an O atom, strong hydrogen bonding exists in ethanol but not in methyl chloride. Hydrogen bonding explains the lower vapor pressure of ethanol compared to methyl chloride.

11.8
a. Zinc, a metal, is a metallic solid.
b. Sodium iodide, an ionic substance, exists as an ionic solid.
c. Silicon carbide, a compound in which carbon and silicon might be expected to form covalent bonds to other carbon and silicon atoms, exists as a covalent network solid.
d. Methane, at room temperature a gaseous molecular compound with covalent bonds, freezes as a molecular solid.

11.9 Only $MgSO_4$ is an ionic solid; C_2H_5OH, CH_4, and CH_3Cl form molecular solids; thus $MgSO_4$ should have the highest melting point. Of the molecular solids, CH_4 has the lowest molecular weight (16.0 amu), and would be expected to have the lowest melting point. Both C_2H_5OH and CH_3Cl have approximately the same molecular weights (46.0 amu vs. 50.5 amu), but C_2H_5OH exhibits strong hydrogen bonding and therefore would be expected to have the higher melting point. The order of increasing melting points is: CH_4, CH_3Cl, C_2H_5OH, and $MgSO_4$.

11.10 The atom at the center of each unit cell belongs entirely to that cell. In addition, each of the four corners of the cell contains one atom, which is shared by a total of four unit cells. Therefore, the corners contribute one whole atom. This is summarized as follows:

$$\frac{\text{"Atoms"}}{\text{Unit cell}} = 1\ \text{central atom} + \left[4\ \text{corners} \times \frac{1/4\ \text{atom}}{1\ \text{corner}}\right] = 2\ \text{atoms}$$

11.11 Use the edge length to calculate the volume of the unit cell. Then use the density to determine the mass of one atom. Divide the molar mass by the mass of one atom.

$$V = (3.509 \times 10^{-10}\ m)^3 = 4.3\underline{2}1 \times 10^{-29}\ m^3$$

$$d = \frac{0.534\ g}{cm^3} \times \left[\frac{100\ cm}{1\ m}\right]^3 = 5.34 \times 10^5\ g/m^3$$

Mass of 1 unit = d x V

$$= (5.34 \times 10^5\ g/m^3) \times (4.321 \times 10^{-29}\ m^3) = 2.3\underline{0}74 \times 10^{-23}\ g$$

There are two atoms in a body-centered cubic unit cell; thus the mass of one lithium atom is:

$$1/2 \times 2.3074 \times 10^{-23}\ g = 1.1\underline{5}37 \times 10^{-23}\ g$$

The known atomic weight of lithium is 6.941 amu, so Avogadro's number is:

$$N_A = \frac{6.941\ g/mol}{1.1537 \times 10^{-23}\ g/atom} = 6.0\underline{1}6 \times 10^{23} = 6.02 \times 10^{23}\ atoms/mol$$

11.12 Use Avogadro's number to convert the molar mass of potassium to the mass per one atom.

$$\frac{39.0983\ g\ K}{1\ mol\ K} \times \frac{1\ mol\ K}{6.022 \times 10^{23}\ atom} = \frac{6.49\underline{2}5 \times 10^{-23}\ g\ K}{atom}$$

There are two K atoms per unit cell; therefore, the mass per unit cell is:

$$\frac{6.49\underline{2}5 \times 10^{-23}\ g\ K}{atom} \times \frac{2\ atoms}{1\ unit\ cell} = \frac{1.29\underline{8}5 \times 10^{-22}\ g}{unit\ cell}$$

The density of 0.856 g/cm^3 is equal to the mass of one unit cell divided by its unknown volume, V. After solving for V, the edge length is found from the cube root of the volume.

$$0.856\ g/cm^3 = \frac{1.29\underline{8}5 \times 10^{-22}\ g}{V}$$

$$V = \frac{1.29\underline{8}5 \times 10^{-22}\ g}{0.856\ g/cm^3} = 1.5\underline{1}7 \times 10^{-22}\ cm^3\ (1.517 \times 10^{-28}\ m^3)$$

$$\text{Edge length} = \sqrt[3]{1.517 \times 10^{-28}\ m^3} = 5.3\underline{3}3 \times 10^{-10} = 5.33 \times 10^{-10}\ m\ (5.33\ Å)$$

ANSWERS TO REVIEW QUESTIONS

11.1 The six different phase transitions, with examples in parentheses, are: melting (snow melting), sublimation (dry ice subliming directly to carbon dioxide gas), freezing (water freezing), vaporization (water evaporating), condensation (dew forming on the ground), gas-solid condensation, or deposition (frost forming on the ground).

11.2 Iodine can be purified by heating in a beaker covered with a dish containiing ice or ice water. Only pure iodine should sublime, crystallizing on the cold bottom surface of the dish above the iodine. The common impurities in iodine do not sublime, nor do they vaporize significantly.

11.3 The vapor pressure of a liquid is the partial pressure of the vapor over the liquid, measured at equilibrium. In molecular terms, vapor pressure involves molecules of a liquid vaporizing from the liquid phase, colliding with any surface above the liquid, and exerting pressure on it. The equilibrium is a dynamic one because molecules of the liquid are continually leaving the liquid phase and returning to it from the vapor phase.

11.4 Steam at 100°C will melt more ice than the same weight of water at 100°C because it contains much more energy in the form of its heat of vaporization. It will transfer this energy to the ice and condense in doing so; obviously water at 100°C cannot condense since it is a liquid.

11.5 The heat of fusion is smaller than the heat of vaporization because melting requires only enough energy for molecules to escape from their sites in the crystal lattice, leaving other molecular attractions intact. In vaporization, sufficient energy must be added to break almost all molecular attractions.

11.6 Evaporation leads to cooling of a liquid because the gaseous molecules require heat to evaporate; as they leave the other liquid molecules, they remove the heat energy required to vaporize them. This leaves less energy in the liquid, whose temperature then drops.

11.7 As the temperature increases for a liquid and its vapor in the closed vessel, the two, which are separated by a meniscus, gradually become identical. The meniscus first becomes fuzzy and then disappears altogether as the temperature reaches the critical temperature. Above this temperature, only the vapor exists.

11.8 A permanent gas can be liquefied only by lowering the temperature below its critical temperature, in addition to compressing the gas.

11.9 The pressure in the cylinder of nitrogen at room temperature (<u>above</u> its critical temperature of -147°C) decreases continuously as gas is released because the number of molecules in the vapor phase, which governs the pressure, decreases continuously. The pressure in the cylinder of propane at room temperature (<u>below</u> its critical temperature) is constant because liquid propane and gaseous propane exist at equilbrium in the cylinder. The pressure will remain constant at the vapor pressure of propane until only gaseous propane remains. At that point, the pressure will decrease until all of the propane is gone.

11.10 The vapor pressure of a liquid depends on the intermolecular forces in the liquid phase, since the ease with which a molecule leaves the liquid phase depends on how strongly it is attracted to the other molecules. If such molecules attract each other strongly, the vapor pressure will be relatively low; if they attract each other weakly, the vapor pressure will be relatively high.

11.11 Surface tension makes a liquid act as though it had a skin because for an object to break through the surface, the surface area must increase. This requires energy, so there is some resistance to the object breaking through the surface.

11.12 London forces, also known as dispersion forces, originate between any two molecules that are weakly attracted to each other by means of small, instantaneous dipoles that occur as a result of the varying positions of the electrons during their movement about their nuclei.

11.13 Hydrogen bonding is a weak to moderate attractive force that exists between a hydrogen atom covalently bonded to a very electronegative atom, X, and a lone pair of electrons on another small, electronegative atom, Y. (X and Y may be same or different atoms.) The hydrogen bonding in water involves a hydrogen atom of one water molecule bonding to a lone pair of electrons on the oxygen atom of another water molecule.

11.14 Molecular substances have relatively low melting points because the forces broken by melting are weak intermolecular attractions in the solid state, not strong bonding attractions.

11.15 A crystalline solid has a well-defined, orderly structure; a noncrystalline solid has a random arrangement of structural units.

11.16 In a face-centered cubic cell, there are atoms at the center of each face of the unit cell, in addition to those at the corners.

11.17 The structure of TlI is a simple cubic lattice for both the metal ions and the anions. Thus the structure consists of two interpenetrating cubic lattices of cation and anion.

11.18 The coordination number of Cs^+ in CsCl is 8; the coordination number of Na^+ in NaCl is 6; and the coordination number of Zn^{2+} in ZnS is 4.

11.19 Starting with the edge length of a cubic crystal, we can calculate the volume of a unit cell by cubing the edge length. Then, knowing the density of the crystalline solid, we can calculate the mass of the atoms in the unit cell. Then the mass of the atoms in the unit cell is divided by the number of atoms in the unit cell, giving the mass of one atom. Dividing the mass of one mole of the crystal by the mass of one atom yields a value for Avogadro's number.

11.20 X rays can strike a crystal, and be reflected, at various angles; at most angles, the reflected waves will be out of phase and will destructively interfere. At certain angles, however, the reflected waves will be in phase and will constructively interfere, giving rise to a diffraction pattern.

SOLUTIONS TO PRACTICE PROBLEMS

Note on significant figures: The final answer is given first with one nonsignificant figure (rightmost significant figure underlined), and is then rounded to the correct number of significant figures. Intermediate answers usually also have at least one nonsignificant figure.

11.21 a. Vaporization
b. Freezing of eggs and sublimation of ice
c. Condensation
d. Gas-solid condensation, deposition
e. Freezing

11.23 Dropping a line from the intersection of a 450 mmHg line with the diethyl ether curve in Figure 11.7 intersects the temperature axis about 0.8 of the distance between 0 and 20°C, giving a boiling point for diethyl ether of about 18°C.

11.25 The total amount of energy provided by the heater in 4.54 min is:

$$4.54 \text{ min} \times \frac{60 \text{ s}}{1 \text{ min}} \times \frac{3.48 \text{ J}}{\text{s}} = 94\underline{7}.9 \text{ J} \quad (0.94\underline{7}9 \text{ kJ})$$

The heat of fusion per mole of I_2 is:

$$\frac{0.9479 \text{ kJ}}{15.5 \text{ g } I_2} \times \frac{2 \times 126.9 \text{ g } I_2}{\text{mol } I_2} = 15.\underline{5}2 = \frac{15.5 \text{ kJ}}{\text{mol } I_2}$$

11.27 The heat absorbed per 10.0 g of alcohol is:

$$10.0 \text{ g alcohol} \times \frac{\text{mol alcohol}}{60.1 \text{ alcohol}} \times \frac{42.1 \text{ kJ}}{\text{mol alcohol}} = 7.0\underline{0}4 = 7.00 \text{ kJ}$$

11.29 Since all of the heat released by freezing the water is used to evaporate the remaining water, you must first calculate the amount of heat released in the freezing:

$$9.31 \text{ g } H_2O \times \frac{\text{mol } H_2O}{18.02 \text{ g } H_2O} \times \frac{6.01 \text{ kJ}}{\text{mol } H_2O} = 3.1\underline{0}505 \text{ kJ}$$

Finally, calculate the mass of H_2O that was vaporized by the 3.10505 kJ of heat:

$$3.10505 \text{ kJ} \times \frac{18.02 \text{ g } H_2O}{\text{mol } H_2O} \times \frac{\text{mol } H_2O}{44.9 \text{ kJ}} = 1.2\underline{4}6 = 1.25 \text{ g } H_2O$$

11.31 Calculate how much heat is released by cooling 50.0 g of H_2O from 55 to 15°C.

$$\text{Heat rel'd} = (64.3 \text{ g})(15°\text{C} - 55°\text{C})\left(\frac{4.18 \text{ J}}{\text{g}\cdot°\text{C}}\right) = -1.0\underline{7}509 \times 10^4 \text{ J} = (-10.\underline{7}509 \text{ kJ})$$

The heat released is used first to melt the ice, then to warm the liquid from 0 to 15°C. Let the mass of ice equal **y** grams. Then for fusion, and for warming, we have:

$$\text{Fusion: } (\mathbf{y} \text{ g } H_2O)\left(\frac{1 \text{ mol } H_2O}{18.02 \text{ g } H_2O}\right)\left(\frac{6.01 \text{ kJ}}{1 \text{ mol } H_2O}\right) = 0.33\underline{3}5\mathbf{y} \text{ kJ}$$

$$\text{Warming: } (\mathbf{y} \text{ g } H_2O)(15°\text{C} - 0°\text{C})\left(\frac{4.18 \text{ J}}{\text{g}\cdot°\text{C}}\right) = 6\underline{2}.70\mathbf{y} \text{ J } (0.06\underline{2}70\mathbf{y} \text{ kJ})$$

Since the total heat required for melting and warming must equal the heat released by cooling, equate the two and solve for **y**.

$$10.7509 \text{ kJ} = 0.3335\mathbf{y} \text{ kJ} + 0.0627\mathbf{y} \text{ kJ} = \mathbf{y}(0.335 + 0.0627) \text{ kJ}$$

$$\mathbf{y} = 10.\underline{7}509 \text{ kJ} \div 0.39\underline{6}3 \text{ kJ} = 27.\underline{1}2 \text{ kJ}$$

Thus 27.1 g of ice were added.

11.33 At the normal boiling point, the vapor pressure of a liquid is 760.0 mmHg. Use the Clausius-Clapeyron equation to find P_2 when P_1 = 760.0 mmHg, T_1 = 334.8 K, and T_2 = 298.1 K.

Use ΔH_{vap} = 31.4 kJ/mol

$$\text{Log } \frac{P_2}{P_1} = \frac{\Delta H_{vap}}{2.303 R} \times \frac{(T_2 - T_1)}{T_2T_1} = \frac{31.4 \times 10^3 \text{ J/mol}}{(2.303)(8.31 \text{ J/(K}\cdot\text{mol))}} \times \frac{(298.1 - 334.8 \text{ K})}{(298.1\text{K})(334.8 \text{ K})}$$

$$\text{Log } \frac{P_2}{P_1} = -0.60\underline{3}3; \ \log P_2 = \log P_1 - 0.6033 = \log(760.0) - 0.6033 = 2.27\underline{7}51$$

$$P_2 = \text{antilog}(2.27751) = 18\underline{9}.4 = 189 \text{ mmHg}$$

11.35 From the Clausius-Clapeyron equation:

$$\Delta H_{vap} = 2.303R \left[\frac{T_2T_1}{(T_2 - T_1)}\right] \left[\log \frac{P_2}{P_1}\right]$$

$$= 2.303[8.31 \text{ J/(K}\cdot\text{mol)}] \left[\frac{(553.1 \text{ K})(524.1 \text{ K})}{(553.1 - 524.1)\text{K}}\right] \left[\log \frac{760.0 \text{ mmHg}}{400.0 \text{ mmHg}}\right]$$

$$= 5.332 \times 10^4 \text{ J/mol}$$

$$= 53.3 \text{ kJ/mol}$$

11.37 The phase diagram for oxygen is shown on the next page. It is plotted from these points: triple point = -219°C, boiling point = -183°C, and finally the critical point = -118°C.

Oxygen phase diagram:

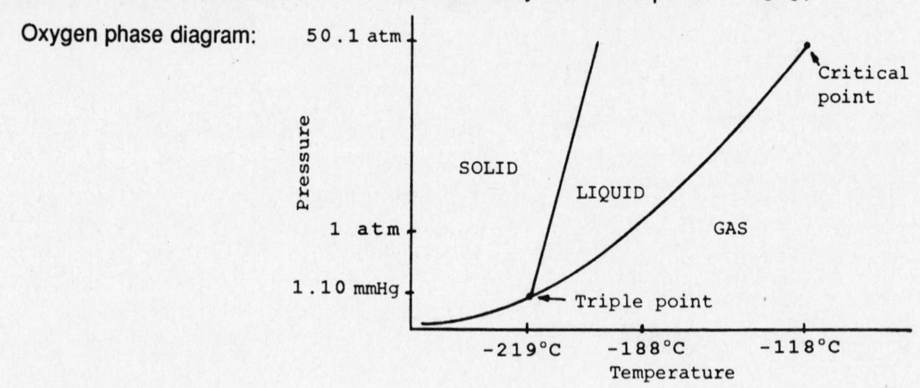

11.39 Liquefied at 25°C: SO_2 and C_2H_2. To liquefy CH_4, lower its temperature below -82°C and then compress it. To liquefy CO, lower its temperature below -140°C and then compress it.

11.41 Br_2 phase diagram:

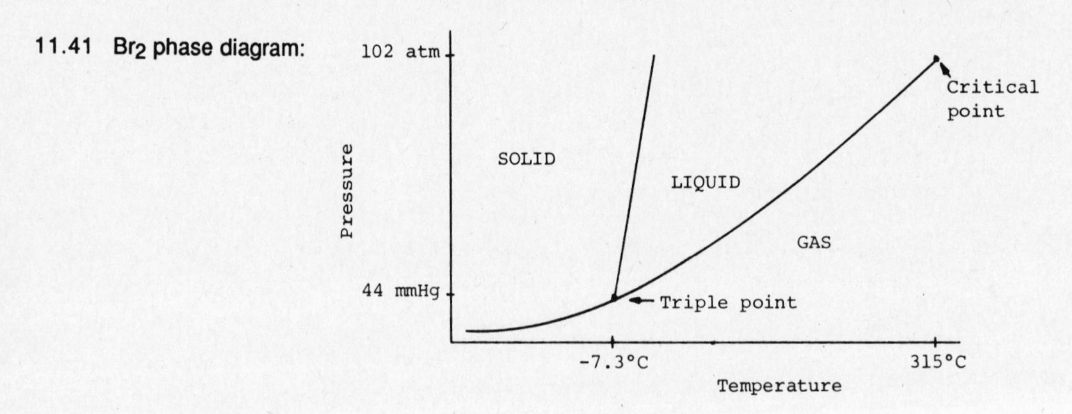

a. Circle "solid." The pressure of 40 mmHg is lower than the pressure at the triple point, so the liquid phase cannot exist.

b. Circle "liquid." The pressure of 400 mmHg is above the triple point so the gas will condense to a liquid.

11.43 Yes, the heats of vaporization of 0.9, 5.6, and 20.4 kJ/mol, for H_2, N_2, and Cl_2, increase in the order of the respective molecular weights of 2.016, 28.02, and 71.0. (London forces increase in order of increasing molecular weight.)

11.45 a. London forces
b. London and dipole-dipole forces, H-bonding
c. London and dipole-dipole forces
d. London forces

11.47 The order is $CCl_4 < SiCl_4 < GeCl_4$ (in order of increasing molecular weight).

11.49 CCl_4 has the lowest vapor pressure because it has the largest molecular weight and thus the greatest intermolecular forces.

11.51 The order of increasing vapor pressure is $HOCH_2CH_2OH$, FCH_2CH_2OH, FCH_2CH_2F. There is no hydrogen bonding in the third molecule; the second molecule can hydrogen-bond only at one end; and the first molecule can hydrogen-bond at both ends for the strongest interaction.

11.53 The order is $CH_4 < C_2H_6 < CH_3OH < CH_2OHCH_2OH$. The weakest forces are the London forces in CH_4 and C_2H_6, which increase with molecular weight. The next strongest interaction is in CH_3OH, which can hydrogen-bond at only one end of the molecule. The strongest interaction is in the last molecule, which can hydrogen-bond at both ends.

11.55 a. Metallic
b. Metallic
c. Covalent
d. Covalent network
e. Molecular

11.57 a. Metallic
b. Covalent network (like diamond)
c. Molecular
d. Molecular

11.59 The order is $(C_2H_5)_2O < C_4H_9OH < KCl < CaO$. Melting points increase in the order of attraction between molecules or ions in the solid state. Hydrogen bonding in C_4H_9OH causes it to melt at a higher temperature than $(C_2H_5)_2O$. Both KCl and CaO are ionic solids with much stronger attraction than the organic molecules. In CaO, the higher charges cause the lattice energy to be higher than in KCl.

11.61 a. Low-melting and brittle
b. High-melting, hard, and brittle
c. Malleable and electrically conducting
d. Hard and high-melting

11.63 a. LiCl b. SiC c. CHI_3 d. Co

11.65 In a simple cubic lattice with one atom at each lattice point, there are atoms only at the corners of unit cells. Each corner is shared by eight unit cells, and there are eight corners per unit cell. Therefore, there is one atom per unit cell.

11.67 Calculate the volume of the unit cell, change density to g/m^3 and then convert volume to mass using density:

$$\text{Volume} = (2.866 \times 10^{-10}\ \text{m})^3 = 2.354 \times 20^{-29}\ \text{m}^3$$

$$\left[\frac{7.87\ \text{g}}{\text{cm}^3}\right] \times \left[\frac{10^2\ \text{cm}}{\text{m}}\right]^3 = 7.87 \times 10^6\ \text{g/m}^3$$

$$\text{Mass of one cell} = (7.87 \times 10^6\ \text{g/m}^3) \times (2.354 \times 10^{-29}\ \text{m}^3) = 1.8\underline{5}26 \times 10^{-22}\ \text{g}$$

Since Fe is a body-centered cubic cell, there are 2 Fe atoms in the cell, and

$$\text{the mass of one Fe} = 1.8526 \times 10^{-22}\ \text{g} \div 2 = 9.2\underline{6}3 \times 10^{-23}\ \text{g}$$

Using the molar mass to calculate the mass of one Fe atom, the agreement is good:

$$\frac{55.85\ \text{g Fe}}{1\ \text{mol Fe}} \times \frac{1\ \text{mol Fe}}{6.022 \times 10^{23}\ \text{Fe atoms}} = 9.27\underline{4}3 \times 10^{-23}\ \text{g/Fe atom}$$

11.69 There are four Cu atoms in the face-centered cubic structure, so the mass of one cell is:

$$4\ \text{Cu atoms} \times \frac{1\ \text{mol Cu}}{6.022 \times 10^{23}\ \text{Cu atoms}} \times \frac{63.5\ \text{g Cu}}{1\ \text{mol Cu}} = 4.2\underline{1}8 \times 10^{-22}\ \text{g}$$

$$\text{Cell volume} = \frac{4.2\underline{1}8 \times 10^{-22}\ \text{g}}{8.93\ \text{g/cm}^3} = 4.7\underline{2}3 \times 10^{-23}\ \text{cm}^3$$

All edges are the same length in a cubic cell, so the edge length, $\underline{l}$, is found:

$$l = \sqrt[3]{V} = \sqrt[3]{4.7\underline{2}3 \times 10^{-23}\ \text{cm}^3} = 3.6\underline{1}4 \times 10^{-8} = 3.61 \times 10^{-8}\ \text{cm}\ (3.61\ \text{Å})$$

11.71 Calculate the volume from the edge length of 4.079 Å (4.079 x 10^{-8} cm) and then use it to calculate the mass of the unit cell:

$$\text{Cell volume} = (4.079 \times 10^{-8}\ \text{cm})^3 = 6.787 \times 10^{-23}\ \text{cm}^3$$

$$\text{Cell mass} = (19.3\ \text{g/cm}^3)(6.878 \times 20^{-23}\ \text{cm}^3) = 1.3\underline{1}0 \times 10^{-21}\ \text{g}$$

Calculate the mass of one gold atom:

$$1\ \text{Au atom} \times \frac{1\ \text{mol Au}}{6.022 \times 10^{23}\ \text{Au atoms}} \times \frac{196.97\ \text{g Au}}{1\ \text{mol Au}} = 3.27\underline{1} \times 10^{-22}\ \text{g Au}$$

$$\frac{1.310 \times 10^{-21}\ \text{g}}{1\ \text{unit cell}} \times \frac{1\ \text{Au atom}}{3.271 \times 10^{-22}\ \text{Au}} = \frac{4.004\ \text{Au atom}}{\text{cell}}\ \text{(= face-cent'd. cubic)}$$

11.73 Calculate volume from the edge (3.165 Å = 3.165×10^{-8} cm); use it to calculate the mass:

$$\text{Cell volume} = (3.165 \times 10^{-8}\ \text{cm})^3 = 3.17\underline{0}5 \times 10^{-23}\ \text{cm}^3$$

For a body-centered cubic lattice, there are two atoms per cell, so their mass is:

$$2\ \text{W atoms} \times \frac{1\ \text{mol W}}{6.022 \times 10^{23}\ \text{W atoms}} \times \frac{183.8\ \text{g W}}{1\ \text{mol W}} = 6.10\underline{4}3 \times 10^{-22}\ \text{g W}$$

$$\text{Density} = \frac{6.1043 \times 10^{-22}\ \text{g W}}{3.1705 \times 10^{-23}\ \text{cm}^3} = 19.2\underline{5}3 = 19.25\ \text{g/cm}^3$$

11.75 Use Avogadro's number to calculated the number of atoms in 1.74 g (= d x 1.000 cm^3):

$$1.74\ \text{g Mg} \times \frac{1\ \text{mol Mg}}{24.305\ \text{g Mg}} \times \frac{6.022 \times 10^{23}\ \text{Mg atoms}}{1\ \text{mol Mg}} = 4.311 \times 10^{22}\ \text{atoms}$$

Since the space occupied by the Mg atoms = 0.741 cm^3, each atom's volume is found:

$$\text{Volume 1 Mg atom} = \frac{0.741\ \text{cm}^3}{4.311 \times 10^{22}\ \text{Mg atoms}} = 1.7\underline{1}9 \times 10^{-23}\ \text{cm}^3$$

$$\text{Volume} = \frac{4\pi r^3}{3} \quad \text{so } r = \sqrt[3]{\frac{3V}{4\pi}}$$

$$r = \sqrt[3]{\frac{3}{4\pi}(1.719 \times 10^{-23}\ \text{cm}^3)} = 1.6\underline{0}1 \times 10^{-8} = 1.60 \times 10^{-8}\ \text{cm}\ (1.60\ \text{Å})$$

11.77 Water vapor condensed directly to solid water (frost) without forming liquid water (Section 11.2). After heating, most of the frost melted to liquid water, which then was vaporized to water vapor. Some of the frost may have sublimed directly to water vapor (Section 11.2).

11.79 From Table 5.6, the vapor pressures are 21.1 mmHg at 23°C and 12.8 mmHg at 15°C. If the moisture did not begin to condense until the air had been cooled to 15°C, then the partial pressure of water in the air at 23°C must have been 12.8 mmHg. The relative humidity is:

$$\%\ \text{relative humidity} = \frac{12.8\ \text{mmHg}}{21.1\ \text{mmHg}} \times 100\% = 60.\underline{6}6 = 60.7\%$$

11.81 After labeling the problem data as below, use the Clausius-Clapeyron equation to obtain ΔH_{vap}, which can then be used to calculate the boiling point.

At T_1 = 299.3 K, P_1 = 100.0 mmHg; at T_2 = 333.8 K, P_2 = 400.0 mmHg

$$\text{Log}\ \frac{400.0}{100.0} = \frac{\Delta H_{vap}}{2.303 \times 8.314\ \text{J/(K}\cdot\text{mol)}} \times \frac{333.8\ \text{K} - 299.3\ \text{K}}{333.8\ \text{K} \times 299.3\ \text{K}}$$

$$0.6021 = \Delta H_{vap}\ (1.804 \times 10^{-5}\ \text{mol/J})$$

$$\Delta H_{vap} = 33.4 \times 10^3\ \text{J/mol}\ \ (33.4\ \text{kJ/mol})$$

Now we use this value of ΔH_{vap} and the following data to calculate the boiling point:

At T_1 = 299.3 K, P_1 = 100.0 mmHg; at T_2 (boiling pt), P_2 = 760 mmHg

$$\text{Log}\ \frac{760.0}{100.0} = \frac{33.4 \times 10^3\ \text{J/mol}}{2.303 \times 8.314\ \text{J/(K}\bullet\text{mol)}} \times \frac{T_2 - 299.3\ \text{K}}{T_2 \times 299.3\ \text{K}}$$

$$0.8808 = 1.744 \times 10^3\ \text{K}\left[\frac{1}{299.3\text{K}} - \frac{1}{T_2}\right]$$

$$\frac{1}{T_2} = \frac{1}{299.3\text{K}} - \frac{0.8808}{1.744 \times 10^3\ \text{K}} = 2.8\underline{3}6 \times 10^{-3}/\text{K}$$

$$T_2 = 35\underline{2}.6 = 353\ \text{K}\ (80^\circ\text{C})$$

11.83 a. As this gas is compressed at 20°C, it will condense into a liquid since 20°C is above the triple point but below the critical point.

b. As this gas is compressed at -70°C, it will condense directly to the solid phase since the temperature of -70°C is below the triple point.

c. As this gas is compressed at 40°C, it will not condense since 40°C is above the critical point.

11.85 In propanol, hydrogen bonding exists between the hydrogen of the OH group and the lone pair of electrons of oxygen of the OH group of an adjacent propanol molecule. For two adjacent propanol molecules, the hydrogen bond may be represented as: C_3H_7-O-H•••O(H)C_3H_7.

11.87 Ethylene glycol molecules are capable of hydrogen bonding to each other whereas pentane molecules do not. The greater intermolecular forces in ethylene glycol are reflected in greater resistance to flow (viscosity) and high boiling point.

11.89 Aluminum (Group IIIA) forms a metallic solid. Silicon (Group IVA) forms a covalent network solid (Section 11.8). Phosphorus (Group VA) forms a molecular solid. Sulfur (Group VIA) forms an amorphous solid.

11.91 a. Lower: KCl. The lattice energy should be lower for ions with lower charge. A lower lattice energy implies a lower melting point.

b. Lower: CCl_4. Both are molecular solids so the compound with the lower molecular weight should have weaker London forces and therefore the lower melting point.

c. Lower: Zn. Melting points for Group IIB metals are lower than for metals near the middle of the transition metal series.

d. Lower: C_2H_5Cl. Ethyl chloride cannot hydrogen-bond, but acetic acid can. The compound with the weaker intermolecular forces has the lower melting point.

11.93 The face-centered cubic structure means that one atom is at each lattice point. All edges are the same length in such a structure, so the volume is found:

$$\text{Volume} = l^3 = (3.839 \times 10^{-8}\ \text{cm})^3 = 5.65\underline{7}9 \times 10^{-23}\ \text{cm}^3$$

$$\text{Mass of unit cell} = dV = (22.42\ \text{g/cm}^3)(5.6579 \times 10^{-23}\ \text{cm}^3) = 1.26\underline{8}5 \times 10^{-21}\ \text{g}$$

There are 4 atoms in a face-centered cubic cell, so:

$$\text{Mass of 1 Ir atom} = \text{mass of unit cell} \div 4 = 1.2685 \times 10^{-21}\ \text{g} \div 4 = 3.17\underline{1}2 \times 10^{-22}\ \text{g}$$

$$\text{Molar mass of Ir} = (3.1712 \times 10^{-22}\ \text{g/Ir atom}) \times (6.022 \times 10^{23}\ \text{Ir atoms/mol})$$

$$= 190.96 = 191.0\ \text{g/mol}\ \ \text{(The atomic weight = 191.0 amu.)}$$

11.95 From problem 11.69, the cell edge length (l) is 3.6$\underline{1}$4 Å. There are four copper-atom radii along the diagonal of a unit-cell face. Since the diagonal square = $l^2 + l^2$ (Pythagorean theorem):

$$4r = \sqrt{2l^2} = \sqrt{2}\,l \ \text{or}\ r = \frac{\sqrt{2}}{4}(3.6\underline{1}4\ \text{Å}) = 1.2\underline{7}7 = 1.28\ \text{Å}$$

11.97 The body diagonal (diagonal passing through the center of the cell) is 4 times the radius, r, of a sphere. Also, from the geometry of the cube and the Pythagorean theorem, the body diagonal equals $\sqrt{3}$ (l), where l is the edge length of the unit cell (see problem 11.97). Thus:

$$4r = \sqrt{3}\,(l),\ \text{or}\ l = 4r/(\sqrt{3})$$

Since the unit cell contains 2 spheres, the volume occupied by the spheres is

$$V_{spheres} = 2 \times \left[(4/3)\,\pi r^3\right],\ \text{and}$$

$$V_{cell} = l^3 = \left[\frac{4r}{\sqrt{3}}\right]^3 = \frac{64r^3}{3\sqrt{3}}$$

Finally, to obtain the % volume of the cell occupied, divide $V_{spheres}$ by V_{cell}:

$$\% V = \frac{V_{spheres}}{V_{cell}} \times 100\% = \frac{2\left[\frac{4\pi r^3}{3}\right]}{\frac{64r^3}{3\sqrt{3}}} \times 100\% = \frac{\pi\sqrt{3}}{8} \times 100\% = 68.01 = 68\%$$

<u>Cumulative-Skills Problems (require skills from previous chapters)</u>

11.99 Use the ideal gas law to calculate n, the number of moles of N_2:

$$N_2\!: \ n = \frac{PV}{RT} = \frac{(745/760)\ \text{atm} \times 5.40\ \text{L}}{[0.082057\ \text{L}\bullet\text{atm/(K}\bullet\text{mol)}] \times 293\ \text{K}} = 0.22\underline{0}1\ \text{mol}$$

C_3H_8O: $n = 0.6149 \text{ g } C_3H_8O \times \frac{1 \text{ mol } C_3H_8O}{60.06 \text{ g } C_3H_8O} = 0.010\underline{2}38 \text{ mol}$

$X_{C_3H_8O} = \frac{0.010238 \text{ mol } C_3H_8O}{(0.010238 \text{ mol} + 0.2201 \text{ mol})} = 0.044\underline{4}4 \text{ mol fraction}$

Partial P $= 0.0444 \times 745 \text{ mmHg} = 33.\underline{1}07 \text{ mmHg} = 33.1 \text{ mmHg}$

Vapor pressure of C_3H_8O = 33.1 mmHg

11.101 Calculate the moles of HCN in 10.0 mL of the solution (density = 0.687 g HCN/mL HCN):

$10.0 \text{ mL HCN} \times \frac{0.687 \text{ g HCN}}{\text{mL HCN}} \times \frac{1 \text{ mol HCN}}{27.03 \text{ g HCN}} = 0.25\underline{4}1 \text{ mol HCN}$

0.2541 mol HCN(l) → 0.2541 mol HCN(g)

(ΔH_f^o = 105 kJ/mol) (ΔH_f^o = 135 kJ/mol)

$\Delta H^o = 0.2541 \text{ mol} \times [135 \text{ kJ/mol} - 105 \text{ kJ/mol}] = 7.\underline{6}23 = 7.6 \text{ kJ}$

11.103 First convert the mass to moles; then multiply by the standard heat of formation to obtain the heat absorbed in vaporizing this mass:

$25.0 \text{ g } P_4 \times \frac{1 \text{ mol } P_4}{123.92 \text{ g } P_4} = 0.20\underline{1}7 \text{ mol } P_4$

$0.20\underline{1}7 \text{ mol } P_4 \times \frac{95.4 \text{ J}}{°\text{C} \cdot \text{mol } P_4} \times (44.1°\text{C} - 25.0°\text{C}) = 36\underline{7}.52 \text{ J } (0.36\underline{7}52 \text{ kJ})$

$2.63 \text{ kJ/mol } P_4 \times 0.2017 \text{ mol } P_4 = 0.53\underline{0}47 \text{ kJ}$

Total heat $= 0.53047 \text{ kJ} + 0.36752 \text{ kJ} = 0.89\underline{7}99 = 0.898 \text{ kJ}$

11.105 Use the ideal gas law to calculate the total number of moles of monomer and dimer:

$n = \frac{PV}{RT} = \frac{(436/760) \text{ atm} \times 1.000 \text{ L}}{[0.082057 \text{ L} \cdot \text{atm/(K} \cdot \text{mol)}] \times 373.75 \text{ K}} = 0.018\underline{7}057 \text{ mol monomer + dimer}$

$(0.0187057 \text{ monomer + dimer}) \times \frac{0.630 \text{ mol dimer}}{1 \text{ mol dimer + monomer}} = 0.011\underline{7}8 \text{ mol dimer}$

$0.0187057 \text{ mol both} - 0.01178 \text{ mol dimer} = 0.006\underline{9}2 \text{ mol monomer}$

Mass dimer $= 0.011\underline{7}8 \text{ mol dimer} \times \frac{120.1 \text{ g dimer}}{\text{mol dimer}} = 1.4\underline{1}4 \text{ g dimer}$

Mass monomer $= 0.00692 \text{ mol monomer} \times \frac{60.05 \text{ g monomer}}{\text{mol monomer}} = 0.4\underline{1}55 \text{ g monomer}$

Density $= \frac{1.414 \text{ g} + 0.4155 \text{ g}}{1.000 \text{ L}} = 1.8\underline{2}9 = 1.83 \text{ g/L vapor}$

CHAPTER 12

SOLUTIONS

SOLUTIONS TO EXERCISES

Note on significant figures: The final answer to all mathematical solutions is given first with one nonsignificant figure (last significant figure underlined) and is then rounded to the correct number of figures. Intermediate answers usually also have at least one nonsignificant figure.

12.1 An example of a solid solution prepared from a liquid and a solid is a dental filling made of liquid mercury and solid silver.

12.2 The C_4H_9OH molecules will be more soluble in water because their OH ends can form hydrogen bonds with water.

12.3 The Na^+ ion has a larger energy of hydration because its ionic radius is smaller, giving Na^+ a more concentrated electric field than K^+.

12.4 Write Henry's law ($S = K_HP$) for 159 mmHg (P_2) and divide it by Henry's law for 1 atm, or 760 mmHg (P_1). Then substitute the experimental values of P_1, P_2, and S_1 to solve for S_2.

$$\frac{S_2}{S_1} = \frac{K_HP_2}{K_HP_1} = \frac{P_2}{P_1}$$

$$S_2 = \frac{P_2S_1}{P_1} = \frac{(159 \text{ mmHg})(0.0404 \text{ g } O_2/L)}{760 \text{ mmHg}} = 8.4\underline{5}2 \times 10^{-3} = 8.45 \times 10^{-3} \text{ g } O_2/L$$

12.5 The mass of HCl in 20.2% HCl (0.202 = fraction of HCl) is

$$0.202 \times 35.0 \text{ g} = 7.0\underline{7}0 = 7.07 \text{ g}$$

The mass of H_2O in 20.2% HCl is

$$35.0 \text{ g solution} - 7.07 \text{ g HCl} = 27.\underline{9}3 = 27.9 \text{ g } H_2O$$

12.6 Calculate the moles of toluene, using its molar mass of 92.14 g/mol:

$$35.6 \text{ g toluene} \times \frac{1 \text{ mol toluene}}{92.14 \text{ g toluene}} = 0.38\underline{6}3 \text{ g toluene}$$

To calculate molality, divide the moles of toluene by the mass in kg of the solvent (C_6H_6):

$$\text{Molality} \times \frac{0.3863 \text{ mol toluene}}{0.125 \text{ kg solvent}} = 3.0\underline{9}04 = 3.09 \text{ m toluene}$$

12.7 The number of moles of toluene = 0.3863 (previous exercise); the number of moles of benzene is

$$125 \text{ g benzene} \times \frac{1 \text{ mol benzene}}{78.11 \text{ g benzene}} = 1.6\underline{0}03 \text{ mol benzene}$$

The total number of moles is 1.6003 + 0.3863 = 1.9$\underline{8}$66, and the mole fractions are:

$$\text{Mol fraction benzene} \times \frac{1.6003 \text{ mol benzene}}{1.9866 \text{ mol}} = 0.80\underline{5}54 = 0.806$$

$$\text{Mol fraction toluene} = \frac{0.3863 \text{ mol toluene}}{1.9866 \text{ mol}} = 0.19\underline{4}4 = 0.194$$

The sum of the mole fractions = 1.000.

12.8 This solution contains 0.120 moles of methanol dissolved in 1.00 kg of ethanol. The number of moles in 1.00 kg of ethanol is:

$$1.00 \times 10^3 \text{ g } C_2H_5OH \times \frac{1 \text{ mol } C_2H_5OH}{46.07 \text{ g } C_2H_5OH} = 21.\underline{7}06 \text{ mol } C_2H_5OH$$

The total number of moles is 21.706 + 0.120 = 21.$\underline{8}$26, and the mole fractions are:

$$\text{Mol fraction } C_2H_5OH = \frac{21.\underline{7}06 \text{ mol } C_2H_5OH}{21.826 \text{ mol}} = 0.99\underline{4}501 = 0.995$$

$$\text{Mol fraction } CH_3OH = \frac{0.12\underline{0} \text{ mol } CH_3OH}{21.826 \text{ mol}} = 0.00\underline{5}49 = 0.005$$

The sum of the mole fractions is 1.000.

12.9 One mole of solution contains 0.250 moles methanol and 0.750 moles ethanol. The mass of this amount of ethanol, the solvent, is

$$0.750 \text{ mol } C_2H_5OH = \frac{46.07 \text{ g } C_2H_5OH}{1 \text{ mol } C_2H_5OH} = 34.\underline{5}5 \text{ g } C_2H_5OH \ (0.03455 \text{ kg})$$

The molality of methanol in the ethanol solvent is

$$\frac{0.250 \text{ mol } CH_3OH}{0.3455 \text{ kg } C_2H_5OH} = 7.2\underline{3}58 = 7.24 \text{ m } CH_3OH$$

12.10 Assume an amount of solution contains one kilogram of water. The mass of urea in this mass is

$$3.42 \text{ mol urea} \times \frac{60.05 \text{ g urea}}{1 \text{ mol urea}} = 20\underline{5}.4 \text{ g urea}$$

The total mass of solution is 205.4 + 1000.0 g = $120\underline{5}.4$ g. The volume and molarity are:

$$\frac{1205.4 \text{ g}}{1.045 \text{ g/mL}} = 115\underline{3}.49 \text{ mL } (1.15\underline{3}49 \text{ L})$$

$$\text{Molarity} = \frac{3.42 \text{ mol urea}}{1.15349 \text{ L solution}} = 2.9\underline{6}49 \text{ mol/L} = 2.96 \text{ M}$$

12.11 Assume a volume equal to 1.000 L of solution. Then

$$\text{Mass of solution} = 1.029 \text{ g/mL} \times (1.000 \times 10^3 \text{ mL}) = 1029 \text{ g}$$

$$\text{Mass of urea} = 2.00 \text{ mol urea} \times \frac{60.05 \text{ g urea}}{1 \text{ mol urea}} = 120.1 \text{ g urea}$$

$$\text{Mass of water} = (1029 - 120.1) \text{ g} = 90\underline{8}.9 \text{ g water } (0.90\underline{8}9 \text{ kg})$$

$$\text{Molality} = \frac{2.00 \text{ mol urea}}{0.9089 \text{ kg solvent}} = 2.2\underline{0}04 = 2.20 \text{ m urea}$$

12.12 Calculate the moles of naphthalene and moles of chloroform:

$$0.515 \text{ g } C_{10}H_8 \times \frac{1 \text{ mol } C_{10}H_8}{128.17 \text{ g } C_{10}H_8} = 0.0040\underline{1}8 \text{ mol } C_{10}H_8$$

$$60.8 \text{ g } CHCl_3 \times \frac{1 \text{ mol } CHCl_3}{119.38 \text{ g } CHCl_3} = 0.50\underline{9}29 \text{ mol } CHCl_3$$

The total number of moles is 0.004018 + 0.50929 = $0.51\underline{3}4$ mol, and the mole fraction of chloroform is:

$$\text{Mol fraction } CHCl_3 = \frac{0.50\underline{9}29 \text{ mol } CHCl_3}{0.5133} = 0.99\underline{2}1$$

Use Raoult's law to calculate the vapor pressure of chloroform:

$$P = P^oX = (156 \text{ mmHg})(0.9921) = 15\underline{4}.7 = 155 \text{ mmHg}$$

12.13 Solve for c_m in the freezing-point depression equation ($\Delta T = K_f c_m$; K_f in Table 12.3):

$$c_m = \frac{\Delta T}{K_f} = \frac{0.150°C}{1.858°C/m} = 0.080\underline{7}3 \text{ m}$$

Use the molal concentration to solve for the mass of ethylene glycol:

$$\frac{0.08073 \text{ m glycol}}{1 \text{ kg solvent}} \times 0.0378 \text{ kg solvent} = 0.0030\underline{5}1 \text{ mol glycol}$$

$$0.003051 \text{ mol glycol} \times \frac{62.1 \text{ g glycol}}{1 \text{ mol glycol}} = 0.18\underline{9}4 = 0.189 \text{ g glycol}$$

12.14 Calculate the moles of ascorbic acid (vit C) from the molality, and then divide the mass of 0.930 g by the number of moles to obtain the molar mass:

$$\frac{0.0555 \text{ mol vit C}}{1 \text{ kg } H_2O} \times 0.0950 \text{ kg } H_2O = 0.0052\underline{7}2 \text{ mol vit C}$$

$$\frac{0.930 \text{ g vit C}}{0.005272 \text{ mol vit C}} = 17\underline{6}.4 = 176 \text{ g/mol}$$

The molecular weight of ascorbic acid, or vitamin C, is 176 amu.

12.15 The molal concentration of white phosphorus is

$$c_m = \frac{\Delta T}{K_f} = \frac{0.159°C}{2.40°C/m} = 0.066\underline{2}5 \text{ m}$$

The number of moles of white phosphorus (P_x) present in this solution is:

$$\frac{0.07725 \text{ mol } P_x}{1 \text{ kg } CS_2} \times 0.0250 \text{ kg } CS_2 = 0.0016\underline{5}6 \text{ mol } P_x$$

The molar mass of white phosphorus equals the mass divided by moles:

$$0.205 \text{ g} \div 0.001656 \text{ mol} = 12\underline{3}.7 = 124 \text{ g/mol}$$

Thus, the molecular weight of P_x is 124 amu. The number of P atoms in the molecule of white phosphorus is obtained by dividing the molecular weight by the atomic weight of P:

$$\frac{124 \text{ amu } P_x}{30.97 \text{ amu P}} = 4.0\underline{0}3 = 4.00$$

Hence the molecular formula is P_4 (x = 4).

12.16 The number of moles of sucrose is

$$5.0 \text{ g sucrose} \times \frac{1 \text{ mol sucrose}}{342.3 \text{ g sucrose}} = 0.01\underline{4}6 \text{ mol sucrose}$$

The molarity of the solution is

$$\frac{0.0146 \text{ mol sucrose}}{0.100 \text{ L}} = 0.1\underline{4}6 \text{ M sucrose}$$

The osmotic pressure, π, is equal to MRT and is calculated:

$$\frac{0.146 \text{ mol sucrose}}{\text{L}} \times \frac{0.0821 \text{ L•atm}}{\text{mol•K}} \times 293 \text{ K} = 3.\underline{5}1 = 3.5 \text{ atm}$$

12.17 The number of ions from each formula units is *i*. Here,

$i = 1 + 2 = 3$

The boiling point elevation is

$$\Delta T_b = K_b c_m = 3 \times \frac{0.521^oC}{m} \times 0.050\ m = 0.07\underline{8}1 = 0.078^oC$$

The boiling point of aqueous $MgCl_2$ is 100.078°C.

12.18 $AlCl_3$ would be most effective in coagulating colloidal sulfur because of the greater magnitude of charge on the Al ion (+3).

ANSWERS TO REVIEW QUESTIONS

12.1 An example of a gaseous solution is air (oxygen dissolved in nitrogen). An example of a liquid solution is any carbonated beverage (carbon dioxide dissolved in water). An example of a solid solution is any gold-silver alloy typically used in jewelry.

12.2 Glycerol is miscible in water because it is similar to water in having -OH groups that form strong hydrogen bonds to other -OH groups in water. Benzene has a limited solubility in water because the benzene molecules must break the strong hydrogen bonds between water molecules and replace them with weaker London forces, an unfavorable situation.

12.3 The two factors are the natural tendency of substances to mix, and the relative strengths of the forces of attraction between species in the pure substances compared to those forces in the solution.

12.4 Octane is immiscible in water because the octane molecules must break the strong hydrogen bonds between water molecules and substitute weaker London forces for them, an unfavorable situation.

12.5 Ionic substances show a wide range of solubilities in water because of the wide variations in lattice energies and hydration energies. (Lattice energies depend on the charges on the ions as well as on the distances between the centers of the charges of the cations and anions.) A substance is very soluble when the hydration energy is much larger than the lattice energy. A substance is very insoluble when the lattice energy is much larger than the hydration energy.

12.6 The sodium ion is attracted to the negative end of a water dipole and dissolves as a hydrated sodium ion. The chloride ion is attracted to the positive end of a water dipole and dissolves as a hydrated chloride ion.

12.7 The usual effect of raising the temperature is to increase the solubility of an ionic compound in water. Typical exceptions are $CaSO_4$ and $Ca(OH)_2$, whose solubility decreases as temperature is increased.

12.8 A salt whose heat of solution is exothermic is $CaSO_4$. A salt whose heat of solution is endothermic is KNO_3.

12.9 Fewer fish can live in the same volume of water in the summer than in the water because fish depend on the amount of dissolved air. The solubility of all gases in the air decreases in the summer because the temperatures are higher.

12.10 A carbonated beverage must be stored in a closed container because the dissolved carbon dioxide has a higher solubility at the higher pressures in a closed container than at the usual atmospheric pressures encountered.

12.11 According to Le Chatelier's principle, a gas is more soluble in a liquid at higher pressures because when the gas dissolves in the liquid, the system decreases in volume, tending to decrease the applied pressure. However, when a solid dissolves in a liquid, there is very little volume change. Thus, pressure has very little effect on the solubility of a solid in a liquid.

12.12 The four ways to express the concentration of a solute in a solution are: (1) molarity, which is moles per liter; (2) mass percentage of solute, which is the percentage by mass of solute contained in a given mass of solution; (3) molality, which is the moles of solute per kilogram of solvent; and (4) mole fraction, which is the moles of the component substance divided by the total moles of solution.

12.13 The vapor pressure of the solvent over the more dilute solution is larger than the vapor pressure of solvent over the more concentrated solution. Thus the more dilute solution loses more solvent by evaporation and becomes more concentrated. Because the cabinet is closed, the excess solvent molecules condense into the more concentrated solution, diluting it. The changes in solvent composition stop when the concentrations of each solution become equal.

12.14 In fractional distillation, the vapor that first appears over a solution will have a greater mole fraction of the more volatile component. If a portion of this is vaporized and condensed, the liquid will be still richer in the more volatile component. After successive distillation stages, eventually the more volatile component will be obtained in pure form (Figure 12.14).

12.15 The boiling point of the solution is higher because the nonvolatile solute lowers the vapor pressure of the solvent. Thus the temperature must be increased to a value greater than the boiling point of the pure solvent to achieve a vapor pressure equal to atmospheric.

12.16 It is possible to prepare drinking water from sea water by freezing the sea water, which then forms almost pure ice. The dissolved salts are left behind in a concentrated solution, which does not freeze because the salts lower the freezing point of the solution. After draining off the concentrated solution, the "sea ice" can be melted and used for drinking.

12.17 One application is the use of ethylene glycol in automobile radiators as antifreeze; the glycol-water mixture usually has a freezing point well below the average temperature low during the winter. A second application is spreading sodium chloride on icy roads in the winter to melt the ice. The ice usually melts because at equilibrium a concentrated solution of NaCl usually freezes at a temperature below that of the roads.

12.18 The lettuce wilts if left too long in a salad dressing containing vinegar and salt because of osmosis. There is a higher salt concentration outside the lettuce membrane than inside, and the water flows through the membrane in order to lower the salt concentration outside to equal the salt concentration inside the lettuce.

12.19 By applying a pressure greater than the osmotic pressure of the ocean water, the natural osmotic flow can be reversed. Then the water solvent flows from the ocean water through a membrane to a more dilute solution or to pure water, leaving behind the salt and other ionic compounds from the ocean in a more concentrated solution.

12.20 Part of the light from the sun is scattered in the direction of an observer by fine particles in the clouds (Tyndall effect), rather than being completely absorbed by the clouds. The scattered light becomes visible against the darker background of dense clouds.

12.21 The examples are: an aerosol: a fog; a foam: whipped cream; an emulsion: mayonnaise; a sol: solid silver chloride dispersed in water; and a gel: fruit jelly.

12.22 Iron(III) hydroxide precipitates because electrons from the negative electrode neutralize the positive charge of the colloidal iron(III) hydroxide. This allows the colloidal particles to approach each other closely enough to aggregate and finally precipitate.

12.23 A Cottrell precipitator apparently works by aggregating the fine particles of smoke (an aerosol type of colloid) into a solid that can be readily removed from the gas. The electrodes conduct the high voltage into the smoke; the high-voltage current apparently removes the excess of electrical charge on the particles that prevents them from aggregating, thus allowing aggregation and removal.

12.24 Each stearic acid molecule orients itself with its acid group (-COOH) into the heavier water below the benzene, and with its hydrocarbon group into the lighter benzene layer on the surface of the water. As the benzene evaporates, the stearic acid molecules are left on the water surface. Each molecule keeps its acid group positioned in the water, so a single layer results.

12.25 Soap removes oil from a fabric by absorbing the oil into the hydrophobic centers of the soap micelles and off the surface of the fabric. Rinsing removes the micelles from contact with the fabric and leaves only water on the fabric, which can then be dried.

12.26 An example of an anionic detergent is sodium lauryl sulfate detergent ($-OSO_3^-$ group). An example of a cationic detergent is the positively charged nitrogen detergents (R_4N^+ group).

SOLUTIONS TO PRACTICE PROBLEMS

Note on significant figures: The final answer is given first with one nonsignificant figure (rightmost significant figure underlined), and is then rounded to the correct number of significant figures. Intermediate answers usually also have at least one nonsignificant figure. Atomic weights are rounded to two decimal places, except for that of hydrogen.

12.27 An example of a liquid solution of a gas in a liquid is household ammonia, which consists of ammonia, NH_3, gas dissolved in water.

12.29 Boric acid would be more soluble in ethanol because this acid is polar and is more soluble in a more polar solvent. It can also hydrogen-bond to ethanol but not to benzene.

12.31 The order of increasing solubility is: H_2O < CH_2OHCH_2OH < $C_{10}H_{22}$. The solubility in non-polar hexane increases with the decreasing polarity of the solute.

12.33 The Ca^{2+} ion has both a greater charge and a smaller ionic radius than K^+, so Ca^{2+} should have a greater energy of hydration.

12.35 The order is: $Ba(IO_3)_2 < Sr(IO_3)_2 < Ca(IO_3)_2 < Mg(IO_3)_2$. The iodate ion is fairly large so the lattice energy for all these iodates should change to a smaller degree than the hydration energy of the cations. Therefore solubility should increase with decreasing cation radius.

12.37 Using Henry's law: S_1 = the solubility at 1.00 atm (P_1), and S_2 = the solubility at 5.50 atm (P_2)

$$S_2 = S_1 \times \frac{P_2}{P_1} = (0.161 \text{ g/100 mL}) \times \frac{5.50 \text{ atm}}{1.00 \text{ atm}} = 0.88\underline{5}5 = 0.886 \text{ g/100 mL}$$

12.39 First calculate the mass of KI in the solution; then calculate the mass of water needed.

$$\text{Mass KI} = 145 \text{ g} \times \frac{2.50 \text{ g KI}}{100 \text{ g soln}} = 3.6\underline{2}50 \text{ g KI}$$

$$\text{Mass } H_2O = 145 \text{ g soln} - 3.625 \text{ g KI} = 14\underline{1}.375 \text{ g } H_2O$$

Dissolve 3.6$\underline{2}$5 g KI in 14$\underline{1}$.3 g of water.

12.41 Multiply the mass of KI by 100 g of soln per 2.50 g KI (reciprocal of percentage).

$$0.258 \text{ g KI} \times \frac{100 \text{ g soln}}{2.50 \text{ g KI}} = 10.\underline{3}2 = 10.3 \text{ g soln}$$

12.43 Convert mass of vanillin to moles, convert mg of ether to kg, and divide for molality.

$$0.0372 \text{ g vanillin} \times \frac{1 \text{ mol vanillin}}{152.2 \text{ g vanillin}} = 2.4\underline{4}4 \times 10^{-4} \text{mol vanillin}$$

$$168.5 \text{ mg ether} \times 1 \text{ kg}/10^6 \text{ mg} = 168.5 \times 10^{-6} \text{ kg ether}$$

$$\text{Molality} = \frac{2.444 \times 10^{-4} \text{ mol vanillin}}{168.5 \times 10^{-6} \text{ kg ether}} = 1.4\underline{5}04 = 1.45 \text{ m vanillin}$$

12.45 Convert mass of fructose to moles and then multiply by 1 kg H_2O per 0.125 mol fructose (the reciprocal of molality).

$$1.75 \text{ g fruct.} \times \frac{1 \text{ mol fruct.}}{180.16 \text{ g fruct.}} \times \frac{1 \text{ kg } H_2O}{0.125 \text{ mol fruct.}} = 0.077\underline{7}08 \text{ kg} \quad (77.7 \text{ g } H_2O)$$

12.47 Convert masses to moles and then calculate the mole fractions.

$$65.0 \text{ g alc.} \times \frac{1 \text{ mol alc.}}{60.09 \text{ g alc.}} = 1.0\underline{8}1 = 1.08 \text{ g mol alc.}$$

$$35.0 \text{ g } H_2O \times \frac{1 \text{ mol } H_2O}{18.02 \text{ g } H_2O} = 1.9\underline{4}2 = 1.94 \text{ g mol } H_2O$$

$$\text{Mol fractions alc.} = \frac{\text{mol alc.}}{\text{total mol}} = \frac{1.0\underline{8}1 \text{ mol}}{3.023 \text{ mol}} = 0.35\underline{7}59 = 0.358$$

$$\text{Mol fractions } H_2O = \frac{\text{mol } H_2O}{\text{total mol}} = \frac{1.942 \text{ mol}}{3.023 \text{ mol}} = 0.64\underline{2}4 = 0.642$$

12.49 In the solution, for every 0.750 mol of NaOCl there is 1.00 kg, or 1.00×10^3 g, H_2O, so:

$$1.00 \times 10^3 \text{ g } H_2O \times \frac{1 \text{ mol } H_2O}{18.02 \text{ g } H_2O} = 55.\underline{4}9 \text{ mol } H_2O$$

$$\text{Total mol} = 55.49 \text{ mol } H_2O + 0.750 \text{ mol NaOCL} = 56.24 \text{ mol}$$

$$\text{Mol fractions NaOCl} = \frac{\text{mol NaOCl}}{\text{mol soln}} = \frac{0.750 \text{ mol}}{56.24 \text{ mol}} = 0.013\underline{3}3 = 0.0133$$

12.51 The total moles of solution = 3.31 mol H_2O + 1.00 mol HCl = 4.31 mol.

$$\text{Mol fractions HCl} = \frac{1 \text{ mol HCl}}{4.31 \text{ mol}} = 0.23\underline{2}01 = 0.232$$

$$3.31 \text{ mol } H_2O \times \frac{18.02 \text{ g } H_2O}{1 \text{ mol } H_2O} \times \frac{1 \text{ kg } H_2O}{10^3 \text{ g } H_2O} = 5.9\underline{6}5 \times 10^{-2} \text{ kg } H_2O$$

$$\text{Molality} = \frac{1.00 \text{ mol HCl}}{5.965 \times 10^{-2} \text{ kg } H_2O} = 16.\underline{7}6 = 16.8 \text{ m}$$

12.53 The mass of 1.000 L of solution is 1.022 kg. In the solution, there are 0.585 moles of $H_2C_2O_4$ (OA) for every 1.0000 kg of water. Convert this number of moles to mass.

$$0.585 \text{ mol OA} \times \frac{90.04 \text{ g OA}}{1 \text{ mol OA}} \times \frac{1 \text{ kg OA}}{10^3 \text{ g OA}} = 0.052\underline{6}7 \text{ kg OA}$$

The total mass of the solution containing 1.000 kg H_2O and 0.585 moles of OA is calculated:

$$\text{Mass} = 1.0000 \text{ kg } H_2O + 0.05267 \text{ kg OA} = 1.052\underline{6}7 \text{ kg}$$

Use this to relate the mass of 1.000 L (1.022 kg) of solution to the amount of solute:

$$1.022 \text{ kg soln} \times \frac{0.585 \text{ mol OA}}{1.05267 \text{ kg soln}} = 0.56\underline{7}9 \text{ mol OA}$$

$$\text{Molarity} = \frac{0.5679 \text{ mol OA}}{1.000 \text{ L soln}} = 0.56\underline{7}8 = 0.568 \text{ M}$$

12.55 In 1.000 L of vinegar, there is 0.763 moles of acetic acid. The total mass of the 1.000 L solution is 1.004 kg. Start by calculating the mass of acetic acid (AA) in the solution.

$$0.763 \text{ mol AA} \times \frac{60.05 \text{ g AA}}{1 \text{ mol AA}} = 45.\underline{8}2 \text{ g AA } (0.045\underline{8}2 \text{ kg AA})$$

The mass of water may be found by difference:

$$\text{Mass } H_2O = 1.004 \text{ kg soln} - 0.04582 \text{ kg AA} = 0.95\underline{8}2 \text{ kg } H_2O$$

$$\text{Molality} = \frac{0.763 \text{ mol TA}}{0.9582 \text{ kg } H_2O} = 0.79\underline{6}2 = 0.796 \text{ m AA}$$

12.57 To find the mole fraction of sucrose, first find the amounts of both sucrose (Suc) and water:

$$20.2 \text{ g Suc} \times \frac{1 \text{ mol Suc}}{342.3 \text{ g Suc}} = 0.059\underline{0}1 \text{ mol Suc}$$

$$60.5 \text{ g } H_2O \times \frac{1 \text{ mol } H_2O}{18.01 \text{ g } H_2O} = 3.3\underline{5}7 \text{ mol } H_2O$$

$$X_{Suc} = \frac{0.05901 \text{ mol Suc}}{(3.357 + 0.05901) \text{ mol}} = 0.017\underline{2}7$$

From Raoult's law, the vapor pressure (P) and lowering (ΔP) are:

$$P = P^o_{H_2O} X_{H_2O} = P^o_{H_2O} (1 - X_{Suc}) = (42.2 \text{ mmHg}) (1 - 0.01727)$$

$$= 41.\underline{4}7 = 41.5 \text{ mmHg}$$

$$\Delta P = P^o_{H_2O} X_{Suc} = (42.2 \text{ mmHg})(0.017\underline{2}7) = 0.72\underline{8}7 = 0.729 \text{ mmHg}$$

12.59 Find the molality of glycerol (Gly) in the solution first:

$$0.152 \text{ g Gly} \times \frac{1 \text{ mol Gly}}{92.095 \text{ g Gly}} = 0.0016\underline{5}04 \text{ mol Gly}$$

$$\text{Molality} = \frac{0.0016\underline{5}04 \text{ mol Gly}}{0.1000 \text{ kg solvent}} = 0.082\underline{5}2 \text{ m}$$

Substitute $K_b = 0.521\ ^oC/m$ and $K_f = 1.858^oC/m$ (Table 12.3) into equations for ΔT_b and ΔT_f:

$$\frac{0.521°C}{m \text{ Gly}} \times 0.08\underline{2}52 \text{ m Gly} = 0.042\underline{9}9°C; \quad \frac{1.858°C}{m \text{ Gly}} \times 0.08252 \text{ m Gly} = 0.15\underline{3}3°C$$

$$T_b = 100.000^oC + 0.04299^oC = 100.04\underline{2}99 = 100.043^oC$$

$$T_f = 0.000^oC - 0.1533^oC = -0.15\underline{3}3 = -0.153^oC$$

12.61 Calculate ΔT_f, the freezing point depression, and, using K_f = 1.858°C/m (Table 12.3), the molality, c_m.

$$\Delta T_f = 0.000°C - (-0.086°C) = 0.086°C$$

$$c_m = \frac{\Delta T_f}{K_f} = \frac{0.086°C}{1.858°C/m} = 0.04\underline{6}28 = 0.046 \text{ m}$$

12.63 Find the moles of unknown solute from the definition of molality:

$$Mol_{solute} = m \times \text{kg solvent} = \frac{0.0698 \text{ mol}}{1 \text{ kg solvent}} \times 0.002135 \text{ kg solvent} = 1.4\underline{9}0 \times 10^{-4} \text{ mol}$$

$$\text{Molar mass} = \frac{0.0182 \text{ g}}{1.490 \times 10^{-4} \text{ mol}} = 12\underline{2}.1 = 122 \text{ g/mol (molecular wt = 122 amu)}$$

12.65 Calculate ΔT_f, the freezing point depression, and, using K_f = 1.858°C/m (Table 12.3), the molality, c_m.

$$\Delta T_f = 26.84°C - 25.70°C = 1.14°C$$

$$c_m = \frac{\Delta T_f}{K_f} = \frac{1.14°C}{8.00°C/m} = 0.14\underline{2}5 \text{ m}$$

Find the moles of solute by rearranging the definition of molality:

$$\text{Mol} = m \times \text{kg solvent} = \frac{0.1425 \text{ mol}}{1 \text{ kg solvent}} \times 103 \times 10^{-6} \text{ kg solvent} = 1.4\underline{6}7 \times 10^{-5} \text{ mol}$$

$$\text{Molar mass} = \frac{2.39 \times 10^{-3} \text{ g}}{1.467 \times 10^{-5} \text{ mol}} = 16\underline{2}.9 = 163 \text{ g/mol}$$

The molecular weight is 163 amu.

12.67 Use the equation for osmotic pressure (π) to solve for the molarity of the solution.

$$M = \frac{\pi}{RT} = \frac{1.47 \text{ mmHg} \times \frac{1 \text{ atm}}{760 \text{ mmHg}}}{(0.0821 \text{ L•atm/mol•K})(21 + 273)} = 8.0\underline{1}3 \times 10^{-5} \text{ mol/L}$$

Now find the number of moles in 106 mL (0.106 L) using the molarity.

$$0.106 \text{ L} \times \frac{8.013 \times 10^{-5} \text{ mol}}{\text{L}} = 8.4\underline{9}4 \times 10^{-6} \text{ mol}$$

$$\text{Molar mass} = \frac{0.582 \text{ g}}{8.494 \times 10^{-6} \text{ mol}} = 6.8\underline{5}1 \times 10^{4} \text{ g/mol}$$

The molecular weight is 6.85 x 10^4 amu.

12.69 Begin by noting that $i = 3$. Then calculate ΔT_f from the product of iK_fc_m:

$$3 \times \frac{1.86\ ^\circ C}{m} \times 0.0085\ m = 0.04\underline{7}4^\circ C$$

The freezing point $= 0.000^\circ C - 0.0474^\circ C = -0.04\underline{7}4 = -0.047^\circ C$.

12.71 Begin by calculating the molarity of $Cr(NH_3)_5Cl_3$.

$$1.40 \times 10^{-2}\ g\ Cr(NH_3)_5Cl_3 \times \frac{1\ mol\ Cr(NH_3)_5Cl_3}{243.5\ g\ Cr(NH_3)_5Cl_3} = 5.7\underline{4}9 \times 10^{-5}\ mol\ Cr(NH_3)_5Cl_3$$

$$\text{Molarity} = \frac{5.7\underline{4}9 \times 10^{-5}\ mol\ Cr(NH_3)_5Cl_3}{0.0250\ L} = 0.0022\underline{9}9\ M$$

Now find the hypothetical osmotic pressure, assuming $Cr(NH_3)_5Cl_3$ does not ionize:

$$\pi = MRT = (2.30 \times 10^{-3}\ M) \times \frac{0.0821\ L\bullet atm}{K\bullet mol} \times 298\ K \times \frac{760\ mmHg}{1\ atm} = 42.\underline{7}7\ mmHg$$

The measured osmotic pressure is greater than the hypothetical osmotic pressure. The number of ions formed per formula unit = ratio of the measured pressure to the hypothetical pressure:

$$i = \frac{119\ mmHg}{42.77\ mmHg} = 2.7\underline{8}2 \cong 3\ \text{ions/form unit}$$

12.73 a. Aerosol (liquid water in air)
b. Sol (solid $Mg(OH)_2$ in liquid water)
c. Foam (air in liquid soap solution)
d. Sol (solid silt in liquid water)

12.75 Since the As_2S_3 particles are negatively charged, the effective coagulation requires a highly charged cation, so $Al_2(SO_4)_3$ is the best choice.

12.77 Using Henry's law [$S_2 = S_1 \times (P_2/P_1)$, where $P_1 = 1.00$ atm], find the solubility of each gas at P_2, its partial pressure. For N_2, $P_2 = 0.800 \times 1.00\ atm = 0.800$ atm; for O_2, $P_2 = 0.200$ atm.

$$N_2\text{: } S_2 \times \frac{P_2}{P_1} = (0.0175\ g/L\ H_2O) \times \frac{0.800\ atm}{1.00\ atm} = 0.014\underline{0}0\ g/L\ H_2O$$

$$O_2\text{: } S_2 \times \frac{P_2}{P_1} = (0.03935\ g/L\ H_2O) \times \frac{0.200\ atm}{1.00\ atm} = 0.0078\underline{6}0\ g/L\ H_2O$$

In 1.00 L of the water, there are 0.0140 g of N_2 and 0.00786 g of O_2. If the water is heated to drive off both dissolved gases, the gas mixture that is expelled will contain 0.0140 g of N_2 and 0.00786 g of O_2. Convert both masses to moles using the molar masses:

$$0.0140\ g\ N_2 \times \frac{1\ mol\ N_2}{28.01\ g\ N_2} = 4.9\underline{9}8 \times 10^{-4}\ mol$$

$$0.00786 \text{ g } O_2 \times \frac{1 \text{ mol } O_2}{32.00 \text{ g } O_2} = 2.4\underline{5}6 \times 10^{-4} \text{ mol}$$

Now calculate the mole fractions of each gas:

$$X_{N_2} = \frac{\text{mol } N_2}{\text{total mol}} = \frac{4.998 \times 10^{-4} \text{ mol}}{(4.998 + 2.456) \times 10^{-4} \text{ mol}} = 0.67\underline{0}51 = 0.671$$

$$X_{O_2} = 1 - X_{N_2} = 1 - 0.67\underline{0}51 = 0.32\underline{9}49 = 0.329$$

12.79 Assume a volume of 1.000 L, whose mass is then 1.024 kg. Use the percent composition given to find the mass of each of the components of the solution.

$$1.024 \text{ kg soln} \times \frac{8.50 \text{ kg } NH_4Cl}{100.00 \text{ kg soln}} = 0.087\underline{0}4 \text{ kg } NH_4Cl$$

$$\text{Mass of } H_2O = 1.024 \text{ kg soln} - 0.08704 \text{ kg } NH_4Cl = 0.9370 \text{ kg } H_2O$$

Convert mass of NH_4Cl and water to moles:

$$87.04 \text{ g } NH_4Cl \times \frac{1 \text{ mol } NH_4Cl}{53.49 \text{ g } NH_4Cl} = 1.6\underline{2}7 \text{ mol } NH_4Cl$$

$$937.9 \text{ g } H_2O \times \frac{1 \text{ mol } H_2O}{18.015 \text{ g } H_2O} = 52.\underline{0}1 \text{ mol } H_2O$$

$$\text{Molarity} = \frac{\text{mol } NH_4Cl}{\text{L soln}} = \frac{1.627 \text{ mol}}{1.00 \text{ L}} = 1.6\underline{2}7 = 1.63 \text{ M}$$

$$\text{Molality} = \frac{\text{mol } NH_4Cl}{\text{kg } H_2O} = \frac{1.627 \text{ mol}}{0.9370 \text{ kg } H_2O} = 1.7\underline{3}6 = 1.74 \text{ m}$$

$$X_{NH_4Cl} = \frac{\text{mol } NH_4Cl}{\text{total moles}} = \frac{1.627 \text{ mol}}{(52.01 + 1.627) \text{ mol}} = 0.030\underline{3}3 = 0.0303$$

12.81 In 1.00 mol of gas mixture, there are 0.43 mol of propane (Pro) and 0.57 mol of butane (But). Calculate the masses of these components first.

$$0.43 \text{ mol Pro} \times \frac{44.1 \text{ g Pro}}{1 \text{ mol Pro}} = 1\underline{9}.0 \text{ g Pro}$$

$$0.57 \text{ mol But} \times \frac{58.12 \text{ g But}}{1 \text{ mol But}} = 3\underline{3}.1 \text{ g But}$$

The mass of 1.00 mol of gas mixture is the sum of the masses of the two components:

$$19.0 \text{ g Pro} + 33.1 \text{ g But} = 52.1 \text{ g mixture}$$

Therefore in 52.1 g of the mixture, there are 19.0 g of propane and 33.1 g of butane. For a sample with a mass of 58 g:

$$58 \text{ g mixt} \times \frac{19.0 \text{ g Pro}}{52.1 \text{ g mixt}} = 2\underline{1}.1 = 21 \text{ g Pro}$$

$$58 \text{ g mixt} \times \frac{33.1 \text{ g But}}{52.1 \text{ g mixt}} = 3\underline{6}.8 = 37 \text{ g But}$$

12.83

$$P_{ED} = P^o{}_{ED}(X_{ED}) = 173 \text{ mmHg } (0.35) = 6\underline{0}.55 \text{ mmHg}$$

$$P_{PD} = P^o{}_{PD}(E_{PD}) = 127 \text{ mmHg } (0.65) = 8\underline{2}.55 \text{ mmHg}$$

$$P = P_{ED} + P_{PD} = 60.55 + 82.55 = 14\underline{3}.1 = 143 \text{ mmHg}$$

12.85 In 1.00 kg of a saturated solution of urea, there are 0.44 kg of urea (a molecular solute) and 0.56 kg of water. First convert the mass of urea to moles.

$$0.44 \times 10^3 \text{ g urea} \times \frac{1 \text{ mol urea}}{60.06 \text{ g urea}} = 7\underline{3}.3 \text{ mol urea}$$

Then find the molality of the urea in the solution:

$$\text{Molality} = \frac{\text{mol urea}}{\text{kg } H_2O} \times \frac{7.33 \text{ mol } CaCl_2}{0.56 \text{ kg}} = 1\underline{3}.1 \text{ m}$$

$$\Delta T_f = K_f c_m = (1.858^oC/m)(13.1 \text{ m}) = 2\underline{4}.3^oC$$

$$T_f = 0.0^oC - 24.3^oC = -2\underline{4}.3 = -24^oC$$

12.87

$$M = \frac{\pi}{RT} = \frac{7.7 \text{ atm}}{(0.0821 \text{ L•atm/mol•K})(37 + 273)K} = 0.3\underline{0}2 = 0.30 \text{ mol/L}$$

12.89 Consider the equation $\Delta T_f = K_f i\, c_m$. For $CaCl_2$, $i = 3$; for glucose $i = 1$. Since $c_m = 0.10$ for both solutions, the product of i and c_m will be larger for $CaCl_2$, as will ΔT_f. The solution of $CaCl_2$ will thus have the lower freezing point.

Cumulative-Skills Problems (require skills from previous chapters)

12.91

$$Na^+(g) + Cl^-(g) \rightarrow NaCl(s) \qquad \Delta H = -787 \text{ kJ/mol}$$

$$NaCl(s) \rightarrow Na^+(aq) + Cl^-(aq) \qquad \Delta H = +4 \text{ kJ/mol}$$

$$Na^+(g) + Cl^-(g) \rightarrow Na^+(aq) + Cl^-(aq) \qquad \Delta H = -783 \text{ kJ/mol}$$

The heat of hydration of Na^+:

$$Na^+(g) + Cl^-(g) \rightarrow Na^+(aq) + Cl^-(aq) \qquad \Delta H = -783 \text{ kJ/mol}$$

$$Cl^-(aq) \rightarrow Cl^-(g) \qquad \Delta H = +338 \text{ kJ/mol}$$

$$Na^+(g) \rightarrow Na^+(aq) \qquad \Delta H = -445 \text{ kJ/mol}$$

12.93 $$15.0\ g\ MgSO_4{\cdot}7H_2O \times \frac{1\ mol}{246.5\ g} = 0.060\underline{8}54\ mol\ MgSO_4{\cdot}7H_2O$$

$$0.060\underline{8}54\ mol\ MgSO_4{\cdot}7H_2O \times \frac{7\ mol\ H_2O}{1\ mol\ hydrate} \times \frac{18.0\ g\ H_2O}{1\ mol\ H_2O} = 7.6\underline{6}7\ g\ H_2O$$

$$kg\ H_2O = (100.0\ g\ H_2O + 7.667\ g\ H_2O) \times \frac{1\ kg\ H_2O}{1000\ g\ H_2O} = 0.107\underline{6}6\ kg\ H_2O$$

$$m = \frac{0.060854\ mol\ MgSO_4}{0.10766\ kg\ H_2O} = 0.56\underline{5}2\ mol/kg = 0.565\ m$$

12.95 $$15.0\ g\ CuSO_4{\cdot}5H_2O \times \frac{1\ mol}{249.7\ g} = 0.060\underline{0}7\ mol\ CuSO_4{\cdot}5H_2O\ \text{or}\ CuSO_4$$

$$100\ g\ soln \times \frac{1\ mL\ soln}{1.167\ g\ soln} \times \frac{1\ L}{1000\ mL} = 0.0856\underline{8}9\ L$$

$$M = \frac{0.060\underline{0}7\ mol\ CuSO_4}{0.085689\ L} = 0.70\underline{1}02 = 0.701\ mol/L$$

12.97 $$0.159°C \times \frac{m}{1.858°C} = 0.085\underline{5}7\ m = \frac{0.08557\ mol\ AA + H^+}{1000\ g\ (\text{or}\ 1000\ mL)\ H_2O}$$

Note that 0.0830 mol AA + mol H^+ = 0.08557 mol/L (AA + H^+)

Mol H^+ = 0.08557 - 0.0830 = 0.002$\underline{5}$7 mol

$$\%\ dissoc = \frac{mol\ H^+}{mol\ AA}(100) = \frac{0.00257\ mol}{0.0830\ mol}(100) = 3.\underline{0}9 = 3.1\%$$

12.99 Calculate the empirical formula first, using the mass of C, O, and H in 1.000 g:

$$1.434\ g\ CO_2 \times \frac{12.00\ g\ C}{44.00\ g\ CO_2} = 0.39\underline{1}09\ g\ C;\ 0.783\ g\ H_2O \times \frac{2.016\ g\ H}{18.016\ g\ H_2O} = 0.087\underline{6}1\ g\ H$$

g O = 1.000 g - 0.39109 g - 0.08761 g = 0.52$\underline{1}$3 g O

Mol C = 0.39109 g C x 1 mol/12.00 g = 0.03259 mol C (lowest integer = 1)

Mol H = 0.08761 g H x 1 mol/1.008 g = 0.08691 mol H (lowest integer = 8/3)

Mol O = 0.5213 g O x 1 mol/16.00 g = 0.03258 mol O (lowest integer = 1)

Therefore the empirical formula is $C_3H_8O_3$. The formula weight from the freezing point is next calculated by first finding the molality:

$0.0894^oC \times (m/1.858^oC) = 0.04811 = (0.048\underline{1}1\ mol/1000\ g\ H_2O)$

$(0.04811\ mol/1000\ g\ H_2O) \times 25.0\ g\ H_2O = 0.0012\underline{0}3\ mol\ (\text{in}\ 25.0\ g\ H_2O)$

Molar mass = M_m = 0.1107 g/0.001203 mol = 92.02 g/mol

Since this is also the formula weight, the molecular formula is $C_3H_8O_3$ also.

CHAPTER 13

CHEMICAL REACTIONS: ACID-BASE AND OXIDATION-REDUCTION CONCEPTS

SOLUTIONS TO EXERCISES

Note on significant figures: The final answer to all mathematical solutions is given first with one nonsignificant figure (last significant figure underlined) and is then rounded to the correct number of figures. Intermediate answers usually also have at least one nonsignificant figure.

13.1 See labels below reaction: $H_2CO_3(aq) + CN^-(aq) \rightleftharpoons HCN(aq) + HCO_3^-(aq)$

$H_2CO_3(aq)$	$CN^-(aq)$	$HCN(aq)$	$HCO_3^-(aq)$
acid	base	acid	base

H_2CO_3 is the proton donor (Bronsted-Lowry acid) on the left and HCN is the proton donor (Bronsted-Lowry acid) on the right. The CN^- and HCO_3^- ions are proton acceptors (Bronsted-Lowry bases). HCN is the conjugate acid of CN^-.

13.2 The $HC_2H_3O_2$ is a stronger acid than H_2S. Also HS^- is a stronger base than $C_2H_3O_2^-$ ion. The equilibrium favors the weaker acid and weaker base; therefore, the reactants are favored.

13.3 a. PH_3 b HI c. H_2SO_3 d. H_3AsO_4 e. HSO_4^-

13.4 Part a involves molecules with all single bonds; part b does not, so bonds are drawn in.

a.

```
     F       H               F   H
     ••      ••              ••  ••
   F : B  + :N : H    →   F : B : N : H
     ••      ••              ••  ••
     F       H               F   H
   Lewis   Lewis
   acid    base
```

b. $:\ddot{O}:^{2-}$ (Lewis base) + $:O::C::O:$ (Lewis acid) $\longrightarrow$ $[:\ddot{O}:C(=O:)(-\ddot{O}:)]^{2-}$

13.5 The oxidation number of K, x_K, = +1 and that of O, x_O, = -2. Letting x_{Cr} equal the oxidation number of the chromium atom, we write an equation and solve for x_{Cr}:

$$2x_K + x_{Cr} + 7x_O = 0$$

$$+2 + 2x_{Cr} - 14 = 0$$

$$x_{Cr} = +6$$

13.6 The oxidation number of O, x_O, = -2. Letting x_{Mn} = the oxidation number of Mn, we write:

$$x_{Mn} + 4x_O = -1$$

$$x_{Mn} - 8 = -1$$

$$x_{Mn} = +7$$

13.7 a. Silver metal is below hydrogen in the activity series and hence will not react with the H^+ ion.

b. Magnesium metal is above silver in the activity series and can reduce the Ag^+ ion to Ag(s).

13.8 Note that the I atom and the N atom change oxidation numbers. Balancing I gives:

1 e^- gained (N: +5 → +4)

$$\overset{0}{I_2} + H\overset{+5}{N}O_3 \rightarrow 2H\overset{+5}{I}O_3 + \overset{+4}{N}O_2 + H_2O$$

10 e^- lost (I: 0 → +5)

Multiply HNO_3 and NO_2 by 10 to balance the number of electrons lost and gained:

10 e^- gained

$$\overset{0}{I_2} + 10H\overset{+5}{N}O_3 \rightarrow 2H\overset{+5}{I}O_3 + 10\overset{+4}{N}O_2 + H_2O$$

10 e^- lost

There are 30 O's on the left and 26 O's on the right, so balance the O atoms by writing a coefficent of 4 in front of H_2O. This also balances the H atoms, with 10 on each side:

$$I_2 + 10HNO_3 \rightarrow 2HIO_3 + 10NO_2 + 4H_2O$$

13.9 The skeleton equation with oxidation numbers written above the equation for Mn and I is:

5 e^- gained (Mn: +7 → +2)

$$\overset{-1}{I^-} + \overset{+7}{MnO_4^-} \rightarrow \overset{+5}{IO_3^-} + \overset{+2}{Mn^{2+}}$$

6 e^- lost (I: −1 → +5)

To balance the number of electrons lost and gained, multiply I^- and IO_3^- by 5, and MnO_4^- and Mn^{2+} by 6. This gives the unbalanced equation:

30 e^- gained

$$5\overset{-1}{I^-} + 6\overset{+7}{MnO_4^-} \rightarrow 5\overset{+5}{IO_3^-} + 6\overset{+2}{Mn^{2+}}$$

30 e^- lost

There are 24 O's on the left and 15 O's on the right, so balance the O atoms by writing $9H_2O$ on the right. To balance the H atoms, then write $18H^+$ on the left, giving:

$$5I^- + 6MnO_4^- + 18H^+ \rightarrow 5IO_3^- + 6Mn^{2+} + 9H_2O$$

13.10 The skeleton equation with oxidation numbers written above the equation for O and Mn is:

3 e^- gained (Mn: +7 → +4)

$$\overset{-1}{H_2O_2} + \overset{+7}{MnO_4^-} \rightarrow \overset{0}{O_2} + \overset{+4}{MnO_2}$$

2 e^- lost per 2 O

To balance the number of electrons lost and gained, multiply H_2O_2 and O_2 by 3, and MnO_4^- and MnO_2 by 2.

6 e^- gained

$$3\overset{-1}{H_2O_2} + 2\overset{+7}{MnO_4^-} \rightarrow 3\overset{0}{O_2} + 2\overset{+4}{MnO_2}$$

6 e^- lost

To balance in basic solution, we first obtain the balanced equation as if it were in acid solution. Then we add OH^- ions to react with all of the H^+ to give H_2O. Since there are 4 extra O's on the left, we add $4H_2O$ on the right; then we add $2H^+$ to the left to balance the H atoms:

$$3H_2O_2 + 2MnO_4^- + 2H^+ \rightarrow 3O_2 + 2MnO_2 + 4H_2O$$

We can convert this reaction in acid solution to basic solution by adding two OH^- to both sides, forming two H_2O molecules on the left. Canceling 2 H_2O molecules gives the final equation.

$$3H_2O_2 + 2MnO_4^- \rightarrow 3O_2 + 2MnO_2 + 2H_2O + 2OH^-$$

13.11 The skeleton equation is: $Zn + NO_3^- \rightarrow Zn^{2+} + NH_4^+$. Splitting into half-reactions gives:

$$Zn \rightarrow Zn^{2+}\ ;\ NO_3^- \rightarrow NH_4^+$$

The first half-reaction is balanced by adding $2e^-$ to the right to balance the charges.

$$Zn \rightarrow Zn^{2+} + 2e^-$$

The second half-reaction is balanced successively in O and then H atoms:

$$NO_3^- \rightarrow NH_4^+ + 3H_2O;\ NO_3^- + 10H^+ \rightarrow NH_4^+ + 3H_2O$$

Finally, noting the left side has a net charge of +9 and the right side a charge of +1, the half-reaction is balanced by adding $8e^-$ to the left to balance the charges.

$$NO_3^- + 10H^+ + 8e^- \rightarrow NH_4^+ + 3H_2O$$

To combine, multiply the first half reaction by 4 so that e's cancel upon adding half-reactions.

$$4Zn \rightarrow 4Zn^{2+} + 8e^-$$

$$NO_3^- + 10H^+ + 8e^- \rightarrow NH_4^+ + 3H_2O$$

$$NO_3^- + 10H^+ + 4Zn \rightarrow 4Zn^{2+} + NH_4^+ + 3H_2O$$

13.12 To obtain the number of equivalents of NaOH, multiply the normality by the volume in liters (25.89 mL is equivalent to 0.02589 L).

$$0.02589\ L \times \frac{0.1156\ eq\ NaOH}{L} = 0.00299\underline{2}8\ eq\ NaOH$$

By definition, the equivalents of citric acid (CA) and NaOH must be equal, so

$$\text{Equivalents CA} = \text{equivalents NaOH} = 0.00299\underline{2}8\ eq$$

The equivalent mass of citric acid is its mass per equivalent; therefore

$$\text{Equivalent mass of CA} = \frac{0.1916\ g\ CA}{0.0029928\ eq} = 64.0\underline{1}8 = 64.02\ g/eq$$

Since citric acid is triprotic, it contains 3 eq per mole. Its molar mass is the product of the equivalent mass and the number of equivalents per mole:

$$\text{Molar mass} = 3\ eq/mol \times 64.018\ g/eq = 192.\underline{0}5 = 192.0\ g/mol$$

13.13 Calculate the equivalents of $K_2Cr_2O_7$ from the normality and the volume in liters (28.3 mL is equivalent to 0.0283 L).

$$0.0283\ L \times \frac{0.0795\ eq\ K_2Cr_2O_7}{L} = 0.0022\underline{4}98\ eq\ K_2Cr_2O_7$$

By definition, the equivalents of $K_2Cr_2O_7$ and SO_2 must be equal, so

Equivalents SO_2 = equivalents $K_2Cr_2O_7$ = 0.0022$\underline{4}$98 eq

The SO_2 is oxidized as $H_2SO_3 + H_2O \rightarrow H_2SO_4 + 2H^+ + 2e^-$, so

1 mol SO_2 = 2 eq SO_2

The mass of SO_2 may be calculated using the molar mass of 64.06 g/mol:

$$0.0022\underline{4}98 \text{ eq } SO_2 \times \frac{1 \text{ mol } SO_2}{2 \text{ eq } SO_2} \times \frac{64.06 \text{ g } SO_2}{\text{eq } SO_2} = 0.072\underline{0}6 = 0.0721 \text{ g } SO_2$$

ANSWERS TO REVIEW QUESTIONS

13.1 An Arrhenius acid is a substance that when dissoved in water increases the concentration of the hydrogen ion. For example, 1 mol of an acid such as HCl supplies 1 mol of H^+ to an aqueous solution. An Arrhenius base is a substance that when dissolved in water increases the concentration of the hydroxide ion. For example, 1 mol of a base such as NaOH supplies 1 mol of OH^- to an aqueous solution.

13.2 a. Weak b. Weak c. Strong d. Strong e. Weak f. Weak

13.3 Thermochemical evidence for the Arrhenius concept comes from the fact that when 1 mol of any strong acid (1 mol H^+) is neutralized with 1 mol of any strong base (1 mol OH^-), the heat of neutralization released is always the same (ΔH^o = -55.90 kJ). See answer to 13.1.

13.4 A Bronsted-Lowry acid is a molecule or ion that donates a proton (proton donor), whereas a base is a molecule or ion that accepts a proton (proton acceptor). Example:

$$H_3O^+(aq) + NH_3(aq) \rightarrow NH_4^+(aq) + H_2O(l)$$

where H_3O^+ is the acid and NH_3 is the base.

13.5 The conjugate acid of a base is a species that differs from the base by one proton. For example, H_2CO_3 is the conjugate acid of the base HCO_3^-, even though both contain H.

13.6 An equation in which $H_2PO_3^-$ acts as a Bronsted-Lowry acid is:

$$H_2PO_3^-(aq) + OH^-(aq) \rightarrow HPO_3^{2-}(aq) + H_2O(l)$$

An equation in which it acts as a Bronsted-Lowry base is:

$$H_2PO_3^-(aq) + H_3O^+(aq) \rightarrow H_3PO_3(aq) + H_2O(l)$$

13.7 Four ways the Bronsted-Lowry concept enlarges on the Arrhenius concept are:

(1) a base is a proton acceptor, of which the OH^- ion is only one example;
(2) ions, as well as molecules, can act as acids and bases;
(3) acid-base reactions can also occur in nonaqueous solution; and
(4) some species can act as either acids or bases, depending on the other reactant.

13.8 An equilibrium involving a base and a relatively weaker acid on one side, and a base and a relatively stronger acid on the other side, always favors the weaker-acid side because the proton bonds more strongly to the weaker acid.

13.9 Since acetic acid is a weaker acid than formic acid, its conjugate base holds its proton more strongly than formic acid does. Thus the acetate ion, $C_2H_3O_2^-$, is a stronger base than the formate ion, CHO_2^-.

13.10 Two important factors that determine the strength of an acid are the polarity of the bond to which the H atom is attached, and the strength of the bond. An increase in the polarity of the bond makes it easier to remove the proton, and thus increases the acid strength. An increase in the strength of the bond, or how tightly the proton is held, makes it more difficult to remove the proton, and thus decreases acid strength. The strength of the bond depends on the size of the atom, so that larger atoms have weaker bonds, whereas smaller atoms have stronger bonds.

13.11 According to the Lewis concept, an acid is an electron-pair acceptor and a base is an electron-pair donor. An example is:

$$\underset{\text{acid}}{Ag^+(aq)} + \underset{\text{base}}{2(:NH_3)} \rightarrow Ag(NH_3)_2^+(aq)$$

13.12 In oxidation, at least one atom in a substance loses electrons so that its oxidation number increases. In reduction, at least one atom in a substance gains electrons, so that its oxidation number decreases.

13.13 $$\underset{\text{reducing agent}}{Fe(s)} + \underset{\text{oxidizing agent}}{2AgNO_3(aq)} \rightarrow 2Ag^+(aq) + Fe(NO_3)_2(aq)$$

13.14 A displacement reaction is an oxidation-reduction reaction in which a free element appears to displace another element in a compound.

13.15 In an acid-base reaction, an equivalent of an acid is the mass of the acid that yields 1 mol of H^+; an equivalent of a base is the mass of the base that accepts 1 mol of H^+. In an oxidation-reduction reaction, an equivalent of an oxidizing or reducing agent is the mass that supplies or reacts with 1 mole of electrons.

13.16 To prepare a 0.10 N solution of barium hydroxide, add 0.10 of an equivalent of $Ba(OH)_2$ to 1L of water. Since the molar mass of $Ba(OH)_2$ is 171.4 g, one equivalent of $Ba(OH)_2$ would be equivalent to 171.4/2, or 85.7 g. Therefore 8.57 g of $Ba(OH)_2$ is added to 1 L.

SOLUTIONS TO PRACTICE PROBLEMS

Note on significant figures: The final answer to all mathematical solutions is given first with one nonsignificant figure (last significant figure underlined) and is then rounded to the correct number of figures. Intermediate answers usually also have at least one nonsignificant figure.

13.17 The ionization equation is $\underset{\text{acid}}{HSO_3^-} + \underset{\text{base}}{H_2O} \rightleftharpoons \underset{\text{acid}}{H_3O^+} + \underset{\text{base}}{SO_3^{2-}}$

13.19 a. SO_4^{2-} b. HS^- c. HPO_4^{2-} d. NH_3

13.21 a. HCN b. H_2CO_3 c. $HSeO_4^-$ d. HPO_4^{2-}

13.23 a. $H_2C_2O_4$ and $HC_2O_4^-$ are conjugates, and ClO^- and HClO are conjugates. $H_2C_2O_4$ and HClO are the acids; ClO^- and $HC_2O_4^-$ are the bases.
b. HPO_4^{2-} and $H_2PO_4^-$ are conjugates, and NH_3 and NH_4^+ are conjugates. NH_4^+ and $H_2PO_4^-$ are the acids; HPO_4^{2-} and NH_3 are the bases.
c. H_2O and OH^- are conjugates, and SO_4^{2-} and HSO_4^- are conjugates. H_2O and HSO_4^- are the acids; SO_4^{2-} and OH^- are the bases.
d. $Fe(H_2O)_6^{2+}$ and $Fe(H_2O)_5OH^+$are conjugates, and H_2O and H_3O^+ are conjugates. $Fe(H_2O)_6^{2+}$ and H_3O^+ are the acids; H_2O and $Fe(H_2O)_5OH^+$ are the bases.

13.25 The reaction is $HSO_4^- + ClO^- \rightarrow HClO + SO_4^{2-}$. According to Table 13.1, HClO is a weaker acid than HSO_4^-. Since the equilibrium for this type of reaction favors formation of the weaker acid (or weaker base), the reaction occurs to a significant extent.

13.27 a. NH_4^+ is a weaker acid than H_3PO_4, so the left-hand species are favored at equilibrium.
b. HCN is a weaker acid H_2S, so the left-hand species are favored at equilibrium.
c. H_2O is a weaker acid than HCO_3^-, so the right-hand species are favored at equilibrium.
d. H_2O is a weaker acid than $Al(H_2O)_6^{3+}$, so the right-hand species are favored at equilibrium.

13.29 Trichloroacetic acid is the stronger acid because in general the equilibrium favors the formation of the weaker acid, which is formic acid in this case.

13.31 a. H_2S is stronger because acid strength decreases with increasing anion charge for polyprotic acid species.
b. H_2SO_3 is stronger because for a series of oxyacids, acid strength increases with increasing electronegativity.
c. HBr is stronger because Br is more electronegative than Se. Within a period, acid strength increases as electronegativity increases.
d. HIO_4 is stronger because acid strength increases with the number of oxygen atoms bonded to the central atom.
e. H_2S is stronger because within a group, acid strength increases with the increasing size of the central atom in binary acids.

13.33 For a, the completed equation is $CO_2 + OH^- \rightarrow HCO_3^-$. For b, the completed equation is $AlCl_3 + Cl^- \rightarrow AlCl_4^-$. For a, the CO_2 is a Lewis acid and the OH^- ion is a Lewis base. For b, the $AlCl_3$ is a Lewis acid and the Cl^- ion is a Lewis base. The Lewis formulas are all shown on the next page.

a. :O: ‖ C ‖ :O: + [:Ö:H]⁻ ⟶ [H:Ö:C(=O:)—:Ö:]⁻

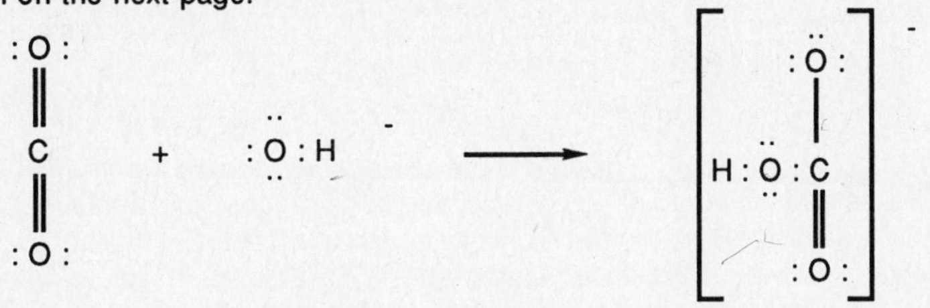

b. :Cl—Al(—Cl:)—Cl: + [:Cl:]⁻ ⟶ [:Cl—Al(—Cl:)(—Cl:)—Cl:]⁻

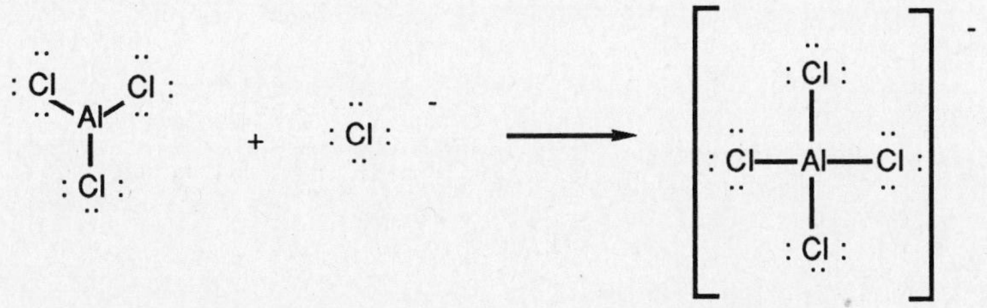

13.35 a. Each water molecule donates a pair of electrons to chromium(III), making the water molecule a Lewis base and the Cr^{3+} ion a Lewis acid.
b. The oxygen atom in $(C_2H_5)O$ donates a pair of electrons to the boron atom in BF_3, making $(C_2H_5)O$ a Lewis base and the BF_3 molecule a Lewis acid.

13.37 The equation is $H_2S + HOCH_2CH_2NH_2 \rightarrow HOCH_2CH_2NH_3^+ + HS^-$. The H_2S is a Lewis acid and $HOCH_2CH_2NH_2$ is a Lewis base. The hydrogen ion from H_2S accepts a pair of electrons from the N atom in $HOCH_2CH_2NH_2$.

13.39 a. Because all 3 O's = a total of -6, both Ga's = +6; thus the oxidation number of Ga = +3.
b. Because both O's = a total of -4, the oxidation number of Pb = +4.
c. Because the 4 O's = a total of -8 and K = +1, the oxidation number of Br = +7.
d. Because the 4 O's = a total of -8 and the 2 K's = +2, the oxidation number of Mn = +6.

13.41 a. Because the charge of -1 = $[x_N + 2$ (from 2 H's)], x_N must equal -3.
b. Because the charge of -1 = $[x_I - 6$ (from 3 O's)], x_I must equal +5.
c. Because the charge of -1 = $[x_{Al} - 8$ (4 O's) + 4 (4 H's)], x_{Al} must equal +3.
d. Because the charge of -1 = $[x_P - 8$ (4 O's) + 2 (2 H's)], x_P must equal +5.

13.43 a. From the list of common polyatomic anions in Table 2.5, the formula of the ClO_3 anion must be ClO_3^-. Thus the oxidation state of Mn is Mn^{2+} (see also Table 13.4). For the oxidation state of Cl in ClO_3^-, the charge of -1 must equal $x_{Cl} - 6$, so x_{Cl} must equal +5.
b. From the list of common polyatomic anions in Table 2.5, the formula of the CrO_4 anion must be CrO_4^{2-}. Thus the oxidation state of Fe is Fe^{3+}. For the oxidation state of Cr in CrO_4^{2-}, the charge of -2 must equal $x_{Cr} - 8$, so x_{Cr} must equal +6.

c. From the list of common polyatomic anions in Table 2.5, the formula of the Cr_2O_7 anion must be $Cr_2O_7^{2-}$. Thus the oxidation state of Hg is Hg^{2+}. For the oxidation state of Cr in $Cr_2O_7^{2-}$, the charge of -2 must equal $2x_{Cr}$ - 14, so x_{Cr} must equal +6.

d. From the list of common polyatomic anions in Table 2.5, the formula of the PO_4 anion must be PO_4^{3-}. Thus the oxidation state of Co is Co^{2+}. For the oxidation state of P in PO_4^{3-}, the charge of -3 must equal x_P - 8, so x_P must equal +5.

13.45 a. Phosphorus changes from an oxidation number of 0 in P_4 to +5 in P_4O_{10}, losing electrons and acting as a reducing agent. Oxygen changes from an oxidation number of 0 in O_2 to -2 in P_4O_{10}, gaining electrons and acting as an oxidizing agent.

b. Cobalt changes from an oxidation number of 0 in Co(s) to +2 in $CoCl_2$, losing electrons and acting as a reducing agent. Chlorine changes from an oxidation number of 0 in Cl_2 to -1 in $CoCl_2$, gaining electrons and acting as an oxidizing agent.

13.47 a. Al: changes from oxidation number 0 to +3; Al is the reducing agent.
F: changes from oxidation number 0 to -1; Fe_2 is the oxidizing agent.

b. Hg: changes from oxidation state +2 to 0; Hg^{2+} is the oxidizing agent.
N: changes from oxidation state +3 to +5; NO_2^- is the reducing agent.

13.49 $Ni(s) \rightarrow Ni^{2+} + 2e^-$ (oxidation); $Cu^{2+} + 2e^- \rightarrow Cu(s)$ (reduction)

13.51 a. No reaction occurs; Pb is below Mg in the activity series (Table 13.2).
b. Reaction occurs; Zn is above Ag in the activity series (Table 13.2).
c. Reaction occurs; Cr is above H_2 in the activity series.
d. No reaction occurs; Au is below H_2 in the activity series.

13.53 a. Sulfite (+4 sulfur) can only be oxidized to the last higher oxidation state of +6 (sulfate ion). Permanganate ion (+7 manganese) can be reduced, in theory, to +4, +2, and 0 (metal) oxidation states, but in acid solution it tends to be reduced to Mn^{2+}.

b. Copper metal can be oxidized, in theory, to the +1 or +2 oxidation states, but in nitric acid solution it tends to be oxidized to Cu^{2+}. Nitrate ion (+5 nitrogen) can be reduced in theory to the +4, +3, +2, +1, or 0 (N_2) oxidation states, but in dilute acid it tends to be reduced to +2 as NO(g).

13.55 a. The oxidizing agent is HNO_3, and the reducing agent is HI. Note that I and N atoms change oxidation numbers. Balancing I and N gives

-3 (from HNO_3 to NO)

$$\overset{-1}{2HI} + \overset{+5}{HNO_3} \rightarrow \overset{0}{I_2} + \overset{+2}{NO}$$

+2 (from 2HI to I_2)

Multiply HNO_3 and NO by 2, and 2HI and I_2 by 3, so that the absolute values of the changes in oxidation number are equal:

$$\overset{-1}{6HI} + 2H\overset{+5}{N}O_3 \rightarrow 3\overset{0}{I_2} + 2\overset{+2}{N}O$$

(-6 from +5 to +2; +6 from -1 to 0)

Balance the O atoms by writing $4H_2O$ on the right, which also balances the H atoms.

$$6HI + 2HNO_3 \rightarrow 3I_2 + 2NO + 4H_2O$$

b. The oxidizing agent is H_2SO_4, and the reducing agent is Ag. Note that the Ag and the S change oxidation numbers. Balancing the S and the Ag gives

$$2\overset{0}{Ag} + 2H_2\overset{+6}{S}O_4 \rightarrow \overset{+1}{Ag_2}SO_4 + \overset{+4}{S}O_2$$

(-2 from +6 to +4; +2 from 0 to +1)

Since the absolute changes in oxidation numbers are the same, balance the oxygens.

$$2Ag + 2H_2SO_4 \rightarrow Ag_2SO_4 + SO_2 + 2H_2O$$

c. The oxidizing agent is $KMnO_4$ and the reducing agent is $MnCl_2$. The Mn in both is changed to the Mn in MnO_2. Balancing Mn's and writing the changes in oxidation numbers gives

$$\overset{+2}{Mn}Cl_2 + K\overset{+7}{Mn}O_4 + KOH \rightarrow 2\overset{+4}{Mn}O_2 + KCl$$

(-3 from +7 to +4; +2 from +2 to +4)

Multiplying each Mn species so that the absolute values of the changes in oxidation numbers are equal gives

$$3\overset{+2}{Mn}Cl_2 + 2K\overset{+7}{Mn}O_4 + KOH \rightarrow 5\overset{+4}{Mn}O_2 + KCl$$

(-6 from +7 to +4; +6 from +2 to +4)

There are 6Cl's on the left and 1Cl on the right. Balance the Cl with 6KCl. Then balance the K with 4KOH on the left, followed by balancing O and H with $2H_2O$ on the right.

$$3MnCl_2 + 2KMnO_4 + 4KOH \rightarrow 5MnO_2 + 6KCl + 2H_2O$$

d. The oxidizing agent is H_2O_2 and the reducing agent is $Cr(OH)_3$. Balancing O gives these changes in oxidation numbers:

$$\overset{+3}{Cr(OH)_3} + \overset{-1}{H_2O_2} + KOH \rightarrow \overset{+6}{K_2CrO_4} + \overset{-2}{2H_2O}$$

(change for O: −2; change for Cr: +3)

Cross-multiplying by 2 and 3 so that the absolute values of the changes in oxidation number are equal gives

$$\overset{+3}{2Cr(OH)_3} + \overset{-1}{3H_2O_2} + KOH \rightarrow \overset{+6}{2K_2CrO_4} + \overset{-2}{6H_2O}$$

(change for O: −6; change for Cr: +6)

Balance the 4K on the right with 4KOH; finish by writing $8H_2O$ on the right.

$$2Cr(OH)_3 + 3H_2O_2 + 4KOH \rightarrow 2K_2CrO_4 + 8H_2O$$

e. $HClO_3$ is both the oxidizing agent and the reducing agent. Balance the Cl by writing $2HClO_3$. Now write the change in oxidation number for one $HClO_3$ forming $HClO_4$ and one $HClO_3$ forming ClO_2:

$$\overset{+5}{2HClO_3} \rightarrow \overset{+7}{HClO_4} + \overset{+4}{ClO_2}$$

(change to ClO_2: −1; change to $HClO_4$: +2)

Adjust coefficients so the absolute values of changes in oxidation number are the same:

$$\overset{+5}{3HClO_3} \rightarrow \overset{+7}{HClO_4} + \overset{+4}{2ClO_2}$$

(change to ClO_2: −2; change to $HClO_4$: +2)

Add H_2O on the right to balance the oxygens and the hydrogens:

$$3HClO_3 \rightarrow HClO_4 + 2ClO_2 + H_2O$$

13.57 a. The oxidizing agent is H_3AsO_4 and the reducing agent is Zn. Using oxidation numbers, the total absolute change in oxidation numbers is 8; using the half-reaction method, the overall number of electrons involved is 8. The balanced equation is:

$$H_3AsO_4 + 4Zn + 8HNO_3 \rightarrow AsH_3 + 4Zn(NO_3)_2 + 4H_2O$$

b. The oxidizing agent is O_2 and the reducing agent is $SnCl_2$. Using oxidation numbers, the total absolute change in oxidation numbers is 4; using the half-reaction method, the overall number of electrons involved is 4. The balanced equation is:

$$2SnCl_2 + O_2 + 8HCl \rightarrow 2H_2SnCl_6 + 2H_2O$$

c. The oxidizing agent and the reducing agent is K_2MnO_4 (it reacts with itself). Using oxidation numbers, the total absolute change in oxidation numbers is 2; using the half-reaction method, the overall number of electrons involved is 2. The balanced equation is:

$$3K_2MnO_4 + 2H_2O \rightarrow MnO_2 + 2KMnO_4 + 4KOH$$

d. The oxidizing agent is H_2SO_4 and the reducing agent is HI. Using oxidation numbers, the total absolute change in oxidation numbers is 8; using the half-reaction method, the overall number of electrons involved is 8. The balanced equation is:

$$H_2SO_4 + 8HI \rightarrow H_2S + 4I_2 + 4H_2O$$

e. The oxidizing agent is PbO_2 and the reducing agent is $MnSO_4$. Using oxidation numbers, the total absolute change in oxidation numbers is 10; using the half-reaction method, the overall number of electrons involved is 10. The balanced equation is:

$$2MnSO_4 + 5PbO_2 + 3H_2SO_4 \rightarrow 2HMnO_4 + 5PbSO_4 + 2H_2O$$

13.59 The half-reaction method will be used. The balanced half-reactions of the oxidizing agent and reducing agent will be given as obtained by this method. The next equation will be the sum of the half-reactions after each has been adjusted to have the same number of electrons (the electrons will not appear). If this is not the ultimate (final) equation, it will be labeled the penultimate equation, and it will then be simplified to the ultimate (final) equation.

a. $Cr_2O_7^{2-} + 14H^+ + 6e^- \rightarrow 2Cr^{3+} + 7H_2O$

$C_2O_4^{2-} \rightarrow 2CO_2 + 2e^-$

Ultimate (final) equation—no simplification needed:

$$Cr_2O_7^{2-} + 3C_2O_4^{2-} + 14H^+ \rightarrow 2Cr^{3+} + 6CO_2 + 7H_2O$$

b. $NO_3^- + 4H^+ + 3e^- \rightarrow NO + 2H_2O$

$Cu \rightarrow Cu^{2+} + 2e^-$

Ultimate (final) equation—no simplification needed:

$$2NO_3^- + 8H^+ + 3Cu \rightarrow 2NO + 3Cu^{2+} + 4H_2O$$

c. $MnO_2 + 4H^+ + 2e^- \rightarrow Mn^{2+} + 2H_2O$

$HNO_2 + H_2O \rightarrow NO_3^- + 3H^+ + 2e^-$

Penultimate equation:

$MnO_2 + HNO_2 + H_2O + 4H^+ \rightarrow Mn^{2+} + NO_3^- + 2H_2O + 3H^+$

Ultimate (final) equation:

$MnO_2 + HNO_2 + H^+ \rightarrow Mn^{2+} + NO_3^- + H_2O$

d. $PbO_2 + SO_4^{2-} + 4H^+ + 2e^- \rightarrow PbSO_4 + 2H_2O$

$Mn^{2+} + 4H_2O \rightarrow MnO_4^- + 8H^+ + 5e^-$

Penultimate equation:

$5PbO_2 + 2Mn^{2+} + 5SO_4^{2-} + 20H^+ + 8H_2O \rightarrow 5PbSO_4 + 2MnO_4^- + 16H^+ + 10H_2O$

Ultimate (final) equation:

$5PbO_2 + 2Mn^{2+} + 5SO_4^{2-} + 4H^+ \rightarrow 5PbSO_4 + 2MnO_4^- + 2H_2O$

e. $Cr_2O_7^{2-} + 14H^+ + 6e^- \rightarrow 2Cr^{3+} + 7H_2O$

$HNO_2 + H_2O \rightarrow NO_3^- + 3H^+ + 2e^-$

Penultimate equation:

$Cr_2O_7^{2-} + 3HNO_2 + 14H^+ + 3H_2O \rightarrow 2Cr^{3+} + 3NO_3^- + 7H_2O + 9H^+$

Ultimate (final) equation:

$Cr_2O_7^{2-} + 3HNO_2 + 5H^+ \rightarrow 2Cr^{3+} + 3NO_3^- + 4H_2O$

13.61 The balanced half-reactions of the oxidizing agent and reducing agent will be given as obtained by the half-reaction method, as if the reaction occurred in acid solution. The next equation will be the sum of the half-reactions after adjusting the number of electrons to be equal (the electrons will not appear). If this is not the ultimate (final) equation, it will be labeled the penultimate equation, which will be simplified to the ultimate (final) equation. Then OH^- ions will be added to both sides of the equation so that OH^- will react with the H^+ to give H_2O.

a. $H_2O_2 + 2H^+ + 2e^- \rightarrow 2H_2O$

$Mn^{2+} + 2H_2O \rightarrow MnO_2 + 4H^+ + 2e^-$

Penultimate equation:

$H_2O_2 + Mn^{2+} + 2H^+ + 2H_2O \rightarrow MnO_2 + 2H_2O + 4H^+$

Ultimate (final) equation (in acid):

$H_2O_2 + Mn^{2+} \rightarrow MnO_2 + 2H^+$

Ultimate (final) equation (after adding $2OH^-$ to both sides):

$H_2O_2 + Mn^{2+} + 2OH^- \rightarrow MnO_2 + 2H_2O$

b. $MnO_4^- + 4H^+ + 3e^- \rightarrow MnO_2 + 2H_2O$

$NO_2^- + H_2O \rightarrow NO_3^- + 2H^+ + 2e^-$

Penultimate equation:

$2MnO_4^- + 3NO_2^- + 8H^+ + 3H_2O \rightarrow 2MnO_2 + 3NO_3^- + 4H_2O + 6H^+$

Ultimate equation (in acid):

$2MnO_4^- + 3NO_2^- + 2H^+ \rightarrow 2MnO_2 + 3NO_3^- + H_2O$

Ultimate equation (after adding $2OH^-$ to both sides and canceling H_2O):

$2MnO_4^- + 3NO_2^- + H_2O \rightarrow 2MnO_2 + 3NO_3^- + 2OH^-$

c. $ClO_3^- + 2H^+ + e^- \rightarrow ClO_2 + H_2O$

$Mn^{2+} + 2H_2O \rightarrow MnO_2 + 4H^+ + 2e^-$

Penultimate equation:

$2ClO_3^- + Mn^{2+} + 4H^+ + 2H_2O \rightarrow 2ClO_2 + MnO_2 + 2H_2O + 4H^+$

Ultimate equation (same in acid or in base):

$2ClO_3^- + Mn^{2+} \rightarrow 2ClO_2 + MnO_2$

d. $MnO_4^- + 4H^+ + 3e^- \rightarrow MnO_2 + 2H_2O$

$NO_2 + H_2O \rightarrow NO_3^- + 2H^+ + e^-$

Penultimate equation:

$MnO_4^- + 3NO_2 + 4H^+ + 3H_2O \rightarrow MnO_2 + 3NO_3^- + 2H_2O + 6H^+$

Ultimate equation (in acid):

$MnO_4^- + 3NO_2 + H_2O \rightarrow MnO_2 + 3NO_3^- + 2H^+$

Ultimate equation (after adding $2OH^-$ to both sides and canceling H_2O):

$$MnO_4^- + 3NO_2 + 2OH^- \rightarrow MnO_2 + 3NO_3^- + H_2O$$

e. $Cl_2 + 2e^- \rightarrow 2Cl^-$

$Cl_2 + 6H_2O \rightarrow 2ClO_3^- + 12H^+ + 10e^-$

Ultimate equation (in acid after dividing coefficients by 2):

$$3Cl_2 + 3H_2O \rightarrow 5Cl^- + ClO_3^- + 6H^+$$

Ultimate equation (after adding $6OH^-$ to both sides and canceling $3H_2O$):

$$3Cl_2 + 6OH^- \rightarrow 5Cl^- + ClO_3^- + 3H_2O$$

13.63 The half-reaction method will be used. The balanced half-reactions of the oxidizing agent and reducing agent will be given as obtained by this method. The next equation will be the sum of the half-reactions after each has been adjusted to have the same number of electrons (the electrons will not appear). If this is not the ultimate (final) equation, it will be labeled the penultimate equation, and it will then be simplified to the ultimate (final) equation.

a. $NO_3^- + 2H^+ + e^- \rightarrow NO_2 + H_2O$

$8H_2S \rightarrow S_8 + 16H^+ + 16e^-$

Penultimate equation:

$$16NO_3^- + 8H_2S + 32H^+ \rightarrow 16NO_2 + S_8 + 16H_2O + 16H^+$$

Ultimate (final) equation:

$$16NO_3^- + 8H_2S + 16H^+ \rightarrow 16NO_2 + S_8 + 16H_2O$$

b. $NO_3^- + 4H^+ + 3e^- \rightarrow NO + 2H_2O$

$Cu \rightarrow Cu^{2+} + 2e^-$

Ultimate (final) equation—no simplification needed:

$$2NO_3^- + 8H^+ + 3Cu \rightarrow 2NO + 3Cu^{2+} + 4H_2O$$

c. $MnO_4^- + 8H^+ + 5e^- \rightarrow Mn^{2+} + 4H_2O$

$SO_2 + 2H_2O \rightarrow SO_4^{2-} + 4H^+ + 2e^-$

Penultimate equation:

$$2MnO_4^- + 5SO_2 + 16H^+ + 10H_2O \rightarrow 2Mn^{2+} + 5SO_4^{2-} + 8H_2O + 20H^+$$

Ultimate equation:

$$2MnO_4^- + 5SO_2 + 2H_2O \rightarrow 2Mn^{2+} + 5SO_4^{2-} + 4H^+$$

d. $Bi(OH)_3 + 3e^- \rightarrow Bi + 3OH^-$

$Sn(OH)_3^- + 3OH^- \rightarrow Sn(OH)_6^{2-} + 2e^-$

Penultimate equation:

$$2Bi(OH)_3 + 3Sn(OH)_3^- + 9OH^- \rightarrow 2Bi + 3Sn(OH)_6^{2-} + 6OH^-$$

Ultimate equation:

$$2Bi(OH)_3 + 3Sn(OH)_3^- + 3OH^- \rightarrow 2Bi + 3Sn(OH)_6^{2-}$$

13.65 The balanced half-reactions of the oxidizing agent and reducing agent will be given as obtained by the half-reaction method, as if the reaction occurred in acid solution. The next equation will be the sum of the half-reactions after adjusting the number of electrons to be equal (the electrons will not appear). If this is not the ultimate (final) equation, it will be labeled the penultimate equation, which will be simplified to the ultimate (final) equation. For basic solution, OH^- will be added to both sides of the equation to react with the H^+ to give H_2O.

a. $MnO_4^- + 4H^+ + 3e^- \rightarrow MnO_2 + 2H_2O$

$I^- + 3H_2O \rightarrow IO_3^- + 6H^+ + 6e^-$

Ultimate equation (in acid):

$$2MnO_4^- + I^- + 2H^+ \rightarrow 2MnO_2 + IO_3^- + H_2O$$

Ultimate equation (after adding $8OH^-$ and canceling $4H_2O$):

$$2MnO_4^- + I^- + H_2O \rightarrow 2MnO_2 + IO_3^- + 2OH^-$$

b. $Cr_2O_7^{2-} + 14H^+ + 6e^- \rightarrow 2Cr^{3+} + 7H_2O$

$2Cl^- \rightarrow Cl_2 + 2e^-$

Ultimate equation:

$$Cr_2O_7^{2-} + 6Cl^- + 14H^+ \rightarrow 2Cr^{3+} + 3Cl_2 + 7H_2O$$

c. $NO_3^- + 4H^+ + 3e^- \rightarrow NO + 2H_2O$

$S_8 + 16H_2O \rightarrow 8SO_2 + 32H^+ + 32e^-$

Penultimate equation:

$$32NO_3^- + 3S_8 + 128H^+ + 48H_2O \rightarrow 32NO + 24SO_2 + 64H_2O + 96H^+$$

Ultimate (final) equation:

$32NO_3^- + 3S_8 + 32H^+ \rightarrow 32NO + 24SO_2 + 16H_2O$

d. $MnO_4^- + 4H^+ + 3e^- \rightarrow MnO_2 + 2H_2O$

$H_2O_2 \rightarrow O_2 + 2H^+ + 2e^-$

Penultimate equation:

$2MnO_4^- + 3H_2O_2 + 2H^+ \rightarrow 2MnO_2 + 3O_2 + 4H_2O$

Ultimate equation:

$2MnO_4^- + 3H_2O_2 \rightarrow 2MnO_2 + 3O_2 + 2H_2O + 2OH^-$

e. $2NO_3^- + 12H^+ + 10e^- \rightarrow N_2 + 6H_2O$

$Zn \rightarrow Zn^{2+} + 2e^-$

Ultimate equation:

$2NO_3^- + 5Zn + 12H^+ \rightarrow N_2 + 5Zn^{2+} + 6H_2O$

13.67 In each case, divide the molar mass by number of moles of H^+ or OH^- that react per mole of substance.

a. $\text{Equiv mass} = \frac{98.00 \text{ g } H_3PO_4}{1 \text{ mol } H_3PO_4} \times \frac{1 \text{ mol } H_3PO_4}{3 \text{ eq } H_3PO_4} = 32.6\underline{66} = 32.67 \text{ g/eq}$

b. Equiv mass = 23.95 g/eq (1 mol OH^- per mol LiOH, so same value as molar mass)

c. $\text{Equiv mass} = \frac{58.34 \text{ g } Mg(OH)_2}{1 \text{ mol } Mg(OH)_2} \times \frac{1 \text{ mol } Mg(OH)_2}{2 \text{ eq } Mg(OH)_2} = 29.17 \text{ g/eq}$

d. Equiv mass = 60.05 g/eq (1 mol H^+ per mol $HC_2H_3O_2$, so same value as molar mass)

13.69 Use the equivalent mass of 98.08/2, or 49.04 g/eq, for H_2SO_4, to convert the mass of 1.68 g to equivalents. Divide by the volume of 0.145 L (145 mL) to find normality.

$$N = \frac{1.68 \text{ g } H_2SO_4}{0.145 \text{ L}} \times \frac{1 \text{ eq } H_2SO_4}{49.04 \text{ g } H_2SO_4} = 0.23\underline{6}2 = 0.236 \text{ eq/L}$$

13.71 Use the equivalent mass of 171.38/2, or 85.69 g/eq, for $Ba(OH)_2$, to convert the mass of 2.56 g to equivalents. Divide by the volume of 0.135 L (135 mL) to find normality.

$$N = \frac{2.56 \text{ g } Ba(OH)_2}{0.135 \text{ L}} \times \frac{1 \text{ eq } Ba(OH)_2}{85.69 \text{ g } Ba(OH)_2} = 0.22\underline{1}2 = 0.221 \text{ eq/L}$$

13.73 By definition, equivalents of $Ba(OH)_2$ = equivalents of HCl, so first find the equivalents of HCl using N x V. Divide equivalents by the volume of 0.0250 L (25.0 mL) of $Ba(OH)_2$ to find normality. Then convert normality to molarity using 2.000 eq/mol.

$$\text{Eq } Ba(OH)_2 = \text{eq HCl} = 0.150 \text{ eq/L} \times 0.0453 \text{ L} = 0.0067\underline{9}5 \text{ eq}$$

$$N = \frac{0.006795 \text{ eq } Ba(OH)_2}{0.0250 \text{ L}} = 0.27\underline{1}8 = 0.272 \text{ eq/L}$$

$$M = \frac{0.006795 \text{ eq } Ba(OH)_2}{0.0250 \text{ L}} \times \frac{1 \text{ mol } Ba(OH)_2}{2 \text{ eq } Ba(OH)_2} = 0.13\underline{5}9 = 0.136 \text{ mol/L}$$

13.75 Equivalents of $Mg(OH)_2$ = equivalents of HCl, so find the equivalents of HCl using N x V. Multiply equivalents of $Mg(OH)_2$ by 58.34/2, or 29.17 g/eq, to find the mass of $Mg(OH)_2$.

$$\text{Eq } Mg(OH)_2 = \text{eq HCl} = 0.2056 \text{ eq/L} \times 0.0391 \text{ L} = 0.0080\underline{3}8 \text{ eq}$$

$$\text{Mass } Mg(OH)_2 = 0.008038 \text{ eq } Mg(OH)_2 \times \frac{29.17 \text{ g } Mg(OH)_2}{1 \text{ eq } Mg(OH)_2} = 0.23\underline{4}4 = 0.234 \text{ g}$$

13.77 Divide each molar mass by the number of electrons in the half-reaction in Table 13.5.

a. $KMnO_4$ ($5e^-$): 158.0 g/mol x 1 mol/5 eq = $31.6\underline{0}0$ = 31.60 g/eq

b. $K_2Cr_2O_7$ ($6e^-$): 294.2 g/mol x 1 mol/6 eq = $49.0\underline{3}3$ = 49.03 g/eq

c. $SnCl_2$ ($2e^-$): 189.6 g/mol x 1 mol/2eq = $94.8\underline{0}0$ = 94.80 g/eq

d. Na_2SO_3 ($2e^-$): 126.04 g/mol x 1 mol/2eq = $63.02\underline{0}0$ = 63.020 g/eq

13.79 Use the equivalent mass of 158.0/5, or 31.60 g/eq, for $KMnO_4$, to convert the 6.58 g mass to equivalents. Divide by the volume of 0.125 L (125 mL) to find normality.

$$N = \frac{6.58 \text{ g } KMnO_4}{0.125 \text{ L}} \times \frac{1 \text{ eq } KMnO_4}{31.60 \text{ g } KMnO_4} = 1.6\underline{6}6 = 1.67 \text{ eq/L}$$

13.81 Equivalents of $FeCl_2$ = equivalents of $KMnO_4$, so find the equivalents of $KMnO_4$ using N x V. Divide equivalents by the volume of 0.0250 L (25.0 mL) of $FeCl_2$ to find N and convert to M.

$$\text{Eq } FeCl_2 = \text{eq } KMnO_4 = 0.150 \text{ eq/L} \times 0.0392 \text{ L} = 0.00658\underline{8}0 \text{ eq}$$

$$N = \frac{0.005880 \text{ eq } FeCl_2}{0.0250 \text{ L}} = 0.23\underline{5}2 = 0.235 \text{ eq/L}$$

$$M = \frac{0.005880 \text{ eq } FeCl_2}{0.0250 \text{ L}} \times \frac{1 \text{ mol } FeCl_2}{1 \text{ eq } FeCl_2} = 0.23\underline{5}2 = 0.235 \text{ mol/L}$$

13.83 Equivalents of H_2O_2 = equivalents of $KMnO_4$, so find the eq of $KMnO_4$ using N x V. Multiply the eq of H_2O_2 by the equivalent mass of 34.04/2, or 17.01 g/eq, of H_2O_2 to convert to mass. Divide by the mass of the solution (12.5).

$$\text{Eq } H_2O_2 = \text{eq } KMnO_4 = 0.5045 \text{ eq/L} \times 0.0393 \text{ L} = 0.019\underline{8}2 \text{ eq}$$

$$\text{Mass } H_2O_2 = 17.01 \text{ g/eq} \times 0.01982 \text{ eq} = 0.33\underline{7}1 \text{ g}$$

$$\text{Mass \%} = \frac{0.3371 \text{ g } H_2O_2}{12.5 \text{ g}} \times 100\% = 2.6\underline{9}6 = 2.70\%$$

13.85 a. BaO is a base: $BaO + H_2O \rightarrow Ba^{2+} + 2OH^-$

b. H_2S is an acid: $H_2S + H_2O \rightarrow H_3O^+ + HS^-$

c. CH_3NH_2 is a base: $CH_3NH_2 + H_2O \rightarrow CH_3NH_3^+ + OH^-$

d. SO_2 is an acid: $SO_2 + 2H_2O \rightarrow H_3O^+ + HSO_3^-$

13.87 a. $H_2O_2(aq) + S^{2-}(aq) \rightarrow HO_2^-(aq) + HS^-(aq)$

b. $HCO_3^-(aq) + OH^-(aq) \rightarrow CO_3^{2-}(aq) + H_2O(l)$

c. $NH_4^+(aq) + CN^-(aq) \rightarrow NH_3(aq) + HCN(aq)$

d. $H_2PO_4^-(aq) + OH^-(aq) \rightarrow HPO_4^{2-}(aq) + H_2O(aq)$

13.89 The order is: $H_2S < H_2Se < HBr$. H_2Se is stronger than H_2S because, within a group, acid strength increases with increasing size of the central atom in binary acids. HBr is a strong acid, whereas the others are weak acids.

13.91 a. The ClO^- ion is a Bronsted base and water is a Bronsted acid. The complete chemical equation is: $ClO^-(aq) + H_2O(l) \rightleftharpoons HClO(aq) + OH^-(aq)$. The equilibrium does not favor the products because ClO^- is a weaker base than OH^-. In Lewis language, a proton from H_2O acts as a Lewis acid by sharing a pair of electrons on the oxygen of ClO^-.

$$H^+ + [:\ddot{\underset{..}{Cl}}:\ddot{\underset{..}{O}}:]^- \rightarrow H:\ddot{\underset{..}{Cl}}:\ddot{\underset{..}{O}}:$$

b. The NH_2^- ion is a Bronsted base and NH_4^+ is a Bronsted acid. The complete chemical equation is: $NH_4^+ + NH_2^- \rightarrow 2NH_3$. The equilibrium favors the products because the reactants form the solvent, a weakly ionized molecule. In Lewis language, the proton from NH_4^+ acts as a Lewis acid by sharing a pair of electrons on the nitrogen of NH_2^-.

$$H^+ + \left[\begin{matrix} :\ddot{\underset{..}{N}}:H \\ H \end{matrix}\right]^- \rightarrow \begin{matrix} H:\ddot{\underset{..}{N}}:H \\ H \end{matrix}$$

13.93 a. This is an oxidation-reduction reaction. The Cl_2 causes the oxidation of Br^- in HBr and is the oxidizing agent, and the HBr causes the reduction of the Cl_2 and is the reducing agent.

b. This is an acid-base reaction. HBr is the acid and $Ca(OH)_2$ is the base.

c. This is an acid-base reaction. $NaHSO_4$ is the acid and NaCN is the base.

d. This is an oxidation-reduction reaction. NaClO is both the oxidizing and reducing agent.

13.95 The balanced half-reactions of the oxidizing agent and reducing agent will be given as obtained by the half-reaction method. The next equation will be the sum of the half-reactions after adjusting the number of electrons to be equal. If this is not the ultimate (final) equation, it will be labeled the penultimate equation, which will be simplified to the ultimate equation. For basic solution, OH^- will be added to both sides of the equation to react with the H^+ to give H_2O.

a. $MnO_4^- + 4H^+ + 3e^- \rightarrow MnO_2 + 2H_2O$

$8S^{2-} \rightarrow S_8 + 16e^-$

Ultimate (final) equation: (in acid)

$16MnO_4^- + 24S^{2-} + 64H^+ \rightarrow 16MnO_2 + 3S_8 + 32H_2O$

Ultimate (final) equation (after adding $64OH^-$ to both sides and canceling $32H_2O$):

$16MnO_4^- + 24S^{2-} + 32H_2O \rightarrow 16MnO_2 + 3S_8 + 64OH^-$

b. $IO_3^- + 6H^+ + 6e^- \rightarrow I^- + 3H_2O$

$HSO_3^- + H_2O \rightarrow SO_4^{2-} + 3H^+ + 2e^-$

Penultimate equation:

$IO_3^- + 3HSO_3^- + 6H^+ + 3H_2O \rightarrow I^- + 3SO_4^{2-} + 3H_2O + 9H^+$

Ultimate equation:

$IO_3^- + 3HSO_3^- \rightarrow I^- + 3SO_4^{2-} + 3H^+$

c. $CrO_4^{2-} + 4H^+ + 3e^- \rightarrow Cr(OH)_4^-$

$Fe(OH)_2 + OH^- \rightarrow Fe(OH)_3 + e^-$

Penultimate equation:

$CrO_4^{2-} + 3Fe(OH)_2 + 4H^+ + 3OH^- \rightarrow Cr(OH)_4^- + 3Fe(OH)_3$

Ultimate equation: (in acid)

$CrO_4^{2-} + 3Fe(OH)_2 + H^+ + 3H_2O \rightarrow Cr(OH)_4^- + 3Fe(OH)_3$

Ultimate equation (after adding OH^- to both sides and combining H_2O's):

$CrO_4^{2-} + 3Fe(OH)_2 + 4H_2O \rightarrow Cr(OH)_4^- + 3Fe(OH)_3 + OH^-$

d. $Cl_2 + 2e^- \rightarrow 2Cl^-$

$Cl_2 + 2H_2O \rightarrow 2ClO^- + 4H^+ + 2e^-$

Ultimate equation (in acid after dividing coefficients by 2):

$Cl_2 + H_2O \rightarrow Cl^- + ClO^- + 2H^+$

Ultimate equation (after adding $2OH^-$ to both sides and canceling H_2O):

$Cl_2 + 2OH^- \rightarrow Cl^- + ClO^- + H_2O$

13.97 The half-reaction method can be used to find the electron change for each half-reaction in acid solution:

$O_2 + 4H^+ + 4e^- \rightarrow 2H_2O$

$Fe(OH)_2 + H_2O \rightarrow Fe(OH)_3 + H^+ + e^-$

Penultimate equation:

$O_2 + 4Fe(OH)_2 + 4H^+ + 4H_2O \rightarrow 2H_2O + 4Fe(OH)_3 + 4H^+$

Ultimate equation (for either acid or base):

$O_2 + 4Fe(OH)_2 + 2H_2O \rightarrow 4Fe(OH)_3$

13.99 Since there is a two-electron oxidation of arsenic(III) to arsenic(V), 2 moles of e^- are lost for each mole of H_3AsO_3. Thus the equivalent mass of H_3AsO_3 is the molar mass divided by 2: 125.94/2 = 62.97 g/eq. The number of equivalents of H_3AsO_3 is:

$0.840 \text{ g } H_3AsO_3 \times (1 \text{ eq } H_3AsO_3/62.97 \text{ g } H_3AsO_3) = 0.01334 \text{ eq } H_3AsO_3$

At the endpoint, eq H_3AsO_3 = eq I_2; the normality of the 25.4 ml (0.0254 L) of I_2 is:

$$N = \frac{\text{eq } I_2}{\text{L } I_2} = \frac{0.01334 \text{ eq } I_2}{0.0254 \text{ L}} = 0.52\underline{5}1 = 0.525 \text{ eq/L}$$

Cumulative-Skills Problems (require skills from previous chapters)

13.101 For $(HO)_mYO_n$ acids, acid strength increases with n, regardless of the number of OH's. The structure of H_3PO_4 is $(HO)_3PO$; since H_3PO_3 and H_3PO_4 have the same acidity, H_3PO_3 must also have n = 1; thus m = 2. This leaves one H, which must bond to phosphorus, giving a structure of $(HO)_2(O)PH$. Assuming that only 2H's react with NaOH, the mass of NaOH that reacts with 1.00 g of H_3PO_3(PA) is calculated:

$$1.00 \text{ g PA} \times \frac{1 \text{ mol PA}}{81.994 \text{ g PA}} \times \frac{2 \text{ mol NaOH}}{1 \text{ mol PA}} \times \frac{40.00 \text{ g NaOH}}{1 \text{ mol NaOH}} = 0.97\underline{5}6 = 0.976 \text{ g NaOH}$$

13.103 BF_3 acts as a Lewis acid, accepting an electron pair from NH_3: $BF_3 + :NH_3 \rightarrow F_3B:NH_3$. The NH_3 acts as a Lewis base in donating an electron pair to BF_3. When 10.0 g of each are mixed, the BF_3 is the limiting reagent since it has the higher formula weight. The mass of $BF_3:NH_3$ formed is:

$$10.00 \text{ g } BF_3 \times \frac{1 \text{ mol } BF_3}{67.81 \text{ g } BF_3} \times \frac{1 \text{ mol } BF_3:NH_3}{1 \text{ mol } BF_3} \times \frac{84.84 \text{ g } BF_3:NH_3}{1 \text{ mol } BF_3:NH_3} = 12.\underline{5}1 = 12.5 \text{ g}$$

13.105 Since two moles of $Na_2S_2O_3$ lose two electrons (to form one mole $Na_2S_4O_6$), one mole loses one electron and the normality = the molarity = 0.1381 N. The equivalents of $Na_2S_2O_3$ in the 38.2 mL (0.0382 L) are:

$$0.0382 \text{ L } Na_2S_2O_3 \times \frac{0.1381 \text{ eq } Na_2S_2O_3}{\text{L}} = 0.0052\underline{7}54 \text{ eq } Na_2S_2O_3$$

By definition, eq $Na_2S_2O_3$ = eq I_2 = eq KIO_3. Since one mole of KIO_3 gains a total of six electrons in being reduced first to I_2, then to I^-, the equivalent mass of KIO_3 is 214.02/6, or 35.67 g/eq. Use this to calculate g KIO_3, which is divided by the sample mass of 0.4381 g to find percentage:

$$\% = \frac{0.0052754 \text{ eq } KIO_3}{0.4381 \text{ g}} \times \frac{35.67 \text{ g } KIO_3}{1 \text{ eq } KIO_3} \times 100\% = 42.\underline{9}52 = 43.0\%$$

13.107 Using eq $Mg(OH)_2$ = eq $MgCl_2$, and eq $Al(OH)_3$ = eq $AlCl_3$, calculate the total equivalents of both bases from the volume of 48.5 mL (0.0485 L) of HCl and the normality of 0.187 eq HCl/L. Then write two equations in two unknowns.

$$0.0485 \text{ L HCl} \times 0.187 \text{ eq HCl/L} = 0.0090\underline{6}95 \text{ eq } [Mg(OH)_2 + Al(OH)_3]$$

(1) $0.0090695 - \text{eq } Mg(OH)_2 = \text{eq } Al(OH)_3$

(2) $(95.21 \text{ g } MgCl_2/2 \text{ eq}) \times \text{eq } Mg(OH)_2 + (133.34 \text{ g } AlCl_3/3 \text{ eq}) \times \text{eq } Al(OH)_3 = 0.420 \text{ g}$

Substituting equation 1 into equation 2 for eq $Al(OH)_3$:

$$47.605 \text{ eq } Mg(OH)_2 + 44.447[0.0090695 - \text{eq } Mg(OH)_2] = 0.420$$

$$3.158 \text{ eq } Mg(OH)_2 + 0.4031 = 0.420$$

$$\text{Eq } Mg(OH)_2 = 0.0169 \div 3.158 = 0.0053\underline{5}14 \text{ eq}$$

$$\text{Eq } Al(OH)_3 = 0.0090695 - \text{eq } Mg(OH)_2 = 0.0037181 \text{ eq}$$

$$\text{g } Mg(OH)_2 = 0.0053514 \text{ eq} \times (58.32 \text{ g/2 eq}) = 0.15\underline{6}04 \text{ g } Mg(OH)_2$$

$$\text{g } Al(OH)_3 = 0.0037181 \text{ eq} \times (77.99 \text{ g/3 eq}) = 0.096\underline{6}6 \text{ g } Al(OH)_3$$

Since $Mg(OH)_2$ and $Al(OH)_3$ are the only active ingredients, the $Mg(OH)_2$ percentage is:

$$\% \; Mg(OH)_2 = \frac{0.15604 \text{ g } Mg(OH)_2}{(0.15604 \text{ g } + \; 0.09666 \text{ g})} \times 100\%$$

$$= 61.\underline{7}49 = 61.7\% \text{ of active ingredients}$$

13.109 Rxn: $Ca^{2+}(aq) + 2OH^-(aq) + 2H^+(aq) + 2Cl^-(aq) \rightarrow Ca^{2+}(aq) + 2Cl^-(aq) + 2H_2O(l)$.

ΔH^o = [-(-542.96) - 2(-229.94) - 0 - 2(-167.46) + (-542.96) + 2(-167.46) + 2(-285.84)] kJ

ΔH^o = -111.80 kJ/mol $Ca(OH)_2(aq)$

$$34.5 \text{ kJ} \times \frac{1 \text{ mol } Ca(OH)_2}{111.80 \text{ kJ}} \times \frac{74.10 \text{ g } Ca(OH)_2}{1 \text{ mol } Ca(OH)_2} = 22.\underline{8}6 = 22.9 \text{ g}$$

CHAPTER 14

RATES OF REACTION

SOLUTIONS TO EXERCISES

Note on significant figures: The final answer to all mathematical solutions is given first with one nonsignificant figure (last significant figure underlined) and is then rounded to the correct number of figures. Intermediate answers usually also have at least one nonsignificant figure.

14.1 Rate of formation of NO_2F = $\Delta[NO_2F]/\Delta t$. Rate of reaction of NO_2 = $-\Delta[NO_2F]$. Divide each rate by the coefficient of the corresponding substance:

$$\frac{1}{2}\frac{\Delta[NO_2F]}{\Delta t} = -\frac{1}{2}\frac{\Delta[NO_2]}{\Delta t} \text{ or } \frac{\Delta[NO_2F]}{\Delta t} = -\frac{\Delta[NO_2]}{\Delta t}$$

14.2 $$\text{Rate} = -\frac{\Delta[I^-]}{\Delta t} = -\frac{[0.00101\text{ M} - 0.00169\text{ M}]}{8.00\text{ s} - 2.00\text{ s}} = 1.1\underline{3} \times 10^{-4} = 1.13 \times 10^{-4}\text{ M/s}$$

14.3 The order with respect to CO is zero, and with respect to NO_2 is 2. The overall order = 2, the sum of the exponents in the rate law.

14.4 By comparing experiments 1 and 2, we see that the rate is quadrupled when the $[NO_2]$ is doubled. Thus the reaction is second order in NO_2, and the rate law is:

$$\text{Rate} = k[NO_2]^2$$

The rate constant may be found by substituting experimental values into the rate-law expression. Using values from Experiment 1:

$$k = \frac{\text{Rate}}{[NO_2]^2} = \frac{7.1 \times 10^{-5}\text{ mol/(L}\cdot\text{s)}}{(0.010\text{ mol/L})^2} = 0.7\underline{1}0 = 0.71\text{ L/(mol}\cdot\text{s)}$$

14.5 a. For $[N_2O_5]$ after 6.00×10^2 s, use the first-order rate law and solve for the concentration at time t:

$$\text{Log}\frac{[N_2O_5]_t}{[1.65 \times 10^{-2}\text{ M}]} = \frac{-(4.8 \times 10^{-4}\text{/s})(6.00 \times 10^2\text{ s})}{2.303}$$

$$\text{Log } [N_2O_5]_t = \log(1.65 \times 10^{-2} \text{ M}) - \frac{(4.8 \times 10^{-4}/\text{s})(6.00 \times 10^{2} \text{ s})}{2.303}$$

$$\text{Log } [N_2O_5]_t = -1.78\underline{2}5 - 0.1\underline{2}505 = -1.9\underline{0}76$$

$$[N_2O_5]_t = 1.\underline{2}37 \times 10^{-2} = 0.012 \text{ mol/L}$$

b. Use the first-order rate law and solve for time t. Let the initial concentration be 100.0%, and the concentration at time t be 10.0%.

$$\text{Log } \frac{[10.0\%]}{[100.0\%]} = \frac{-(4.8 \times 10^{-4}/\text{s})\ t}{2.303}$$

$$t = \frac{-(2.303)(\log 0.100)}{4.80 \times 10^{-4}/\text{s}} = 4.7\underline{9}7 \times 10^{3} \text{ s or } 1.33 \text{ hr}$$

14.6 Use the expression relating the half-life and the rate constant to calculate the half-life, $t_{1/2}$.

$$t_{1/2} = \frac{0.693}{k} = \frac{0.693}{9.2/\text{s}} = 0.07\underline{5}3 = 0.075 \text{ s}$$

By definition, the half-life is the amount of time it takes to decrease the amount of substance present by one-half. Thus, it takes $0.07\underline{5}3$ s for concentration to decrease by 50%, and another $0.07\underline{5}3$ s for the concentration to decrease by 50% of the remaining 50% (to 25% left), for a total of $0.15\underline{0}6$, or 0.151 s.

14.7 Solve for E_a by substituting the given values into the two-temperature Arrhenius equation:

$$\text{Log } \frac{2.14 \times 10^{-2}}{1.05 \times 10^{-3}} = \frac{E_a}{2.303 \times 8.31 \text{ J/(K·mol)}} \left[\frac{1}{759 \text{ K}} - \frac{1}{836 \text{ K}}\right]$$

$$1.30\underline{9}20 = \frac{E_a}{19.\underline{1}38 \text{ J/(K·mol)}} \left[1.2\underline{1}35 \times 10^{-4}\right]$$

$$E_a = \frac{1.30920 \times 19.138 \text{ J/(K·mol)}}{1.2135 \times 10^{-4} \text{ K}} = 2.0\underline{6}1 \times 10^{5} = 2.06 \times 10^{5} \text{ J/mol}$$

Solve for the rate constant, k_2, at 865 K by using the same equation, and using $E_a = 2.061 \times 10^5$ J/mol:

$$\text{Log } \frac{k_2}{2.14 \times 10^{-2}/(M^{1/2}·s)} = \frac{2.065 \times 10^{5} \text{ J/mol}}{2.303 \times 8.31 \text{ J/(K·mol)}} \left[\frac{1}{836 \text{ K}} - \frac{1}{865 \text{ K}}\right] = 0.43\underline{2}71$$

Now calculate the antilog of 0.43271:

$$\frac{k_2}{2.14 \times 10^{-2}/(M^{1/2}·s)} = 2.7\underline{0}84$$

$$k_2 = 5.7\underline{9}6 \times 10^{-2} = 5.80 \times 10^{-2}/(M^{1/2} · s)$$

14.8 The net chemical equation is the overall sum of the two elementary reactions:

$$H_2O_2 + I^- \rightarrow H_2O + IO^-$$

$$H_2O_2 + IO^- \rightarrow H_2O + O_2 + I^-$$

$$2H_2O_2 \rightarrow 2H_2O + O_2$$

The IO^- is an intermediate; I^- is a catalyst. Neither of these appears in the net equation.

14.9 The reaction is bimolecular because it is an elementary reaction that involves two molecules.

14.10 For $NO_2 + NO_2 \rightarrow N_2O_4$, the rate law is: rate $= k[NO_2]^2$.
(The rate must be proportional to the concentration of both reactants.)

14.11 The slow step is the rate-determining step. Therefore, the rate law predicted by the mechanism given is:

$$\text{Rate} = k_1 [H_2O_2][I^-]$$

14.12 According to the rate-determining (slow) step, the rate law is:

$$\text{Rate} = k_2[NO_3][NO]$$

Because NO_3 is an intermediate, it cannot appear in the overall equation. We can eliminate it by substituting a mathematical equality for it. Note that $[NO_3]$ appears in the first step, a fast equilibrium step. At equilibrium, the ratio k_1/k_2 is equal to the ratio of product over reactants:

$$\frac{k_1}{k_{-1}} = \frac{[NO_3]}{[NO][O_2]}$$

Rearranging to obtain an equality for $[NO_3]$ gives:

$$[NO_3] = k_1/k_{-1}\ [NO][O_2]$$

Substituting the right-hand side into the rate law for $[NO_3]$ gives:

$$\text{Rate} = k_2(k_1/k_{-1})\ [O_2][NO]^2$$

The $k_2(k_1/k_{-1})$ product is equal to the observed rate constant, k. Note that k_2 is a contribution from step 2, whereas the (k_1/k_{-1}) ratio is a contribution from step 1.

ANSWERS TO REVIEW QUESTIONS

14.1 The four variables are: (1) the concentration of each reactant, (2) the concentration of catalyst, if present, (3) the temperature, and (4) the surface area of any solid reactant or solid catalyst.

14.2 The rate of reaction of HBr with O_2 can be defined as the decrease in HBr concentration or the increase in Br_2 product formed over the time interval Δt:

$$\text{Rate} = -\frac{1}{4}\frac{\Delta[HBr]}{\Delta t} = \frac{1}{2}\frac{\Delta[Br_2]}{\Delta t}$$

14.3 Two physical properties that might be used to determined the rate of a reaction are: (1) color, for reactions involving a colored reactant or product, and (2) pressure, for reactions involving a change in moles of gas. More generally, any type of electromagnetic radiation that is absorbed by a reactant or product may be used; for example, ultraviolet radiation.

14.4 Consider the general rate law for the reaction of A and B to give D and E, with solid catalyst C, written as C(s):

$$\text{Rate} = k[A]^m[[B]^n[C(s)]]^p$$

(The exponents m, n, and p are the orders of the individual reactants and catalyst.) Assuming that m and n are positive numbers, the rate law predicts that increasing the concentrations of either A or B will increase the rate of reaction. In addition, the rate will be increased if the concentration of any soluble catalyst is increased. In this case, the catalyst is a solid and the rate will be increased by increasing the surface area of the catalyst - making it as finely divided as possible. Finally, increasing the temperature will be reflected by an increase in the rate constant, k, and thus the rate of the reaction.

14.5 The text's discussion of reaction order gives for the reaction $2NO + 2H_2 \rightarrow N_2 + 2H_2O$, its experimental rate law as:

$$\text{Rate} = k[NO]^2[H_2]$$

Thus the reaction is only first order in H_2, not second order as might be thought from the overall balanced equation. Thus this exponent in the rate law has no relationship to the coefficient in the balanced equation.

14.6 The rate law for this reaction is

$$\text{Rate} = k[I^-][H_3AsO_4][H^+]$$

The overall order is third order; that is, 1 + 1 + 1 = 3.

14.7 When the rate is quadrupled by doubling the concentration of one reactant, the order with respect to that reactant must be two:

$$\text{New rate} = 2^m = 4/1, \text{ and thus } m = 2$$

14.8 The doubling of the concentration of a reactant whose order = 0.50 (one half-order) will cause an increase in the new rate of 1.41 times the old rate:

$$\text{New rate} = 2^{0.50}(\text{old rate}) = \sqrt{2}\ (\text{old rate}) = 1.41(\text{old rate})$$

14.9 If the half-life for the formation of B(g) and C(g) from A(g) is 25 s, then the time for A(g) to decrease to 1/4 its initial value will be double the half-life (2 x $t_{1/2}$), or 50 s. The time for A(g) to decrease to 1/8 its initial value will be 3 x $t_{1/2}$, or 75 s.

14.10 The order cannot be first order because the half-life of a first-order reaction is constant over the time of the reaction. However, the half-life of a second-order reaction depends inversely on the concentration of reactants and should increase as the concentrations decrease. Thus the reaction must be second order since it cannot be first order.

14.11 For a collision to result in a reaction (1) the colliding reactant molecules must have a total energy greater than the activation energy, and (2) the reactant molecules must collide with the proper orientation to form the activated complex and products.

14.12 The potential energy diagram for the exothermic reaction of A and B to give activated complex $AB^{\ddagger}$ and products C and D is given below.

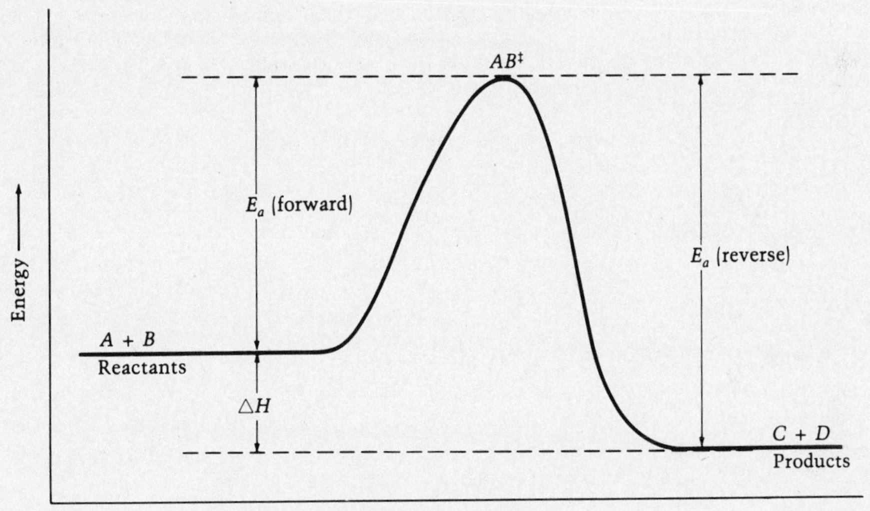

14.13 The activated complex for the reaction of NO_2 with NO_3 to give NO, NO_2, and O_2 has the structure below (dashed lines = bonds about to form or break):

O-N---O---O---N(-O)$_2$

14.14 The Arrhenius equation expressed with the base e is: $k = Ae^{-Ea/RT}$. With the base 10, the Arrhenius equation is: $k = A\,10^{-Ea/(2.303RT)}$. The A term is the frequency factor and is equal to the product of p and Z from collision theory. The term p is the fraction of collisions with reactant molecules properly oriented and Z is the frequency of collisions. Thus A is the number of collisions with the molecules properly oriented. The E_a term is the activation energy, the minimum energy of collision required for two molecules to react. The R term is the gas constant and T is the absolute temperature.

14.15 In the reaction of $NO_2(g)$ with $CO(g)$, an example of an intermediate is the temporary formation of NO_3 from the reaction of two NO_2 molecules in the first step:

$$NO_2 + NO_2 \rightarrow NO_3 + NO$$

$$NO_3 + CO \rightarrow NO_2 + CO_2$$

14.16 It is generally impossible to predict the rate law from the equation alone because most reactions consists of several elementary steps whose combined result is summarized in the rate law. If these elementary steps are unknown, the rate law cannot be predicted.

14.17 The mechanism cannot be $2NO_2Cl \rightarrow 2NO_2 + Cl_2$ because the reaction is first order in NO_2Cl. The order in NO_2Cl would have to reflect the total number of molecules (2) for the proposed mechanism, but it does not.

14.18 The characteristic of the rate-determining step in a mechanism is that it is, relatively speaking, the slowest step of all of the elementary steps (even though it may occur in seconds). Thus the rate of disappearance of reactant(s) is limited by the rate of this step.

14.19 For the rate of decomposition of N_2O_4 : Rate $= k_1[N_2O_4]$

For the rate of formation of N_2O_4: Rate $= k_{-1}[NO_2]^2$

At equilibrium the rates are equal, so:

$$k_1[N_2O_4] = k_{-1}[NO_2]^2$$

$$[N_2O_4] = (k_{-1}/k_1)\,[NO_2]^2$$

14.20 A catalyst operates by providing a pathway (mechanism) that occurs faster than the uncatalyzed pathway (mechanism) of the reaction. The catalyst is not consumed because after reacting in an early step, it is regenerated in a later step.

14.21 In physical adsorption, molecules adhere to a surface through <u>weak</u> intermediate forces, whereas in chemisorption, the molecules adhere to the surface by <u>stronger</u> chemical bonding.

14.22 In the first step of catalytic hydrogenation of ethylene, the ethylene and hydrogen molecules diffuse to the catalyst surface and undergo chemisorption. Then the electrons of ethylene form temporary bonds to the metal catalyst and the hydrogen molecule breaks into two hydrogen atoms. The hydrogen atoms next migrate to an ethylene held in position on the metal catalyst surface, forming ethane. Finally, because it cannot bond to the catalyst, the ethane diffuses away from the surface.

SOLUTIONS TO PRACTICE PROBLEMS

Note on significant figures: The final answer to all mathematical solutions is given first with one nonsignificant figure (last significant figure underlined) and is then rounded to the correct number of digits. Intermediate answers usually also have at least one nonsignificant figure.

14.23 For the reaction $2NO_2 \rightarrow 2NO + O_2$, the rate of decomposition of NO_2 and the rate of formation of O_2 are, respectively:

$$\text{Rate} = -\Delta[NO_2]/\Delta t, \text{ and rate} = \Delta[O_2]/\Delta t$$

To relate the two rates, divide each rate by the coefficient of the corresponding substance in the chemical equation and equate them.

$$-\frac{1}{2}\frac{\Delta[NO_2]}{\Delta t} = \frac{\Delta[O_2]}{\Delta t}$$

14.25 For the reaction $5Br^- + BrO_3^- + 6H^+ \rightarrow 3Br_2 + 3H_2O$, the rate of decomposition of Br^- and the rate of decomposition of BrO_3^- are, respectively:

$$\text{Rate} = -\Delta[Br^-]/\Delta t, \text{ and rate} = -\Delta[BrO_3^-]/\Delta t$$

To relate the two rates, divide each rate by the coefficient of the corresponding substance in the chemical equation and equate them.

$$\frac{1}{5}\frac{\Delta[Br^-]}{\Delta t} = \frac{\Delta[BrO_3^-]}{\Delta t}$$

14.27 $$\text{Rate} = -\frac{\Delta[NH_4NO_2]}{\Delta t} = -\frac{[0.432\text{ M} - 0.500\text{ M}]}{3.00\text{ hr} - 0.00\text{ hr}} = 0.02\underline{2}7 = 0.023\text{ M/hr}$$

14.29 $$\text{Rate} = -\frac{\Delta[\text{Azo.}]}{\Delta t} = -\frac{[0.0129\text{ M} - 0.0150\text{ M}]}{10.00\text{ m} - 0.00\text{ m}} \times \frac{1\text{ m}}{60\text{ s}} = 3.\underline{5}0 \times 10^{-6} = 3.5 \times 10^{-6}\text{ M/s}$$

14.31 If the rate law is rate = $k[H_2S][Cl_2]$, the order with respect to H_2S is one (first order), and the order with respect to Cl_2 is also one (first order). The overall order is 1 + 1 = 2, second order.

14.33 If the rate law is rate = $k[MnO_4^-][H_2C_2O_4]$, the order with respect to MnO_4^- is one (first order), the order with respect to $H_2C_2O_4$ is one (first order), and the order with respect to H^+ is zero. The overall order is 2, second order.

14.35 The reaction rate doubles when the concentration of CH_3NNCH_3 is doubled, so the reaction is first order in azomethane. The rate equation should have the form:

$$\text{Rate} = k[CH_3NNCH_3]$$

Substituting values for the rate and concentration yields a value for k:

$$k = \frac{\text{rate}}{[\text{Azo.}]} = \frac{2.8 \times 10^{-6}\ \text{M/s}}{1.13 \times 10^{-2}\ \text{M}} = 2.\underline{4}7 \times 10^{-4} = 2.5 \times 10^{-4}\ \text{/s}$$

14.37 Doubling [NO] quadruples the rate, so the reaction is second order in NO. Doubling $[H_2]$ doubles the rate, so the reaction is first order in H_2. The rate law should have the form:

$$\text{Rate} = k[\text{NO}]^2[\text{H}_2]$$

Substituting values for the rate and concentration yields a value for k:

$$k = \frac{\text{rate}}{[\text{NO}]^2[\text{H}_2]} = \frac{2.6 \times 10^{-5}\ \text{M/s}}{[6.4 \times 10^{-3}\ \text{M}]^2[2.2 \times 10^{-3}]} = 2.\underline{8}8 \times 10^2 = 2.9 \times 10^2/\text{M}^2\text{s}$$

14.39 By comparing experiments 2 and 1, we see that tripling $[ClO_2]$ increases the rate ninefold; that is, $3^m = 9$, so $m = 2$ (and the reaction is second order in ClO_2). From experiments 3 and 2, we see that tripling $[OH^-]$ triples the rate, so the reaction is first order in OH^-. The rate law is:

$$\text{Rate law} = k[\text{ClO}_2]^2[\text{OH}^-]$$

Substituting values for the rate and concentrations yields a value for k:

$$k = \frac{\text{rate}}{[\text{ClO}_2]^2[\text{OH}^-]} = \frac{0.0248\ \text{M/s}}{[0.060\ \text{M}]^2[0.030\ \text{M}]} = 2.\underline{2}9 \times 10^2 = 2.3 \times 10^2/\text{M}^2\text{s}$$

14.41 Let $[SO_2Cl_2]_o = 0.0248$ M and $[SO_2Cl_2]_t$ = the concentration after 4.5 hr. Substituting these and $k = 2.2 \times 10^{-5}$/s into the first-order rate equation gives:

$$\text{Log}\ \frac{[\text{SO}_2\text{Cl}_2]_t}{[0.0248\ \text{M}]} = \frac{-(2.2 \times 10^{-5}/\text{s})\left(4.5\ \text{h} \times \frac{3600\ \text{s}}{1\ \text{h}}\right)}{2.303} = -0.1\underline{5}47$$

Taking the antilogs of both sides gives:

$$\frac{[\text{SO}_2\text{Cl}_2]_t}{[0.0248\ \text{M}]} = 0.7\underline{0}03$$

$$[\text{SO}_2\text{Cl}_2] = 0.7003 \times [0.0248\ \text{M}] = 1.\underline{7}3 \times 10^{-2} = 1.7 \times 10^{-2}\ \text{M}$$

14.43 First find the rate constant, k, by substituting experimental values into the first-order rate equation. Let $[EtCl]_o = 0.00100$ M, $[EtCl]_t = 0.00067$ M, and t = 155 s. Solving for k yields:

$$k = \frac{-2.303 \log \frac{[0.00067\ \text{M}]_t}{[0.00100\ \text{M}]_o}}{155\ \text{s}} = 2.5\underline{8}4 \times 10^{-3}/\text{s}$$

Now let $[EtCl]_t$ = the concentration after 256 s; $[EtCl]_o$ again = 0.00100 M, and use the value of k of $1.5\underline{6}4 \times 10^{-3}$/s to calculate $[EtCl]_t$.

$$\text{Log}\ \frac{[EtCl]_t}{[0.00100\ M]} = \frac{-(2.584 \times 10^{-3}/s)(256\ s)}{2.303} = -0.28\underline{7}2$$

Converting both sides to antilogs gives:

$$\frac{[EtCl]_t}{[0.00100\ M]} = 0.51\underline{6}1$$

$$[EtCl]_t = 0.5161 \times [0.00100\ M] = 5.1\underline{6}1 \times 10^{-4} = 5.16 \times 10^{-4}\ M$$

14.45 For a first-order reaction, divide 0.693 by the rate constant to find the half-life:

$$t_{1/2} = 0.693/(6.3 \times 10^{-4}/s) = 1.\underline{1}0 \times 10^3 = 1.1 \times 10^3\ s\ (1\underline{8}.3\ m)$$

$$t_{25\%\ left} = t_{1/4\ left} = 2 \times t_{1/2} = 2 \times 1.1 \times 10^3\ s = 2.2 \times 10^3\ s\ (3\underline{6}.6\ m)$$

$$t_{12.5\%\ left} = t_{1/8\ left} = 3 \times t_{1/2} = 3 \times 1.1 \times 10^3\ s = 3.3 \times 10^3\ s\ (55\ m)$$

14.47 For a first-order reaction, divide 0.693 by the rate constant to find the half-life:

$$t_{1/2} = 0.693/(2.0 \times 10^{-6}/s) = 3.\underline{4}65 \times 10^5\ s\ (9\underline{6}.25\ \text{or}\ 96\ hr)$$

$$t_{25\%\ left} = t_{1/4\ left} = 2 \times t_{1/2} = 2 \times 96.25\ hr = 1\underline{9}2.5\ hr$$

$$t_{12.5\%\ left} = t_{1/8\ left} = 3 \times t_{1/2} = 3 \times 96.25\ hr = 2\underline{8}8.75\ hr$$

$$t_{6.25\%\ left} = t_{1/16\ left} = 4 \times t_{1/2} = 4 \times 96.25\ hr = 3\underline{8}5.0\ hr$$

$$t_{3.125\%\ left} = t_{1/32\ left} = 5 \times t_{1/2} = 5 \times 96.25\ hr = 4\underline{8}1.25\ hr$$

14.49 Use the first-order rate equation and solve for time t. Let $[Cr^{3+}]_o$ = 100.0%; then the concentration at time t, $[Cr^{3+}]_t$ = (100.0% - 90.0%, or 10.0%). Use k = $2.0 \times 10^{-6}/s$.

$$\text{Log}\ \frac{[10.0\%]}{[100.0\%]} = \frac{-(2.0 \times 10^{-6}/s)\ t}{2.303}$$

$$t = \frac{-(2.303)(\text{Log}\ 0.100)}{2.0 \times 10^{-6}/s} = 1.\underline{1}51 \times 10^6\ s,\ \text{or}\ 3.2 \times 10^2\ hr$$

14.51 For the first-order plot, follow Figure 14.9 and plot log $[ClO_2]$ vs. the time in seconds. The data used for plotting are:

t, min	log $[ClO_2]$
0.00	-3.321
1.00	-3.365
2.00	-3.408
3.00	-3.452

The plot requires a graph with too many lines to be reproduced here, but it yields a straight line, demonstrating that the reaction is first order in [ClO_2]. The slope of the line may be calculated from the difference between the last point and the first point:

$$\text{Slope} = \frac{[-3.452 - (-3.321)]}{[3.00 - 0.00 \text{ s}]} = \frac{-0.043\underline{6}66}{\text{s}}$$

Just as the slope, m, was obtained for the plot in Figure 14.9, we can also equate m to -k/2.303 and calculate k as follows:

$$k = -2.303\ \underline{m} = -2.303 \times (-0.04366/\text{s}) = 0.10056 = 0.101/\text{s}$$

14.53 The potential-energy diagram is below. Since the activation energy for the forward reaction is +10 kJ and ΔH^o = -200 kJ, the activation energy for the reverse reaction is +210 kJ.

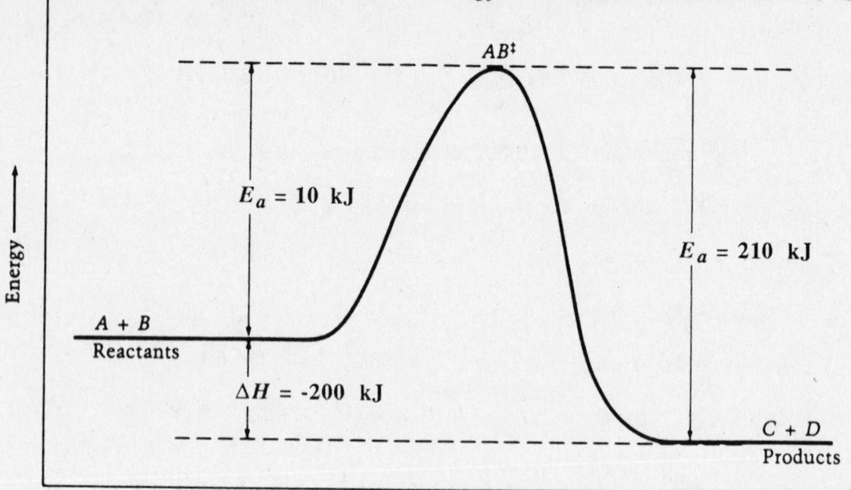

14.55 Solve the two-temperature Arrhenius equation (Section 14.6) for E_a by substituting T_1 = 308 K (from 35°C), $k_1 = 1.4 \times 10^{-4}$/s, T_2 = 318 K (from 45°C), and $k_2 = 5.0 \times 10^{-4}$/s :

$$\text{Log}\ \frac{5.0 \times 10^{-4}}{1.4 \times 10^{-4}} = \frac{E_a}{2.303 \times 8.31\ \text{J/(K•mol)}} \left[\frac{1}{308\ \text{K}} - \frac{1}{318\ \text{K}}\right]$$

Rearranging E_a to the left side and calculating [1/308 - 1/318] gives:

$$E_a = \frac{(2.303 \times 8.31\ \text{J/K}) \times \log\left[\frac{5.0 \times 10^{-4}}{1.4 \times 10^{-4}}\right]}{0.00010209/\text{K}} = 1.\underline{0}37 \times 10^5 = 1.0 \times 10^5\ \text{J/mol}$$

To find the rate at 55°C (328 K), we use the first equation but let k_2 in the numerator be unknown, and solve for k_2:

$$\text{Log}\ \frac{k_2}{1.4 \times 10^{-4}/\text{s}} = \frac{1.037 \times 10^5\ \text{J/mol}}{2.303 \times 8.31\ \text{J/(K•mol)}} \left[\frac{1}{308\ \text{K}} - \frac{1}{328\ \text{K}}\right] = 1.072$$

$$\frac{k_2}{1.4 \times 10^{-4}/s} = 1.\underline{1}80 \times 10^1$$

$$k_2 = 1.180 \times 10^1 \times (1.4 \times 10^{-4}/s) = 1.6\underline{5}2 \times 10^{-3} = 1.7 \times 10^{-3}/s$$

14.57 Since the rate constant is proportional to the rate of a reaction, tripling the rate at 25°C also means that the rate constant at 25°C is tripled. Thus, $k_{35} = 3k_{25}$, and the latter can be substituted for k_{35} in the Arrhenius equation:

$$\text{Log } \frac{3k_{25}}{k_{25}} = \frac{E_a}{2.303 \times 8.31\ \text{J/(K}\cdot\text{mol)}} \left[\frac{1}{298\ \text{K}} - \frac{1}{308\ \text{K}}\right]$$

$$0.4771 = (5.693 \times 10^{-6}\ \text{mol/J})\ E_a$$

$$E_a = 0.4771 \div (5.693 \times 10^{-6}\ \text{mol/J}) = 8.3\underline{8}04 \times 10^4\ \text{J/mol, or } 83.8\ \text{kJ/mol}$$

14.59 For plotting log k vs. 1/T, the data below are used:

k	log k	1/T (1/K)
0.527	-0.2781	1.686×10^{-3}
0.776	-0.1101	1.658×10^{-3}
1.121	0.0496	1.631×10^{-3}
1.607	0.2060	1.605×10^{-3}

The plot requires a graph with too many lines to be reproduced here, but it yields a straight line. The slope of the line is calculated from the difference between the last and the first points:

$$\text{Slope} = \frac{[0.2060 - (-0.2781)]}{[1.605 \times 10^{-3} - 1.686 \times 10^{-3}]/\text{K}} = -59\underline{7}6\ \text{K}$$

Since the slope = $-E_a/(2.303\ R)$, we can solve for E_a, using R = 8.31 J/K:

$$\frac{-E_a}{2.303 \times 8.31\ \text{J/K}} = -59\underline{7}6\ \text{K}$$

$$E_a = 5976\ \text{K} \times 2.303 \times 8.31\ \text{J/K} = 1.1\underline{4}3 \times 10^5\ \text{J, or } 114\ \text{kJ}$$

14.61 The $NOCl_2$ is a reaction intermediate, being produced in the first reaction and consumed in the second. The overall reaction is the sum of the two elementary reactions:

$$NO + Cl_2 \rightarrow NOCl_2$$

$$NOCl_2 + NO \rightarrow 2NOCl$$

$$2NO + Cl_2 \rightarrow 2NOCl$$

14.63 a. bimolecular b. bimolecular c. unimolecular d. termolecular

14.65 a. Only O_3 occurs on the left side of the equation, so the rate law is

$$\text{Rate} = k[O_3]$$

b. Both $NOCl_2$ and NO occur on the left side of the equation, so the rate law is

$$\text{Rate} = k[NOCl_2][NO]$$

14.67 Step 1 of the isomerization of cyclopropane, C_3H_6, is slow, so the rate law for the overall reaction will be the rate law for this step, with k_1 = to k, the overall rate constant:

$$\text{Rate} = k[C_3H_6]^2$$

14.69 Step 2 of this reaction is slow, so the rate law for the overall reaction would appear to be the rate law for this step:

$$\text{Rate} = k_2[I\ ]^2[H_2]$$

However, the rate law includes an intermediate, the I atom, and cannot be used unless the intermedate is eliminated. This can only be done using an equation for step 1. At equibrium, we can write the following equality for step 1:

$$k_1[I_2] = k_{-1}[I\ ]^2$$

Rearranging, and then substituting for the $[I\ ]^2$ term:

$$[I\ ]^2 = [I_2]\frac{k_1}{k_{-1}}$$

$$\text{Rate} = k_2\left[k_1/k_{-1}\right][I_2][H_2] = k[I_2][H_2] \qquad (\text{k = the measured rate constant})$$

14.71 The Cl_2 molecule is the catalyst. It is consumed in the first reaction and regenerated in the third reaction. It speeds up the reaction by providing a pathway with a lower activation energy than that of a reaction pathway involving no Cl_2. The overall reaction is obtained by adding the three steps together:

$$2N_2O \rightarrow 2N_2 + O_2$$

Chlorine, Cl^2, is added to the mixture to give the catalytic activity.

14.73 All rates of reaction are calculated by dividing the decrease in concentration by the difference in times; hence only the setup for the first rate (after 10 minutes) is given below. This setup is:

$$\text{Rate}_{10\ \text{min}} = -\frac{(1.29 - 1.50) \times 10^{-2}\ \text{M}}{(10 - 0)\ \text{min}} \times \frac{1\ \text{min}}{60\ \text{s}} = 3.\underline{5}0 \times 10^{-6} = 3.5 \times 10^{-6}\ \text{M/s}$$

A summary of the times and rates is given in the table:

Time, min	Rate
10	$3.\underline{5}0 \times 10^{-6}$ = 3.5×10^{-5} M/s
20	$3.\underline{1}6 \times 10^{-6}$ = 3.2×10^{-5} M/s
30	$2.\underline{5}0 \times 10^{-6}$ = 2.5×10^{-5} M/s

14.75 The calculation of the average concentration, and the division of the rate by this average concentration, is the same for all three time intervals. Thus only the setup for the first interval is given:

$$k_{10\,min} = \frac{rate}{ave.\ conc.} = \frac{3.50 \times 10^{-6}\ M/s}{\frac{(1.50 + 1.29) \times 10^{-2}\ M}{2}} = 2.\underline{5}08 \times 10^{-4}/s$$

A summary of the times, rate constants, and average rate constant is given in the table below.

Time	Rate	k
10 min	3.50×10^{-6} M/s	$2.\underline{5}08 \times 10^{-4}$/s
20 min	3.17×10^{-6} M/s	$2.\underline{6}52 \times 10^{-4}$/s
30 min	2.50×10^{-6} M/s	$2.\underline{4}39 \times 10^{-4}$/s
-----	---- average k =	$2.\underline{5}33 \times 10^{-4}$ = 2.5×10^{-4}/s

14.77 Use the first-order rate equation: $k = 1.26 \times 10^{-4}/s$, the initial methyl acetate $[MA]_o$ = 100%, and $[MA]_t$ = (100% - 85%, or 15%).

$$t = \frac{-(2.303)\left(\log \frac{15\%}{100\%}\right)}{1.26 \times 10^{-4}/s} = 1.5\underline{0}59 \times 10^4 = 1.51 \times 10^4\ s$$ (use 15% as exact no.)

14.79 Use $k = 1.26 \times 10^{-4}/s$ and substitute into the $t_{1/2}$ equation:

$$t_{1/2} = \frac{0.693}{k} = \frac{0.693}{1.26 \times 10^{-4}/s} = 5.5\underline{0}0 \times 10^3 = 5.50 \times 10^3\ s\ (1.52\ hr)$$

14.81 First find the rate constant from the rearranged first-order rate equation, substituting the initial concentration of $[comp]_o$ = 0.0350 M, and the $[comp]_t$ = 0.0250 M.

$$k = \frac{-(2.303)\left(\log \frac{0.0250\ M}{0.0350\ M}\right)}{65\ s} = 5.\underline{1}8 \times 10^{-3}/s$$

Now arrange the first-order rate equation to solve for $[comp]_t$; substitute the above value of k, again using $[comp]_o$ = 0.0350 M.

$$\text{Log}\ \frac{[comp]_t}{[0.0350\ M]_o} = \frac{-(5.18 \times 10^{-3}/s)(98\ s)}{2.303} = -0.2\underline{2}04$$

Taking antilogarithms of both sides gives:

$$\frac{[comp]_t}{[0.0350\ M]_o} = 0.6\underline{0}2$$

$$[comp]_t = 0.602 \times [0.0350\ M]_o = 0.02108 = 0.021\ M$$

14.83 The log [CH_3NNCH_3] and time data for the plot are tabulated below.

t, min	logCH_3NNCH_3]
0	-1.824
10	-1.889
20	-1.959
30	-2.022

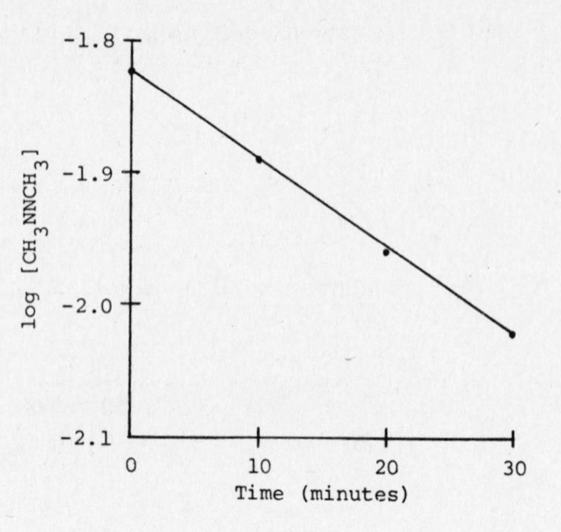

From the graph, the slope, m, is calculated:

$$m = \frac{-2.022 - (-1.824)}{(30 - 0)\ \text{min}} = -6.\underline{6}0 \times 10^{3}/\text{min}$$

Since the slope also = k/2.303, we solve for k by equating the two right-hand terms:

$$k = -(-6.60 \times 10^{-3}/\text{min})(2.303)(1\ \text{m}/60\ \text{s})$$

$$k = 2.\underline{5}3 \times 10^{-4} = 2.5 \times 10^{-4}/\text{s}$$

14.85 Rearrange the two-temperature Arrhenius equation to solve for E_a in J, using k_1 = 0.498 M/s at T_1 = 592 K (319°C) and k_2 = 1.81 M/s at 627 K (354°C). Assume ($1/T_1 - 1/T_2$) has three significant figures.

$$E_a = \frac{(2.303 \times 8.31\ \text{J/K}) \times \log\left[\frac{1.81}{0.498}\right]}{\left[\frac{1}{592\ \text{K}} - \frac{1}{627\ \text{K}}\right]} = 1.1\underline{3}7 \times 10^{5} = 1.14 \times 10^{5}\ \text{J/mol}$$

To obtain A, rearrange the log form of the one-temperature Arrhenius equation; substitute the value of E_a obtained above and use k_1 = 0.498 M/s at T_1 = 592 K.

$$\text{Log}\ A = \log 0.498 + \frac{1.137 \times 10^{5}\ \text{J}}{2.303 \times 8.31\ \text{J/K} \times 592\ \text{K}} = 9.\underline{7}3$$

$$A = \underline{5}.37 \times 10^{9} = 5 \times 10^{9}$$

To obtain k at 383°C (656 K), also use the log form of the one-temperature Arrhenius equation:

$$\text{Log}\ k = 9.73 - \frac{1.137 \times 10^{5}\ \text{J}}{2.303 \times 8.31\ \text{J/K} \times 656\ \text{K}} = 0.\underline{6}73$$

$$k = 4.709 = 5\ \text{M/s}$$

14.87 If the reaction occurs in one step, the coefficients of NO_2 and CO in this elementary reaction are each 1, so the rate law should be:

$$\text{Rate} = k[NO_2][CO]$$

14.89 The slow first step determines the observed rate, so the overall rate constant, k, should be equal to k_1 for the first step and the rate law should be:

$$\text{Rate} = k_1[NO_2Br] = k[NO_2Br]$$

14.91 The slow step determines the observed rate; assuming k_2 is the rate constant for the second step, the rate law would appear to be:

$$\text{Rate} = k_2[NH_3][HCNO]$$

However, this rate law includes two intermediate substances that are neither reactants nor the products. The rate law cannot be used unless both are eliminated. This can only be done using an equation from step 1. At equilibrium in step 1, we can write the following equality, assuming k_1 and k_{-1} are the rate constants for the forward and back reactions, respectively:

$$k_1[NH_4^+][CNO^-] = k_{-1}[NH_3][HCNO]$$

Rearranging and then substituting for the $[NH_3][HCNO]$ product gives:

$$[NH_3][HCNO] = (k_1/k_{-1})[NH_4^+][CNO^-]$$

$$\text{Rate} = k_2(k_1/k_{-1})[NH_4^+][CNO^-] = k[NH_4^+][CNO^-] \qquad \text{(k = measured rate constant)}$$

Cumulative-Skills Problems (require skills from other chapters)

14.93 Using the first-order rate law, the initial rate of decomposition is given by:

$$\text{Rate} = k[H_2O_2] = (7.40 \times 10^{-4}/s) \times (1.50\ M\ H_2O_2) = 1.110 \times 10^{-3}\ M\ H_2O_2/s$$

The heat liberated per second per mol of H_2O_2 can be found by first calculating the standard enthalpy of the decomposition of 1 mol of H_2O_2:

	$H_2O_2(l)$	$\rightarrow$	$H_2O(l)$	+	$1/2O_2(g)$
ΔH^o_f =	-187.8 kJ/mol		-285.84 kJ/mol		0 kJ/mol

For the reaction, the standard enthalpy change is:

$$\Delta H^o = -285.84\ kJ/mol - (-187.8\ kJ/mol) = -98.\underline{0}4\ kJ/mol\ H_2O_2$$

The heat liberated per second is:

$$\frac{98.04\ kJ}{mol\ H_2O_2} \times \frac{1.110 \times 10^{-3}\ mol\ H_2O_2}{L \cdot s} \times 2.00\ L = 0.21\underline{7}65 = 0.218\ kJ/s$$

14.95 Use the ideal gas law (P/RT = n/V) to calculate the mol/L of each gas:

$$[O_2] = \frac{(345/760)\ atm}{0.082057\ L \cdot atm/(K \cdot mol) \times 612\ K} = 0.0090\underline{3}9\ mol\ O_2/L$$

$$[NO] = \frac{(155/760)\ atm}{0.082057\ L \cdot atm/(K \cdot mol) \times 612\ K} = 0.0040\underline{6}1\ mol\ NO/L$$

The rate of decrease of NO is:

$$\frac{1.16 \times 10^{-5}\ L^2}{mol^2 \cdot s} \times \left(\frac{4.061 \times 10^{-3}\ mol}{L}\right) \times \frac{9.039 \times 10^{-3}\ mol}{L} = 1.7\underline{2}9 \times 10^{-12}\ mol/(L \cdot s)$$

The rate of decrease in atm/s is found by multiplying by RT:

$$\frac{1.729 \times 10^{-12}\ mol}{L \cdot s} \times \frac{0.082057\ L \cdot atm}{K \cdot mol} \times 612\ K = 8.6\underline{8}2 \times 10^{-11}\ atm/s$$

The rate of decrease in mmHg/s is:

$$8.682 \times 10^{-11}\ atm/s \times (760\ mmHg/atm) = 6.598 \times 10^{-8} = 6.60 \times 10^{-8}\ mmHg/s$$

CHAPTER 15

CHEMICAL EQUILIBRIUM; GASEOUS REACTIONS

SOLUTIONS TO EXERCISES

Note on significant figures: The final answer to all mathematical solutions is given first with one nonsignificant figure (last significant figure underlined) and is then rounded to the correct number of figures. Intermediate answers usually also have at least one nonsignificant figure.

15.1 Use the "table" approach, giving the starting, change, and equilibrium number of moles of each.

Amt. (Mol)	$CO(g)$	+	$H_2O(g)$ $\rightleftharpoons$	$CO_2(g)$	+	$H_2(g)$
Starting	1.00		1.00	0		0
Change	-x		-x	+x		+x
Equilibrium	(1.00 - x)		(1.00 - x)	x		x = 0.59

Since we are given that x = 0.59 in the statement of the problem, we can use that to calculate the equilibrium amounts of the reactants and products:

Equilibrium amount CO = 1.00 - 0.59 = 0.41 mol

Equilibrium amount H_2O = 1.00 - 0.59 = 0.41 mol

Equilibrium amount CO_2 = x = 0.59 mol

Equilibrium amount H_2 = x = 0.59 mol

15.2 For the equation $2NO_2 + 7H_2 \rightarrow 2NH_3 + 4H_2O$, the expression for the equilibrium constant, K_c, is:

$$K_c = \frac{[NH_3]^2[H_2O]^4}{[NO_2]^2[H_2]^7}$$

Note that each concentration term is raised to a power equal to that of its coefficient in the chemical equation.

For the equation $NO_2 + 7/2H_2 \rightarrow NH_3 + 2H_2O$, the expression for the equilibrium constant, K_c, is:

$$K_c = \frac{[NH_3][H_2O]^2}{[NO_2][H_2]^{7/2}}$$

Note again the correspondence between the power and coefficient for each molecule.

15.3 The chemical equation for the reaction is:

$$CO(g) + H_2O(g) \rightleftharpoons H_2(g) + CO_2(g)$$

The expression for the equilibrium constant for this reaction is:

$$K_c = \frac{[CO_2][H_2]}{[CO][H_2O]}$$

We obtain the concentration of each substance by dividing the moles of substance by its volume. The equilibrium concentrations are as follows: [CO] = 0.057 M, $[H_2O]$ = 0.057 M, $[CO_2]$ = 0.043 M, and $[H_2]$ = 0.043 M. Substituting these values into the equation for the equilibrium constant gives:

$$K_c = \frac{(0.043)(0.043)}{(0.057)(0.057)} = 5.\underline{6}9 \times 10^{-1} = 5.7 \times 10^{-1}$$

15.4 Use the "table" approach, giving the starting, change, and equilibrium concentrations of each by dividing moles by volume in liters.

Conc. (M)	$2H_2S(g)$	$\rightleftharpoons$	$2H_2(g)$	+	$S_2(g)$
Starting	0.0100		0		0
Change	-2x		+2x (= 0.00285)		+x
Equilibrium	0.0100 - 2x		2x		x

Since the problem stated that 0.00285 M H_2 was formed, we can use the 0.00285 M to calculate the other concentrations. The S_2 molarity should be one-half that, or 0.00143 M, and the H_2S molarity should be 0.0100 - 0.00285, or 0.007$\underline{1}$5 M. Substituting into the equilibrium expression gives:

$$K_c = \frac{[H_2]^2[S_2]}{[H_2S]^2} = \frac{(0.00285)^2(0.00143)}{(0.00715)^2} = 2.\underline{2}7 \times 10^{-4} = 2.3 \times 10^{-4}$$

15.5 Use the expression that relates K_c to K_p:

$$K_p = K_c (RT)^{\Delta n}$$

The Δn term is the sum of the coefficients of the gaseous products minus the sum of the coefficients of the gaseous reactants. In this case, $\Delta n = (2 - 1) = 1$, and K_p is:

$$K_p = (3.26 \times 10^{-2})(0.0821 \times 464)^1 = 1.2\underline{4}1 = 1.24$$

15.6 For a heterogeneous equilibrium, the concentration terms for liquids and solids are omitted because such concentrations are constant at a given temperature and are incorporated into the measured value of K_c. For this case, K_c is thus defined:

$$K_c = \frac{[Ni(CO)_4]}{[CO]^4}$$

15.7 Since the equilibrium constant is very large (>10^4), the equilibrium mixture will contain mostly products. Rearrange the K_c expression to solve for $[NO_2]$.

$$[NO_2] = \sqrt{K_c[O_2][NO]^2} = \sqrt{4.0 \times 10^{13}(0.50)(0.50)^2} = 2.\underline{23} \times 10^6 = 2.2 \times 10^6$$

15.8 First divide moles by volume in liters to convert to molar concentrations, giving 0.00015 M CO_2 and 0.010 M CO. Substitute these values into the reaction quotient and calculate Q.

$$Q = \frac{[CO]^2}{[CO_2]} = \frac{(0.010)^2}{(0.00015)} = 6.\underline{66} \times 10^{-1} = 6.7 \times 10^{-1}$$

Since Q = 0.67 and is less than K_c, the reaction will go to the right, forming more CO.

15.9 Rearrange the K_c expression, and substitute for K_c (= 0.0415) and the given moles to solve for moles of PCl_5.

$$[PCl_5] = \frac{[PCl_3][Cl_2]}{K_c} = \frac{(0.020)(0.020)}{0.0415} = 9.\underline{63} \times 10^{-3} = 9.6 \times 10^{-3} \text{ mol}$$

Since the volume is 1.00 L, the concentration of PCl_5 is 0.0096 M.

15.10 Use the "table" approach, giving the starting, change, and equilibrium number of moles of each.

Amt. (Mol)	$H_2(g)$	+	$I_2(g)$	$\rightleftharpoons$	$2HI(g)$
Starting	0.500		0.500		0
Change	-x		-x (= 0.00285)		+2x
Equilibrium	0.500 - x		0.500 - x		2x

Substitute the equilibrium concentrations into the expression for K_c (= 49.7).

$$K_c = \frac{[HI]^2}{[H_2][I_2]};\quad 49.7 = \frac{(2x)^2}{(0.500 - x)(0.500 - x)} = \frac{(2x)^2}{(0.500 - x)^2}$$

Taking the square root of both sides of the right-hand equation and solving for x gives:

$$\pm 7.05 = \frac{2x}{(0.500 - x)} \text{ or } \pm 7.05(0.500 - x) = 2x$$

Using the positive root, x = 0.390
Using the negative root, x = 0.698 (this must be rejected because 0.698 is greater than the 0.500 starting number of moles).

Substituting x = 0.390 mol into the last line of the table to solve for equilibrium concentrations gives these amounts: 0.11 mol H_2, 0.11 mol I_2, and 0.78 mol HI.

15.11 Use the "table" approach for starting, change, and equilibrium concentrations of each species.

Conc. (M)	$PCl_5(g)$	$\rightleftharpoons$	$PCl_3(g)$	+	$Cl_2(g)$
Starting	1.00		0		0
Change	-x		+x		+x
Equilibrium	1.00 - x		x		x

Substitute the equilibrium concentration expressions from the table into the equilibrium equation and solve for x, using the quadratic formula.

$$[PCl_3][Cl_2] = K_c \times [PCl_5] = 0.0211 \times (1.00 - x)$$

$$x^2 + 0.0211x - 0.0211 = 0$$

$$x = \frac{-0.0211 \pm \sqrt{(0.0211)^2 - 4(-0.0211)}}{2} = \frac{-0.0211 \pm 0.2913}{2}$$

x = -0.15591 (impossible; reject), or x = 0.13509 = 0.135 M (logical)

Solve for the equilibrium concentrations using 0.135 M: $[PCl_5]$ = 0.86 M, $[Cl_2]$ = 0.135 M, and $[PCl_3]$ = 0.135 M.

15.12 a. Increasing the pressure will cause a net reaction to occur from right to left, and more $CaCO_3$ will form.

b. Increasing the concentration of hydrogen will cause a net reaction to occur from right to left, forming more Fe and H_2O.

15.13 a. Since there are equal numbers of moles of gas on each side of the equation, increasing the pressure will not increase the amount of product.

b. Since the reaction increases the number of moles of gas, increasing the pressure will decrease the amount of product.

c. Since the reaction decreases the number of moles of gas, increasing the pressure will increase the amount of product.

15.14 Since this is an endothermic reaction and absorbs heat, high temperatures will be more favorable to the production of carbon monoxide.

15.15 Since this is an endothermic reaction and absorbs heat, high temperatures will give the best yield of carbon monoxide. Since the reaction increases the number of moles of gas, decreasing the pressure will increase the yield also.

ANSWERS TO REVIEW QUESTIONS

15.1 A reasonable graph showing the decrease in concentration $N_2O_4(g)$ and the increase in concentration of $NO_2(g)$ is shown below:

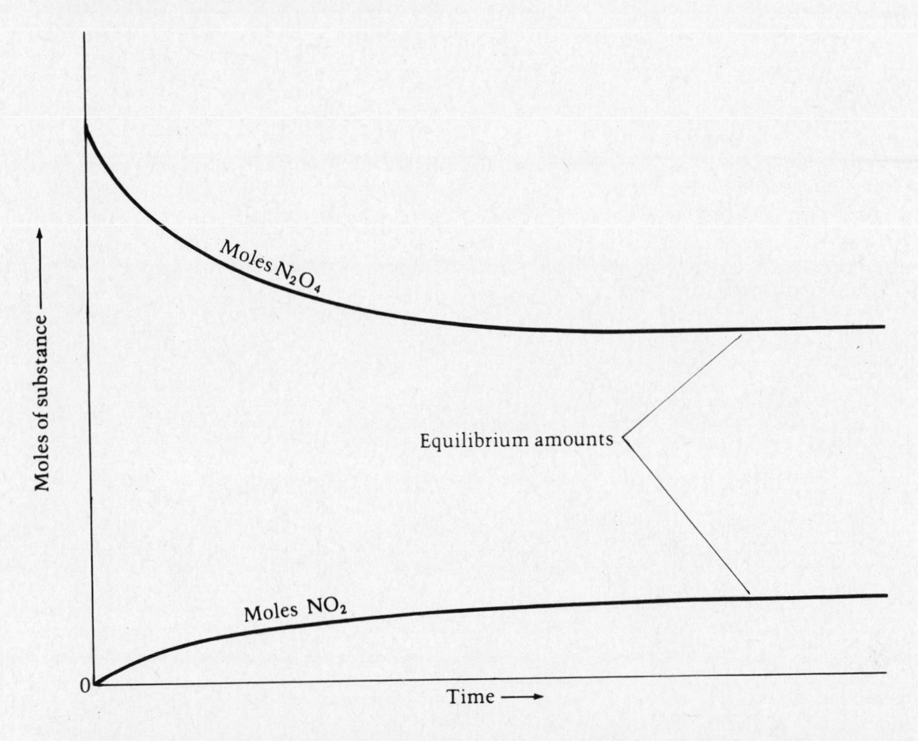

At first the concentration of N_2O_4 is large, and the rate of the forward reaction is large, but then as the concentration of N_2O_4 decreases, the rate of the forward reaction decreases. In contrast, the concentration of NO_2 builds up from zero to a low concentration. Thus the initial rate of the reverse reaction is zero, but steadily increases as the concentration of NO_2 increases. Eventually the two rates become equal when the reaction reaches equilibrium. This is a dynamic equilibrium because both the forward and reverse reactions are occurring at all times even though there is no net change in concentration at equilibrium.

15.2 The 1.0 mol of $H_2(g)$ and 1.0 mol of $I_2(g)$ in the first mixture reaches equilibrium when the amounts of reactants decrease to 0.50 mol each and when the amount of product increases to 1.0 mol. The total number of moles at the reactants at the start is 2.0 mol, which is the same number of moles as in the second mixture, the 2.0 mol of HI which is to be allowed to come to equilibrium. The second mixture should produce the same number of moles of H_2, I_2, and HI at equilibrium because if the total number of moles is constant, it should not matter from which direction an equilibrium is approached.

15.3 The equilibrium constant for a gaseous reaction can be written using partial pressures instead of concentrations because all the reactants and products are in the same vessel. Therefore at constant temperature, the pressure, **P**, is proportional to the concentration, **n/V**. (The ideal gas law says that **P** = **(n/V)**/RT.)

15.4 The addition of reactions 1 and 2 yields reaction 3:

(Reaction 1) $HCN + OH^- \rightleftharpoons CN^- + H_2O$

(Reaction 2) $H_2O \rightleftharpoons H^+ + OH^-$

(Reaction 3) $HCN \rightleftharpoons H^+ + CN^-$

The rule states that if a given equation can be obtained from the sum of other equations, the equilibrium constant for the given equation equals the product of the other equilibrium constants. Thus K for reaction 3 is found:

$$K = K_1 \times K_2 = (4.9 \times 10^4) \times (1.0 \times 10^{-14}) = 4.9 \times 10^{-10}$$

15.5 a. Homogeneous equilibrium. NO, O_2, and NO_2 are all gases and exist in a single phase.

b. Heterogeneous equilibrium. Both $Cu(NO_3)_2$ and CuO are solids whereas the others are gases, and hence there are two phases present.

c. Homogeneous equilibrium. N_2O, N_2, and O_2 are all gases and exist in a single phase.

d. Heterogeneous equilibrium. Two are solids and two gases, so two phases are present.

15.6 Pure liquids and solids can be ignored in an equilibrium expression because their concentrations are constant and in effect are incorporated into the value of K at a given temperature.

15.7 The main qualitative information obtainable from the magnitude of the equilibrium constant is the relative amounts of reactants and products at equilibrium. If K is large (greater than 10^2 to 10^4, depending on the type of reaction), mostly products are present at equilibrium. If K is near unity (10^{-1} to 10^1), the amounts of reactants and products are close to being equal, or equal (K = 1), at equilibrium. If K is small (less than 10^{-2} to 10^{-4}, depending on the type of reaction), mostly reactants are present at equilibrium.

15.8 The reaction quotient is the product of the concentrations of the products divided by the product of the concentrations of the reactants. It has the same form as the equilibrium constant expression, but the concentrations used are not necessarily those at equilibrium. Comparing it with the equilibrium constant permits predictions on the direction of a reaction.

15.9 Note that the total number of moles of products is 2.00, the same as the total number of moles in the example. This mixture of products should produce the same number of moles of CO, H_2O, CO_2, and H_2 at equilibrium because if the total number of moles is constant, it should not matter from which direction an equilibrium is approached. Thus the concentrations are: 0.0114 M CO, 0.0114 M H_2O, 0.0086 M CO_2, and 0.0086 M H_2.

15.10 The equilibrium composition of a mixture can be altered by removing one or more of the products from the reaction vessel, adding more of one or more of the reactants to the reaction vessel, changing the partial pressure of one or more gases reactants and/or products, and changing the temperature. (Adding a catalyst will not alter the composition.)

15.11 The platinum acts as a catalyst in this reaction by providing a surface to which the H_2 and O_2 molecules can bond and form new bonds. The platinum has no effect on the equilibrium composition of the mixture, although it greatly increases the rate of reaching equilibrium.

15.12 Some reaction mixtures can form two or more sets of products, one of which may not form in the absence of a catalyst because its rate is almost zero compared to the rate of formation of the other set of products. For example, NH_3 and O_2 can form either NO and H_2O, or N_2 and H_2O. In the absence of a catalyst such as platinum, NH_3 burns in O_2 as follows:

$$4NH_3(g) + 3O_2(g) \rightarrow 2N_2(g) + 6H_2O(g)$$

Adding a platinum catalyst increases the rate of the reaction below from near zero to a rate much larger than that of the reaction above:

$$4NH_3(g) + 5O_2(g) \rightarrow 4NO(g) + 6H_2O(g)$$

(This is not an equilibrium situation, however; at equilibrium the NO dissociates to N_2 and O_2.)

15.13 The reaction below readily reaches equilibrium from either direction.

$$Fe_3O_4(s) + 4H_2(g) \rightleftharpoons 3Fe(s) + 4H_2O(g)$$

In a stream of H_2, the excess H_2(reactant) forces the reaction to the right in accordance with LeChatelier's principle. In a stream of $H_2O(g)$, the excess H_2O (product) forces the reaction to the left.

15.14 The yield of ammonia from nitrogen and hydrogen can be improved by (1) removing the ammonia by liquefying it, (2) increasing the nitrogen concentration, (3) increasing the total pressure on the mixture since the moles of gas decrease during the reaction, and (4) lowering the temperature since heat is evolved in the reaction (ΔH^o is neg). Each shifts the reaction to the right according to Le Chatelier's principle.

SOLUTIONS TO PRACTICE PROBLEMS

Note on significant figures: The final answer to all mathematical solutions is given first with one nonsignificant figure (last significant figure underlined) and is then rounded to the correct number of figures. Intermediate answers usually also have at least one nonsignificant figure.

15.15 Use the "table" approach, giving the starting, change, and equilibrium number of moles of each.

Amt. (Mol)	$PCl_5(g)$	$\rightleftharpoons$	$PCl_3(g)$	+	$Cl_2(g)$
Starting	1.500		0		0
Change	-x		+x		+x
Equilibrium	1.500 - x		x (= 0.203)		x

Hence the equilibrium amounts are: 1.297 mol PCl_5, 0.203 mol PCl_3, and 0.203 mol Cl_2.

15.17 Use the "table" approach, giving the starting, change, and equilibrium number of moles of each.

Amt. (Mol)	$CO(g)$	+	$2H_2(g)$	$\rightleftharpoons$	$CH_3OH(g)$
Starting	0.1500		0.3000		0
Change	-x		-2x		+x
Equilibrium	0.1500 - x (= 0.1187)		0.3000 - 2x		x

Since 0.1500 - x = 0.1187, x = 0.0313. Therefore the amounts of substances at equilibrium are: 0.1187 mol CO, 0.2374 mol H_2, and 0.0313 mol CH_3OH.

15.19 a. $K_c = \dfrac{[PCl_5]}{[PCl_3][Cl_2]}$ b. $K_c = \dfrac{[O_3]^2}{[O_2]^3}$

c. $K_c = \dfrac{[NO]^2[Cl_2]}{[NOCl]^2}$ d. $K_c = \dfrac{[N_2]^2[H_2O]^6}{[NH_3]^4[O_2]^3}$

15.21 The reaction is: $CH_4(g) + 2H_2S(g) \rightarrow CS_2(g) + 4H_2(g)$

15.23 The equilibrium constant expressions when the reaction is halved and then reversed are:

Halved: $K_c = \dfrac{[N_2]^{1/2}[H_2O]}{[NO][H_2]}$ reversed: $K_c = \dfrac{[NO][H_2]}{[N_2]^{1/2}[H_2O]}$

15.25 Since K_c = 1.84 for $2HI \rightarrow H_2 + I_2$, the value of K_c for $H_2 + I_2 \rightarrow 2HI$ must be the reciprocal of K_c for the first reaction. Mathematically, this can be shown as follows:

Forward: $K_c(f) = \dfrac{[H_2][I_2]}{[HI]^2} = 1.84$ reverse: $K_c = \dfrac{[HI]^2}{[H_2][I_2]} = \dfrac{1}{\dfrac{[H_2][I_2]}{[HI]^2}} = \dfrac{1}{K_c(f)}$

Thus for the reverse reaction K_c is calculated:

$$K_c = 1 \div 1.84 = 5.4\underline{3}4 \times 10^{-1} = 5.43 \times 10^{-1}$$

15.27 First calculate the molar concentrations of each of the compounds in the equilibrium:

$[PCl_5]$ = 0.0158 mol PCl_5 ÷ 5.00 L = $0.0031\underline{6}0$ M

$[PCl_3]$ = 0.0185 mol PCl_3 ÷ 5.00 L = $0.0037\underline{0}0$ M

$[Cl_2]$ = 0.0870 mol Cl_2 ÷ 5.00 L = $0.017\underline{4}0$ M

Now substitute these into the expression for K_c:

$$K_c = \frac{0.003160}{(0.003700)(0.01740)} = 49.\underline{0}8 = 49.1/M$$

15.29 Substitute the following concentrations into the K_c expression:

$[CH_3OH] = 0.0313 \text{ mol} \div 1.50 \text{ L} = 0.020\underline{8}6 \text{ M}$

$[CO] = 0.1187 \text{ mol} \div 1.50 \text{ L} = 0.079\underline{1}3 \text{ M}$

$[H_2] = 0.2374 \text{ mol} \div 1.50 \text{ L} = 0.15\underline{8}2$

$$K_c = \frac{(0.02086)}{(0.07913)(0.1582)^2} = 1.0\underline{5}3 \times 10^1 = 1.05 \times 10^1/M^2$$

15.31 For each mole of NOBr that reacts, (1.000 - 0.094 = 0.906) mol remains. Starting with 2.00 mol NOBr, 2 x 0.906 mol NOBr, or 1.812 mol NOBr, remains. Since the volume is 1.00 L, the concentration of NOBr at equilibrium is 1.812 M. Next assemble a table of starting, change, and equilibrium concentrations:

Conc. (M)	$2NOBr(g)$	$\rightleftharpoons$	$2NO(g)$	+	$Br_2(g)$
Starting	2.00		0		0
Change	-2x		+2x		+x
Equilibrium	2.00 - 2x (= 1.812)		2x		x

Since 2.00 - 2x = 1.812, x = 0.094 M. Therefore, the equilibrium concentrations are: [NOBr] = 1.812 M, [NO] = 0.188 M, and $[Br_2]$ = 0.094 M.

$$K_c = \frac{[NO]^2[Br_2]}{[NOBr]^2} = \frac{(0.188 \text{ M})^2(0.094 \text{ M})}{(1.812 \text{ M})^2} = 1.01 \times 10^{-1} = 1.0 \times 10^{-3} \text{ M}$$

15.33 a. $K_p = \frac{P_{HBr}^2}{P_{H_2}P_{Br_2}}$ b. $K_p = \frac{P_{CH_4}P_{H_2S}^2}{P_{CS_2}P_{H_2}^4}$

c. $K_p = \frac{P_{H_2O}^2P_{Cl_2}^2}{P_{HCl}^4P_{O_2}}$ d. $K_p = \frac{P_{CH_3OH}}{P_{CO}P_{H_2}^2}$

15.35 There are 3 mol of gaseous product for every 5 mol of gaseous reactant, so $\Delta n = 3 - 5 = -2$. Using this, we calculate K_p from K_c:

$$K_p = K_c(RT)^{\Delta n} = 0.28\,(0.0821 \times 1173)^{-2} = 3.\underline{0}19 \times 10^{-5} = 3.0 \times 10^{-5}$$

15.37 For each 1 mol of gaseous product, there are 1.5 mol of gaseous reactants; thus $\Delta n = 1 - 1.5 = -0.5$. Using this, we calculate K_c from K_p:

$$K_c = \frac{K_p}{(RT)^{\Delta n}} = \frac{6.55}{(0.0821 \times 900)^{-0.5}} = 6.55 \times (0.0821 \times 900)^{1/2} = 56.\underline{3}03 = 56.3$$

15.39 a. $K_c = \frac{[CO]}{[CO_2]}$ b. $K_c = \frac{[CO_2]}{[CO]}$

c. $K_c = \frac{[CO_2]}{[SO_2][O_2]^{1/2}}$ d. $K_c = [Pb^{2+}][I^-]^2$

15.41 a. Nearly complete; K_c is very large (>10^2 to 10^4), indicating nearly complete reaction.

b. Not complete; K_c is very small (<10^{-2} to 10^{-4}), indicating very little reaction.

15.43 K_c is extremely small, indicating very little reaction at room temperature. Since the decomposition of HF yields equal amounts of H_2 and F_2, at equilibrium $[H_2] = [F_2]$. So for the decomposition of HF:

$$K_c = \frac{[H_2][F_2]}{[HF]^2} = \frac{[H_2]^2}{[HF]^2}$$

$$[H_2] = (K_c)^{1/2}[HF] = (1.0 \times 10^{-95})^{1/2} (1.0 \text{ mol}) = 3.\underline{1}6 \times 10^{-48} = 3.2 \times 10^{-48} \text{ mol}$$

This result does agree with what is expected from the very small magnitude of K_c.

15.45 Calculate Q, the reaction quotient, and compare it to the equilibrium constant. If Q is larger, the reaction goes to the left, and vice versa. For all, Q is found by combining these terms:

$$Q = \frac{[CS_2][H_2]^4}{[CH_4][H_2S]^2}$$

a. $Q = \frac{(1.51)(1.08)^4}{(1.15)(1.20)^2} = 1.2\underline{4}05$ (<3.59; reaction goes to right)

b. $Q = \frac{(0.90)(1.78)^4}{(1.07)(1.20)^2} = 5.8\underline{6}3$ (>3.59; reaction goes to left)

c. $Q = \frac{(1.10)(1.68)^4}{(1.10)(1.49)^2} = 3.5\underline{8}8$ (=3.59; reaction is at equilibrium)

d. $Q = \frac{(1.25)(1.75)^4}{(1.45)(1.29)^2} = 4.8\underline{5}8$ (>3.59; reaction goes to left)

15.47 Calculate Q, the reaction quotient, and compare it to the equilibrium constant. If it is larger, the reaction will go to the right, and vice versa. Q is found by combining these terms:

$$Q = \frac{[CH_3OH]}{[CO][H_2]^2} = \frac{(0.020)}{(0.10)(0.10)^2} = 2.\underline{0}0 \times 10^1 \quad (>10.5;\ \text{reaction goes to left})$$

15.49 Substitute into the expression for K_c and solve for $[COCl_2]$:

$$K_c = 1.23 \times 10^3 = \frac{[COCl_2]}{[CO][Cl_2]} = \frac{(COCl_2)}{(0.012)(0.025)}$$

$$[COCl_2] = (1.23 \times 10^3)(0.012)(0.025) = 0.3\underline{6}9 = 0.37 \text{ M}$$

15.51 Divide moles of substance by the volume of 5.0 L to obtain concentration. The starting concentrations are 3.0×10^{-4} M for both $[I_2]$ and $[Br_2]$. Assemble a table of starting, change, and equilibrium concentrations:

Conc. (M)	$I_2(g)$	+	$Br_2(g)$	⇌	$2IBr(g)$
Starting	3.0×10^{-4}		3.0×10^{-4}		0
Change	-x		-x		+2x
Equilibrium	3.0×10^{-4} - x		3.0×10^{-4} - x		2x

Substituting into the equilibrium expression gives:

$$K_c = 1.2 \times 10^2 = \frac{[IBr]^2}{[I_2][Br_2]} = \frac{(2x)^2}{(3.0 \times 10^4 - x)(3.0 \times 10^4 - x)}$$

Taking the square root of both sides:

$$1\underline{0}.95 = \frac{(2x)}{(3.0 \times 10^{-4} - x)}$$

Rearranging and simplifying the right side gives:

$$(3.0 \times 10^{-4} - x) = \frac{(2x)}{10.95} = (0.1\underline{8}2\ x)$$

$$x = 2.\underline{5}3 \times 10^{-4} \text{ M}$$

Thus, $[I_2] = [Br_2] = 4.\underline{7}0 \times 10^{-5} = 4.7 \times 10^{-5}$ M, and $[IBr] = 5.\underline{0}6 \times 10^{-4} = 5.1 \times 10^{-4}$ M.

15.53 Divide moles of substance by the volume of 5.0 L to obtain concentration. The starting concentrations are 0.10 M both for $[PCl_3]$ and $[Cl_2]$. Assemble a table of starting, change, and equilibrium concentrations:

Conc. (M)	$PCl_3(g)$	+	$Cl_2(g)$	⇌	$PCl_5(g)$
Starting	0.10		0.10		0
Change	-x		-x		+x
Equilibrium	0.10 - x		0.10 - x		x

Substituting into the equilibrium expression for K_c gives:

$$K_c = 49 = \frac{[PCl_5]}{[PCl_3][Cl_2]} = \frac{x}{(0.10 - x)(0.10 - x)}$$

Rearranging and solving for x yields:

$$49(0.10 - x)^2 = 49(x^2 - 0.20x + 0.010) = x$$

$$49x^2 - 10.8x + 0.49 = 0 \text{ (quadratic eqn)}$$

Using the solution to the quadratic:

$$x = \frac{10.8 \pm \sqrt{(10.8)^2 - 4(49)(0.49)}}{2(49)}$$

$x = 0.157$ (impossible; reject), or $x = 0.06\underline{3}89 = 0.064$ M (logical)

Thus, $[PCl_3] = [Cl_2] = 0.036$ M, and $[PCl_5] = 0.064$ M; the vessel contains 0.18 mol PCl_3, 0.18 mol Cl_2, and 0.32 mol PCl_5.

15.55 Divide moles of substance by the volume of 10.00 L to obtain concentration. The starting concentrations are 0.1000 M for [CO] and 0.3000 M for $[H_2]$. Assemble a table of starting, change, and equilibrium concentrations:

Conc. (M)	CO(g)	+	$3H_2(g)$	⇌	$CH_4(g)$	+	$H_2O(g)$
Starting	0.1000		0.3000		0		0
Change	-x		-3x		+x		+x
Equilibrium	0.1000 - x		0.3000 - x		x		x

Substituting into the equilibrium expression for K_c gives:

$$K_c = 3.92 = \frac{[CH_4][H_2O]}{[CO][H_2]^3} = \frac{x^2}{(0.1000 - x)[3(0.1000 - x)]^3} = \frac{x^2}{27(0.1000 - x)^4}$$

Multiplying both sides by 27 and taking the square root of both sides gives:

$$10.29 = \frac{x}{(0.1000 - x)^2} \quad \text{or} \quad 10.29x^2 - 3.058x + 0.1029 = 0$$

Using the solution to the quadratic equation:

$$x = \frac{3.058 \pm \sqrt{(-3.058)^2 - 4(10.29)(0.1029)}}{2(10.29)}$$

$x = 0.2585$ (can't be > 0.1000, so reject), or $x = 0.038\underline{6}8 = 0.0387$ M (use)

Thus: $[CO] = 0.0613$ M, $[H_2] = 0.1839$ M, $[CH_4] = 0.0387$ M, and $[H_2O] = 0.0387$ M.

15.57 a. Forward direction b. Reverse direction

15.59 a. A pressure increase has no effect because number of moles of reactants equals that of products.

b. A pressure increase has no effect because number of moles of reactants equals that of products.

c. A pressure increase causes the reaction to go to the left because the number of moles of reactants is less than that of the products.

15.61 The fraction would not increase because an increase in temperature decreases the amounts of products of an exothermic reaction.

15.63 The value of ΔH^o is calculated from the ΔH^o_f values below each substance in the reaction:

$2NO_2(g)$ +	$7H_2(g)$ ⇌	$2NH_3(g)$ +	$4H_2O(g)$
2(33.2)	7(0)	2(-45.9)	4(-241.8)

$$\Delta H^o = -967.2 + (-91.8) - 66.4 = -1125.4 \text{ kJ/2 mol } NO_2$$

The equilibrium constant will decrease with temperature because raising the temperature of an exothermic reaction will cause the reaction to go farther to the left.

15.65 Since the reaction is exothermic, the formation of products will be favored by low temperatures. Since there are more molecules of gaseous products than gaseous reactants, the formation of products will be favored by low pressures.

15.67 Substitute the concentrations into the equilibrium expression to calculate K_c.

$$K_c = \frac{[CH_3OH]}{[CO][H_2]^2} = \frac{(0.015)}{(0.096)(0.191)^2} = 4.\underline{2}8 = 4.3$$

15.69 Assume 100.00 g of gas: 90.55 g are CO, and 9.45 g are CO_2. The moles of each are:

$$90.55 \text{ g CO} \times \frac{1 \text{ mol CO}}{28.01 \text{ g CO}} = 3.2\underline{3}28 \text{ mol}; \; 9.45 \text{ g } CO_2 \times \frac{1 \text{ mol } CO_2}{44.01 \text{ g } CO_2} = 0.2147 \text{ mol } CO_2$$

Total moles of gas = (3.2328 + 0.2147) mol = 3.44$\underline{7}$5 mol. Use the ideal gas law to convert to the volume of gaseous solution:

$$V = \frac{nRT}{P} = \frac{(3.4475 \text{ mol})(0.082057 \text{ L·atm/(K·mol)})(850 + 273 \text{ K})}{1.000 \text{ atm}} = 317.\underline{6}9 \text{ L}$$

The concentrations are:

$$[CO] = \frac{3.2328 \text{ mol CO}}{317.69 \text{ L}} = 0.010\underline{1}76 \text{ M}; \; [CO_2] = \frac{0.2147 \text{ mol } CO_2}{317.69 \text{ L}} = 6.7\underline{5}8 \times 10^{-4} \text{ M}$$

Find K_c by substituting into the equilibrium expression:

$$K_c = \frac{[CO]^2}{[CO_2]} = \frac{(0.10\underline{1}76)^2}{(6.758 \times 10^{-4})} = 0.15\underline{3}2 = 0.153 \text{ M}$$

15.71 After calculating the concentrations after mixing, calculate Q, the reaction quotient, and compare it with K_c.

$$[N_2] = [H_2] = 1.00 \text{ mol} \div 2.00 \text{ L} = 0.500 \text{ M}$$

$$[NH_3] = 2.00 \text{ mol} \div 2.00 \text{ L} = 1.00 \text{ M}$$

$$Q = \frac{[NH_3]^2}{[N_2][H_2]^3} = \frac{(1.00)^2}{(0.500)(0.500)^3} = 16.\underline{0}0 = 16.0/\text{M}^2$$

Since Q is greater than K_c, the reaction will go in the reverse direction (to the left) to reach equilibrium.

15.73 To calculate the concentrations after mixing, the volume can be assumed to be 1.00 L, or symbolized as V. Since the volumes in the numerator and denominator cancel each other, they do not matter. We will assume a 1.00 L volume and calculate Q, the reaction quotient, and compare it with K_c.

$$[CO] = [H_2O] = [CO_2] = [H_2] = 1.00 \text{ mol/1.00 L}$$

$$Q = \frac{[ICO_2][H_2]}{[CO][H_2O]} = \frac{(1.00 \text{ mol/1.00 L})(1.00 \text{ mol/1.00 L})}{(1.000 \text{ mol/1.00 L})(1.00 \text{ mol/1.00 L})} = 1.\underline{0}0$$

Since Q is greater than K_c, the reaction will go in the reverse direction (left) to reach equilibrium.

15.75 Assemble a table of starting, change, and equilibrium concentrations, letting 2x = the change in [HBr].

Conc. (M)	$2HBr(g)$	$\rightleftharpoons$	$H_2(g)$	+	$Br_2(g)$
Starting	0.010		0		0
Change	-2x		+x		+x
Equilibrium	0.010 - 2x		x		x

$$K_c = 0.016 = \frac{[H_2][Br_2]}{[HBr]^2} = \frac{(x)(x)}{(0.010 - 2x)^2}$$

$$0.1\underline{2}6 = \frac{(x)}{(0.010 - 2x)}$$

$$1.26 \times 10^{-3} - (2.52 \times 10^{-1})x = x$$

$$x = (1.26 \times 10^{-3}) \div 1.252 = 1.\underline{0}06 \times 10^{-3} = 1.0 \times 10^{-3} \text{ M}$$

Therefore, [HBr] = 0.008 M, or 0.008 mol, $[H_2]$ = 0.0010 M, or 0.0010 mol, and $[Br_2]$ = 0.0010 M, or 0.0010 mol.

15.77 The starting concentration of $COCl_2$ = 1.00 mol ÷ 25.00 L = 0.0400 M. Assemble a table of starting, change, and equilibrium concentrations.

Conc. (M)	$COCl_2(g)$	⇌	$Cl_2(g)$	+	$CO(g)$
Starting	0.0400		0		0
Change	-x		+x		+x
Equilibrium	0.0400 - x		x		x

Substituting into the equilibrium expression for K_c gives:

$$K_c = 8.05 \times 10^{-4} = \frac{[CO]\,[Cl_2]}{[COCl_2]} = \frac{(x)(x)}{(0.0400 - x)}$$

Rearranging and solving for x yields:

$$3.22 \times 10^{-5} - (8.05 \times 10^{-4})x - x^2 = 0$$

$$x^2 + (8.05 \times 10^{-4})x - 3.22 \times 10^{-5} = 0 \text{ (quadratic equation)}$$

Using the solution to the quadratic:

$$x = \frac{-8.05 \times 10^{-4} \pm \sqrt{(8.05 \times 10^{-4})^2 - 4(1)(-3.22 \times 10^{-5})}}{2}$$

$x = -6.09 \times 10^{-3}$ (impossible; reject), or $x = 5.2\underline{8}6 \times 10^{-3}$ (logical; use)

% dissoc = change ÷ starting (100%) = 0.005286 ÷ 0.0400 x 100% = $13.\underline{2}1$ = 13.2%

15.79 Using 1.00 mol/10.00 L or 0.100 M, and 4.00 mol/10.00 L or 0.400 M, for the respective starting concentrations for CO and H_2, assemble a table of starting, change, and equilibrium concentrations.

Conc. (M)	$CO(g)$	+	$3H_2(g)$	⇌	$CH_4(g)$ +	$H_2O(g)$
Starting	0.100		0.400		0	0
Change	-x		-3x		+x	+x
Equilibrium	0.100 - x		0.400 -3 x		x	x

Substituting into the equilibrium expression for K_c gives:

$$K_c = 3.92 = \frac{[CH_4][H_2O]}{[CO][H_2]^3} = \frac{x^2}{(0.100 - x)(0.400 - 3x)^3}$$

$$f(x) = \frac{x^2}{(0.100 - x)(0.400 - 3x)^3}$$

Since K_c is > 1 (>50% reaction), choose x = 0.05 (about half of CO reacting) and use that for the first entry in the table of x, f(x), and interpretations:

x	f(x)	interpretation
0.05	3.20	x > 0.05
0.06	8.45	x < 0.06
0.055	5.18	x < 0.055
0.0525	4.07	x < 0.0525 (but close)
0.052	3.87	f(x) of 3.87 ≅ 3.92

At equilibrium concentrations and moles are: CO: 0.048 M and 0.48 mol; H_2: 0.244 M and 2.44 mol; CH_4: 0.052 M and 0.52 mol; and H_2O: 0.052 M and 0.52 mol.

15.81 The dissociation is endothermic.

15.83 For $N_2 + 3H_2 \rightarrow 2NH_3$, K_p is defined in terms of pressures as:

$$K_p = \frac{P_{NH_3}^{\ 2}}{P_{N_2}P_{H_2}^{\ 3}}$$

But by the ideal gas law, where [i] = mol/L:

$$P_i = (n_iRT)/V \quad \text{or} \quad P_i = [\,i\,]RT$$

Substituting the right-hand equality into the K_p expression gives:

$$K_p = \frac{[[NH_3]RT]^2}{[[N_2]RT][[H_2]RT]^3} = \frac{[NH_3]^2}{[N_2][H_2]^3}(RT)^{-2}$$

$$K_p = K_c\,(RT)^{-2} \quad \text{or} \quad K_c = K_p\,(RT)^2$$

Cumulative-Skills Problems (require skills from previous chapters)

15.85 For $Sb_2S_3(s) + 3H_2(g) \rightleftharpoons 2Sb(s) + 3H_2S(g)$ [$+3Pb^{2+} \rightarrow 3PbS(s) + 6H^+$]:

Starting M of $H_2(g)$ = 0.0100 mol ÷ 2.50 L = 0.00400 M H_2

1.029 g PbS ÷ 239.26 g PbS/mol H_2S = 4.30$\underline{0}$7 mol H_2S

4.3007 mol H_2S ÷ 2.50 L = [1.7203×10^{-3}] = M of H_2S

Conc. (M)	$3H_2(g)$	+	$Sb_2S_3(s)$	⇌	$3H_2S(g)$ + $2Sb(s)$
Starting	0.00400				0
Change	-0.0017203				+0.0017203
Equilibrium	0.0022797				0.0017203

Substituting into the equilibrium expression for K_c gives:

$$K_c = \frac{[H_2S]^3}{[H_2]^3} = \frac{(0.0017203)^3}{(0.0022797)^3} = 4.2\underline{9}7 \times 10^{-1} = 4.30 \times 10^{-1}$$

15.87 For $PCl_5(g) \rightleftharpoons PCl_3(g) + Cl_2(g)$:

Starting M of PCl_5 = 0.0100 mol ÷ 2.00 L = 0.00500 M

Conc. (M)	$PCl_5(g)$	$\rightleftharpoons$	$PCl_3(g)$	+	$Cl_2(g)$
Starting	0.00500		0		0
Change	-x		+x		+x
Equilibrium	0.00500 - x		x		x

Substituting into the equilibrium expression for K_c gives:

$$K_c = 4.15 \times 10^{-2} = \frac{[PCl_3]\,[Cl_2]}{[PCl_5]} = \frac{(x)(x)}{(0.00500 - x)}$$

$$x^2 + (4.15 \times 10^{-2})x - 2.075 \times 10^{-4} = 0 \text{ (quadratic)}$$

Solving the quadratic gives x = -4.60×10^{-2} (impossible) and x = 4.51×10^{-3} M (use)

Total M of gas = 0.00451 + 0.00451 + 0.00049 = 0.0095$\underline{1}$0 M

P = (n/V)RT = (0.009510 M)(0.082057 L•atm/K•mol)(523 K) = 0.40$\underline{8}$1 = 0.408 atm

CHAPTER 16

ACID-BASE EQUILIBRIA

SOLUTIONS TO EXERCISES

Note on answers to equilibrium calculations: The rounded answer (with significant figures) is always given first. After the stepwise solution, the numerical answer is given again, but with one nonsignificant figure. (The last significant figure is indicated by underlining it.)

16.1 A 0.125 M solution of $Ba(OH)_2$, a strong base, ionizes completely to yield 0.125 M Ba^{2+} ion and 2 x 0.125 M, or 0.250 M, OH^- ion. Use the K_w equation to calculate the $[H^+]$.

$$[H^+] = \frac{K_w}{[OH^-]} = \frac{1.00 \times 10^{-14}}{(0.250)} = 4.00\underline{0} \times 10^{-14} = 4.00 \times 10^{-14} \text{ M}$$

16.2 The K_w equation gives 1.0×10^{-9} M $[H^+]$, which is basic (<1.00×10^{-7} M of a neutral solution):

$$[H^+] = \frac{K_w}{[OH^-]} = \frac{1.00 \times 10^{-14}}{(1.0 \times 10^{-5})} = 1.\underline{0}0 \times 10^{-9} = 1.0 \times 10^{-9} \text{ M (basic)}$$

16.3 Calculate the negative log of the $[H^+]$:

$$-\text{Log } [H^+] = -\log (0.045) = 1.3\underline{4}6 = 1.35$$

Note: the number of places after the decimal point in the pH = the no. of sig figs of $[H^+]$.

16.4 Calculate the pOH of 0.025 M OH^- and then subtract from 14.00 to find pH:

$$\text{pOH} = -\log [OH^-] = -\log (0.025) = 1.6\underline{0}2$$

$$\text{pH} = 14.00 - 1.602 = 12.3\underline{9}7 = 12.40$$

16.5 Since pH = 3.16, by definition the log $[H^+]$ = -3.16. Enter this on the calculator and convert to the antilog (number) of -3.16:

$$\text{Antilog } (-3.16) = 10^{-3.16} = 6.\underline{9}1 \times 10^{-4} = 6.9 \times 10^{-4} \text{ M}$$

16.6 Find the pOH by subtracting the pOH from 14.00. Then enter -3.40 on the calculator to convert to the antilog (number) corresponding to -3.40.

$$pOH = 14.00 - 10.6 = 3.\underline{4}0$$

$$\text{Antilog } (-3.40) = 10^{-3.40} = \underline{3}.98 \times 10^{-4} = 4 \times 10^{-4} \text{ M}$$

16.7 Rounded answer: $K_a = 1.4 \times 10^{-4}$ (correct sig figs). To solve, assemble a table of starting, change, and equilibrium concentrations. Use HL as the symbol for lactic acid.

Conc. (M)	HL	⇌	H^+	+	L^-
Starting	0.025		0		0
Change	-x		+x		+x
Equilibrium	0.025 - x		x		x

The value of x equals the value of the molarity of the H^+ ion, which can be obtained from the pH: $[H^+]$ = antilog (-pH) = antilog (-2.75) = $1.\underline{7}8 \times 10^{-3}$ M. After using x to substitute into the equilibrium expression, substitute this value for x and solve for K_a:

$$K_a = \frac{[H^+][L^-]}{[HL]} = \frac{[x]^2}{(0.025 - x)} = \frac{(1.78 \times 10^{-3})^2}{0.0232} = 1.\underline{36} \times 10^{-4} \text{ (rounded answer above)}$$

$$\text{Degree of ionization} = \frac{1.78 \times 10^{-3}}{0.025} = 0.071 \text{ (7.1\%)}$$

16.8 Rounded answer: $[H^+] = [C_2H_3O_2^-] = 1.3 \times 10^{-3}$ M; pH = 2.89. To solve, assemble a table of starting, change, and equilibrium concentrations. Use HAc as a symbol for acetic acid.

Conc. (M)	HAc	⇌	H^+	+	Ac^-
Starting	0.10		0		0
Change	-x		+x		+x
Equilibrium	0.10 - x		x		x

Write the equilibrium expression in terms of chemical symbols, then substitute x and (0.10 - x):

$$\frac{[H^+][Ac^-]}{[HAc]} = K_a = \frac{[x]^2}{(0.10 - x)} = 1.7 \times 10^{-5}$$

Solve the equation for x, assuming that x is much smaller than 0.10, so that $(0.10 - x) \cong 0.10$. Solve by taking the square root of $1.7 \times 10^{-5} \times (0.10)$.

$$\frac{[x]^2}{(0.10)} \cong 1.7 \times 10^{-5}; \quad x^2 \cong 1.7 \times 10^{-5} (0.10) = 1.7 \times 10^{-6}$$

$$x = 1.\underline{3}03 \times 10^{-3} \text{ M} = [H^+] = [Ac^-]$$

Check to make sure that the assumption that $(0.10 - x) \cong 0.10$ is valid:

$$0.10 - 1.3 \times 10^{-3} = 0.0987, \text{ or } \cong 0.10 \text{ to 2 sig figs}$$

The pH of the solution = $-\log (1.303 \times 10^{-3})$ = 2.8<u>8</u>505 (rounded answer above)

$$\text{Degree of ionization} = \frac{1.303 \times 10^{-3}}{0.10} = 0.013 \ (1.3\%)$$

16.9 Rounded answer: $[H^+]$ = 5.8×10^{-4} M; pH = 3.24 (correct sig figs). To solve, assemble a table of starting, change, and equilibrium concentrations. Let HPy symbolize pyruvic acid.

Conc. (M)	HPy	$\rightleftharpoons$	H^+	+	Py^-
Starting	0.0030		0		0
Change	-x		+x		+x
Equilibrium	0.0030 - x		x		x

Write the equilibrium expression in terms of chemical symbols, then substitute the terms x and (0.0030 - x):

$$\frac{[H^+][Py^-]}{[HPy]} = K_a = \frac{[x]^2}{(0.0030 - x)} = 1.4 \times 10^{-4}$$

In this case, x cannot be neglected compared to 0.0030 M. (If it is neglected, the calculated $[H^+]$ is 6.4×10^{-4}, which when subtracted from 0.0030 yields 0.00236, a significant change.) The <u>quadratic formula</u> must be used. Reorganize the above equilibrium expression into the form $ax^2 + bx + c = 0$, and substitute for a, b, and c in the quadratic formula.

$$x^2 + 1.4 \times 10^{-4} - 4.20 \times 10^{-7} = 0$$

$$x = \frac{-1.4 \times 10^{-4} \pm \sqrt{1.96 \times 10^{-8} + 1.68 \times 10^{-6}}}{2} = -7.00 \times 10^{-5} \pm 6.52 \times 10^{-4}$$

Using the + root: $x = [H^+] = 5.\underline{8}2 \times 10^{-4}$ M

pH = $-\log (5.82 \times 10^{-4})$ = 3.2<u>3</u>507 (rounded answer above)

16.10 Rounded answer: $[H^+]$ = 0.051 M; pH = 1.29; $[SO_3^{2-}]$ = 6.3×10^{-8} M. To solve, note that $K_{a1} = 1.3 \times 10^{-2} > K_{a2} = 6.3 \times 10^{-8}$, and hence the second ionization and K_{a2} can be neglected. Assemble a table of starting, change, and equilibrium concentrations:

Conc. (M)	H_2SO_3	$\rightleftharpoons$	H^+	+	HSO_3^-
Starting	0.25		0		0
Change	-x		+x		+x
Equilibrium	0.25 - x		x		x

Write the equilibrium expression in terms of chemical symbols, then substitute x and (0.25 - x):

$$\frac{[H^+][HSO_3^-]}{[H_2SO_3]} = K_{a1} = \frac{[x]^2}{(0.25 - x)} = 1.3 \times 10^{-2} \text{ gives } x^2 + 0.013x - 0.00325 = 0$$

In this case, x cannot be neglected compared to 0.25 M. (If it is neglected, the calculated $[H^+]$ is 5.7×10^{-2}, which when subtracted from 0.25 yields 0.193, a significant change.) The quadratic formula must be used. Reorganize the above equilibrium expression into the form $ax^2 + bx + c = 0$, and substitute for a, b, and c in the quadratic formula.

$$x = \frac{-1.3 \times 10^{-2} \pm \sqrt{1.69 \times 10^{-4} + 1.30 \times 10^{-2}}}{2} = -6.50 \times 10^{-3} \pm 5.738 \times 10^{-2}$$

Using the + root, $x = [H^+] = 5.\underline{0}88 \times 10^{-2}$ M

$$pH = -\log (5.088 \times 10^{-2}) = 1.2\underline{9}3$$

To calculate $[SO_3^{2-}]$, we note from the equilibrium concentrations in the table above that $[H^+] = [HSO_3^-]$ (= 5.088×10^{-2} M). In exact terms, $[H^+] = (0.051 + y)$ from the ionization of HSO_3^-, and $[HSO_3^-] = (0.051 - y)$ from ionization. Substituting into the K_{a2} expression gives:

$$\frac{[H^+][SO_3^{2-}]}{[HSO_3^-]} = K_{a2} = \frac{(0.051 + y)(SO_3^{2-})}{(0.051 - y)} = 6.3 \times 10^{-8}$$

Assuming that y is much smaller than 0.051, we see that the (0.051 + y) cancels the (0.051 - y) term, leaving:

$$[SO_3^{2-}] \cong 6.3 \times 10^{-8} \text{ M}$$

16.11 Rounded answer: $K_b = 3.3 \times 10^{-6}$ (correct sig figs). To solve, convert the pH to $[OH^-]$:

$$pOH = 14.00 - pH = 14.00 - 9.84 = 4.16$$

$$[OH^-] = \text{antilog } (-4.16) = 6.\underline{9}2 \times 10^{-5} \text{ M}$$

Using the symbol Qu for quinine, assemble a table of starting, change, and equilibrium concentrations.

Conc. (M)	Qu +	H_2O	$\rightleftharpoons$	HQu^+	+	OH^-
Starting	0.0015			0		0
Change	-x			+x		+x
Equilibrium	0.0015 - 6.92×10^{-5}			6.92×10^{-5}		6.92×10^{-5}

Write the equilibrium expression in terms of chemical symbols, then substitute the terms after subtracting in the denominator, and solve for K_b:

$$K_b = \frac{[HQu^+][OH^-]}{[Qu]} = \frac{(6.92 \times 10^{-5})^2}{(0.00143)} = 3.\underline{3}48 \times 10^{-6} \text{ (rounded answer above)}$$

16.12 Rounded answer: $[H^+] = 5.3 \times 10^{-12}$ M. To solve, assemble a table of starting, change, and equilibrium concentrations:

Conc. (M)	NH_3 +	H_2O	$\rightleftharpoons$	NH_4^+ +	OH^-
Starting	0.20			0	0
Change	-x			+x	+x
Equilibrium	0.20 - x			x	x

Write the equilibrium expression in terms of chemical symbols, then substitute the terms and solve for $[OH^-]$ and then $[H^+]$:

$$\frac{[NH_4^+][OH^-]}{[NH_3]} \cong \frac{(x)^2}{(0.20)}$$

$x = [OH^-] \cong 1.89 \times 10^{-3}$ M (note that x is negligible compared to 0.20)

$[H^+] = 1.00 \times 10^{-14} \div 1.89 \times 10^{-3} = 5.\underline{2}9 \times 10^{-12}$ M (rounded answer above)

16.13 a. NH_4NO_3 yields an acidic solution because NH_4^+ hydrolyzes in water to form the H_3O^+ ion (+ NH_3); the NO_3^- ion does not hydrolyze.

b. KNO_3 does not change the pH (7.00) of neutral water because neither ion hydrolyzes.

c. $Al(NO_3)_3$ yields an acidic solution because Al^{3+} hydrolyzes in water to form the H_3O^+ ion; the NO_3^- ion does not hydrolyze.

16.14 a. Calculate K_b of the F^- ion from the K_a of its conjugate acid, HF:

$$K_b = K_w \div K_a = 1.00 \times 10^{-14} \div 6.8 \times 10^{-4} = 1.\underline{4}7 \times 10^{-11} = 1.5 \times 10^{-11}$$

b. Calculate K_a of $C_6H_5NH_2^+$ from K_b of its conjugate base, $C_6H_5NH_2$:

$$K_a = K_w \div K_b = 1.00 \times 10^{-14} \div (4.2 \times 10^{-10}) = 2.\underline{3}8 \times 10^{-5} = 2.4 \times 10^{-5}$$

16.15 Rounded answer: $[OH^-] = 1.6 \times 10^{-6}$ M; pH = 8.19. Assemble the usual table, letting $[Ben^-]$ equal the equilibrium concentration of the benzoate anion (the only ion that hydrolyzes). Then calculate K_b of the Ben^- ion from K_a of its conjugate acid, HBen. Assume x is much smaller than the 0.015 M concentration in the denominator and solve for x in the numerator of the equilibrium expression. Finally, calculate pOH from the $[OH^-]$ and pH from the pOH.

Conc. (M)	Ben^- +	H_2O	$\rightleftharpoons$	HBen +	OH^-
Starting	0.015			0	0
Change	-x			+x	+x
Equilibrium	0.015 - x			x	x

$$K_b = K_w \div K_a = 1.00 \times 10^{-14} \div (6.3 \times 10^{-5}) = 1.\underline{5}8 \times 10^{-10}$$

$$\frac{[HBen][OH^-]}{[Ben^-]} \cong \frac{(x)^2}{(0.015)} \cong 1.58 \times 10^{-10}$$

$x = [OH^-] \cong 1.\underline{5}39 \times 10^{-6}$ M (note that x is negligible compared to 0.015)

$pOH = -\log [OH^-] = -\log (1.539 \times 10^{-6}) = 5.8\underline{12}$

$pH = 14.00 - 5.812 = 8.1\underline{88} = 8.19$

16.16 Rounded answer: $[CHO_2^-] = 8.5 \times 10^{-5}$ M; degee of ionization = 8.5×10^{-4} (0.085%). Assemble the usual table, using starting $[H^+] = 0.20$ M from 0.20 M HCl, and letting HFo symbolized $HCHO_2$. Assume x is negligible compared to 0.10 M and 0.20 M, and solve for x in the numerator. Calculate the degree of ionization from $[CHO_2^-]$.

Conc. (M)	HFo	$\rightleftharpoons$	H^+	+	Fo^-
Starting	0.10		0.20		0
Change	-x		+x		+x
Equilibrium	0.10 - x		0.20 + x		x

$$\frac{[H^+][Fo^-]}{[HFo]} = K_a = \frac{(0.20 + x)(x)}{(0.10 - x)} = 1.7 \times 10^{-4}$$

$$\frac{(0.20)(x)}{(0.10)} \cong 1.7 \times 10^{-4}; \; x = [Fo^-] \cong 8.\underline{5}0 \times 10^{-5} \text{ M}$$

Degee of ionization = $8.50 \times 10^{-5} \div 0.10 = 8.50 \times 10^{-4}$ (0.085%)

16.17 Rounded answer: $[H^+] = 2.4 \times 10^{-4}$ M; pH = 3.63. Assemble the usual table, using a starting $[CHO_2^-]$ of 0.018 M from 0.018 M $NaCHO_2$, and symbolizing $HCHO_2$ as HFo and the CHO_2^- anion as Fo^-. Assume x is negligible compared to 0.025 M and 0.018 M, and solve for the x in the numerator.

Conc. (M)	HFo	$\rightleftharpoons$	H^+	+	Fo^-
Starting	0.025		0		0.018
Change	-x		+x		+x
Equilibrium	0.025 - x		x		0.018 + x

$$\frac{[H^+][Fo^-]}{[HFo]} = K_a = \frac{(x)(0.018 + x)}{(0.025 - x)} = 1.7 \times 10^{-4}$$

$$\frac{(x)(0.018)}{(0.025)} \cong 1.7 \times 10^{-4}; \; x = [H^+] \cong 2.\underline{3}6 \times 10^{-4} \text{ M}$$

$pH = -\log [H^+] = -\log (2.36 \times 10^{-4}) = 3.6\underline{2}7$

16.18 Rounded answer: $[H^+] = 5.5 \times 10^{-5}$ M; pH = 5.26. Find the mol/L of HAc ($HC_2H_3O_2$) and the mol/L of Ac^- ($C_2H_3O_2^-$), and assemble the usual table. Substitute the equilibrium concentrations into the equilibrium expression; then assume x is negligible compared to the starting concentrations of both HAc and Ac^-. Solve for x in the numerator of the equilibrium expression, and calculate the pH from this value.

Total volume = 0.030 L + 0.070 L = 0.100 L

(0.15 mol HAc/L) x 0.030 L = 0.0045 mol HAc ($\div$ 0.100 L total volume = 0.045 M)

(0.20 mol Ac^-/L) x 0.070 L = 0.0140 mol Ac^- (÷ 0.100 L total volume = 0.14 M)

Now substitute these starting concentrations into the usual table.

Conc. (M)	HAc	⇌	H^+	+	Ac^-
Starting	0.045		0		0.14
Change	-x		+x		+x
Equilibrium	0.045 - x		x		0.14 + x

$$\frac{[H^+][Ac^-]}{[HAc]} = K_a = \frac{(x)(0.045 + x)}{(0.14 - x)} = 1.7 \times 10^{-5}$$

$$\frac{(x)(0.045)}{(0.14)} \cong 1.7 \times 10^{-5};\ x = [H^+] \cong 5.\underline{46} \times 10^{-6}\ M$$

$$pH = -\log [H^+] = -\log (5.46 \times 10^{-6}) = 5.2\underline{62}$$

16.19 Rounded answer: $[OH^-] = 6.8 \times 10^{-11}$; pH = 3.83. The OH^- ion reacts with the $HCHO_2$ (HFo) to form additional CHO_2^- (Fo^-) plus H_2O. Calculate the stoichiometric amounts of NaOH; then subtract the moles of NaOH from the moles of HFo. Add the resulting moles of Fo^- to the 0.018 starting moles of Fo^- already present.

(0.10 mol NaOH/L) x 0.0500 L = 0.0050 mol NaOH (reacts with 0.0050 mol HFo)

Mol HFo left = (0.025 - 0.0050) mol = 0.020 mol HFo

Total mol Fo^- now present = 0.018 orig + 0.0050 formed from NaOH = 0.023 mol Fo^-

$$[HFo] = \frac{0.020\ mol\ HFo}{1.050\ L\ soln} = 0.01\underline{9}0\ M;\ [Fo^-] = \frac{0.023\ mol\ Fo^-}{1.050\ L\ soln} = 0.02\underline{1}9\ M$$

Now account for the ionization of HFo to Fo^- at equilbrium by assembling the usual table. Assume x is negligible compared to 0.0190 M and 0.0219 M, and solve the equilibrium expression for x in the numerator. Calculate the pH from x, the $[H^+]$.

Conc. (M)	HFo	⇌	H^+	+	Fo^-
Starting	0.0190		0		0.0290
Change	-x		+x		+x
Equilibrium	0.0190 - x		x		0.0290 + x

$$\frac{[H^+][Fo^-]}{[HFo]} = K_a = \frac{(x)(0.0219 + x)}{(0.0190 - x)} = 1.7 \times 10^{-4}$$

$$\frac{(x)(0.0219)}{(0.0190)} \cong 1.7 \times 10^{-4};\ x = [H^+] \cong 1.\underline{47} \times 10^{-4}\ M$$

$$pH = -\log [H^+] = -\log (1.47 \times 10^{-4}) = 3.8\underline{31}$$

16.20 Rounded answer: $[H^+]$ = 0.025 M; pH = 1.60. All the OH^- (from the NaOH) reacts with the H^+ from HCl. Calculate the stoichiometric amounts of OH^- and H^+ and subtract the mol of OH^- from the mol of H^+. Then divide the remaining H^+ by the total volume of 0.025 L + 0.015 L, or 0.040 L, to find the $[H^+]$. Then calculate the pH.

Mol H^+ = (0.10 mol HCl/L) x 0.025 L HCl = 0.0025 mol H^+

Mol OH^- = (0.10 mol NaOH/L) x 0.015 L NaOH = 0.0015 mol OH^-

Mol H^+ left = (0.0025 - 0.0015) mol H^+ = 0.0010 mol H^+

$[H^+]$ = 0.0010 mol H^+ ÷ 0.040 L total volume = 0.025 M

pH = -log $[H^+]$ = -log (0.025) = $1.6\underline{0}2$

16.21 Rounded answer: $[OH^-]$ = 9.4×10^{-7} M; pH = 7.97. At the equivalence point, equal molar amounts of HF and NaOH react to form a solution of NaF. Start by calculating the moles of HF. Use this to calculate the volume of NaOH needed to neutralize all of the HF (and use the moles of HF as the moles of F^- formed at the equivalence point). Add the volume of NaOH to the original 0.025 L to find the total volume of solution.

(0.10 mol HF/L) x 0.025 L = 0.0025 mol HF

Volume NaOH = 0.0025 mol HF ÷ (0.15 mol NaOH/L) = $0.01\underline{6}6$ L

Total volume = 0.0166 L + 0.025 L HF soln = 0.0416 L

$[F^-]$ = (0.0025 mol F^- from HF) ÷ 0.0416 L = $0.06\underline{0}09$ M

Since the F^- hydrolyzes to OH^- and HF, use this to calculate the $[OH^-]$. Start by calculating the hydrolysis constant of F^- from the K_a of its conjugate acid, HF. Then assemble the usual table of concentrations, assume x is negligible, and calculate $[OH^-]$ and pH.

$K_b = K_w \div K_a = 1.00 \times 10^{-14} \div (6.8 \times 10^{-4}) = 1.\underline{4}7 \times 10^{-11}$

Conc. (M)	F^-	+	H_2O	$\rightleftharpoons$	HF	+	OH^-
Starting	0.06009				0		0
Change	-x				+x		+x
Equilibrium	0.06009 - x				x		x

$$\frac{[HF][OH^-]}{[F^-]} = K_b = \frac{(x)^2}{(0.06009 - x)} = 1.47 \times 10^{-11}$$

$$\frac{(x)^2}{(0.06009)} \cong 1.47 \times 10^{-11};\ x = [OH^-] \cong 9.\underline{3}9 \times 10^{-7} \text{ M}$$

pOH = -log $[OH^-]$ = -log (9.39×10^{-7}) = $6.0\underline{2}6$

pH = 14.00 - 6.026 = $7.9\underline{7}4$

16.22 Rounded answer: $[H^+]$ = 6.5×10^{-6} M; pH = 5.19. At the equivalence point, equal molar amounts of NH_3 and HCl react to form a solution of NH_4Cl. Start by calculating the moles of NH_3. Use this to calculate the volume of HCl needed to neutralize all of the NH_3 (and use the moles of NH_3 as the moles of NH_4^+ formed at the equivalence point). Add the volume of HCl to the original 0.035 L to find the total volume of solution.

$$(0.20 \text{ mol } NH_3/L) \times 0.035 \text{ L} = 0.00700 \text{ mol } NH_3$$

$$\text{Volume HCl} = 0.00700 \text{ mol } NH_3 \div (0.12 \text{ mol HCl/L}) = 0.0583 \text{ L}$$

$$\text{Total volume} = 0.035 \text{ L} + 0.0583 \text{ L } NH_3 \text{ soln} = 0.0933 \text{ L}$$

$$[NH_4^+] = (0.00700 \text{ mol } NH_4^+ \text{ from } NH_3) \div 0.0933 \text{ L} = 0.0750 \text{ M}$$

Since the NH_4^+ hydrolyzes to H_3O^+ and NH_3, use this to calculate the $[H^+]$. Start by calculating the hydrolysis constant of NH_4^+ from the K_b of its conjugate base, NH_3. Then assemble the usual table of concentrations, assume x is negligible, and calculate $[H^+]$ and pH.

$$K_a = K_w \div K_b = 1.00 \times 10^{-14} \div (1.8 \times 10^{-5}) = 5.56 \times 10^{-11}$$

Conc. (M)	NH_4^+ +	H_2O	$\rightleftharpoons$	NH_3	+	H_3O^+
Starting	0.0750			0		0
Change	-x			+x		+x
Equilibrium	0.0750 - x			x		x

$$\frac{[NH_3][H_3O^+]}{[NH_4^+]} = K_a = \frac{(x)^2}{(0.0750 - x)} = 5.56 \times 10^{-10}$$

$$\frac{(x)^2}{(0.0750)} \cong 5.56 \times 10^{-10}; \ [H^+] \cong 6.46 \times 10^{-6} \text{ M}$$

$$\text{pH} = -\log [H^+] = -\log (6.46 \times 10^{-6}) = 5.189$$

ANSWERS TO REVIEW QUESTIONS

16.1 The self-ionization of water is the reaction of two water molecules in which a proton is transferred from one molecule to the other to form H_3O^+ and OH^- ions. At 25°C, the K_w expression is: $K_w = [H_3O^+][OH^-] = 1.00 \times 10^{-14}$.

16.2 The pH = -log $[H^+]$ of an aqueous solution. Measure pH by using electrodes and a pH meter, or by interpolating the pH from the color changes of a series of acid-base indicators.

16.3 For a neutral solution, $[H^+] = [OH^-]$; thus the $[H^+]$ of a neutral solution at 37°C is the square root of K_w at 37°C:

$$[H^+] = \sqrt{2.5 \times 10^{-14}} = 1.58 \times 10^{-7} \text{ M}; \ \text{pH} = -\log (1.58 \times 10^{-7}) = 6.801 = 6.80$$

16.4 Since pH + pOH = pK_w at any temperature:

$$pH + pOH = -\log(2.5 \times 10^{-14}) = 13.60$$

16.5 The equation is: $HCN(aq) \rightleftharpoons H^+(aq) + CN^-(aq)$. The equilibrium expression is:

$$K_a = \frac{[H^+][CN^-]}{[HCN]}$$

16.6 HCN is the weakest acid. Its K_a of 4.9×10^{-10} is < K_a of 1.7×10^{-5} of $HC_2H_3O_2$; $HClO_4$ is a strong acid, of course.

16.7 Both methods involve direct measurement of the concentrations of the hydrogen ion and the anion of the weak acid and calculation of the concentration of the unionized acid. All concentrations are substituted into the K_a expression to obtain a value for K_a. The first method: measure the electrical conductivity of a solution of the weak acid. The conductivity is proportional to the concentration of the hydrogen ion and anion. The second method: measure the pH of a known starting concentration of weak acid. Convert the pH to $[H^+]$, which will be equal to the [anion].

16.8 The degree of ionization of a weak acid decreases as the concentration of the acid added to the solution increases. Compared to low concentrations, at high concentrations there is less water for each weak acid molecule to react with the weak acid as it ionizes:

$$HA(aq) + H_2O(l) \rightleftharpoons H_3O^+(aq) + A^-(aq)$$

16.9 We can neglect x if K_a/M is $\leq 10^{-3}$. In this case, $K_a/M = (6.8 \times 10^{-2} \div 0.010\ M) = 6.8 \times 10^{-2}$, and x cannot be neglected in the (0.010 - x) term. This says the degree of ionization is significant.

16.10 The ionization of the first H^+: $H_2PHO_3(aq) \rightleftharpoons H^+(aq) + HPHO_3^-(aq)$.

The ionization of the second H^+: $HPHO_3^-(aq) \rightleftharpoons H^+(aq) + PHO_3^-(aq)$.

$$K_{a1} = \frac{[H^+][HPHO_3^-]}{[H_2PHO_3]}; \quad K_{a2} = \frac{[H^+][PHO_3^{2-}]}{[HPHO_3^-]}$$

16.11 The concentration of the -2 anion in a solution of only the diprotic acid molecule is equal approximately to the value of K_{a2}. For the diprotic acid oxalic acid, it is approximately true that the $[H^+] = [HC_2O_4^-]$. This can be shown by substituting $[H^+]$ for $[HC_2O_4^-]$ in the K_{a2} expression:

$$K_{a2} = 5.1 \times 10^{-5} = \frac{[H^+][C_2O_4^{2-}]}{[HC_2O_4^-]} = \frac{[H^+][C_2O_4^{2-}]}{[H^+]} = [C_2O_4^{2-}]$$

16.12 The ionization of aniline: $C_6H_5NH_2(aq) + H_2O(l) \rightleftharpoons C_6H_5NH_3^+(aq) + OH^-(aq)$.

$$K_b = \frac{[C_6H_5NH_3^+][OH^-]}{[C_6H_5NH_2]}$$

16.13 The CH_3NH_2 molecule is the strongest base because its ionization constant of 4.4×10^{-4} is larger than the respective K_b's of 1.8×10^{-5} for NH_3 and 4.2×10^{-10} for $C_6H_5NH_2$.

16.14 A solution of anilinium chloride should be acidic because the $C_6H_5NH_3^+$ ion is a conjugate acid of a weak base; thus it should hydrolyze to form H_3O^+ ion, making the solution acidic. (The Cl^- ion will not hydrolyze because it is the conjugate base of a strong acid.) The equation is:

$$C_6H_5NH_3^+(aq) + H_2O(l) \rightleftharpoons C_6H_5NH_2(aq) + H_3O^+(aq)$$

$$K_h \text{ or } K_b = \frac{[C_6H_5NH_2^+][H_3O^+]}{[C_6H_5NH_3^+]} = \frac{K_w}{K_b} \text{ (where } K_b \text{ is for } C_6H_5NH_2)$$

16.15 The common-ion effect refers to the equilibrium shift that occurs when an ion common to the equilibrium is added to a solution in which the equilibrium has been established. An example is the addition of H^+ (as HCl) to the equilibrium below:

$$HF(aq) \rightleftharpoons H^+(aq) + F^-(aq)$$

The additional H^+ displaces the equilibrium composition to the left.

16.16 When CH_3NH_3Cl is added to a 0.10 M CH_3NH_2 solution, the pH will drop below 11.8 because of the common-ion effect from the $CH_3NH_3^+$ cation. The composition of the CH_3NH_2 equilibrium involves the $CH_3NH_3^+$ cation; thus adding more of it will shift the equilibrium composition back toward CH_3NH_2, using up OH^- and making the solution less basic (more acidic).

16.17 A buffer is a solution that is able to resist significant changes in pH when limited amounts of acid or base (or water) are added to it. A buffer must contain a weak acid and its conjugate base (weak). Strong acids or strong bases do not make effective buffers for pH control in either direction. An example of a buffer pair is the H_2CO_3 and $NaHCO_3$ pair (blood buffer).

16.18 Buffer capacity is the amount of acid or base the buffer can react with before exhibiting a significant pH change. A low-capacity buffer is a solution containing just 0.010 mol of HF and 0.010 mol of F^- per liter; a high-capacity buffer is a solution containing 1.0 mol of HF and 1.0 mol of F^- per liter.

16.19 During the titration of a weak base by a strong acid, the pH at the start is at a high value (about 10). It slowly decreases as strong acid is added until the equivalence pH (from 5 to 6) is reached. (The equivalence point is the point at which a stoichiometric amount of strong acid has been added to the weak base). After the equivalence point, the pH decreases below pH 7, and finally levels off about pH 2.

16.20 Either thymol blue or phenolphthalein would be a suitable indicator for an equivalence point pH of 8.0. The color change with either would occur after the equivalence point has been reached and not before.

SOLUTIONS TO PRACTICE PROBLEMS

Note on answers to equilibrium calculations: The rounded answer (with significant figures) is always given first. After the stepwise solution, the numerical answer is given again, but with one nonsignificant figure. (The last significant figure is indicated by underlining it.)

16.21 a. $[H^+] = 0.25$ M; $[OH^-] = 4.0 \times 10^{-14}$ M.

b. $[OH^-] = 1.25$ M; $[H^+] = 8.\underline{0}0 \times 10^{-15} = 8.0 \times 10^{-15}$ M.

c. $[OH^-] = 0.0070$ M; $[H^+] = 1.\underline{4}2 \times 10^{-12} = 1.4 \times 10^{-12}$ M.

d. $[H^+] = 2.5$ M; $[OH^-] = 4.0 \times 10^{-15}$ M.

16.23 The $[H^+] = 0.050$ M (HCl is a strong acid); using K_w, the $[OH^-] = 2.0 \times 10^{-13}$ M.

16.25 Since the $Sr(OH)_2$ forms $2OH^-$ per formula unit, the $[OH^-] = 2 \times 0.0050 = 0.010$ M.

$$[H^+] = K_w \div [OH^-] = 1.0 \times 10^{-14} \div 0.010 \text{ M} = 1.\underline{0}0 \times 10^{-12} = 1.0 \times 10^{-12} \text{ M}$$

16.27 a. Basic; 5×10^{-9} M $H^+ < 1.0 \times 10^{-7}$ M H^+

b. Acidic; 2×10^{-5} M $H^+ > 1.0 \times 10^{-7}$ M H^+

c. Neutral; 1×10^{-7} M H^+ is the same as 1.0×10^{-7} M H^+ of neutral water

d. Acidic; 2×10^{-6} M $H^+ > 1.0 \times 10^{-7}$ M H^+

16.29 The $[H^+]$ calculated below is $< 1.0 \times 10^{-7}$ M, so the solution is acidic.

$$[H^+] = K_w \div [OH^-] = 1.0 \times 10^{-14} \div (1.5 \times 10^{-9} \text{ M}) = 6.\underline{6}6 \times 10^{-6} = 6.7 \times 10^{-6} \text{ M}$$

16.31 a. Basic ($11.2 > 7.0$) b. Neutral ($7.0 = 7.0$) c. Acidic ($1.2 < 7.0$) d. Acidic ($6.1 < 7.0$)

16.33 a. Acidic ($3.5 < 7.0$) b. Neutral ($7.0 = 7.0$) c. Basic ($9.0 > 7.0$) d. Acidic ($5.5 < 7.0$)

16.35 Record the same number of places after the decimal point in the pH as the number of significant figures in the $[H^+]$.

a. $-\text{Log}(1.0 \times 10^{-8}) = 8.00$

b. $-\text{Log}(5.0 \times 10^{-12}) = 11.3\underline{0}1 = 11.30$

c. $-\text{Log}(7.5 \times 10^{-3}) = 2.1\underline{2}4 = 2.12$

d. $-\text{Log}(6.35 \times 10^{-9}) = 8.19\underline{7}2 = 8.197$

16.37 Record the same number of places after the decimal point in the pH as the number of significant figures in the $[H^+]$.

$$-\text{Log}\ (7.5 \times 10^{-3}) = 2.1\underline{2}49 = 2.12$$

16.39 First convert the $[OH^-]$ to $[H^+]$ using the K_w equation. Then find the pH, recording the same number of places after the decimal point in the pH as the number of significant figures in the $[H^+]$.

$$[H^+] = K_w \div [OH^-] = 1.0 \times 10^{-14} \div (0.0040) = 2.\underline{5}0 \times 10^{-12}\ M$$

$$-\text{Log}\ (2.50 \times 10^{-12}) = 11.6\underline{0}2 = 11.60$$

16.41 From the definition pH = -log $[H^+]$, -pH = log $[H^+]$, so enter the negative value of the pH on the calculator and use the inverse and log keys (or 10^x) key to find the antilog of -pH:

$$\text{Log}\ [H^+] = -pH = -5.12$$

$$[H^+] = \text{antilog}\ (-5.12) = 10^{-5.12} = 7.\underline{5}8 \times 10^{-6} = 7.6 \times 10^{-6}\ M$$

16.43 From the definition pH = -log $[H^+]$, -pH = log $[H^+]$, so enter the negative value of the pH on the calculator and use the inverse and log keys (or 10^x) key to find the antilog of -pH. Then use the K_w equation to calculate $[OH^-]$ from $[H^+]$.

$$\text{Log}\ [H^+] = -pH = -11.63$$

$$[H^+] = \text{antilog}\ (-11.63) = 10^{-11.63} = 2.\underline{3}4 \times 10^{-12}\ M$$

$$[OH^-] = K_w \div [H^+] = 1.0 \times 10^{-14} \div (2.34 \times 10^{-12}) = 4.\underline{2}7 \times 10^{-3} = 4.3 \times 10^{-3}\ M$$

16.45 First calculate the molarity of the OH^- ion from the mass of NaOH. Then convert the $[OH^-]$ to $[H^+]$ using the K_w equation. Then find the pH, recording the same number of places after the decimal point in the pH as the number of significant figures in the $[H^+]$.

$$\frac{5.80\ \text{g NaOH}}{1.00\ \text{L}} \times \frac{1\ \text{mol NaOH}}{40.01\ \text{g NaOH}} = \frac{0.1450\ \text{mol NaOH}}{\text{L}} = 0.14\underline{5}0\ M\ OH^-$$

$$[H^+] = K_w \div [OH^-] = 1.0 \times 10^{-14} \div (0.1450) = 6.\underline{8}96 \times 10^{-14}\ M$$

$$pH = -\log [H^+] = -\log (6.896 \times 10^{-14}) = 13.1\underline{6}14 = 13.16$$

16.47 Figure 16.4 shows that the methyl red indicator is yellow at pH values above about 5.5 (slightly past the midpoint of the range for methyl red). Bromthymol blue is yellow at pH's up to about 6.5 (slightly below the midpoint of the range for bromthymol blue). Therefore the pH of the solution is between 5.5 and 6.5, and the solution is acidic. (Note: normal rain has a pH of 5.6, as indicated in the related topic discussion of acid rain.)

16.49 a. $HNO_2(aq) + H_2O(l) \rightleftharpoons H_3O^+(aq) + NO_2^-(aq)$; $HNO_2(aq) \rightleftharpoons H^+(aq) + NO_2^-(aq)$

$$K_a = \frac{[H^+][NO_2^-]}{[HNO_2]}$$

b. $HClO(aq) + H_2O(l) \rightleftharpoons H_3O^+(aq) + ClO^-(aq)$; $HClO(aq) \rightleftharpoons H^+(aq) + ClO^-(aq)$

$$K_a = \frac{[H^+][ClO^-]}{[HClO]}$$

c. $HCN(aq) + H_2O(l) \rightleftharpoons H_3O^+(aq) + CN^-(aq)$; $HCN(aq) \rightleftharpoons H^+(aq) + CN^-(aq)$

$$K_a = \frac{[H^+][CN^-]}{[HCN]}$$

d. $HCHO_2(aq) + H_2O(l) \rightleftharpoons H_3O^+(aq) + CHO_2^-(aq)$; $HCHO_2(aq) \rightleftharpoons H^+(aq) + CHO_2^-(aq)$

$$K_a = \frac{[H^+][CHO_2^-]}{[HCHO_2]}$$

16.51 Rounded answer: $K_a = 1.9 \times 10^{-6}$ (correct sig figs). To solve, assemble a table of starting, change, and equilibrium concentrations. Start by converting pH to $[H^+]$:

$$[H^+] = \text{antilog}(-pH) = \text{antilog}(-3.21) = 6.\underline{1}6 \times 10^{-4} \text{ M.}$$

Conc. (M)	HN_3	$\rightleftharpoons$	H^+	+	N_3^-
Starting	0.20		0		0
Change	-x		+x		+x
Equilibrium	0.20 - x		x		x

The value of x equals the value of the molarity of the H^+ ion, which was obtained above from the pH: $[H^+] = 6.\underline{1}6 \times 10^{-4}$ M. After using x to substitute into the equilibrium expression, substitute this value for x and solve for K_a:

$$K_a = \frac{[H^+][N_3^-]}{[HN_3]} = \frac{[x]^2}{(0.20 - x)} = \frac{(6.16 \times 10^{-4})^2}{0.20} = 1.\underline{8}9 \times 10^{-6} \text{ (rounded answer above)}$$

16.53 Rounded answer: degree of ionization = 0.00015; pH = 5.42. To solve, assemble a table of starting, change, and equilibrium concentrations. Use HBo as the symbol for boric acid, and Bo- as the symbol for $B(OH)_4^-$.

Conc. (M)	HBo	$\rightleftharpoons$	H^+	+	Bo^-
Starting	0.025		0		0
Change	-x		+x		+x
Equilibrium	0.025 - x		x		x

The value of x equals the value of the molarity of the H^+ ion, which can be obtained from the equilibrium expression. Substitute into the equilibrium expression, and solve for x.

$$\frac{[H^+][Bo^-]}{[HBo]} = K_a = \frac{[x]^2}{(0.025 - x)} = 5.9 \times 10^{-10}$$

Solve the equation for x, assuming that x is much smaller than 0.025, so that $(0.025 - x) \cong 0.025$. Solve by taking the square root of $5.9 \times 10^{-10} \times (0.025)$.

$$\frac{[x]^2}{(0.025)} \cong 5.9 \times 10^{-10};\ x^2 \cong 5.9 \times 10^{-10}\,(0.025) = 1.\underline{4}75 \times 10^{-11}$$

$$x = 3.\underline{8}4 \times 10^{-6}\ M = [H^+];\ \text{degree ionization} = (3.84 \times 10^{-6}/0.025) = 0.0001\underline{5}3$$

Check to make sure that the assumption that $(0.025 - x) \cong 0.025$ is valid:

$$0.025 - 3.84 \times 10^{-6} = 0.02499, \text{ or } \cong 0.025 \text{ to 2 sig figs.}$$

$$pH = -\log [H^+] = -\log (3.84 \times 10^{-6}) = 5.4\underline{1}56 \text{ (rounded answer above)}$$

16.55 Rounded answer: $[H^+] = [C_6H_4NH_2COO^-] = 1.0 \times 10^{-3}$ M. To solve, assemble a table of starting, change, and equilibrium concentrations. Use HPaba as a symbol for p-aminobenzoic acid (PABA), and $Paba^-$ as the symbol for the -1 anion.

Conc. (M)	HPaba	⇌	H^+	+	$Paba^-$
Starting	0.050		0		0
Change	-x		+x		+x
Equilibrium	0.050 - x		x		x

Write the equilibrium expression in terms of chemical symbols, then substitute the terms x and (0.050 - x):

$$\frac{[H^+][Paba^-]}{[HPaba]} = K_a = \frac{[x]^2}{(0.050 - x)} = 2.2 \times 10^{-5}$$

Technically, x should not be neglected compared to 0.050 M. If it is neglected, the calculated $[H^+]$ is 1.04×10^{-3}, which when subtracted from 0.050 yields 0.0489, a significant change. However, the quadratic formula gives the same value of x within two, but not three, significant figures: 1.02×10^{-3} M. Therefore we neglect x and avoid the quadratic:

$$x^2 = 2.2 \times 10^{-5} \times 0.050 = 1.1 \times 10^{-6}$$

$$x = [H^+] = [Paba^-] \cong 1.\underline{0}4 \times 10^{-3}\ M \text{ (rounded answer above)}$$

16.57 Rounded answer: $[HC_2H_3O_2] = 0.26$ M. To solve, first convert the pH to $[H^+]$, which also equals the $[C_2H_3O_2^-]$, here symbolized as $[Ac^-]$. Then assemble the usual table and substitute into the equilibrium constant expression to solve for the $[HC_2H_3O_2]$, here symbolized as [HAc].

$$[H^+] = \text{antilog}(-2.68) = 2.\underline{0}89 \times 10^{-3}\ M$$

Conc. (M)	HAc	$\rightleftharpoons$	H^+	+	Ac^-
Starting	x		0		0
Change	-2.089×10^{-3}		$+2.089 \times 10^{-3}$		$+2.089 \times 10^{-3}$
Equilibrium	$(x - 2.089 \times 10^{-3})$		2.089×10^{-3}		2.089×10^{-3}

Write the equilibrium expression in terms of chemical symbols, and then substitute the x and the $(x - 2.089 \times 10^{-3})$ terms into the expression:

$$\frac{[H^+][Ac^-]}{[HAc]} = K_a = \frac{(2.089 \times 10^{-3})^2}{(x - 0.002089)} = 1.7 \times 10^{-5}$$

Solve the equation for x, assuming that 0.002089 is much smaller than x, so that we can say that $(x - 0.002089) \cong (x)$. Solve by rearranging to:

$$(x) = [HAc] \cong \frac{(2.089 \times 10^{-3})^2}{1.7 \times 10^{-5}} \cong 0.2\underline{5}6 \text{ M}$$

Adding 0.00289 to 0.256 does indeed give 0.26 to two significant figures.

16.59 Rounded answer: $[H^+] = 4.9 \times 10^{-3}$ M; pH = 2.31. To solve, assemble the usual table of starting, change, and equilibrium concentrations of HF and F^- ion.

Conc. (M)	HF	$\rightleftharpoons$	H^+	+	F^-
Starting	0.040		0		0
Change	-x		+x		+x
Equilibrium	0.040 - x		x		x

Write the equilibrium expression in terms of chemical symbols, then substitute the terms x and (0.040 - x):

$$\frac{[H^+][F^-]}{[HF]} = K_a = \frac{[x]^2}{(0.040 - x)} = 6.8 \times 10^{-4}$$

In this case, x cannot be neglected compared to 0.040 M. (If it is neglected, subtracting the calculated $[H^+]$ from 0.040 yields a significant change.) The quadratic formula must be used. Reorganize the above equilibrium expression into the form $ax^2 + bx + c = 0$, and substitute for a, b, and c in the quadratic formula.

$$x^2 + 6.8 \times 10^{-4} - 2.72 \times 10^{-5} = 0$$

$$x = \frac{-6.8 \times 10^{-4} \pm \sqrt{(6.8 \times 10^{-4})^2 + (4)(2.72 \times 10^{-5})}}{2}$$

$$x = \frac{-6.8 \times 10^{-4} \pm \sqrt{1.092 \times 10^{-4}}}{2}$$

$$x = 9.768 \times 10^{-3} \div 2 = 4.\underline{8}8 \times 10^{-3}$$

$$\text{pH} = -\log(4.88 \times 10^{-3}) = 2.3\underline{1}1 \text{ (rounded answer above)}$$

16.61 Rounded answer: $[H^+] = 3.61 \times 10^{-1}$ M. To solve, assemble the usual table of starting, change, and equilibrium concentrations of $(NO_2)_2C_6H_3CO_2H$, symbolized as HDin, and the $(NO_2)_2C_6H_3CO_2^-$ ion, symbolized as Din^-.

Conc. (M)	HDin	⇌	H^+	+	Din^-
Starting	2.00		0		0
Change	-x		+x		+x
Equilibrium	2.00 - x		x		x

Write the equilibrium expression in terms of chemical symbols, then substitute the terms x and (2.00 - x):

$$\frac{[H^+][Din^-]}{[HDin]} = K_a = \frac{[x]^2}{(2.00 - x)} = 7.94 \times 10^{-2}$$

In this case, x cannot be neglected compared to 2.00 M. (If it is neglected, subtracting the calculated $[H^+]$ from 2.00 yields a significant change.) The quadratic formula must be used. Reorganize the above equilibrium expression into the form $ax^2 + bx + c = 0$, and substitute for a, b, and c in the quadratic formula.

$$x^2 + 7.94x\ 10^{-2} - 1.588 \times 10^{-1} = 0$$

$$x = \frac{-7.94 \times 10^{-2} \pm \sqrt{(7.94 \times 10^{-2})^2 + (4)(1.588 \times 10^{-1})}}{2}$$

$$x = \frac{-7.94 \times 10^{-2} \pm \sqrt{6.41504 \times 10^{-1}}}{2}$$

$$x = 7.215 \times 10^{-1} \div 2 = 3.6\underline{0}76 \times 10^{-1}$$

(pH not required but value of pH = $0.44\underline{2}7$)

16.63 Rounded answers: (a) $[H^+] = 3.7 \times 10^{-3}$ M; (b) $[C_8H_4O_4^{2-}] = K_{a2} = 3.9 \times 10^{-6}$ M. The $C_8H_4O_4^{2-}$ ion will be symbolized as the $[Ph^{2-}]$ ion. To solve, note that $K_{a1} = 1.2 \times 10^{-3} > K_{a2} = 3.9 \times 10^{-6}$, and hence the second ionization and K_{a2} can be neglected. Assemble a table of starting, change, and equilibrium concentrations; let $H_2Ph = H_2C_8H_4O_4$ and $HPh = HC_8H_4O_4^-$.

Conc. (M)	H_2Ph	⇌	H^+	+	HPh^-
Starting	1.50		0		0
Change	-x		+x		+x
Equilibrium	1.50 - x		x		x

Write the equilibrium expression in terms of chemical symbols, then substitute x and (0.25 - x):

$$\frac{[H^+][HPh^-]}{[H_2Ph]} = K_{a1} = \frac{[x]^2}{(0.015 - x)} = 1.2 \times 10^{-3}$$

In this case, x cannot be neglected in the (0.015 M - x) term. Solving the quadratic gives:

$$x = (7.3697 \times 10^{-3}) \div 2$$

$x = [H^+] \cong 3.\underline{6}84 \times 10^{-3}$ M (rounded answer above)

Since $[HPh^-]$ is $\cong [H^+]$, these terms cancel in the K_{a2} expression. This reduces to the equality that the concentration of the -2 phthalate ion, $[Ph^{2-}]$, = K_{a2} = 3.9×10^{-6} M.

16.65 The equation is: $C_2H_5NH_2(aq) + H_2O(l) \rightleftharpoons C_2H_5NH_3^+(aq) + OH^-(aq)$. The K_b expression is:

$$K_b = \frac{[C_2H_5NH_3^+][OH^-]}{[C_2H_5NH_2]}$$

16.67 Rounded answer: $K_b = 3.2 \times 10^{-5}$ (correct sig figs). To solve, convert the pH to $[OH^-]$:

$$pOH = 14.00 - pH = 14.00 - 11.34 = 2.66$$

$$[OH^-] = \text{antilog}(-2.66) = 2.\underline{1}88 \times 10^{-3}\ M$$

Using the symbol EtN for ethanolamine, assemble a table of starting, change, and equilibrium concentrations.

Conc. (M)	EtN	+	H_2O	$\rightleftharpoons$	$HEtN^+$	+	OH^-
Starting	0.15				0		0
Change	-x				+x		+x
Equilibrium	$0.15 - 2.188 \times 10^{-3}$				2.188×10^{-3}		2.188×10^{-3}

Write the equilibrium expression in terms of chemical symbols, then substitute the terms and solve for K_b:

$$K_b = \frac{[HEtN^+][OH^-]}{[EtN]} = \frac{(2.188 \times 10^{-3})^2}{(0.15 - 0.002188)} = 3.\underline{2}3 \times 10^{-5} \text{ (rounded answer above)}$$

16.69 Rounded answer: $[OH^-] = 5.7 \times 10^{-3}$ M; pH = 11.76. To solve, assemble a table of starting, change, and equilibrium concentrations:

Conc. (M)	CH_3NH_2	+	H_2O	$\rightleftharpoons$	$CH_3NH_3^+$	+	OH^-
Starting	0.080				0		0
Change	-x				+x		+x
Equilibrium	0.080 - x				x		x

Write the equilibrium expression in terms of chemical symbols, then substitute the terms and the value of K_b:

$$\frac{[CH_3NH_3^+][OH^-]}{[CH_3NH_2]} = K_b = \frac{(x)^2}{(0.080 - x)} = 4.4 \times 10^{-4}$$

In this case, x cannot be neglected compared to 0.080 M. (If it is neglected, subtracting the calculated $[H^+]$ from 0.080 yields a significant change.) The quadratic formula must be used.

Reorganize the above equilibrium expression into the form $ax^2 + bx + c = 0$, and substitute for a, b, and c in the quadratic formula.

$$x^2 + 4.4 \times 10^{-4}\ x - 3.52 \times 10^{-5} = 0$$

$$x = \frac{-4.4 \times 10^{-4} \pm \sqrt{(4.4 \times 10^{-4})^2 + (4)(3.52 \times 10^{-5})}}{2}$$

$$x = \frac{-4.4 \times 10^{-4} \pm \sqrt{1.4099 \times 10^{-4}}}{2}$$

$x = 1.143 \times 10^{-2} \div 2 = 5.\underline{7}17 \times 10^{-3} = [OH^-]$ (rounded answer above)

$pOH = -\log(5.717 \times 10^{-3}) = 2.2\underline{4}28$

$pH = 14.00 - 2.2428 = 11.7\underline{5}72$ (rounded answer above)

16.71 a. No hydrolysis because the iodide ion is the anion of a strong acid.

b. Hydrolysis occurs: equation: $CHO_2^- + H_2O \rightleftharpoons HCHO_2 + OH^-$. Equilibrium expression:

$$K_b = \frac{K_w}{K_a} = \frac{[HCHO_2][OH^-]}{[CHO_2^-]}$$

c. Hydrolysis occurs: equation: $CH_3NH_3^+ + H_2O \rightleftharpoons H_3O^+ + CH_3NH_2$. Equilibrium expression:

$$K_a = \frac{K_w}{K_b} = \frac{[H_3O^+][CH_3NH_2]}{[CH_3NH_3^+]}$$

d. Hydrolysis occurs: equation: $IO^- + H_2O \rightleftharpoons HIO + OH^-$. Equilibrium expression:

$$K_b = \frac{K_w}{K_a} = \frac{[HIO][OH^-]}{[IO^-]}$$

16.73 Acid ionization is: $Zn(H_2O)_6^{2+}(aq) + H_2O(l) \rightleftharpoons Zn(H_2O)_5(OH)^+(aq) + H_3O^+(aq)$

16.75 a. $Fe(NO_3)_3$ is a salt of a weak base, $Fe(OH)_3$, and a strong acid, HNO_3, so it would be expected to be acidic. Fe^{3+} is not in Group IA or IIA, so it would be expected to form a metal hydrate ion that hydrolyzes to form an acidic solution.

b. Na_2CO_3 is a salt of a strong base, NaOH, and the anion of a weak acid, HCO_3^-, so it would be expected to be basic.

c. $Ca(CN)_2$ is a salt of a strong base, $Ca(OH)_2$, and a weak acid, HCN, so it would be expected to be basic.

d. NH_4ClO_4 is a salt of a weak base, NH_3, and a strong acid, $HClO_4$, so it would be expected to be acidic.

16.77 a. Both ions hydrolyze:

$$NH_4^+ + H_2O \rightleftharpoons NH_3 + H_3O^+ \qquad C_2H_3O_2^- + H_2O \rightleftharpoons HC_2H_3O_2 + OH^-$$

Calculate the hydrolysis constants of each so as to compare them:

$$NH_4^+ \text{ as an acid: } K_a = \frac{K_w}{K_b} = \frac{1.0 \times 10^{-14}}{1.8 \times 10^{-5}} = 5.\underline{55} \times 10^{-10}$$

$$C_2H_3O_2^- \text{ as a base: } K_b = \frac{K_w}{K_a} = \frac{1.0 \times 10^{-14}}{1.7 \times 10^{-5}} = 5.\underline{88} \times 10^{-10}$$

Since the hydrolysis constant (K_b) for the hydrolysis of $C_2H_3O_2^-$ is slightly larger than the hydrolysis constant (K_a) for the hydrolysis of NH_4^+, the solution will be slightly basic, but close to pH 7.0.

b. Both ions hydrolyze:

$$C_6H_5NH_3^+ + H_2O \rightleftharpoons C_6H_5NH_2 + H_3O^+; \; C_2H_3O_2^- + H_2O \rightleftharpoons HC_2H_3O_2 + OH^-$$

Calculate the hydrolysis constants of each so as to compare them:

$$C_6H_5NH_3^+ \text{ as an acid: } K_a = \frac{K_w}{K_b} = \frac{1.0 \times 10^{-14}}{4.2 \times 10^{-10}} = 2.\underline{38} \times 10^{-5}$$

$$C_2H_3O_2^- \text{ as a base: } K_b = \frac{K_w}{K_a} = \frac{1.0 \times 10^{-14}}{1.7 \times 10^{-5}} = 5.\underline{88} \times 10^{-10}$$

Since the hydrolysis constant (K_a) for the hydrolysis of $C_6H_5NH_3^+$ is larger than the hydrolysis constant (K_b) for the hydrolysis of $C_2H_3O_2^-$, the solution will be acidic, and significantly less than pH 7.0.

16.79 a. The reaction is: $NO_2^- + H_2O \rightleftharpoons HNO_2 + OH^-$, and the hydrolysis constant, K_b, is obtained by dividing K_w by the K_a of the conjugate acid, HNO_2:

$$K_b = K_w \div K_a = 1.0 \times 10^{-14} \div 4.5 \times 10^{-4} = 2.\underline{22} \times 10^{-11} = 2.2 \times 10^{-11}$$

b. The reaction is: $C_5H_5NH^+ + H_2O \rightleftharpoons H_3O^+ + C_5H_5N$, and the hydrolysis constant, K_a, is obtained by dividing K_w by the K_b of the conjugate base, C_5H_5N:

$$K_a = K_w \div K_b = 1.0 \times 10^{-14} \div 1.4 \times 10^{-9} = 7.\underline{14} \times 10^{-6} = 7.1 \times 10^{-6}$$

16.81 Rounded answer: $[OH^-] = [CH_3CH_2CO_2H] \cong 4.4 \times 10^{-6}$ M; pH = 8.64. Assemble the usual table, letting $[Pr^-]$ equal the equilibrium concentration of the propionate anion (the only ion that hydrolyzes). Then calculate K_b of the Pr^- ion from K_a of its conjugate acid, HPr. Assume x is much smaller than the 0.025 M concentration in the denominator and solve for x in the numerator of the equilibrium expression. Finally, calculate pOH from the $[OH^-]$ and pH from the pOH.

Conc. (M)	Pr^- + H_2O	$\rightleftharpoons$	HPr	+	OH^-
Starting	0.025		0		0
Change	-x		+x		+x
Equilibrium	0.025 - x		x		x

$K_b = K_w \div K_a = 1.0 \times 10^{-14} \div (1.3 \times 10^{-5}) = 7.\underline{6}9 \times 10^{-10}$

$$\frac{[HPr][OH^-]}{[Pr^-]} \cong \frac{(x)^2}{(0.025)} \cong 7.\underline{6}9 \times 10^{-10}$$

$x = [OH^-] = [Pr^-] \cong 4.\underline{3}8 \times 10^{-6}$ M (note that x is negligible compared to 0.025)

$pOH = -\log [OH^-] = -\log (4.38 \times 10^{-6}) = 5.3\underline{5}8$

$pH = 14.00 - 5.358 = 8.6\underline{4}2$ (rounded answer above)

16.83 Rounded answer: $[H^+] = [C_5H_5N] \cong 1.0 \times 10^{-3}$ M; pH = 2.99. Assemble the usual table, letting $[PyNH^+]$ equal the equilibrium concentration of the pryridinium cation (the only ion that hydrolyzes). Then calculate K_a of the $PyNH^+$ ion from K_b of its conjugate base, PyN. Assume x is much smaller than the 0.15 M concentration in the denominator and solve for x in the numerator of the equilibrium expression. Finally, calculate pH from the $[H^+]$.

Conc. (M)	$PyNH^+$ + H_2O	$\rightleftharpoons$	H_3O^+	+	PyN
Starting	0.15		0		0
Change	-x		+x		+x
Equilibrium	0.15 - x		x		x

$K_a = K_w \div K_b = 1.00 \times 10^{-14} \div 1.4 \times 10^{-9} = 7.\underline{1}4 \times 10^{-6}$

Write the equilibrium expression in terms of chemical symbols, then substitute the terms and solve for K_b:

$$\frac{[PyN][H_3O^+]}{[PyNH^+]} \cong \frac{(x)^2}{(0.15)} \cong 7.\underline{1}4 \times 10^{-6}$$

$x = [H^+] \cong 1.\underline{0}3 \times 10^{-3}$ M (note that x is negligible compared to 0.15)

$pH = -\log [H^+] = -\log (1.03 \times 10^{-3}) = 2.9\underline{8}7$ (rounded answer above)

16.85 Rounded answers: degree of ionization: a. 0.029 (2.9%) b. 0.0065 (0.65%). To solve, assemble a table of starting, change, and equilibrium concentrations for each part. For each part, assume x is much smaller than the 0.80 M starting concentration of HF. Then solve for x in the numerator of each equilibrium expression by solving the product of 6.8×10^{-4} and the other terms.

a.

Conc. (M)	HF	$\rightleftharpoons$	H^+	+	F^-
Starting	0.80		0		0
Change	-x		+x		+x
Equilibrium	0.80- x		x		x

$$\frac{[H^+][F^-]}{[HF]} = K_a \cong \frac{(x)^2}{(0.80)} \cong 6.8 \times 10^{-4}$$

$x^2 = 6.8 \times 10^{-4} \times 0.80$ M

$x \cong 2.\underline{3}3 \times 10^{-2}$ M $= [H^+]$ (rounded answer above)

$0.80 - 2.33 \times 10^{-2} = 0.7767$, or $\cong 0.78$ to 2 sig figs. (Borderline case; quadratic gives a $[H^+] = 2.298$ or 2.30×10^{-2} M, not much different)

Degree of ionization $= 2.3 \times 10^{-2} \div 0.80 = 0.0287$ (2.87%)

b.

Conc. (M)	HF	⇌	H^+	+	F^-
Starting	0.80		0.10		0
Change	-x		+x		+x
Equilibrium	0.80- x		0.10 + x		x

Assuming x is negligible compared to 0.10 and to 0.80, we substitute into the equilibrium expression (0.10) for $[H^+]$ from 0.10 M HCl and (0.80) from the HF:

$$\frac{[H^+][F^-]}{[HF]} = K_a \cong \frac{(0.10)(x)}{(0.80)} \cong 6.8 \times 10^{-4}$$

$$x \cong \frac{(0.80)6.8 \times 10^{-4}}{(0.10)} \cong 5.\underline{4}4 \times 10^{-3}$$

$0.80 - 5.44 \times 10^{-3} = 0.794$, or $\cong 0.79$ to 2 sig figs. [Using just $(0.10 + x)$ and the quadratic gives $x = 5.17 \times 10^{-3}$.]

Degree of ionization (quadratic) $= 5.17 \times 10^{-3} \div 0.80 = 0.006\underline{4}6$ (0.6$\underline{4}$6%)

16.87 Rounded answer: pH = 3.17; ($[H^+] = 6.8 \times 10^{-4}$ M). Assemble the usual table, using a starting NO_2^- of 0.10 M from 0.10 M KNO_2, and a starting HNO_2 of 0.15 M. Assume x is negligible compared to 0.10 M and 0.15 M, and solve for the x in the numerator.

Conc. (M)	HNO_2	⇌	H^+	+	NO_2^-
Starting	0.15		0		0.10
Change	-x		+x		+x
Equilibrium	0.15 - x		x		0.10 + x

$$\frac{[H^+][NO_2^-]}{[HNO_2]} = K_a = \frac{(x)(0.10 + x)}{(0.15 - x)} = 4.5 \times 10^{-4}$$

$$\frac{(x)(0.10)}{(0.15)} \cong 4.5 \times 10^{-4};\ x = [H^+] \cong 6.\underline{7}5 \times 10^{-4} \text{ M}$$

pH $= -\log [H^+] = -\log (6.75 \times 10^{-4}) = 3.1\underline{7}06$ (rounded answer above)

16.89 Rounded answer: pH = 10.47 ([OH^-] = 2.9×10^{-4} M). Assemble the usual table, using a starting $CH_3NH_3^+$ of 0.15 M from 0.15 M CH_3NH_3Cl, and a starting CH_3NH_2 of 0.10 M. Assume x is negligible compared to 0.15 M and 0.10 M, and solve for the x in the numerator.

Conc. (M)	CH_3NH_2	$+ H_2O$	$\rightleftharpoons$	OH^-	+	$CH_3NH_3^+$
Starting	0.10			0		0.20
Change	-x			+x		+x
Equilibrium	0.10 - x			x		0.20 + x

$$\frac{[CH_3NH_3^+][OH^-]}{[CH_3NH_2]} = K_b = \frac{(0.15 + x)(x)}{(0.10 - x)} = 4.4 \times 10^{-4}$$

$$\frac{(0.15)(x)}{(0.10)} \cong 4.4 \times 10^{-4};\ x = [OH^-] \cong 2.\underline{9}3 \times 10^{-4}\ M$$

$$pOH = -\log [OH^-] = -\log (2.93 \times 10^{-4}) = 3.5\underline{3}2$$

$$pH = 14.00 - pOH = 14.00 - 3.532 = 10.4\underline{6}8$$ (rounded answer above)

16.91 Rounded answer: pH = 3.45 ([H^+] = 3.5×10^{-4} M). Find the mol/L of HF, and the mol/L of F^-, and assemble the usual table. Substitute the equilibrium concentrations into the equilibrium expression; then assume x is negligible compared to the starting concentrations of both HF and F^-. Solve for x in the numerator of the equilibrium expression, and calculate the pH from this value.

Total volume = 0.045 L + 0.035 L = 0.080 L

(0.10mol HF/L) x 0.035 L = 0.0035 mol HF (÷ 0.080 L total volume = 0.04375 M)

(0.15 mol F^-/L) x 0.045 L = 0.00675 mol F^- (÷ 0.080 L total volume = 0.084375 M)

Now substitute these starting concentrations into the usual table:

Conc. (M)	HF	$\rightleftharpoons$	H^+	+	F^-
Starting	0.04375		0		0.084375
Change	-x		+x		+x
Equilibrium	0.04375 - x		x		0.084375 + x

$$\frac{[H^+][F^-]}{[HF]} = K_a = \frac{(x)(0.084375 + x)}{(0.04375 - x)} = 6.8 \times 10^{-4}$$

$$\frac{(x)(0.084375)}{(0.04375)} \cong 6.8 \times 10^{-4};\ x = [H^+] \cong 3.\underline{5}2 \times 10^{-4}\ M$$

$$pH = -\log [H^+] = -\log (3.52 \times 10^{-4}) = 3.4\underline{5}3$$

16.93 Rounded answer: pH before HCl: 9.26; pH after HCl: 9.09. First use the 0.10 M NH_3 and 0.10 M NH_4^+ to calculate the $[OH^-]$ and pH before HCl is added. Assemble a table of starting, change, and equilibrium concentrations. Assume x is negligible compared to 0.10 M and substitute the approximate concentrations into the equilibrium expression.

Conc. (M)	NH_3 +	H_2O	$\rightleftharpoons$	NH_4^+	+	OH^-
Starting	0.10			0.10		0
Change	-x			+x		+x
Equilibrium	0.10 - x			0.10 + x		x

$$\frac{[NH_4^+][OH^-]}{[NH_3]} = K_b \cong \frac{(0.10)(x)}{(0.10)} \cong 1.8 \times 10^{-5}$$

$x = [OH^-] \cong 1.8 \times 10^{-5}$ M (note that x is negligible compared to 0.10)

$[H^+] = 1.00 \times 10^{-14} \div 1.8 \times 10^{-5} = 5.555 \times 10^{-10}$ M

pH before HCl added = 9.2553 (rounded answer above)

Now calculate the pH after the 0.012 L (12 mL) of 0.20 M HCl is added by noting that the H^+ ion reacts with the NH_3 to form additional NH_4^+. Calculate the stoichiometric amount of HCl; then subtract the moles of HCl from the moles of NH_3. Add the resulting moles of NH_4^+ to the 0.0125 starting moles of NH_4^+ in the 0.125 L of buffer.

(0.20 mol HCl/L) x 0.012 L = 0.0024 mol HCl (reacts with 0.0024 mol NH_3)

mol NH_3 left = (0.0125 - 0.0024) mol = 0.0101 mol NH_3

total mol NH_4^+ now present = 0.0125 orig + 0.0024 formed from HCl = 0.0149 mol NH_4^+

$$[NH_3] = \frac{0.0101 \text{ mol } NH_3}{0.137 \text{ L soln}} = 0.0737 \text{ M}; \quad [NH_4^+] = \frac{0.0149 \text{ mol } NH_4^+}{0.137 \text{ L soln}} = 0.108 \text{ M}$$

Now account for the ionization of NH_3 to NH_4^+ and OH^- at equilibrium by assembling the usual table. Assume x is negligible compared to 0.0737 M and 0.108 M, and solve the equilibrium expression for x in the numerator. Calculate the pH from x, the $[OH^-]$.

Conc. (M)	NH_3	+	H_2O	$\rightleftharpoons$	NH_4^+	+	OH^-
Starting	0.0737				0.108		0
Change	-x				+x		+x
Equilibrium	0.0737 - x				0.108 + x		x

$$\frac{[NH_4^+][OH^-]}{[NH_3]} = K_b \cong \frac{(0.108)(x)}{(0.0737)} \cong 1.8 \times 10^{-5}$$

$x = [OH^-] \cong 1.22 \times 10^{-5}$ M (note that x is negligible compared to 0.10)

$[H^+] = 1.00 \times 10^{-14} \div 1.22 \times 10^{-5} = 8.19 \times 10^{-10}$ M

pH before HCl added = 9.086 (rounded answer above)

16.95 Use the Henderson-Hasselbalch equation, where $pK_a = -\log K_a$.

$$pH = -\log K_a + \log \frac{[\text{buff base}]}{[\text{buff acid}]} = -\log (1.4 \times 10^{-3}) + \log \frac{[0.15\ M]}{[0.10\ M]} = 3.0\underline{2}99 = 3.03$$

16.97 Calculate the K_a of the pyridinium ion from the K_b of pyridine, and then use the Henderson-Hasselbalch equation, where $pK_a = -\log K_a$.

$$K_a = K_w \div K_b = 1.00 \times 10^{-14} \div 1.4 \times 10^{-9} = 7.\underline{1}4 \times 10^{-6}$$

$$pH = -\log K_a + \log \frac{[\text{buff base}]}{[\text{buff acid}]} = -\log (7.14 \times 10^{-6}) + \log \frac{[0.15\ M]}{[0.10\ M} = 5.3\underline{2}2 = 5.32$$

16.99 Symbolize acetic acid as HAc and sodium acetate as Na^+Ac^-. Use the Henderson-Hasselbalch equation to find the log of $[Ac^-]/[HAc]$. Then solve for $[Ac^-]$ and for moles of NaAc in the 2.0 L of solution.

$$pH = -\log K_a + \log \frac{[Ac^-]}{[HAc]} = 5.00 = -\log (1.7 \times 10^{-5}) + \log \frac{[Ac^-]}{[0.10\ M]}$$

$$5.00 = 4.770 + \log [Ac^-] + \log 1/[0.10]$$

$$\log [Ac^-] = 5.00 - 4.770 - 1.00 = -0.770$$

$$[Ac^-] = 0.1\underline{6}98\ M$$

$$\text{mol NaAc} = 0.1698\ \text{mol/L} \times 2.0\ L = 0.3\underline{3}96 = 0.34\ \text{mol}$$

16.101 Rounded answer: pH = 2.0 ($[H^+]$ = 0.01 M). All the OH^- (from the NaOH) reacts with the H^+ from HCl. Calculate the stoichiometric amounts of OH^- and H^+ and subtract the mol of OH^- from the mol of H^+. Then divide the remaining H^+ by the total volume of 0.020 L + 0.025 L, or 0.045 L, to find the $[H^+]$. Then calculate the pH.

$$\text{Mol } H^+ = (0.10\ \text{mol HCl/L}) \times 0.025\ L\ HCl = 0.0025\ \text{mol } H^+$$

$$\text{Mol } OH^- = (0.10\ \text{mol NaOH/L}) \times 0.020\ L\ NaOH = 0.0020\ \text{mol } OH^-$$

$$\text{Mol } H^+ \text{ left} = (0.0025 - 0.0020)\ \text{mol } H^+ = 0.0005\ \text{mol } H^+$$

$$[H^+] = 0.0005\ \text{mol } H^+ \div 0.045\ L \text{ total volume} = 0.0\underline{1}1\ M$$

$$pH = -\log [H^+] = -\log (0.011) = 1.\underline{9}54$$

16.103 Rounded answer: pH = 8.59 ($[OH^-]$ = 3.9×10^{-6} M). Use HBen to symbolize benzoic acid and Ben^- to symbolize the benzoate anion. At the equivalence point, equal molar amounts of HBen and NaOH react to form a solution of NaBen. Start by calculating the moles of HBen. Use this to calculate the volume of NaOH needed to neutralize all of the HBen (and use the moles of HBen as the moles of Ben^- formed at the equivalence point). Add the volume of NaOH to the original 0.050 L to find the total volume of solution.

Mol HBen = 1.24 g HBen ÷ (122.1 g HBen/mol HBen) = 0.01016 mol HBen

Volume NaOH = 0.01016 mol HBen ÷ (0.180 mol NaOH/L) = 0.05644 L

Total volume = 0.05644 L + 0.050 L HBen soln = 0.10644 L

[Ben^-] = (0.01016 mol Ben^- from HBen) ÷ 0.10644 L = 0.09545 M

Since the Ben^- hydrolyzes to OH^- and HBen, use this to calculate the [OH^-]. Start by calculating the hydrolysis constant of Ben^- from the K_a of its conjugate acid, HBen. Then assemble the usual table of concentrations, assume x is negligible, calculate [OH^-] and pH.

$K_b = K_w \div K_a = 1.0 \times 10^{-14} \div (6.3 \times 10^{-5}) = 1.\underline{5}9 \times 10^{-10}$

Conc. (M)	Ben^- + H_2O	⇌	HBen	+	OH^-
Starting	0.09545		0		0
Change	-x		+x		+x
Equilibrium	0.09545 - x		x		x

$$\frac{[HBen][OH^-]}{[Ben^-]} = K_b = \frac{(x)^2}{(0.09545 - x)} = 1.59 \times 10^{-10}$$

$$\frac{(x)^2}{(0.09545)} \cong 1.59 \times 10^{-10};\ x = [OH^-] \cong 3.\underline{8}95 \times 10^{-6}$$

$pOH = -\log [OH^-] = -\log (3.895 \times 10^{-6}) = 5.4\underline{0}94$

$pH = 14.00 - 5.4094 = 8.5\underline{9}05$

16.105 Rounded answer: pH = 5.97 ([H^+] = 1.1×10^{-6} M). Use EtN to symbolize ethylamine and $EtNH^+$ to symbolize the ethylammonium cation. At the equivalence point, equal molar amounts of EtN and HCl react to form a solution of EtNHCl. Start by calculating the moles of EtN. Use this to calculate the volume of HCl needed to neutralize all of the EtN (and use the moles of EtN as the moles of $EtNH^+$ formed at the equivalence point). Add the volume of HCl to the original 0.032 L to find the total volume of solution.

(0.087 mol EtN/L) x 0.032 L = 0.00278 mol EtN

Volume HCl = 0.00278 mol EtN ÷ (0.15 mol HCl/L) = 0.0185 L

Total volume = 0.0185 L + 0.032 L EtN soln = 0.0505 L

[$EtNH^+$] = (0.00278 mol $EtNH^+$ from EtN) ÷ 0.0505 L = 0.0550 M

Since the $EtNH^+$ hydrolyzes to H_3O^+ and EtN, use this to calculate the [H^+]. Start by calculating the hydrolysis constant of $EtNH^+$ from the K_b of its conjugate base, EtN. Then assemble the usual table of concentrations, assume x is negligible, and calculate [H^+] and pH.

$K_a = K_w \div K_b = 1.0 \times 10^{-14} \div (4.7 \times 10^{-4}) = 2.\underline{1}3 \times 10^{-11}$

Conc. (M)	$EtNH^+$	+	H_2O	$\rightleftharpoons$	EtN	+	H_3O^+
Starting	0.0550				0		0
Change	-x				+x		+x
Equilibrium	0.0550 - x				x		x

$$\frac{[EtN][H^+]}{[EtNH^+]} = K_a = \frac{(x)^2}{(0.0550 - x)} = 2.13 \times 10^{-11}$$

$$\frac{(x)^2}{(0.0550)} \cong 2.13 \times 10^{-11}; \; x = [H^+] \cong 1.08 \times 10^{-6} \text{ M}$$

$$pH = -\log [H^+] = -\log (1.08 \times 10^{-6}) = 5.9\underline{6}56$$

16.107 Rounded answer: pH = 9.08; ($[OH^-]$ = 1.2×10^{-5} M). Calculate the stoichiometric amounts of NH_3 and HCl, which forms NH_4^+. Then divide NH_3 and NH_4^+ by the total volume of 0.500 L + 0.200 L, or 0.700 L, to find the starting concentrations. Calculate the $[OH^-]$, the pOH, and the pH.

Mol NH_3 = (0.10 mol NH_3/L) x 0.500 L = 0.0500 mol NH_3

Mol HCl = (0.25 mol HCl/L) x 0.200 L = 0.0300 mol HCl (→ 0.0300 mol NH_4^+)

Mol NH_3 left = 0.0500 mol - 0.0300 mol HCl = 0.0200 mol NH_3

$0.02\underline{0}0$ mol NH_3 ÷ 0.700 L = 0.0286 M NH_3

$0.03\underline{0}0$ mol NH_4^+ ÷ 0.700 L = 0.0429 M NH_4^+

Conc. (M)	NH_3	+	H_2O	$\rightleftharpoons$	NH_4^+	+	OH^-
Starting	0.0286				0.0429		0
Change	-x				+x		+x
Equilibrium	0.0286 - x				0.0429 + x		x

$$\frac{[NH_4^+][OH^-]}{[NH_3]} = K_b \cong \frac{(0.0429)(x)}{(0.0286)} \cong 1.8 \times 10^{-5}$$

$x = [OH^-] \cong 1.\underline{2}0 \times 10^{-5}$ M (note that x is negligible compared to 0.10)

$[H^+] = 1.0 \times 10^{-14} \div 1.20 \times 10^{-5} = 8.\underline{3}3 \times 10^{-10}$ M

pH = $9.0\underline{7}9$ (rounded answer above)

16.109 Rounded answer: K_a = 1.1×10^{-3} (with correct sig figs). To solve, assemble a table of starting, change, and equilibrium concentrations. Use HSal to symbolize salicylic acid, and Sal^- for the anion. Start by converting pH to $[H^+]$:

$[H^+]$ = antilog (-pH) = antilog (-2.43) = 3.72×10^{-3} M

Starting M of HSal = (2.2 g ÷ 183 g/mol) ÷ 1.00 L = $0.01\underline{5}9$ M

Conc. (M)	HSal	$\rightleftharpoons$	H^+	+	Sal^-
Starting	0.0159		0		0
Change	-x		+x		+x
Equilibrium	0.0159 - x		x		x

The value of x equals the value of the molarity of the H^+ ion, which is 3.72×10^{-3} M. Substitute into the equilibrium expression to find K_a:

$$K_a = \frac{[H^+][Sal^-]}{[HSal]} = \frac{[x]^2}{(0.0159 - x)} = \frac{(3.72 \times 10^{-3})^2}{(0.0159 - 0.00372)} = 1.\underline{1}3 \times 10^{-3} \text{ (see above)}$$

16.111 Rounded answer: $K_a = 1.1 \times 10^{-2}$ (with correct sig figs). To solve, assemble a table of starting, change, and equilibrium concentrations. Start by converting pH to $[H^+]$:

$$[H_+] = \text{antilog}(-pH) = \text{antilog}(-1.73) = 1.86 \times 10^{-2} \text{ M}$$

Starting M of HSO_4^- = 0.050 M

Conc. (M)	HSO_4^-	$\rightleftharpoons$	H^+	+	SO_4^{2-}
Starting	0.050		0		0
Change	-x		+x		+x
Equilibrium	0.050 - x		x		x

The value of x equals the value of the molarity of the H^+ ion, which is 1.86×10^{-2} M. Substitute into the equilibrium expression to find K_a:

$$K_{a2} = \frac{[H^+][SO_4^{2-}]}{[HSO_4^-]} = \frac{[x]^2}{(0.050 - x)} = \frac{(1.86 \times 10^{-2})^2}{(0.050 - 0.0186)} = 1.\underline{1}01 \times 10^{-2} \text{ (see above)}$$

16.113 Rounded answer: $[H^+] = 2.1 \times 10^{-10}$ M; pH = 9.68. The K_{a2} of 4.8×10^{-11} is much smaller than the hydrolysis constant for HCO_3^- hydrolyzing to H_2CO_3 and OH^-. This constant is calculated from K_{a1} of H_2CO_3, the conjugate acid of HCO_3^-:

$$K_b = K_w \div K_{a1} = 1.0 \times 10^{-14} \div 4.3 \times 10^{-7} = 2.32 \times 10^{-8}$$

Conc. (M)	HCO_3^-	+	H_2O	$\rightleftharpoons$	H_2CO_3	+	OH^-
Starting	0.10				0		0
Change	-x				+x		+x
Equilibrium	0.10 - x				x		x

$$\frac{[H_2CO_3][OH^-]}{[HCO_3^-]} \cong \frac{(x)^2}{(0.10)} \cong 2.\underline{3}2 \times 10^{-8}$$

$$x = [OH^-] \cong 4.81 \times 10^{-5} \text{ M}$$

$$pOH = -\log[OH^-] = -\log(4.81 \times 10^{-5}) = 4.317$$

$$pH = 14.00 - 4.317 = 9.683 \text{ (rounded answer above)}$$

Note that since the hydrolysis constant of HCO_3^- is larger than K_{a2}, its hydrolysis is more extensive than its ionization.

16.115 For the base ionization (hydrolysis) of CN^- to $HCN + OH^-$, the base ionization constant is:

$$K_w \div K_a = 1.0 \times 10^{-14} \div 4.9 \times 10^{-10} = 2.\underline{0}4 \times 10^{-5} = 2.0 \times 10^{-5}$$

For the base ionization (hydrolysis) of CO_3^{2-} to $HCO_3^- + OH^-$, the base ionization constant is calculated from the ionization constant (K_{a2}) of HCO_3^-, the conjugate acid of CO_3^{2-}:

$$K_w \div K_a = 1.00 \times 10^{-14} \div 4.8 \times 10^{-11} = 2.\underline{0}8 \times 10^{-4} = 2.1 \times 10^{-4}$$

Because the constant of CO_3^{2-} is larger, it is the stronger base.

16.117 Rounded answer: $[H^+] = 1.4 \times 10^{-3}$ M; pH = 2.84. Assume Al^{3+} is $Al(H_2O)_6^{3+}$. Assemble the usual table to calculate the $[H^+]$ and pH. Use the usual equilibrium expression.

Conc. (M)	$Al(H_2O)_6^{3+}$	+	H_2O	$\rightleftharpoons$	H_3O^+	+	$Al(H_2O)_5(OH)^{2+}$
Starting	0.15				0		0
Change	-x				+x		+x
Equilibrium	0.15 - x				x		x

$K_a = 1.4 \times 10^{-5}$

$$\frac{[Al(H_2O)_5(OH)^{2+}][H_3O^+]}{[Al(H_2O)_6^{3+}]} \cong \frac{(x)^2}{(0.15)} \cong 1.\underline{4} \times 10^{-5}$$

$x = [H^+] \cong 1.\underline{4}49 \times 10^{-3}$ M (note that x is negligible compared to 0.15)

$pH = -\log [H^+] = -\log (1.449 \times 10^{-3}) = 2.8\underline{3}8$ (rounded answer above)

16.119 Rounded answer: $[H^+] = 6.9 \times 10^{-4}$ M; pH = 3.16. Calculate the concentrations of the tartaric acid (H_2Tar) and the hydrogen tartrate ion ($HTar^-$).

$(11.0 \text{ g } H_2Tar \div 150.1 \text{ g/mol}) \div 1.00 \text{ L} = 0.07328 \text{ M } H_2Tar$

$(20.0 \text{ g } KHTar^- \div 188.2 \text{ g/mol}) \div 1.00 \text{ L} = 0.1063 \text{ M } [HTar^-]$

Conc. (M)	H_2Tar	$\rightleftharpoons$	H^+	+	$HTar^-$
Starting	0.07328		0		0.1063
Change	-x		+x		+x
Equilibrium	0.07328 - x		x		0.1063 + x

Neglecting x compared to 0.07328 and 0.1063 and substituting into the K_{a1} expression:

$$\frac{(x)(0.1063)}{(0.07328)} \cong 1.0 \times 10^{-3}; \quad x = [H^+] \cong 6.\underline{8}9 \times 10^{-4} \text{ M}$$

$pH = -\log [H^+] = -\log (6.89 \times 10^{-4}) = 3.1\underline{6}1$

16.121 Use the Henderson-Hasselbalch equation, where $[H_2CO_3]$ = the buffer acid, $[HCO_3^-]$ = the buffer base, and K_{a1} of carbonic acid is the ionization constant.

$$pH = -\log K_a + \log \frac{[HCO_3^-]}{[H_2CO_3]} = 7.40 = -\log (4.3 \times 10^{-7}) + \log \frac{[HCO_3^-]}{[H_2CO_3]}$$

$$\text{Log} \frac{[HCO_3^-]}{[H_2CO_3]} = 7.40 - 6.366 = 1.0\underline{3}4$$

$$[HCO_3^-]/[H_2CO_3] = 1\underline{0}.81/1 = 11/1$$

16.123 Rounded answer: $[H^+] = 4 \times 10^{-3}$ M; pH = 2.4. All the OH^- (from the NaOH) reacts with the H^+ from HCl. Calculate the stoichiometric amounts of OH^- and H^+ and subtract the mol of OH^- from the mol of H^+. Then divide the remaining H^+ by the total volume of 0.456 L + 0.285 L, or 0.741 L, to find the $[H^+]$. Then calculate the pH.

$$\text{Mol } H^+ = (0.10 \text{ mol HCl/L}) \times 0.456 \text{ L HCl} = 0.0456 \text{ mol } H^+$$

$$\text{Mol } OH^- = (0.15 \text{ mol NaOH/L}) \times 0.285 \text{ L NaOH} = 0.0428 \text{ mol } OH^-$$

$$\text{Mol } H^+ \text{ left} = (0.0456 - 0.0428) \text{ mol } H^+ = 0.0028 \text{ mol } H^+$$

$$[H^+] = 0.0028 \text{ mol } H^+ \div 0.741 \text{ L total volume} = 0.00\underline{3}8 \text{ M}$$

$$pH = -\log [H^+] = -\log (0.0038) = 2.42$$

16.125 Rounded answer: $[H^+] = 7.8 \times 10^{-4}$ M; pH = 3.11. Use BzN to symbolize benzylamine and $BzNH^+$ to symbolize the benzylammonium cation. At the equivalence point, equal molar amounts of BzN and HCl react to form a solution of BzNHCl. Start by calculating the moles of BzN. Use this to calculate the volume of HCl needed to neutralize all of the BzN (and use the moles of BzN as the moles of $BzNH^+$ formed at the equivalence point). Add the volume of HCl to the original 0.025 L to find the total volume of solution.

$$(0.025 \text{ mol BzN/L}) \times 0.065 \text{ L} = 0.00162 \text{ mol BzN}$$

$$\text{Volume HCl} = 0.00162 \text{ mol BzN} \div (0.050 \text{ mol HCl/L}) = 0.0324 \text{ L}$$

$$\text{Total volume} = 0.025 \text{ L} + 0.0324 \text{ L} = 0.05\underline{7}4 \text{ L}$$

$$[BzNH^+] = (0.00162 \text{ mol } BzNH^+ \text{ from BzN}) \div 0.0574 \text{ L} = 0.02\underline{8}2 \text{ M}$$

Since the $BzNH^+$ hydrolyzes to H_3O^+ and BzN, use this to calculate the $[H^+]$. Start by calculating the hydrolysis constant of $BzNH^+$ from the K_b of its conjugate base, BzN. Then assemble the usual table of concentrations, assume x is negligible, and calculate $[H^+]$ and pH.

$$K_a = K_w \div K_b = 1.00 \times 10^{-14} \div (4.7 \times 10^{-10}) = 2.\underline{1}3 \times 10^{-15}$$

Conc. (M)	$BzNH^+$	+	H_2O	$\rightleftharpoons$	BzN	+	H_3O^+
Starting	0.0282				0		0
Change	-x				+x		+x
Equilibrium	0.0282 - x				x		x

$$\frac{[BzN][H^+]}{[BzNH^+]} = K_a = \frac{(x)^2}{(0.0282 - x)} = 2.13 \times 10^{-5}$$

$$\frac{(x)^2}{(0.0282)} \cong 2.13 \times 10^{-5};\ x = [H^+] \cong 7.\underline{7}502 \times 10^{-4}\ M$$

$$pH = -\log [H^+] = -\log (7.7502 \times 10^{-4}) = 3.1\underline{1}06$$

16.127 a. $[H^+] \cong 0.100$ M

b. Rounded answer: $[H^+] = 0.11$ M. The 0.100 M H_2SO_4 ionizes to 0.100 M H^+ and 0.100 M HSO_4^-. Assemble the usual table and substitute into the K_{a2} equilibrium expression for H_2SO_4. Solve the resulting quadratic.

Conc. (M)	HSO_4^-	+	H_2O	⇌	SO_4^{2-}	+	H^+
Starting	0.100				0		0.100
Change	-x				+x		+x
Equilibrium	0.100 - x				x		0.100 + x

$$x^2 + 0.111x - 1.10 \times 10^{-3} = 0$$

$$x = \frac{-0.111 \pm \sqrt{(0.111)^2 + (4)(1.10 \times 10^{-3})}}{2}$$

$$x = \frac{-0.111 \pm \sqrt{1.6721 \times 10^{-2})}}{2}$$

$$x = 1.8308 \times 10^{-2} \div 2 = 9.\underline{1}54 \times 10^{-3};\ [H^+] = 0.10\underline{0} + 0.009\underline{1}54 = 0.109$$

Cumulative-Skills Problems (require skills from previous chapters)

16.129 The equilibrium constant for the reaction $HCN + SO_4^{2-} \rightarrow CN^- + HSO_4^-$ is found by adding reactions 1 and 2 and multiplying their respective equilibrium constants:

1: $HCN \rightleftharpoons H^+ + CN^-$ $\quad K_a = 4.9 \times 10^{-10}$

2: $SO_4^{2-} + H^+ \rightleftharpoons HSO_4^{2-}$ $\quad K = 1/K_{a2} = 1/1.1 \times 10^{-2}$

$SO_4^{2-} + HCN \rightleftharpoons HSO_4^{2-} + CN^-$ $\quad K = (4.9 \times 10^{-10})/(1.1 \times 10^{-2}) = 4.5 \times 10^{-8}$

16.131 Use the pH to calculate $[H^+]$, and then use the K_a of 1.7×10^5 and the K_a expression to calculate the molarity of acetic acid (HAc), assuming ionization is negligible. Convert molarity to mass percentage using the formula weight of 60.05 g/mol of HAc.

$$[H^+] = \text{antilog}(-2.45) = 3.\underline{5}48 \times 10^{-3}\ M$$

Write the equilibrium expression in terms of chemical symbols, and then substitute the x and the (0.003548 M) terms into the expression:

$$\frac{[H^+][Ac^-]}{[HAc]} = K_a \cong \frac{(0.003548)^2}{(x)} \cong 1.7 \times 10^{-5}$$

Solve the equation for x, assuming that 0.003548 is much smaller than x.

$$x = (0.003548)^2 \div 1.7 \times 10^{-5} \cong [HAc] \cong 0.7\underline{4}04 \text{ M}$$

$$0.7404 \text{ mol HAc/L} \times (60.05 \text{ g/mol}) \times (1 \text{ L}/1090 \text{ g})(100\%) = 4.\underline{0}78 = 4.1\% \text{ HAc}$$

16.133 Find the $[H^+]$ and $[C_2H_3O_2^-]$ by solving for the approximate $[H^+]$, noting that x is much smaller than the starting 0.92 M of acetic acid. The usual table is used but is not shown, and only the final setup for $[H^+]$ is shown. Use K_f = 1.858°C/m for the constant for water.

$$[H^+] = [C_2H_3O_2^-] = (1.7 \times 10^{-5} \times 0.92\text{M})^{1/2} \cong 0.003\underline{9}54 \text{ M}$$

Total molarity of acid + ions = 0.92 + 0.003954 = 0.9$\underline{2}$39

Molality = m = 0.9239 mol ÷ (1.000 L x 0.953 kg solv/L) = 0.9695 m

Freezing point = $-\Delta T_f = -K_f c_m$ = (-1.858°C/m) x 0.9695 m = -1.801 = -1.8°C

16.135 The $[H^+]$ = -antilog (-4.35) = 4.$\underline{4}$6 x 10^{-5} M. Noting that 0.465 L of 0.0941 M NaOH will produce 0.043756 mol of acetate, Ac^-, ion. Rearranging the K_a expression for acetic acid, HAc and Ac^-, and canceling the volume in the mol/L of each, we obtain:

$$\frac{[HAc]}{[Ac^-]} = \frac{[H^+]}{K_a} = \frac{4.46 \times 10^{-5}}{1.7 \times 10^{-5}} = \frac{2.62}{1.00} \cong \frac{x \text{ mol HAc}}{0.043756 \text{ mol Ac}^-}$$

x = 0.1$\underline{1}$46 mol HAc

Total mol HAc added = 0.1146 + 0.043756 = 0.1583 mol HAc

mol/L of pure HAc needed = 1049 g HAc/L x (1 mol HAc/60.05 g) = 17.4$\underline{6}$7 mol/L

L of pure HAc needed = 0.1583 mol HAc x (L/17.467 mol) = 0.009$\underline{0}$62 L (9.1 mL)

CHAPTER 17

SOLUBILITY AND COMPLEX-ION EQUILIBRIA

SOLUTIONS TO EXERCISES

Note on answers to equilibrium calculations: The rounded answer (with significant figures) is always given first. After the stepwise solution, the numerical answer is given again, but with one nonsignificant figure. (The last significant figure is indicated by underlining it.)

17.1 The equilibria and solubility product expressions are:

a. $BaSO_4(s) \rightleftharpoons Ba^{2+}(aq) + SO_4^{2-}(aq)$; $K_{sp} = [Ba^{2+}][SO_4^{2-}]$

b. $Fe(OH)_3(s) \rightleftharpoons Fe^{3+}(aq) + 3OH^-(aq)$; $K_{sp} = [Fe^{3+}][OH^-]^3$

c. $Ca_3(PO_4)_2(s) \rightleftharpoons 3Ca^{2+}(aq) + 2PO_4^{3-}(aq)$; $K_{sp} = [Ca^{2+}]^3[PO_4^{3-}]^2$

17.2 Rounded answer: $K_{sp} = 1.8 \times 10^{-10}$ (correct sig figs). Calculate the molar solubility. Then assemble the usual concentration table and substitute from it the equilibrium concentrations into the equilibrium expression. (Because no concentrations can be given for solid AgCl, dashes are written; in later problems such spaces will be left blank.)

$$\frac{1.9 \times 10^{-3}\ g}{L} \times \frac{1\ mol}{143\ g} = 1.\underline{3}3 \times 10^{-5}\ M$$

Conc. (M)	AgCl(s)	$\rightleftharpoons$	Ag^+	+	Cl^-
Starting	- -		0		0
Change	- -		$+1.33 \times 10^{-5}$		$+1.33 \times 10^{-5}$
Equilibrium	- -		1.33×10^{-5}		1.33×10^{-5}

$K_{sp} = [Ag^+][Cl^-] = (1.33 \times 10^{-5})(1.33 \times 10^{-5}) = 1.\underline{7}68 \times 10^{-10}$ (rounded answer above)

17.3 Rounded answer: $K_{sp} = 4.5 \times 10^{-36}$. Calculate the molar solubility. Then assemble the usual concentration table and substitute from it into the equilibrium expression. (Because no concentrations can be given for solid $Pb_3(AsO_4)_2$, dashes are given.)

$$\frac{3.0 \times 10^{-5}\ g}{L} \times \frac{1\ mol}{899\ g} = 3.\underline{3}4 \times 10^{-8}\ M$$

Conc. (M)	$Pb_3(AsO_4)_2(s)$	$\rightleftharpoons$	$3Pb^{2+}$	+	$2AsO_4^{3-}$
Starting	- -		0		0
Change	- -		$+3(3.34 \times 10^{-8})$		$+2(3.34 \times 10^{-8})$
Equilibrium	- -		$3(3.34 \times 10^{-8})$		$2(3.34 \times 10^{-8})$

$$K_{sp} = [Pb^{2+}]^3[AsO_4^{3-}]^2 = (3 \times 3.34 \times 10^{-8})^3(2 \times 3.34 \times 10^{-8})^2 = 4.\underline{4}89 \times 10^{-36}$$

(rounded answer above)

17.4 Rounded answer: 0.67 g $CaSO_4$/L. Assemble the usual concentration table. Let x = the molar solubility of $CaSO_4$. When x mol $CaSO_4$ dissolves in one liter of solution, x mol Ca^{2+} and x mol SO_4^{2-} form.

Conc. (M)	$CaSO_4(s)$	$\rightleftharpoons$	Ca^{2+}	+	SO_4^{2-}
Starting	- -		0		0
Change	- -		+x		+x
Equilibrium	- -		x		x

Substitute the equilibrium concentrations into the equilibrium expression and solve for x. Then convert to g $CaSO_4$ per L.

$$[Ca^{2+}][SO_4^{2-}] = K_{sp}$$

$$(x)(x) = (x)^2 = 2.4 \times 10^{-5}$$

$$x = \sqrt{(2.4 \times 10^{-5})} = 4.\underline{8}9 \times 10^{-3} \text{ M (rounded answer above)}$$

$$\frac{4.9 \times 10^{-3} \text{ mol}}{\text{L}} \times \frac{136 \text{ g}}{1 \text{ mol}} = 0.6\underline{6}64 \text{ g/L}$$

17.5 Rounded answers: a. 6.3×10^{-3} M; b. 4.4×10^{-5} M.

a. Let x = the molar solubility of BaF_2. Assemble the usual concentration table and substitute from the table into the equilibrium expression.

Conc. (M)	$BaF_2(s)$	$\rightleftharpoons$	Ba^{2+}	+	$2F^-$
Starting			0		0
Change			+x		+2x
Equilibrium			x		2x

$$[Ba^{2+}][F^-]^2 = K_{sp}$$

$$(x)(2x)^2 = 4x^3 = 1.0 \times 10^{-6}$$

$$\sqrt[3]{\frac{1.0 \times 10^{-6}}{4}} = 6.\underline{2}99 \times 10^{-3} \text{ M}$$

b. At the start, before any BaF_2 dissolves, the solution contains 0.15 M F^-. At equilibrium, x mol of solid BaF_2 dissolves to yield x mol Ba^{2+} and 2x mol F^-. Assemble the usual concentration table, and substitute the equilibrium concentrations into the equilibrium expression. As an approximation, assume x is negligible compared to 0.15 M F^-.

Conc. (M)	$BaF_2(s)$	$\rightleftharpoons$	Ba^{2+}	+	$2F^-$
Starting			0		0.15
Change			+x		+2x
Equilibrium			x		0.15 + 2x

$$[Ba^{2+}][F^-]^2 = K_{sp}$$

$$(x)(0.15 + 2x)^2 \cong (x)(0.15)^2 \cong 1.0 \times 10^{-6}$$

$$x \cong \frac{1.0 \times 10^{-6}}{(0.15)^2} \cong 4.\underline{4}44 \times 10^{-5} \text{ M}$$

Note that subtracting 2x from 0.15 M will not change it (to two significant figures), so that 2x is negligible compared to 0.15 M. The solubility of 4.4×10^{-5} M in 0.15 M NaF is lower than the solublility of 6.3×10^{-3} M in pure water.

17.6 Rounded answer: $Q = 8.5 \times 10^{-5}$ (> K_{sp} of 2.4×10^{-5}, so precipitation occurs at equilibrium). Calculate the ion product, Q, after evaporation assuming no precipitation has occurred, and compare it with the K_{sp}.

$$Q = [Ca^{2+}][SO_4^{2-}]$$

$$Q = (2 \times 0.0052)(2 \times 00.0041) = 8.528 \times 10^{-5} \text{ M}^2 \text{ (> } K_{sp}\text{, so precipitation occurs)}$$

The solution is supersaturated before equilibrium is reached. At equilibrium, precipitation occurs and the solution is saturated.

17.7 Rounded answer: $Q = 8.5 \times 10^{-9}$ (< K_{sp}, so no precipitation occurs). Calculate the concentrations of Pb^{2+} and SO_4^{2-} assuming no precipitation. Use a total volume of 0.456 L + 0.255 L, or 0.711 L.

$$(Pb^{2+}) = \frac{\frac{0.00016 \text{ mol}}{\text{L}} \times 0.255 \text{ L}}{0.711 \text{ L}} = 5.\underline{7}4 \times 10^{-5} \text{ M}$$

$$(SO_4^{2-}) = \frac{\frac{0.00023 \text{ mol}}{\text{L}} \times 0.456 \text{ L}}{0.711 \text{ L}} = 1.\underline{4}8 \times 10^{-4} \text{ M}$$

Calculate the ion product and compare it to K_{sp}.

$$Q = (Pb^{2+})(SO_4^{2-}) = (5.74 \times 10^{-5})(1.48 \times 10^{-4}) = 8.\underline{4}9 \times 10^{-9} \text{ M}^2$$

Since Q is less than the K_{sp} of 1.7×10^{-8}, no precipitation occurs and the solution is unsaturated.

17.8 Rounded answers: 3.6×10^{-11} M; 7.3×10^{-4} % remaining. Because K_{sp} is small, almost all the Pb^{2+} and CrO_4^{2-} react until one ion, the limiting reactant, is essentially completely precipitated. It will be seen below that this is Pb^{2+}. Assuming complete precipitation of Pb^{2+} ion, the amount of CrO_4^{2-} left is calculated:

$$(0.0010 \text{ mol } CrO_4^{2-}/L) \times 0.50 \text{ L} = 5.0 \times 10^{-4} \text{ mol } CrO_4^{2-} \quad (= 5.0 \times 10^{-4} \text{ M in } 1.00 \text{ L})$$

$$- \ (0.00001 \text{ mol } Pb^{2+}/L) \times 0.50 \text{ L} = 5.0 \times 10^{-6} \text{ mol } Pb^{2+} \quad (= 5.0 \times 10^{-6} \text{ M in } 1.00 \text{ L})$$

$$= 4.\underline{9}5 \times 10^{-4} \text{ mol } CrO_4^{2-} \quad (4.\underline{9}5 \times 10^{-4} \text{ M in } 1.00 \text{ L})$$

At equilibrium, a small amount of $PbCrO_4$ will dissolve, producing an unknown concentration, x, of Pb^{2+} ion. Assemble the usual table for exact equilibrium concentrations. Then assume that x is negligible compared to 4.95×10^{-4} M and do an approximate calculation of the Pb^{2+} concentration.

Conc. (M)	$PbCrO_4(s)$	$\rightleftharpoons$	Pb^{2+}	+	CrO_4^{2-}
Starting			0		4.95×10^{-4}
Change			+x		+x
Equilibrium			x		$4.95 \times 10^{-4} + x$

$$[Pb^{2+}][CrO_4^{2-}] = K_{sp}$$

$$(x)(4.95 \times 10^{-4} + x) \cong (x)(4.95 \times 10^{-4}) \cong 1.8 \times 10^{-14}$$

$$x \cong \frac{1.8 \times 10^{-14}}{4.95 \times 10^{-4}} \cong 3.\underline{6}36 \times 10^{-11} \text{ M} \cong [Pb^{2+}]$$

The percentage of Pb^{2+} remaining in solution is

$$3.636 \times 10^{-11} \div (5.0 \times 10^{-6}) \times 100\% = 7.27 \times 10^{-4} \%$$

17.9 The solubility of AgCN would increase as the pH decreases because the increasing concentration of H^+ would react with the CN^- to form the weakly ionized acid HCN. As CN^- is removed, more AgCN dissolves to replace the cyanide:

$$AgCN(s) \rightleftharpoons Ag^+(aq) + CN^-(aq) \ [+ H^+ \rightarrow HCN]$$

In the case of AgCl, the chloride ion is the conjugate base of a strong acid and would therefore not be affected by any amount of hydrogen ion.

17.10 Rounded answer: pH range = 0.0 to 2.52. Find the higher end of the pH range by calculating the minimum concentration of S^{2-} required to prevent precipitation of the most soluble metal sulfide. In this case, this is FeS because its K_{sp} of 6×10^{-18} is much larger than the K_{sp} of 6×10^{-36} for CuS. For FeS, this is:

$$[Fe^{2+}][S^{2-}] = K_{sp}$$

$$(0.050)[S^{2-}] = 1.2 \times 10^{-16}$$

$[S^{2-}] = 1.2 \times 10^{-16}$ M

Substituting into the overall equilibrium ($K_{a1}K_{a2}$) expression for H_2S, and assuming 0.10 M for saturated H_2S, we solve for $[H^+]$:

$$\frac{[H^+]^2[S^{2-}]}{[H_2S]} = 1.1 \times 10^{-20}$$

$[H^+]^2 = (1.1 \times 10^{-20})(0.10\ M) \div 1.2 \times 10^{-16} = 9.16 \times 10^{-6}$; $[H^+] = 3.\underline{0}2 \times 10^{-3}$ M

Higher end of pH range $= -\log(3.02 \times 10^{-3}) = 2.5\underline{1}8 = 2.52$

Find the lower end of the pH range by calculating the sulfide ion concentration required to just begin precipitation of CuS, using the same calculations.

$[Cu^{2+}][S^{2-}] = K_{sp}$

$(0.050)[S^{2-}] = 6 \times 10^{-36}$

$[S^{2-}] = 1.2 \times 10^{-34}$ M

$$\frac{[H^+]^2[S^{2-}]}{[H_2S]} = 1.1 \times 10^{-20}$$

$[H^+]^2 = (1.1 \times 10^{-20})(0.10\ M) \div 1.2 \times 10^{-34} = 9.16 \times 10^{12}$; $[H^+] = 3.\underline{0}2 \times 10^{6}$ M

Theoretical lower end of pH range $= -\log(3.02 \times 10^{6}) = -6.4$ (practical end: pH = 0.0).

17.11 Rounded answer: 1.2×10^{-9} M. Since $K_f = 4.8 \times 10^{12}$, and since the starting concentration of NH_3 is much larger than that of the Cu^{2+} ion, we can make a rough assumption that most of the copper(II) is converted to $Cu(NH_3)_4^{2+}$ ion. This ion then dissociates slightly to give a small concentration of Cu^{2+} and additional NH_3. The amount of NH_3 remaining at the start after reacting with 0.015 M Cu^{2+} is

$[0.100\ M - (4 \times 0.015\ M)] = 0.040$ M starting NH_3

Assemble the usual concentration table using this starting concentration for NH_3 and asssuming that the starting concentration of Cu^{2+} is zero.

Conc. (M)	$Cu(NH_3)_4^{2+}$	$\rightleftharpoons$	Cu^{2+}	+	$4NH_3$
Starting	0.015		0		0.040
Change	$-x$		$+x$		$+4x$
Equilibrium	$0.015 - x$		x		$0.040 + 4x$

Even though this reaction is the opposite of the equation for the formation constant, the formation-constant expression can be used. Simply substitute all exact equilibrium concentrations into the formation-constant expression; then simplify the exact equation by assuming that x is negligible compared to 0.015 and 4x is negligible compared to 0.040.

$$K_f = \frac{[Cu(NH_3)_4^{2+}]}{[Cu^{2+}][NH_3]^4} = \frac{(0.015 - x)}{(x)(0.040 + 4x)^4} \cong \frac{(0.015)}{(x)(0.040)^4} \cong 4.8 \times 10^{12}$$

Rearrange and solve for x:

$$x \cong (0.015) \div (4.8 \times 10^{12} \times (0.040)^4) \cong 1.\underline{2}2 \times 10^{-9}\ M \cong [Cu^{2+}]$$

17.12 Rounded answers: Q = 3.3×10^{-21}; no precipitation. Start by calculating the $[Ag^+]$ in equilibrium with the $Ag(CN)_2^-$ formed from Ag^+ and CN^-. Then use the $[Ag^+]$ to decide whether or not AgI will precipitate, by calculating the ion product and comparing it with the K_{sp} of 8.3×10^{-17} for AgI. Assume that all of the 0.0045 M Ag^+ reacts with CN^- to form 0.0045 M $Ag(CN)_2^-$, and calculate the remaining CN^-. Use these as starting concentrations for the usual concentration table.

[0.20 M KCN - (2 x 0.0045 M)] = 0.191 M starting CN^-

Conc. (M)	$Ag(CN)_2^-$	$\rightleftharpoons$	Ag^+	+	$2CN^-$
Starting	0.0045		0		0.191
Change	-x		+x		+2x
Equilibrium	0.0045 - x		x		0.191 + 2x

Even though this reaction is the opposite of the equation for the formation constant, the formation-constant expression can be used. Simply substitute all exact equilibrium concentrations into the formation-constant expression; then simplify the exact equation by assuming that x is negligible compared to 0.0045 and 2x is negligible compared to 0.191.

$$K_f = \frac{[Ag(CN)_2^-]}{[Ag^+][CN^-]^2} = \frac{(0.0045 - x)}{(x)(0.191 + 2x)^2} \cong \frac{(0.0045)}{(x)(0.191)^2} \cong 5.6 \times 10^{18}$$

Rearrange and solve for x:

$$x \cong (0.0045) \div (5.6 \times 10^{18} \times (0.191)^2) \cong 2.\underline{2}02 \times 10^{-20}\ M \cong [Ag^+]$$

Now calculate the ion product for AgI:

$$Q = [Ag^+][I^-] = (2.20 \times 10^{-20})(0.15) = 3.\underline{3}0 \times 20^{-21}\ M^2$$

Since Q is less than the K_{sp} of 8.3×10^{-17}, no precipitate will form and the solution is unsaturated.

17.13 Rounded answer: 0.44 M. Using the rule from Chapter 14, obtain the overall equilibrium constant for this reaction from the product of the individual equilibrium constants of the two individual equations whose sum gives this equation.

$$AgBr(s) \rightleftharpoons Ag^+(aq) + Br^-(aq) \qquad K_{sp} = 5.0 \times 10^{-13}$$

$$Ag^+(aq) + 2S_2O_3^{2-}(aq) \rightleftharpoons Ag(S_2O_3)_2^{3-}(aq) \qquad K_f = 2.9 \times 10^{13}$$

$$AgBr(s) + 2S_2O_3^{2-}(aq) \rightleftharpoons Ag(S_2O_3)_2^{3-}(aq) + Br^-(aq) \qquad K_c = K_{sp} \times K_f = 14.5$$

Assemble the usual table, using 1.0 M as the starting concentration of $S_2O_3^{2-}$ and x as the unknown concentration of $Ag(S_2O_3)_2^{3-}$ formed.

Conc. (M)	AgBr(s) +	$2S_2O_3^{2-}$	⇌	$Ag(S_2O_3)_2^{3-}$	+	Br^-
Starting		1.0		0		0
Change		-2x		+x		+x
Equilibrium		1.0 - 2x		x		x

The equilibrium-constant expression can now be used. Simply substitute all exact equilibrium concentrations into the equilibrium constant expression; it is not necessary to simplify the exact equation because the solution can be obtained without using the quadratic.

$$K_c = \frac{[Ag(S_2O_3)_2^{3-}][Br^-]}{[S_2O_3^{2-}]^2} = \frac{(x)^2}{(1.0 - 2x)^2} = 14.5$$

Take the square root of both sides of the two right-hand terms, and solve for x:

$$\frac{(x)}{(1.0 - 2x)} = 3.\underline{8}08$$

$$x = 3.808\ (1.0 - 2x)$$

$$7.62x + x = 3.808$$

$x = 0.4\underline{4}17$ = molar solubility of AgBr in 1.0 M $Na_2S_2O_3$.

ANSWERS TO REVIEW QUESTIONS

17.1 The solubility equation is: $Ni(OH)_2(s) \rightleftharpoons Ni^{2+}(aq) + 2OH^-(aq)$. If the molar solubility of $Ni(OH)_2$ = x molar, the concentrations of the ions in the solution must be x M Ni^{2+} and 2x M OH^-. Substituting into the equilibrium expression gives:

$$K_{sp} = [Ni^{2+}][OH^-] = (x)(2x)^2 = 4x^3$$

17.2 Calcium sulfate is less soluble in a solution containing sodium sulfate because the increase in sulfate from the sodium sulfate causes the equilibrium composition in the equation below to shift to the left:

$$CaSO_4(s) \rightleftharpoons Ca^{2+}(aq) + SO_4^{2-}(aq)$$

The result is a decrease in both the calcium ion, and the calcium sulfate, concentrations.

17.3 Substitute the 0.10 M concentration of chloride into the solubility-constant expression and solve for $[Ag^+]$:

$$[Ag^+] = \frac{K_{sp}}{[Cl^-]} = \frac{1.8 \times 10^{-10}}{0.10} = 1.\underline{8}0 \times 10^{-9} = 1.8 \times 10^{-9}\ M$$

17.4 To predict whether PbI_2 will precipitate or not when lead nitrate and potassium iodide are mixed, the concentrations of Pb^{2+} and I^- after mixing would have to be calculated first (if the concentrations are not known or are not given). Then the value of Q, the ion product, would have to be calculated for PbI_2. Finally Q would have to be compared with the value of K_{sp}. If Q is greater than K_{sp}, then a precipitate will form at equilibrium. If Q is $\leq$ than K_{sp}, no precipitate will form.

17.5 Barium fluoride, normally insoluble in water, dissolves in dilute hydrochloric acid because the fluoride ion, once it forms, reacts with the hydrogen ion to form weakly ionized HF:

$$BaF_2(s) \rightleftharpoons Ba^{2+}(aq) + 2F^-(aq) \quad [+ 2H^+ \rightarrow 2HF(aq)]$$

17.6 Metal ions such as Pb^{2+} and Zn^{2+} are separated by controlling the $[S^{2-}]$ in a solution of saturated H_2S by means of adjusting the pH correctly. Because the K_{sp} of 2.5×10^{-27} of PbS is smaller than the K_{sp} of 1.1×10^{-21} for ZnS, the pH can be adjusted to make the $[S^{2-}]$ just high enough to precipitate PbS, without precipitating ZnS.

17.7 When NaCl is first added to a solution of $Pb(NO_3)_2$, a precipitate of $PbCl_2$ forms. As more NaCl is added, the excess chloride reacts further with the insoluble $PbCl_2$, forming soluble complex ions of $PbCl_3^-$ and $PbCl_4^{2-}$:

$$Pb^{2+}(aq) + 2Cl^-(aq) \rightleftharpoons PbCl_2(s)$$

$$PbCl_2(s) + Cl^-(aq) \rightleftharpoons PbCl_3^-(aq)$$

$$PbCl_3^-(aq) + Cl^-(aq) \rightleftharpoons PbCl_4^{2-}(aq)$$

17.8 When a small amount of NaOH is added to a solution of $Al_2(SO_4)_3$, a precipitate of $Al(OH)_3$ forms at first. As more NaOH is added, the excess hydroxide ion reacts further with the insoluble $Al(OH)_3$, forming a soluble complex ion of $Al(OH)_4^-$.

17.9 The Ag^+, Cu^{2+}, and Ni^{2+} ions can be separated in two steps: (1) add HCl to precipitate just the Ag^+ as AgCl, leaving the others in solution. (2) after pouring the solution away from the precipitate, add 0.3 M HCl and H_2S to precipitate only the CuS away from the Ni^{2+} ion, whose sulfide is soluble under these conditions.

17.10 By controlling the pH through the appropriate buffer, one can control the $[CO_3^{2-}]$ using the equilibrium reaction:

$$H^+(aq) + CO_3^{2-}(aq) \rightleftharpoons HCO_3^-(aq)$$

Calcium carbonate is much more insoluble than magnesium carbonate, and thus can precipitate in weakly basic solution whereas magnesium carbonate will not. Magnesium carbonate will precipitate only in highly basic solution. (There is the possibility that $Mg(OH)_2$ might precipitate, but the $[OH^-]$ is too low for this to occur.)

SOLUTIONS TO PRACTICE PROBLEMS

Note on answers to equilibrium calculations: **The rounded answer (with significant figures) is always given first. After the stepwise solution, the numerical answer is given again, but with one nonsignificant figure. (The last significant figure is indicated by underlining it.)**

17.11 a. Soluble (Gr IA salts are soluble)

b. Insoluble (carbonates are generally insoluble)

c. Insoluble ($PbSO_4$ is an insoluble sulfate)

d. Soluble (all ammonium salts are soluble)

17.13 a. $K_{sp} = [Ba^{2+}][CrO_4^{2-}]$

b. $K_{sp} = [Fe^{2+}][OH^-]^2$

c. $K_{sp} = [Pb^{2+}]^3[AsO_4^{3-}]^2$

d. $K_{sp} = [Ag^+]^2[CrO_4^{2-}]$

17.15 Rounded answer: $K_{sp} = 9.3 \times 10^{-10}$ (correct sig figs). Calculate molar solubility, assemble the usual table, and substitute from it the equilibrium concentrations into the equilibrium expression. (Dashes are given for $AgBrO_3(s)$; in later problems such dashes are omitted.)

$$\frac{7.2 \times 10^{-3}\ g}{L} \times \frac{1\ mol}{236\ g} = 3.\underline{0}5 \times 10^{-5}\ M$$

Conc. (M)	$AgBrO_3(s)$	$\rightleftharpoons$	Ag^+	+	BrO_3^-
Starting	--		0		0
Change	--		$+3.05 \times 10^{-5}$		$+3.05 \times 10^{-5}$
Equilibrium	--		3.05×10^{-5}		3.05×10^{-5}

$K_{sp} = [Ag^+][BrO_3^-] = (3.05 \times 10^{-5})(3.05 \times 10^{-5}) = 9.\underline{3}02 \times 10^{-10}$ (rounded ans above)

17.17 Rounded answer: $K_{sp} = 1.3 \times 10^{-7}$ (correct sig figs). Calculate the molar solubility. Then assemble the usual concentration table and substitute from it the equilibrium concentrations into the equilibrium expression.

$$\frac{0.13\ g}{L} \times \frac{1\ mol}{413\ g} = 3.\underline{1}5 \times 10^{-3}\ M$$

Conc. (M)	$Cu(IO_3)_2(s)$	$\rightleftharpoons$	Cu^{2+}	+	$2IO_3^-$
Starting			0		0
Change			$+3.15 \times 10^{-3}$		$+2 \times 3.15 \times 10^{-3}$
Equilibrium			3.15×10^{-3}		$2 \times 3.15 \times 10^{-3}$

$K_{sp} = [Cu^{2+}][IO_3^-]^2 = (3.15 \times 10^{-3})(2 \times 3.15 \times 10^{-3})^2 = 1.\underline{2}502 \times 10^{-7}$

17.19 Rounded answer: $K_{sp} = 1.8 \times 10^{-11}$ (correct sig figs). Calculate the pOH from pH and then convert pOH to $[OH^-]$. Then assemble the usual concentration table and substitute from it the equilibrium concentrations into the equilibrium expression.

$$pOH = 14.00 - 10.52 = 3.48$$

$$[OH^-] = \text{antilog}(-pOH) = \text{antilog}(-3.48) = 3.\underline{3}11 \times 10^{-4}\ M$$

$$[Mg^{2+}] = [OH^-] \div 2 = 3.311 \times 10^{-4} \div 2 = 1.655 \times 10^{-4}\ M$$

Conc. (M)	$Mg(OH)_2(s)$ ⇌	Mg^{2+}	+	$2OH^-$
Starting		0		0
Change		$+1.655 \times 10^{-4}$		$+3.311 \times 10^{-4}$
Equilibrium		1.655×10^{-4}		3.311×10^{-4}

$$K_{sp} = [Mg^{2+}][OH^-]^2 = (1.655 \times 10^{-4})(3.311 \times 10^{-4})^2 = 1.\underline{8}14 \times 10^{-11}$$

17.21 Rounded answers: 5.0×10^{-4} M and 0.092 g $SrSO_4$/L. Assemble the usual concentration table. Let x = the molar solubility of $SrSO_4$. When x mol $SrSO_4$ dissolves in one liter of solution, x mol Sr^{2+} and x mol SO_4^{2-} form.

Conc. (M)	$SrSO_4(s)$	⇌	Sr^{2+}	+	SO_4^{2-}
Starting			0		0
Change			+x		+x
Equilibrium			x		x

Substitute the equilibrium concentrations into the equilibrium expression and solve for x. Then convert to g $SrSO_4$ per L.

$$[Sr^{2+}][SO_4^{2-}] = K_{sp}$$

$$(x)(x) = (x)^2 = 2.5 \times 10^{-7}$$

$$x = \sqrt{(2.5 \times 10^{-7})} = 5.\underline{0}0 \times 10^{-4}\ M \text{ (rounded answer above)}$$

$$\frac{5.0 \times 10^{-4}\ \text{mol}}{L} \times \frac{184\ g}{1\ \text{mol}} = 0.09\underline{2}0\ g/L$$

17.23 Rounded answer: 1.9×10^{-3} M. Let x = the molar solubility of PbF_2. Assemble the usual concentration table and substitute from the table into the equilibrium expression.

Conc. (M)	$PbF_2(s)$	⇌	Pb^{2+}	+	$2F^-$
Starting			0		0
Change			+x		+2x
Equilibrium			x		2x

$$[Pb^{2+}][F^-]^2 = K_{sp}$$

$$(x)(2x)^2 = 4x^3 = 2.7 \times 10^{-8}$$

$$x = \sqrt[3]{\frac{2.7 \times 10^{-8}}{4}} = 1.\underline{8}8 \times 10^{-3}\ M$$

17.25 Rounded answer: 3.1×10^{-4} g/L. Let x = the molar solubility of $SrSO_4$. At the start, before any $SrSO_4$ dissolves, the solution contains 0.15 M SO_4^{2-}. At equilibrium, x mol of solid $SrSO_4$ dissolves to yield x mol Sr^{2+} and x mol SO_4^{2-}. Assemble the usual concentration table, and substitute the equilibrium concentrations into the equilibrium expression. As an approximation, assume x is negligible compared to 0.15 M SO_4^{2-}.

Conc. (M)	$SrSO_4(s)$	$\rightleftharpoons$	Sr^{2+}	+	SO_4^{2-}
Starting			0		0.15
Change			+x		+x
Equilibrium			x		0.15 + x

$$[Sr^{2+}][SO_4^{2-}] = K_{sp}$$

$$(x)(0.15 + x) \cong (x)(0.15) \cong 2.5 \times 10^{-7}$$

$$x \cong \frac{2.5 \times 10^{-7}}{0.15} \cong 1.\underline{6}6 \times 10^{-6} \text{ M}$$

$$\frac{1.66 \times 10^{-6} \text{ mol}}{\text{L}} \times \frac{184 \text{ g}}{1 \text{ mol}} = 3.\underline{0}54 \times 10^{-4} \text{ g/L}$$

Note that subtracting x from 0.15 M will not change it (to two significant figures) so that x is negligible compared to 0.15 M.

17.27 Rounded answer: 1.1×10^{-6} g/L. Calculate the value of K_{sp} from the solubility, using the concentration table. Then, using the common ion calculation, assemble another concentration table. Use 0.020 M NaF as the starting concentration of F^- ion. Substitute the equilibrium concentrations from the table into the equilibrium expression. As an approximation, assume that 2x is negligible compared to 0.020 M F^- ion.

$$\frac{0.0076 \text{ g}}{\text{L}} \times \frac{1 \text{ mol}}{62.3 \text{ g}} = 1.\underline{2}2 \times 10^{-4} \text{ M}$$

Conc. (M)	$MgF_2(s)$	$\rightleftharpoons$	Mg^{2+}	+	$2F^-$
Starting			0		0
Change			$+1.22 \times 10^{-4}$		$+2 \times 1.22 \times 10^{-4}$
Equilibrium			1.22×10^{-4}		$2 \times 1.22 \times 10^{-4}$

$$K_{sp} = [Mg^{2+}][F^-]^2 = (1.22 \times 10^{-4})(2 \times 1.22 \times 10^{-4})^2 = 7.26 \times 10^{-12}$$

Now use K_{sp} to calculate the molar solubility of MgF_2.

Conc. (M)	$MgF_2(s)$	$\rightleftharpoons$	Mg^{2+}	+	$2F^-$
Starting			0		0.020
Change			+x		+2x
Equilibrium			x		0.020 + 2x

$$[Mg^{2+}][F^-]^2 = K_{sp}$$

$$(x)(0.020 + 2x)^2 \cong (x)(0.020)^2 = 7.\underline{2}6 \times 10^{-12}$$

$$x \cong \frac{7.26 \times 10^{-12}}{(0.020)^2} \cong 1.\underline{8}15 \times 10^{-8} \text{ M}$$

$$\frac{1.815 \times 10^{-8} \text{ mol}}{\text{L}} \times \frac{62.3 \text{ g}}{1 \text{ mol}} = 1.\underline{1}3 \times 10^{-6} \text{ g/L}$$

17.29 Rounded answer: 0.40 g/L. The concentration table follows.

	MgC_2O_4	$\rightleftharpoons$	Mg^{2+}	+	$C_2O_4^{2-}$
Start			0		0.020
Change			+x		+x
Equilibrium			x		0.020 + x

The equilibrium equation is

$$K_{sp} = [Mg^{2+}][C_2O_4^{2-}]$$

$$8.5 \times 10^{-5} = x(0.020 + x)$$

$$x^2 + 0.020x - 8.5 \times 10^{-5} = 0$$

Solving the quadratic equation gives

$$x = \frac{-0.020 \pm \sqrt{(0.020)^2 + 4(8.5 \times 10^{-5})}}{2} = 0.00\underline{3}6$$

The solubility in grams per liter is

$$0.0036 \frac{\text{mol}}{\text{L}} \times 112 \frac{\text{g}}{\text{mol}} = 0.04\underline{0}3 \text{ g/L} = 0.04 \text{ g/L}$$

17.31 Rounded answer: $Q = 2.5 \times 10^{-8}$ (> K_{sp} of 1.8×10^{-14}, so precipitation occurs at equilibrium). Calculate the ion product, Q, after preparation of the solution and assuming no precipitation has occurred. Compare it with the K_{sp}.

$$Q = [Pb^{2+}][CrO_4^{2-}]$$

$$Q = (5.0 \times 10^{-4})(5.0 \times 10^{-5}) = 2.\underline{5}0 \times 10^{-8} \text{ M}^2 \quad (> K_{sp} \text{ of } 1.8 \times 10^{-14})$$

The solution is supersaturated before equilibrium is reached. At equilibrium, precipitation occurs and the solution is saturated.

17.33 Rounded answer: $Q = 2.5 \times 10^{-12}$ (< K_{sp}, so no precipitation occurs). Calculate the concentrations of Mg^{2+} and OH^- assuming no precipitation. Use a total volume of 1.0 L + 1.0 L, or 2.0 L. (Note that the concentrations are halved when the volume is doubled.)

$$(Mg^{2+}) = \frac{\frac{0.0020 \text{ mol}}{\text{L}} \times 1.0 \text{ L}}{2.0 \text{ L}} = 1.\underline{0}0 \times 10^{-3} \text{ M}$$

$$(OH^-) = \frac{\frac{0.00010 \text{ mol}}{L} \times 1.0 \text{ L}}{2.0 \text{ L}} = 5.\underline{0}0 \times 10^{-5} \text{ M}$$

Calculate the ion product and compare it to K_{sp}.

$$Q = (Mg^{2+})(OH^-)^2 = (1.00 \times 10^{-3})(5.00 \times 10^{-5})^2 = 2.\underline{5}0 \times 10^{-12} \text{ M}^3$$

Since Q is less than the K_{sp} of 1.8×10^{-11}, no precipitation occurs and the solution is unsaturated.

17.35 Rounded answer: $Q = 1.4 \times 10^{-9}$ (< K_{sp}, so no precipitation occurs). Calculate the concentrations of Ba^{2+} and F^- assuming no precipitation. Use a total volume of 0.045 L + 0.075 L, or 0.120 L.

$$(Ba^{2+}) = \frac{\frac{0.0015 \text{ mol}}{L} \times 0.045 \text{ L}}{0.120 \text{ L}} = 5.\underline{6}25 \times 10^{-4} \text{ M}$$

$$(F^-) = \frac{\frac{0.0025 \text{ mol}}{L} \times 0.075 \text{ L}}{0.120 \text{ L}} = 1.\underline{5}6 \times 10^{-3} \text{ M}$$

Calculate the ion product and compare it to K_{sp}.

$$Q = (Ba^{2+})(F^-)^2 = (5.625 \times 10^{-4})(1.56 \times 10^{-3})^2 = 1.\underline{3}6 \times 10^{-9} \text{ M}^3$$

Since Q is less than the K_{sp} of 1.0×10^{-6}, no precipitation occurs and the solution is unsaturated.

17.37 Rounded answer: 1.8×10^{-3} mol of $CaCl_2$. A mixture of $CaCl_2$ and K_2SO_4 can only precipitate $CaSO_4$ since KCl is soluble. Use the K_{sp} expression to calculate the $[Ca^{2+}]$ needed to just begin precipitating the 0.030 M SO_4^{2-} (essentially a saturated solution). Then convert to moles.

$$[Ca^{2+}][SO_4^{2-}] = K_{sp} = 2.4 \times 10^{-5}$$

$$(Ca^{2+}) = \frac{K_{sp}}{[SO_4^{2-}]} = \frac{2.4 \times 10^{-5}}{2.0 \times 10^{-2}} = 1.\underline{2}0 \times 10^{-3} \text{ M}$$

The number of moles in 1.5 L of this calcium-containing solution is:

$$1.5 \text{ L} \times 1.20 \times 10^{-3} \text{ mol/L} = 1.\underline{8}0 \times 10^{-3} \text{ mol } Ca^{2+} = 1.\underline{8}0 \times 10^{-3} \text{ mol } CaCl_2$$

17.39 Rounded answers: 6.6×10^{-6} M; 0.013 % remaining. Because K_{sp} is small, almost all the Ag^+ and CrO_4^{2-} react until one ion, the limiting reactant, is essentially completely precipitated. It will be seen below that this is CrO_4^{2-}. Assuming complete precipitation of Ag^+ ion, the amount of CrO_4^{2-} left can be calculated: subtract half the moles of Ag^+ from the moles of CrO_4^{2-} to find the moles of CrO_4^{2-} unprecipitated (2 Ag^+ are used to form Ag_2CrO_4).

$(0.10 \text{ mol } CrO_4^{2-}/L) \times 0.025 \text{ L} = 2.\underline{5}0 \times 10^{-3} \text{ mol } CrO_4^{2-}$ $(= 5.\underline{0} \times 10^{-2} \text{ M in } 0.050 \text{ L})$

$- (0.5 \times 0.10 \text{ mol } Ag^+/L) \times 0.025 \text{ L} = 1.\underline{2}5 \times 10^{-3} \text{ mol } Ag_2CrO_4$ $(= 2.\underline{5} \times 10^{-2} \text{ M in } 0.050 \text{ L})$

$= \text{unprecipitated } CrO_4^{2-} = 1.\underline{2}5 \times 10^{-3} \text{ mol } CrO_4^{2-}$ $(2.\underline{5} \times 10^{-2} \text{ M in } 0.050 \text{ L})$

At equilibrium, a small amount of Ag_2CrO_4 will dissolve, producing an unknown concentration, x, of Ag^+ ion. Assemble the usual table for exact equilibrium concentrations. Then assume that x is negligible compared to 2.5×10^{-2} M and do an approximate calculation of the Ag^+ concentration.

Conc. (M)	$Ag_2CrO_4(s)$	$\rightleftharpoons$	$2Ag^+$	+	CrO_4^{2-}
Starting			0		0.0250
Change			+2x		+x
Equilibrium			2x		0.0250 + x

$[Ag^+]^2[CrO_4^{2-}] = K_{sp}$

$(2x)^2(0.0250 + x) \cong (2x)^2(0.0250) \cong 1.1 \times 10^{-12}$

$$[Ag^+] = 2x \cong \sqrt{\frac{1.1 \times 10^{-12}}{0.0250}} \cong 6.\underline{6}3 \times 10^{-6} \text{ M}$$

To calculate the percentage remaining of the initial Ag^+, we must first calculate the initial concentration of Ag^+.

$$(Ag^+) = \frac{2.50 \times 10^{-3} \text{ mol}}{0.0500 \text{ L}} = 5.\underline{0}0 \times 10^{-2} \text{ M}$$

$\% \; Ag^+ \text{ remaining} = (6.63 \times 10^{-6} \text{ M} \div 5.00 \times 10^{-2}) \times 100\% = 0.01\underline{3}2\%$

17.41 Rounded answer: 6.9×10^{-9} M. Because the $AgNO_3$ solution is relatively concentrated, we will ignore the dilution of the solution of Cl^- and I^- from the addition of $AgNO_3$. The $[Ag^+]$ just as the AgCl begins to preciptate can be calculated from the K_{sp} expression for AgCl, using the $[Cl^-] = 0.015$ M. Therefore for AgCl:

$[Ag^+][Cl^-] = K_{sp}$
$[Ag^+][0.015] = 1.8 \times 10^{-10}$

$$[Ag^+] = \frac{1.8 \times 10^{-10}}{0.015} = 1.\underline{2}0 \times 10^{-8} \text{ M}$$

The $[I^-]$ at this point can be obtained by substituting the above $[Ag^+]$ into the K_{sp} expression for AgI. Therefore, for AgI:

$[Ag^+][I^-] = K_{sp}$

$[1.20 \times 10^{-8}][I^-] = 8.3 \times 10^{-17}$

$$[I^-] = \frac{8.3 \times 10^{-17}}{1.2 \times 10^{-8}} = 6.\underline{9}1 \times 10^{-9} \text{ M}$$

17.43 Net ionic equation: $MgC_2O_4(s) + 2H^+(aq) \rightarrow Mg^{2+}(aq) + H_2C_2O_4(aq)$

17.45 Calculate the value of K for the reaction of H^+ with both the SO_4^{2-} and F^- anions as they form by the slight dissolving of the insoluble salts. These constants are the reciprocals of the K_a values of the conjugate acids of these anions.

$$F^- + H^+ \rightarrow HF\text{: } K = \frac{1}{K_a} = \frac{1}{6.8 \times 10^{-4}} = 1.\underline{4}7 \times 10^3$$

$$SO_4^{2-} + H^+ \rightarrow HSO_4^-\text{: } K = \frac{1}{K_{a2}} = \frac{1}{1.1 \times 10^{-2}} = 9.\underline{0}9 \times 10^1$$

Since K for the fluoride ion is relatively larger, more of BaF_2 will dissolve in acid than $BaSO_4$.

17.47 Rounded answer: pH range = 0 to 0.8. Find the higher end of the pH range by calculating the minimum concentration of S^{2-} required to prevent precipitation of the most soluble metal sulfide. In this case, this is CoS because its K_{sp} of 4×10^{-21} is much larger than the K_{sp} of 1.6×10^{-52} for HgS. For CoS, this is:

$$[Co^{2+}][S^{2-}] = K_{sp}$$

$$(0.10)[S^{2-}] = 4 \times 10^{-21}$$

$$[S^{2-}] = \underline{4}.0 \times 10^{-20} \text{ M}$$

Substituting into the overall equilibrium ($K_{a1}K_{a2}$) expression for H_2S, and assuming 0.10 M for saturated H_2S, we solve for $[H^+]$:

$$\frac{[H^+]^2[S^{2-}]}{[H_2S]} = 1.1 \times 10^{-20}$$

$$[H^+]^2 = (1.1 \times 10^{-20})(0.10 \text{ M}) \div 4.0 \times 10^{-20} = 2.75 \times 10^{-2};\ [H^+] = \underline{1}.658 \times 10^{-1} \text{ M}$$

$$\text{Higher end of pH range} = -\log(1.658 \times 10^{-1}) = 0.\underline{7}803$$

Find the lower end of the pH range by calculating the sulfide ion concentration required to just begin precipitation of HgS, using the same calculations.

$$[Hg^{2+}][S^{2-}] = K_{sp}$$

$$(0.10)[S^{2-}] = 1.6 \times 10^{-52}$$

$$[S^{2-}] = 1.6 \times 10^{-51} \text{ M}$$

$$\frac{[H^+]^2[S^{2-}]}{[H_2S]} = 1.1 \times 10^{-20}$$

$[H^+]^2 = (1.1 \times 10^{-20})(0.10\ M) \div 1.6 \times 10^{-51} = 6.875 \times 10^{29}$; $[H^+] = 8.\underline{2}9 \times 10^{14}\ M$

Theoretical lower end of pH = - log (8.29×10^{14}) = -14.91 (practical end: pH = 0.0).

17.49 The equation is: $Ag^+(aq) + 2CN^-(aq) \rightarrow Ag(CN)_2^-(aq)$. The K_f expression is:

$$K_f = \frac{[Ag(CN)_2^-]}{[Ag^+][CN^-]^2} = 5.6 \times 10^{18}$$

17.51 Rounded answer: 5.5×10^{-19} M. Assume that the only $[Ag^+]$ is that in equilibrium with the $Ag(CN)_2^-$ formed from Ag^+ and CN^-. (In other words assume that all of the 0.015 M Ag^+ reacts with CN^- to form 0.015 M $Ag(CN)_2^-$.) Subtract the CN^- that forms 0.015 M $Ag(CN)_2^-$ from the initial 0.100 M CN^-. Use this as starting concentration of CN^- for the usual concentration table.

[0.100 M NaCN - (2 x 0.015 M)] = 0.070 M starting CN^-

Conc. (M)	$Ag(CN)_2^-$	$\rightleftharpoons$	Ag^+	+	$2CN^-$
Starting	0.015		0		0.070
Change	-x		+x		+2x
Equilibrium	0.015 - x		x		0.070 + 2x

Even though this reaction is the opposite of the equation for the formation constant, the formation-constant expression can be used. Simply substitute all exact equilibrium concentrations into the formation-constant expression; then simplify the exact equation by assuming that x is negligible compared to 0.015 and 2x is negligible compared to 0.070.

$$K_f = \frac{[Ag(CN)_2^-]}{[Ag^+][CN^-]^2} = \frac{(0.015 - x)}{(x)(0.070 + 2x)^2} \cong \frac{(0.015)}{(x)(0.070)^2} \cong 5.6 \times 10^{18}$$

Rearrange and solve for x:

$$x \cong (0.015) \div (5.6 \times 10^{18} \times (0.070)^2) \cong 5.\underline{4}6 \times 10^{-19}\ M \cong [Ag^+]$$

17.53 Rounded answers: Q = 2.8×10^{-8}; precipitation occurs at equilibrium. Start by calculating the $[Cd^{2+}]$ in equilibrium with the $Cd(NH_3)_4^{2+}$ formed from Cd^{2+} and NH_3. Then use the $[Cd^{2+}]$ to decide whether or not CdC_2O_4 will precipitate, by calculating the ion product, and comparing it with the K_{sp} of 1.5×10^{-8} for CdC_2O_4. Assume that all of the 0.0020 M Cd^{2+} reacts with NH_3 to form 0.0020 M $Cd(NH_3)_4^{2+}$, and calculate the remaining NH_3. Use these as starting concentrations for the usual concentration table.

[0.10 M NH_3 - (4 x 0.0020 M)] = 0.092 M starting NH_3

Conc. (M)	$Cd(NH_3)_4^{2+}$	$\rightleftharpoons$	Cd^{2+}	+	$4NH_3$
Starting	0.0020		0		0.092
Change	-x		+x		+4x
Equilibrium	0.0020 - x		x		0.092 + 4x

Even though this reaction is the opposite of the equation for the formation constant, the formation-constant expression can be used. Simply substitute all exact equilibrium concentrations into the formation-constant expression; then simplify the exact equation by assuming that x is negligible compared to 0.0020 and 4x is negligible compared to 0.092.

$$K_f = \frac{[Cd(NH_3)_4{}^{2+}]}{[Cd^{2+}][NH_3]^4} = \frac{(0.0020 - x)}{(x)(0.092 + 4x)^4} \cong \frac{(0.0020)}{(x)(0.092)^4} \cong 1.0 \times 10^7$$

Rearrange and solve for x:

$$x \cong (0.0020) \div (1.0 \times 10^7 \times (0.092)^4) \cong 2.\underline{7}9 \times 10^{-6}\ M \cong [Cd^{2+}]$$

Now calculate the ion product for CdC_2O_4:

$$Q = [Cd^{2+}][C_2O_4{}^{2-}] = (2.79 \times 10^{-6})(0.010) = 2.\underline{7}9 \times 10^{-8}\ M^2$$

Since Q is greater than the K_{sp} of 1.5×10^{-8}, the solution is supersaturated before equilibrium. At equilibrium, a precipitate will form and the solution will be saturated.

17.55 Rounded answer: 3.0×10^{-3} M. Using the rule from Chapter 14, obtain the overall equilibrium constant for this reaction from the product of the individual equilibrium constants of the two individual equations whose sum gives this equation.

$$CdC_2O_4(s) \rightleftharpoons Cd^{2+}(aq) + C_2O_4{}^{2-}(aq) \quad K_{sp} = 1.5 \times 10^{-8}$$

$$Cd^{2+}(aq) + 4NH_3(aq) \rightleftharpoons Cd(NH_3)_4{}^{2+}(aq) \quad K_f = 1.0 \times 10^7$$

$$CdC_2O_4(s) + 4NH_3(aq) \rightleftharpoons Cd(NH_3)_4{}^{2+}(aq) + C_2O_4{}^{2-}(aq) \quad K_c = K_{sp} \times K_f = 0.15$$

Assemble the usual table, using 0.10 M as the starting concentration of NH_3 and x as the unknown concentration of $Cd(NH_3)_4{}^{2+}$ formed.

Conc. (M) $CdC_2O_4(s)$ +	$4NH_3$	$\rightleftharpoons$	$Cd(NH_3)_4{}^{2+}$	+	$C_2O_4{}^{2-}$
Starting	0.10		0		0
Change	-4x		+x		+x
Equilibrium	0.10 - 4x		x		x

The equilibrium-constant expression can now be used. Simply substitute all exact equilibrium concentrations into the equilbrium-constant expression.

$$K_c = \frac{[Cd(NH_3)_4{}^{2+}][C_2O_4{}^{2-}]}{[NH_3]^4} = \frac{(x)^2}{(0.10 - 4x)^4} = 0.15$$

Take the square root of both sides of the two right-hand terms, rearrange into a quadratic equation and solve for x:

$$\frac{(x)}{(0.10 - 4x)^2} = 0.3\underline{8}7$$

$$16x^2 - 3.38x + 0.010 = 0$$

$$x = \frac{3.38 \pm \sqrt{(-3.38)^2 - 4(16)(0.010)}}{2(16)} = 0.208 \text{ (too large) and } 3.\underline{0}01 \times 10^{-3}$$

Using the smaller root, $x = 3.\underline{0}01 \times 10^{-3}$ M = theoretical molar solubility of CdC_2O_4.

17.57 The Pb^{2+}, Cd^{2+}, and Sr^{2+} ions can be separated in two steps: (1) add HCl to precipitate just the Pb^{2+} as $PbCl_2$, leaving the others in solution; (2) after pouring the solution away from the precipitate, add 0.3 M HCl and H_2S to precipitate only the CdS away from the Sr^{2+} ion, whose sulfide is soluble under these conditions.

17.59 a. Ag^+ is not possible because no precipitate formed with HCl.
b. Ca^{2+} is possible because no reactions were described involving Ca^{2+}.
c. Mn^{2+} is possible because a precipitate was obtained with basic sulfide ion.
d. Cd^{2+} is not possible because no precipitate was obtained with acidic sulfide solution.

17.61 Rounded answer: 1.3×10^{-4} M. Assemble the usual concentration table. Let x = the molar solubility of $PbSO_4$. When x mol $PbSO_4$ dissolves in one liter of solution, x mol Pb^{2+} and x mol SO_4^{2-} form.

Conc. (M)	$PbSO_4(s)$	$\rightleftharpoons$	Pb^{2+}	+	SO_4^{2-}
Starting			0		0
Change			+x		+x
Equilibrium			x		x

Substitute the equilibrium concentrations into the equilibrium expression and solve for x.

$$[Pb^{2+}][SO_4^{2-}] = K_{sp}$$

$$(x)(x) = (x)^2 = 1.7 \times 10^{-8}$$

$$x = \sqrt{(1.7 \times 10^{-8})} = 1.\underline{3}03 \times 10^{-4} \text{ M (rounded answer above)}$$

17.63 Rounded answers: a. 6.9×10^{-7} M; b. 3.2×10^{-4} g/L. Let x = the molar solubility of Hg_2Cl_2. Assemble the usual concentration table and substitute from the table into the equilibrium expression.

Conc. (M)	$Hg_2Cl_2(s)$	$\rightleftharpoons$	Hg_2^{2+}	+	$2Cl^-$
Starting			0		0
Change			+x		+2x
Equilibrium			x		2x

$$[Hg_2^{2+}][Cl^-]^2 = K_{sp}$$

$$(x)(2x)^2 = 4x^3 = 1.3 \times 10^{-18}$$

$$x = \sqrt[3]{\frac{1.3 \times 10^{-18}}{4}} = 6.\underline{8}7 \times 10^{-7} \text{ M}$$

a. Molar solubility = $6.\underline{8}7 \times 10^{-7}$ M.

b. $$\frac{6.87 \times 10^{-7} \text{ mol}}{\text{L}} \times \frac{472.1 \text{ g } Hg_2Cl_2}{1 \text{ mol}} = 3.\underline{2}4 \text{ g } Hg_2Cl_2/\text{L}$$

17.65 Rounded answers: a. 5.2×10^{-6} M; b. pOH = 4.81. Let x = the molar solubility of $Ce(OH)_3$. Assemble the usual concentration table and substitute from the table into the equilibrium expression.

Conc. (M)	$Ce(OH)_3(s)$	⇌	Ce^{3+}	+	$3OH^-$
Starting			0		0
Change			+x		+3x
Equilibrium			x		3x

$$[Ce^{3+}][OH^-]^3 = K_{sp}$$

$$(x)(3x)^3 = 27x^4 = 2.0 \times 10^{-20}$$

$$x = \sqrt[4]{\frac{2.0 \times 10^{-20}}{27}} = 5.\underline{2}16 \times 10^{-6} \text{ M}$$

a. Molar solubility = $5.\underline{2}16 \times 10^{-6}$ M.

b. $[OH^-] = 3x = 1.\underline{56} \times 10^{-5}$ M; pOH = $4.8\underline{0}54$

17.67 Rounded answer: 0.45 M; 26 g/L. Calculate the pOH from pH, and then convert pOH to $[OH^-]$. Then assemble the usual concentration table and substitute from it the equilibrium concentrations into the equilibrium expression.

$$\text{pOH} = 14.00 - 8.80 = 5.20$$

$$[OH^-] = \text{antilog}(-\text{pOH}) = \text{antilog}(-5.20) = 6.\underline{3}09 \times 10^{-6} \text{ M}$$

Conc. (M)	$Mg(OH)_2(s)$	⇌	Mg^{2+}	+	$2OH^-$
Starting			0		0
Change			+x		$+6.309 \times 10^{-6}$
Equilibrium			x		6.309×10^{-6}

$$K_{sp} = [Mg^{2+}][OH^-]^2 = (x)(6.309 \times 10^{-6})^2 = 1.8 \times 10^{-11}$$

$$x = \frac{1.8 \times 10^{-11}}{(6.309 \times 10^{-6})^2} = 0.4\underline{5}2 \text{ M}$$

$$\frac{0.452 \text{ mol}}{\text{L}} \times \frac{58.3 \text{ g } Mg(OH)_2}{1 \text{ mol}} = 2\underline{6}.3 \text{ g } Mg(OH)_2/\text{L}$$

17.69 Rounded answer: 1.8×10^{-9} M. Let x = the change in M of Mg^{2+}, and 0.10 M = the starting OH^- concentration. Then assemble the usual concentration table and substitute from it the equilibrium concentrations into the equilibrium expression. Assume 2x is negligible compared to 0.10 M and perform an approximate calculation.

Conc. (M)	$Mg(OH)_2(s) \rightleftharpoons$	Mg^{2+}	+	$2OH^-$
Starting		0		0.10
Change		+x		+2x
Equilibrium		x		0.10 + 2x

$$K_{sp} = [Mg^{2+}][OH^-]^2 = (x)(0.10 + 2x)^2 \cong (x)(0.10)^2 \cong 1.8 \times 10^{-11}$$

$$x \cong \frac{1.8 \times 10^{-11}}{(0.10)^2} \cong 1.8\underline{0} \times 10^{-9} \text{ M } Mg(OH)_2$$

17.71 Rounded answer: just slightly greater than 8.0×10^{-3} M. To begin precipitation, we must add just slightly more sulfate ion than that required to give a saturated solution. Use the K_{sp} expression to calculate the $[SO_4^{2-}]$ needed to just begin precipitating the 0.0030 M Ca^{2+}.

$$[Ca^{2+}][SO_4^{2-}] = K_{sp} = 2.4 \times 10^{-5}$$

$$[SO_4^{2-}] = \frac{K_{sp}}{[Ca^{2+}]} = \frac{2.4 \times 10^{-5}}{3.0 \times 10^{-3}} = 8.\underline{0}0 \times 10^{-3} \text{ M}$$

When the sulfate ion concentration slightly exceeds 8.0×10^{-3} M, precipitation begins.

17.73 Rounded answer: Q = 1.0×10^{-5} (< K_{sp}, so the solution is not saturated). Calculate the concentrations of Pb^{2+} and Cl^-. Use a total volume of 3.20 L + 0.80 L, or 4.00 L.

$$[Pb^{2+}] = \frac{\frac{1.25 \times 10^{-3} \text{ mol}}{\text{L}} \times 3.20 \text{ L}}{4.00 \text{ L}} = 1.0\underline{0} \times 10^{-3} \text{ M}$$

$$[Cl^-] = \frac{\frac{5.0 \times 10^{-1} \text{ mol}}{\text{L}} \times 0.80 \text{ L}}{4.00 \text{ L}} = 1.\underline{0}0 \times 10^{-1} \text{ M}$$

Calculate the ion product and compare it to K_{sp}.

$$Q = (Pb^{2+})(Cl^-)^2 = (1.00 \times 10^{-3})(1.00 \times 10^{-1})^2 = 1.\underline{0}0 \times 10^{-5} \text{ M}^3$$

Since Q is less than the K_{sp} of 1.6×10^{-5}, no precipitation occurs and the solution is not saturated.

17.75 Rounded answer: 5.5×10^{-6} g of NaCl. A mixture of $AgNO_3$ and NaCl can only precipitate AgCl since $NaNO_3$ is soluble. Use the K_{sp} expression to calculate the $[Cl^-]$ needed to prepare a saturated solution (just before precipitating the 0.0015 M Ag^+). Then convert to moles, and finally to grams.

$[Ag^+][Cl^-] = K_{sp} = 1.8 \times 10^{-10}$

$[Cl^-] = \dfrac{K_{sp}}{[Ag^+]} = \dfrac{1.8 \times 10^{-10}}{(1.5 \times 10^{-3})} = 1.\underline{2}0 \times 10^{-7}$ M

The number of moles, and grams, in 0.785 L (785 mL) of this chloride-containing solution is:

0.785 L x 1.20×10^{-7} mol/L = $9.\underline{4}2 \times 10^{-8}$ mol Cl^- = $9.\underline{4}2 \times 10^{-8}$ mol NaCl

9.42×10^{-7} mol NaCl x 58.5 g NaCl/1 mol NaCl = $5.\underline{5}1 \times 10^{-6}$ g NaCl

17.77 Rounded answers: 2.1×10^{-4} M; 0.021% remaining. Because K_{sp} is small, almost all the Ba^{2+} and SO_4^{2-} react completely. The stoichiometric calculation follows:

(0.10 mol Ba^{2+}/L) x 0.050 L = $5.\underline{0}0 \times 10^{-3}$ mol Ba^{2+}

\- (0.10 mol SO_4^{2-}/L) x 0.050 L = $5.\underline{0}0 \times 10^{-3}$ mol SO_4^{2-}

= 0 mol Ba^{2+} and SO_4^{2-}

At equilibrium, a small amount of $BaSO_4$ will dissolve, producing an unknown concentration, x, of Ba^{2+} and SO_4^{2-} ions. Assemble the usual table for exact equilibrium concentrations. Then substitute into the K_{sp} equilibrium expression.

Conc. (M)	$BaSO_4(s)$	$\rightleftharpoons$	Ba^{2+}	+	SO_4^{2-}
Starting			0		0
Change			+x		+x
Equilibrium			x		x

$[Ba^{2+}][SO_4^{2-}] = K_{sp}$

$(x)(x) = 1.1 \times 10^{-10}$

$x = \sqrt{(1.1 \times 10^{-10})} = 1.04 \times 10^{-5} \text{ M} = [SO_4^{2-}]$

To calculate the percentage remaining of the initial SO_4^{2-}, we must first calculate the initial concentration of SO_4^{2-} in the 0.100 L after mixing, assuming precipitation has not started.

$(SO_4^{2-}) = \dfrac{\dfrac{0.10 \text{ mol}}{1 \text{ L}} \times 0.0500 \text{ L}}{0.100 \text{ L}} = 5.\underline{0}0 \times 10^{-2}$ M

% SO_4^{2-} not precipitated = (1.04×10^{-5} M ÷ 5.00×10^{-2} M) x 100% = 0.02$\underline{0}$8%

17.79 Rounded answer: 4.7×10^{-2} M. From the magnitude of K_f, assume that Fe^{3+} and SCN^- react essentially completely to form 2.00 M $Fe(SCN)^{2+}$ at equilibrium. Use 2.00 M as the starting concentration of $Fe(SCN)^{2+}$ for the usual concentration table.

Conc. (M)	Fe^{3+}	+	SCN^-	⇌	$FeSCN^{2+}$
Starting	0		0		2.00
Change	+x		+x		-x
Equilibrium	x		x		2.00 - x

Substitute all exact equilibrium concentrations into the formation-constant expression; then simplify the exact equation by assuming that x is negligible compared to 2.00 M.

$$K_f = \frac{[FeSCN^{2+}]}{[Fe^{3+}][SCN^-]} = \frac{(2.00 - x)}{(x)(x)} \cong \frac{(2.00)}{(x)^2} \cong 9.0 \times 10^2$$

Rearrange and solve for x:

$$x \cong \sqrt{\frac{2.00}{9.0 \times 10^2}} \cong 4.\underline{7}1 \times 10^{-2} \cong [Fe^{3+}]$$

Fraction dissociated = $(4.\underline{7}1 \times 10^{-2} \div 2.00) \times 100\% = 0.02\underline{3}5$ (<0.03, so acceptable)

17.81 Rounded answer: 1.4×10^{-2} M. Using the rule from Chapter 14, obtain the overall equilibrium constant for this reaction from the product of the individual equilibrium constants of the two individual equations whose sum gives this equation.

$$AgBr(s) \rightleftharpoons Ag^+(aq) + Br^-(aq) \qquad K_{sp} = 5.0 \times 10^{-13}$$

$$Ag^+(aq) + 2NH_3(aq) \rightleftharpoons Ag(NH_3)_2{}^+(aq) \qquad K_f = 1.7 \times 10^7$$

$$AgBr(s) + 2NH_3(aq) \rightleftharpoons Ag(NH_3)_2{}^{2+}(aq) + Br^-(aq) \quad K_c = K_{sp} \times K_f = 8.\underline{5}0 \times 10^{-6}$$

Assemble the usual table, using 5.0 M as the starting concentration of NH_3 and x as the unknown concentration of $Ag(NH_3)_2{}^+$ formed.

Conc. (M) AgBr(s) +	$2NH_3$	⇌	$Ag(NH_3)_2{}^+$	+	Br^-
Starting	5.0		0		0
Change	-2x		+x		+x
Equilibrium	5.0 - 2x		x		x

The equilibrium-constant expression can now be used. Simply substitute all exact equilibrium concentrations into the equilibrium-constant expression; it will not be necessary to simplify the equation because taking the square root of both sides removes the x^2 term.

$$K_c = \frac{[Ag(NH_3)_2{}^{2+}][Br^-]}{[NH_3]^2} = \frac{(x)^2}{(5.0 - 2x)^2} = 8.\underline{5}0 \times 10^{-6}$$

Take the square root of both sides of the two right-hand terms and solve for x.

$$\frac{(x)}{(5.0 - 2x)} = 2.\underline{9}15 \times 10^{-3}$$

$$x + 5.8x \times 10^{-3}x = 1.\underline{4}575 \times 10^{-2}$$

$x = 1.\underline{4}49 \times 10^{-2}$ M = the molar solubility of AgBr in 5.0 M NH_3.

17.83 Rounded answers: $[C_2O_4^{2-}] = 3.6 \times 10^{-4}$ M; $[Zn^{2+}] = 4.2 \times 10^{-6}$ M; $K_f = 2.6 \times 10^9$. Start by recognizing that because each zinc oxalate produces one oxalate ion, the solubility of zinc oxalate equals the oxalate concentration:

$$ZnC_2O_4(s) + 4NH_3 \rightleftharpoons Zn(NH_3)_4^{2+} + C_2O_4^{2-}$$

Thus the $[C_2O_4^{2-}] = 3.6 \times 10^{-4}$ M. Now calculate the $[Zn^{2+}]$ in equilibrium with the oxalate ion., using the K_{sp} expression for zinc oxalate.

$$K_{sp} = [Zn^{2+}][C_2O_4^{2-}] = 1.5 \times 10^{-9}$$

$$[Zn^{2+}] = \frac{1.5 \times 10^{-9}}{(3.6 \times 10^{-4})} = 4.\underline{1}6 \times 10^{-6} \text{ M}$$

To calculate K_f, the $[Zn(NH_3)_4^{2+}]$ term must be calculated. This can be done by recognizing that the molar solubility of ZnC_2O_4 is the sum of the concentration of Zn^{2+} and $Zn(NH_3)_4^{2+}$ ions:

$$\text{Molar solubility of } ZnC_2O_4 = [Zn^{2+}] + [Zn(NH_3)_4^{2+}]$$

$$3.6 \times 10^{-4} = 4.16 \times 10^{-6} + [Zn(NH_3)_4^{2+}]$$

$$[Zn(NH_3)_4^{2+}] = 3.6 \times 10^{-4} - 4.16 \times 10^{-6} = 3.\underline{5}58 \times 10^{-4} \text{ M}$$

Now the $[NH_3]$ term must be calculated by subtracting the ammonia in $[Zn(NH_3)_4^{2+}]$ from the starting NH_3 of 0.0150 M:

$$[NH_3] = 0.0150 - 4\,[Zn(NH_3)_4^{2+}] = 0.0150 - 4(3.55 \times 10^{-4}) = 0.013\underline{5}7 \text{ M}$$

Solve for K_f by substituting the known concentrations into the K_f expression:

$$K_f = \frac{[Zn(NH_3)_4^{2+}]}{[Zn^{2+}][NH_3]^4} = \frac{3.6 \times 10^{-4}}{(4.\underline{1}6 \times 10^{-6})(0.01357)^4} = 2.\underline{5}52 \times 10^9$$

Cumulative-Skills Problems (require skills from previous chapters)

17.85 Rounded answer: 0.18 M SO_4^{2-}. Using the K_{sp} of 1.1×10^{-21} for ZnS, calculate the $[S^{2-}]$ needed to maintain a saturated solution (without precipitation):

$$[S^{2-}] = \frac{1.1 \times 10^{-21}}{1.5 \times 10^{-4} \text{ M } Zn^{2+}} = 7.\underline{3}3 \times 10^{-18}$$

Next use the overall H_2S ionization expression to calculate the $[H^+]$ needed to achieve this $[S^{2-}]$ level:

$$[H^+] = \sqrt{\frac{1.1 \times 10^{-20}\ (0.10 \text{ M } H_2S)}{7.33 \times 10^{-18} \text{ M } S^{2-}}} = 1.\underline{2}25 \times 10^{-2} \text{ M}$$

Finally calculate the buffer ratio of $[SO_4^{2-}]/[HSO_4^-]$ from the H_2SO_4 K_{a2} expression, where K_{a2} has the value 1.1×10^{-2}:

$$\frac{[SO_4^{2-}]}{[HSO_4^-]} = \frac{K_{a2}}{[H^+]} = \frac{1.1 \times 10^{-2}}{1.225 \times 10^{-2}} = \frac{0.8979}{1.000}$$

If $[HSO_4^-] = 0.20$ M, then

$$[SO_4^{2-}] = 0.8979 \times 0.20 \text{ M} = 0.1\underline{7}95 \text{ M}$$

17.87 Rounded answer: 2.7×10^{-4} M. We begin by solving for $[H^+]$ in the buffer. Ignoring changes in $[HCHO_2]$ as a result of ionization in the buffer, we obtain

$$[H^+] \cong 1.7 \times 10^{-4} \times \frac{0.45 \text{ M}}{0.20 \text{ M}} = 3.\underline{8}25 \times 10^{-4}$$

You should verify that this approximation is valid (you obtain this same result from the Henderson-Hasselbalch equation). The equilibrium for the dissolution of CaF_2 in acidic solution is obtained by subtracting twice the acid ionization of HF from the solubility equilibrium of CaF_2:

$CaF_2(s)$	$\rightleftharpoons$	$Ca^{2+}(aq) + 2F^-(aq)$	K_{sp}
$2H^+(aq) + 2F^-(aq)$	$\rightleftharpoons$	$2HF(aq)$	$1/(K_a)^2$
$2H^+(aq) + CaF_2(s)$	$\rightleftharpoons$	$Ca^{2+}(aq) + 2HF(aq)$	$K_c = K_{sp}/(K_a)^2$

Therefore, $K_c = (3.4 \times 10^{-11})/(6.8 \times 10^{-4})^2 = 7.\underline{3}5 \times 10^{-5}$. In order to solve the equilibrium equation, we require the concentration of HF, which we obtain from the acid ionization constant for HF.

$$K_a = \frac{[H^+][F^-]}{[HF]} \text{ or } 6.8 \times 10^{-4} = (3.825 \times 10^{-4})[F^-]/[HF]$$

$$[F^-]/[HF] = 1.778 \text{ or } [F^-] = 1.\underline{7}78[HF]$$

Let x be the solubility of CaF_2 in the buffer. Then $[Ca^{2+}] = x$ and $[F^-] + [HF] = 2x$. Substituting from the previous equation, we obtain

$$2x = 1.778[HF] + [HF] = 2.778[HF] \text{ or } [HF] = 2x/2.\underline{7}78$$

We can now substitute for $[H^+]$ and $[HF]$ into the equation for K_c.

$$K_c = \frac{[Ca^{2+}][HF]^2}{[H^+]^2} = \frac{x(2x/2.778)^2}{(3.825 \times 10^{-4})^2}$$

$$7.35 \times 10^{-5} = (3.543 \times 10^6)\, x^3$$

$$x^3 = 2.075 \times 10^{-11}$$

$$x = 2.\underline{7}48 \times 10^{-4}$$

17.89 Rounded answers: $[SO_4^{2-}]$ = $[Mg^{2+}]$ = 0.109 M; $[Ba^{2+}]$ = 1.0×10^{-9} M; and finally $[OH^-]$ = 1.3×10^{-5} M. The net ionic equation is

$$Ba^{2+}(aq) + 2OH^-(aq) + Mg^{2+}(aq) + SO_4^{2-}(aq) \rightleftharpoons BaSO_4(s) + Mg(OH)_2(s)$$

Start by calculating the mol/L of each ion after mixing and before precipitation. Use a total volume of 0.0450 + 0.0670 L = 0.112 L.

M of SO_4^{2-} and Mg^{2+} = (0.350 mol/L x 0.0670 L) ÷ 0.112 L = $0.20\underline{9}4$ M

M of Ba^{2+} = (0.250 mol/L x 0.0450 L) ÷ 0.112 L = $0.10\underline{0}4$ M

M of OH^- = (2 x 0.250 mol/L x 0.0450 L) ÷ 0.112 L = $0.20\underline{0}89$ M

Write a table showing precipitation of $BaSO_4$ and $Mg(OH)_2$.

Conc. (M)	Ba^{2+}	+ SO_4^{2-}	+ Mg^{2+}	+ $2OH^-$	$\rightleftharpoons BaSO_4 + Mg(OH)_2$
Starting	0.1004	0.2094	0.2094	0.20089	
Change	-0.1004	-0.1004	-0.1004	-0.20089	
Equil	0.0000	0.1090	0.1090	0.0000	

To calculate $[Ba^{2+}]$, use the K_{sp} expression for $BaSO_4$.

$$[Ba^{2+}] = \frac{1.1 \times 10^{-10}}{0.1090 \text{ M } SO_4^{2-}} = 1.\underline{0}09 \times 10^{-9} \text{ M}$$

$[SO_4^{2-}]$ = $0.10\underline{9}0$ M

Calculate the $[OH^-]$ using the K_{sp} expression for $Mg(OH)_2$:

$$[OH^-] = \sqrt{\frac{1.8 \times 10^{-11}}{0.1090 \text{ M } Mg^{2+}}} = 1.\underline{2}8 \times 10^{-5} \text{ M}$$

$[Mg^{2+}]$ = $0.10\underline{9}0$ M

CHAPTER 18

THERMODYNAMICS AND EQUILIBRIUM

SOLUTIONS TO EXERCISES

Notes on units and significant figures: **The "mol" unit will be omitted from all thermodynamic paramaters such as S°, ΔS°, etc. The final answer to all mathematical solutions is given first with one nonsignificant figure (last significant figure underlined) and is then rounded to the correct number of figures. Intermediate answers usually also have least one nonsignificant figure.**

18.1 Calculate the work, w, done using w = F x d = (mg) x d. Then use w to calculate ΔE.

$$w = (mg) \times d = (2.20 \text{ kg} \times 9.80 \text{ m/s}^2) \times 0.250 \text{ m} = 5.3\underline{9}0 \text{ kg•m}^2/\text{s}^2 = 5.39 \text{ J}$$

$$\Delta U = q + w = (5.3\underline{9}0 \text{ J}) + (-1.505 \text{ J}) = 3.890 = 3.89 \text{ J}$$

18.2 At 1.00 atm and 25°C, the volume occupied by 1.00 mol of any of the gases in the equation is 22.41 L x (298/273) = 24.46 L. Find the change in volume:

$$CH_4(24.46 \text{ L}) + 2O_2(2 \times 24.46 \text{ L}) \rightarrow CO_2(24.46 \text{ L}) + 2H_2O(l)$$

$$\Delta V = 24.26 \text{ L} - (3.00 \times 24.46 \text{ L}) = -48.92 \text{ L}$$

Next calculate the work, w, done by the system in pushing back the atmosphere, using 1.00 atm = 1.01×10^5 Pa, and 1.00 L = 1.00×10^{-3} m^3. Add this to the heat, q_p, at constant P:

$$w = -P\Delta V = -(1.01 \times 10^5 \text{ Pa}) \times (-48.92 \times 10^{-3} \text{ m}^3) = 4.9\underline{4}09 \times 10^3 \text{ J} = 4.94 \text{ kJ}$$

$$\Delta U = q_p + w = (-890.\underline{2} \text{ kJ}) + (+4.9\underline{4}09 \text{ kJ}) = -885.\underline{2}59 = -885.3 \text{ kJ}$$

18.3 When the liquid evaporates, it absorbs heat: ΔH_{vap} = 42.3 kJ/mol at 25°C. The entropy change is:

$$\Delta S = \frac{\Delta H_{vap}}{T} = \frac{42.3 \times 10^3 \text{ J/mol}}{298 \text{ K}} = 14\underline{1}.9 \text{ J/K•mol}$$

The entropy of 1 mole of vapor is calculated using the entropy of 1 mol of liquid (161 J/K):

$$S^\circ = 161 \text{ J/K} + 141.9 \text{ J/K} = 302.9 = 303 \text{ J/K}$$

18.4 a. ΔS^o is positive because there is an increase in moles of gas (Δn_{gas} = +1) from a solid reactant forming a mole of gas. (Entropy increases.)

b. ΔS^o is positive because there is an increase in moles of gas (Δn_{gas} = +1) from a liquid reactant forming a mole of gas. (Entropy increases.)

c. ΔS^o is negative because there is an decrease in moles of gas (Δn_{gas} = -1) from liquid and gaseous reactants forming 2 moles of solid. (Entropy decreases.)

d. ΔS^o is positive because there is an increase in moles of gas (Δn_{gas} = +1) from solid and liquid reactants forming a mole of gas and four moles of an ionic compound. (Entropy increases.)

18.5 The reaction and standard entropies are below. Multiply the S^o values by their stoichiometric coefficients and subtract the entropy of the reactant from the sum of the product entropies.

	$C_6H_{12}O_6$	$\rightarrow$	$2C_2H_5OH(l)$	+	$2CO_2(g)$
S^o:	212		2 x 161		2 x 213.7

$$\Delta S^o = \Sigma nS^o(\text{products}) - S^o(\text{reactant}) = (2 \times 161) + (2 \times 213.7) - 212 = 53\underline{7}.4$$

$$\Delta S^o = 537 \text{ J/K}$$

18.6 The reaction, standard enthalpy changes, and standard entropies are below.

	$CH_4(g)$	+ $2O_2(g)$	$\rightarrow$ $CO_2(g)$	+ $2H_2O(g)$
ΔH_f^o:	-74.9	2 x 0	-393.5	2 x (-241.8)
S^o:	186.1	2 x 205.0	213.7	2 x 188.7

Calculate ΔH^o and ΔS^o for the reaction.

$$\Delta H^o = [-393.5 + 2 \times (-241.8)]\text{kJ} - [-74.9 + 0]\text{kJ} = -802.2 \text{ kJ}$$

$$\Delta S^o = [213.7 + 2 \times 188.7]\text{J/K} - [186.1 + 2 \times 205.0]\text{J/K} = -5.0 \text{ J/K}$$

Now substitute into the equation for ΔG^o in terms of ΔH^o and ΔS^o (as -5.0×10^{-3} kJ/K):

$$\Delta G^o = \Delta H^o - T\Delta S^o = -802.2 \text{ kJ} - (298 \text{ K})(-5.0 \times 10^{-3} \text{ kJ/K}) = -800.\underline{7}1 = -800.7 \text{ kJ}$$

18.7 Write the values of ΔG_f^o multiplied by their stoichiometric coefficients below each formula:

	$CaCO_3(s)$	$\rightarrow$ $CaO(s)$	+ $CO_2(g)$
ΔG_f^o =	-1128.8	-603.5	-394.4

Subtract ΔG_f^o of reactant from that of the products:

$$\Delta G_f^o = [(-603.5) - (-394.4) - (-1128.8)]\text{kJ} = 130.\underline{9} \text{ kJ}$$

18.8 a. C(graphite) + $2H_2(g)$ → $CH_4(g)$

ΔG_f^o: 0 0 -50.8

ΔG^o = [(-50.8) - (0)]kJ = -50.8 kJ (spontaneous reaction)

b. $2H_2(g)$ + $O_2(g)$ → $2H_2O(l)$

ΔG_f^o: 0 0 2 x -237.2

ΔG^o = [(2 x -237.2) - (0)]kJ = -474.$\underline{4}$ kJ (spontaneous reaction)

c. 4HCN(g) + $5O_2(g)$ → $2H_2O(l)$ + $4CO_2(g)$ + $2N_2(g)$

ΔG_f^o: 4 x 125 0 2 x -237.2 4 x (-394.4) 0

ΔG^o = [(2 x -237.2) + 4 x (-394.4) - (4 x 125)]kJ

= -255$\underline{2}$.4 kJ (spontaneous reaction)

d. $Ag^+(aq)$ + $I^-(aq)$ → AgI(s)

ΔG_f^o: 77.1 -51.7 -66.3

ΔG^o = [(-66.3) - (77.1 - 51.7)]kJ = -91.$\underline{7}$ kJ (spontaneous reaction)

18.9 a. $K = P_{CO_2}$ (= K_p)

b. $K = [Pb^{2+}][I^-]$ (= K_{sp})

c. $K = \dfrac{P_{CO_2}}{[H^+][HCO_3^-]}$

18.10 First calculate ΔG^o using the ΔG_f^o values from Table 18.2.

$CaCO_3(s)$ ⇌ CaO(s) + $CO_2(g)$

ΔG_f^o = -1128.8 -603.5 -394.4

Subtract ΔG_f^o of reactant from that of the products:

ΔG_f^o = [(-603.5) - (-394.4) - (-1128.8)]kJ = 130.$\underline{9}$ kJ

Use the equation $\Delta G^o = -2.303\ RT \log K_{th}$ to calculate K_p, equating K to K_p. Use 298 as an exact number and use the 4 sig figs for ΔG^o to assign 4 sig figs to the log K_p:

$$\log K_p = \frac{\Delta G^o}{-2.303\ RT} = \frac{130.9\ \text{kJ}}{(-2.303)(0.008314\ \text{kJ/K})(298\ \text{K})} = -22.9\underline{4}1$$

K_p = 1.$\underline{1}$4 x 10^{-23} = 1.1 x 10^{-23} (only 2 sig figs to the right of the decimal in log K_p)

18.11 First calculate ΔG^o using the ΔG_f^o values in the exercise.

$$Mg(OH)_2(s) \rightarrow Mg^{2+}(aq) + 2OH^-(aq)$$

$$\Delta G_f^o = -933.9 \qquad -456.0 \qquad 2 \times -157.3$$

Subtract ΔG_f^o of reactant from that of the products:

$$\Delta G_f^o = [2 \times (-157.3) + (-456.0) - (-933.9)]\text{kJ} = 163.\underline{3}\ \text{kJ}$$

Use the equation $\Delta G^o = -2.303\ RT \log K$ to calculate K_{sp}, equating K to K_{sp}. Use 298 as an exact number and use the 4 sig figs for ΔG^o to assign 4 sig figs to the log K_{sp}:

$$\log K_{sp} = \frac{\Delta G^o}{-2.303\ RT} = \frac{163.3\ \text{kJ}}{(-2.303)(0.008314\ \text{kJ/K})(298\ \text{K})} = -28.6\underline{1}9$$

$$K_{sp} = 2.\underline{4}04 \times 10^{-29} = 2.4 \times 10^{-29}$$ (2 sig figs to the right of the decimal in log K_{sp})

18.12 From Table 6.2 and 18.1, we have:

$$H_2O(l) \rightarrow H_2O(g)$$

ΔH_f^o: -285.8 -241.8

S^o: 69.9 188.7

Calculate ΔH^o and ΔS^o from these values.

$$\Delta H^o = [-241.8 - (-285.8)]\ \text{kJ} = 44.\underline{0}\ \text{kJ}$$

$$\Delta S^o = [188.7 - 69.9]\ \text{J/K} = 118.8\ \text{J/K}$$

Substitute ΔH^o and ΔS^o into the Gibbs equation:

$$\Delta G^o = \Delta H^o - T\Delta S^o = 44.0\ \text{kJ} - (318\ \text{K})(0.1188\ \text{kJ/K}) = 6.\underline{2}2\ \text{kJ}$$

For this reaction, $K = K_p = P_{H_2O}$. Substitute ΔG^o into the K equation at 318 K:

$$\log K_p = \frac{\Delta G^o}{-2.303\ RT} = \frac{6.22\ \text{kJ}}{(-2.303)(0.008314\ \text{kJ/K})(318\ \text{K})} = -1.\underline{0}22$$

$K = \underline{9}.506 \times 10^{-2} = 1 \times 10^{-1}$ = vapor pressure of H_2O in atm

18.13 First calculate ΔH^o and ΔS^o, using the given ΔH_f^o and S^o values.

$$MgCO_3(s) \rightleftharpoons MgO(s) + CO_2(g)$$

ΔH_f^o = -1112 -601.2 -393.5

ΔS^o = 65.9 26.9 213.7

$$\Delta H^o = [-601.1 - (-393.5 - (-1112))]\ \text{kJ} = 117.3\ \text{kJ}$$

$$\Delta S^o = [(26.9 + 213.7) - 65.9]\ \text{J/K} = 174.7\ \text{J/K}$$

Substitute these values into the expression relating T, ΔH^o, and ΔS^o.

$$T = \frac{\Delta H^o}{\Delta S^o} = \frac{117.3\ kJ}{0.1747\ kJ/K} = 67\underline{1}.4\ K \quad \text{(lower than that for } CaCO_3\text{)}$$

ANSWERS TO REVIEW QUESTIONS

18.1 A spontaneous process is a chemical or physical change that occurs by itself without the continuing intervention of an outside agency. Three examples are: (1) a rock rolling down a hill from the top, (2) heat flowing from a hot object to a cold one, and (3) a piece of iron rusting in moist air. Three examples of a nonspontaneous process are: (1) a rock rolling uphill by itself, (2) heat flowing from a cold object to a warm object, and (3) rust being converted to iron metal.

18.2 A quantity of liquid benzene contains more entropy than the same quantity of frozen benzene because the frozen benzene has an ordered crystalline structure. The liquid has a less-ordered structure in which molecules move freely about.

18.3 The second law of thermodynamics states that the total entropy of a system and its surroundings always increases for a spontaneous process. This is because entropy is produced during any spontaneous or natural process.

18.4 The entropy change for such a system at constant pressure does not equal $\Delta H/T$ because the change must equal $\Delta H/T$ plus the entropy created by the spontaneous process (see previous question). Consider, for example, the decomposition of a solid to a gaseous product.

18.5 The standard entropy of hydrogen gas at 25°C can be obtained from heat measurements. We start with hydrogen at 0 K, where $S^o = 0$, and warm to 298 K (25°C) in small increments. For example, warm from 0.0 K to 2.0 K; calculate ΔS by dividing the heat absorbed by the average temperature (1.0 K), etc.

18.6 To predict the sign of ΔS^o in reactions involving gases, look for a change, Δn_{gas}, in the number of moles of gas. If Δn_{gas} increases, then ΔS^o should be positive; it it decreases, the ΔS^o should be negative. The reason is that the gaseous phase has a higher entropy than the liquid, aqueous, or solid phases.

18.7 Free energy, G, is the difference between the enthalpy of a system and the product of the absolute temperature and the entropy of the system: $G = H - TS$. The free-energy change, $\Delta G = \Delta H - T\Delta S$.

18.8 The standard free-energy change, ΔG^o, occurs when reactants in their standard states are converted to products in their standard states. The standard free energy of formation, ΔG_f^o, is the free-energy change when one mole of substance is formed from its elements in their stable states at 1 atm and at a specified temperature, usually 25°C.

18.9 If ΔG^o for a reaction at 25°C is negative, the equation for the reaction is spontaneous in the direction written. If ΔG^o is positive, the equation for the reaction is nonspontaneous in the direction written.

18.10 In the ideal situation in which a chemical reaction is run so that it produces no entropy, the useful work done is equal to the maximum useful work, w_{max}, or to ΔG.

18.11 When gasoline burns in an automobile engine, the change in free energy shows up as useful work. Gasoline, a mixture of hydrocarbons such as C_8H_{18} or octane, burns to yield energy, gaseous CO_2, and gaseous H_2O.

18.12 A nonspontaneous reaction can be made to occur by coupling it with a spontaneous reaction having a sufficiently negative ΔG^o to furnish the required energy. (The net ΔG^o of the coupled reactions must be negative.)

18.13 As a spontaneous reaction proceeds, the free energy decreases until equilibrium is reached at a minimum ΔG. See the diagram below.

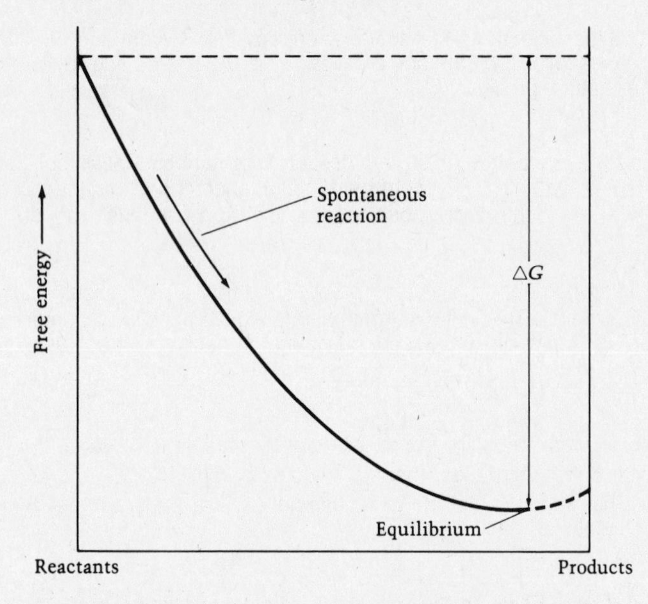

18.14 Since the equilibrium constant is related to ΔH^o and ΔS^o by $-RT \ln K = \Delta H^o - T\Delta S^o$, heat measurements alone can be used to obtain it. The standard enthalpy ΔH^o is the heat of reaction measured at constant pressure. The standard entropy change, ΔS^o, can be calculated from standard entropies, which are obtained from heat capacity data.

18.15 The four combinations are: (1) A negative ΔH^o and positive ΔS^o always gives a negative ΔG^o and a spontaneous reaction. (2) A positive ΔH^o and a negative ΔS^o always gives a positive ΔG^o and a nonspontaneous reaction. (3) A negative ΔH^o and negative ΔS^o may give a negative or a positive ΔG^o. At low temperatures ΔG^o will usually be negative and the reaction spontaneous; at high temperatures, ΔG^o will usually be positive and the reaction nonspontaneous. (4) A positive ΔH^o and a positive ΔS^o may give a negative or a positive ΔG^o. At low temperatures ΔG^o will usually be positive and the reaction nonspontaneous; at high temperatures ΔG^o will usually be negative and the reaction spontaneous.

18.16 We can estimate the temperature at which a nonspontaneous reaction becomes spontaneous by substituting 0 for ΔG^o into the equation $\Delta G^o = \Delta H^o - T\Delta S^o$, and then solving for T using the form $T = \Delta H^o/\Delta S^o$.

SOLUTIONS TO PRACTICE PROBLEMS

Note on significant figures: The final answer to all mathematical solutions is given first with one nonsignificant figure (last significant figure underlined), and is then rounded to the correct number of digits. Intermediate answers usually also have at least one nonsignificant figure.

18.17 The values of q and w are = -65 J and 22 J, respectively.

$$\Delta U = q + w = (-65\ J) + 22\ J = -43\ J.$$

18.19 At 100°C (373 K) and 1 atm (1.01×10^5 Pa), 1.00 mol of $H_2O(g)$ occupies:

$$22.41\ L \times \frac{373\ K}{273\ K} = 30.\underline{6}19\ L\ \ (30.619 \times 10^{-3}\ m^3)$$

The work done by the chemical system in pushing back the atmosphere is:

$$w = -P\Delta V = -(1.01 \times 10^5\ Pa) \times (30.619 \times 10^{-3}\ m^3) = -3.0\underline{9}25 \times 10^3\ J = -3.0\underline{9}25\ kJ$$

$$\Delta U = q_p + w = (40.6\underline{6}\ kJ) + (-3.0925\ kJ) = 37.5\underline{6}7 = 37.57\ kJ$$

18.21 Use the equilibrium relation between ΔS and ΔH_{vap}, as 29.6×10^3 J, at the boiling point:

$$\Delta S = \frac{\Delta H_{vap}}{T} = \frac{29.6 \times 10^3\ J}{334.2\ K} = 8.8\underline{5}6\ J/K$$

18.23 When the liquid condenses, it releases heat: ΔH_{cond} = -37.4 kJ/mol, or -37.4×10^3 J/mol, at 25°C. The entropy change is:

$$\Delta S = \frac{\Delta H_{cond}}{T} = \frac{-37.4 \times 10^3\ J}{298\ K} = -12\underline{5}.503\ J/K$$

The entropy of 1 mole of liquid is calculated using the entropy of 1 mol of vapor (252 J/K):

$$S_{liq} = S_{vap} - \Delta S_{cond} = 252\ J/K + (-125.503\ J/K) = 12\underline{6}.497 = 126\ J/K$$

18.25 a. ΔS^o is negative because there is a decrease in moles of gas ($\Delta n_{gas} = -2$) from 3 moles of gaseous reactants forming 1 mole of gas. (Entropy decreases.)

b. ΔS^o is not predictable because there is no change in moles of gas (Δn_{gas} = 0) from 2 moles of gaseous reactants forming 2 moles of gaseous products.

c. ΔS^o is positive because there is an increase in moles of gas (Δn_{gas} = +1) from 5 moles of gaseous reactants forming 6 moles of gaseous products. (Entropy increases.)

d. ΔS^o is positive because there is an increase in moles of gas (Δn_{gas} = +1) from a solid reactant and 1 mole of gaseous reactant forming 2 moles of gas. (Entropy increases.)

18.27 The reaction and standard entropies are given below. Multiply the S^o values by their stoichiometric coefficients and subtract the entropy of the reactant from the sum of the product entropies. In part b, note that aqueous NaCl exists as $Na^+(aq)$ and $Cl^-(aq)$ ions

a. $2Na(s) + Cl_2(g) \rightarrow 2NaCl(s)$

S^o: 2 x 51.4   223.0   2 x 72.1

$\Delta S^o = S^o(prod) - \Sigma S^o(reac's) = [(2 \times 72.1) - (2 \times 51.4 + 223.0)] = -181.\underline{6}$

ΔS^o = -181.6 J/K

b. $NaCl(s) \rightarrow Na^+(aq) + Cl^-(aq)$

S^o: 72.1   60.2   55.1

$\Delta S^o = \Sigma S^o(prod) - S^o(reac) = [(60.2 + 55.1) - (72.1)] = 43.2$

ΔS^o = 43.2 J/K

c. $CS_2(l) + 3O_2(g) \rightarrow CO_2(g) + 2SO_2(g)$

S^o: 151.0   3 x 205.0   213.7   2 x 248.1

$\Delta S^o = \Sigma S^o(prod's) - \Sigma S^o(reac's)$

$\Delta S^o = [(213.7 + 2 \times 248.1) - (151.0 + 3 \times 205.0)] = -56.\underline{1}$ = -56.1 J/K

d. $2CH_3OH(l) + 3O_2(g) \rightarrow 2CO_2(g) + 4H_2O(g)$

S^o: 2 x 127   3 x 205.0   2 x 213.7   4 x 188.7

$\Delta S^o = \Sigma S^o(prod's) - \Sigma S^o(reac's)$

$\Delta S^o = [(2 \times 213.7 + 4 \times 188.7) - (2 \times 127 + 3 \times 205.0)] = -31\underline{3}.2$ = -313 J/K

18.29 $CH_4(g) + 2O_2(g) \rightarrow CO_2(g) + 2H_2O(l)$

S^o: 186.1   2 x 205.0   213.7   2 x 69.9

$\Delta S^o = S^o(prod's) - \Sigma S^o(reac's)$

$\Delta S^o = [(213.7 + 2 \times 69.9) - (186.1 + 2 \times 205.0)] = -242.\underline{6}$ = -242.6 J/K

18.31 The reaction with standard enthalpies of formation and standard entropies written underneath is

	$2CH_3OH(l)$	+	$3O_2(g)$	→ $2CO_2(g)$	+	$4H_2O(l)$
ΔH_f^o:	2 x (-238.6)		3 x 0	2 x (-393.5)		4 x (-285.840)
ΔS^o:	2 x 127		3 x 205.0	2 x 213.7		4 x 69.940

Calculate ΔH^o and ΔS^o for the reaction.

ΔH^o = [2 x (-393.5) + 4 x (-285.840)]kJ - [2 x (-238.6) + 3 x 0]kJ
= -1453.$\underline{1}$60 kJ = -1453.2kJ

ΔS^o = [2 x 213.7 + 4 x 69.940]J/K - [2 x 127 + 3 x 205.0]J/K = -16$\underline{1}$.84 = -162 J/K

Now substitute into the equation for ΔG^o in terms of ΔH^o and ΔS^o (as -0.16184 kJ/K).

ΔG^o = ΔH^o - $T\Delta S^o$ = -1453.160 kJ - (298.2 K) x (-0.16184 kJ/K)
= -1404.$\underline{8}$99 = -1404.9 kJ

18.33 a. $Na(s) + 1/2Cl_2(g) \rightarrow NaCl(s)$

b. $1/2H_2(g) + C(graphite) + 1/2N_2(g) \rightarrow HCN(l)$

c. $S(rhombic) + O_2(g) \rightarrow SO_2(g)$

d. $P(red) + 3/2H_2(g) \rightarrow PH_3(g)$

18.35 Write the values of ΔG_f^o multiplied by their stoichiometric coefficients below each formula; then subtract ΔG_f^o of reactant from that of the products

a.

	$CH_4(g)$	+ $2O_2(g)$	→	$CO_2(g)$ +	$2H_2O(g)$
ΔG_f^o =	-50.8	2 x 0		-394.4	2 x -228.6

ΔG^o = [(-394.4) + 2(-228.6) - (-50.8 + 0)]kJ = -800.$\underline{8}$ kJ

b.

	$CaCO_3(s)$	+ $2H^+(aq)$	→	$Ca^{2+}(aq)$ +	$H_2O(l)$ +	$CO_2(g)$
ΔG_f^o =	-1128.8	2 x 0		-553.04	-237.192	-394.4

ΔG^o = [(-553.04) + (-237.192) + (-394.4) - (-1128.8 + 0)]kJ = -55.$\underline{8}$32 kJ

ΔG^o = -55.8 kJ

18.37 a. Spontaneous reaction

b. Spontaneous reaction

c. Nonspontaneous reaction

d. Equilibrium mixture-significant amounts of both

e. Nonspontaneous reaction

18.39 Calculate ΔH^o and ΔG^o, using the given ΔH_f^o and ΔG_f^o values.

a.	$Al_2O_3(s)$ +	$2Fe(s)$ →	$Fe_2O_3(s)$ +	$2Al(s)$
ΔH_f^o:	-1676	2 x 0	-825.5	2 x 0
ΔG_f^o:	-1582	2 x 0	-743.6	2 x 0

ΔH^o = [(-825.5) + 0 - (-1676) - 0] kJ = $8\underline{5}0.5$ = 850 kJ

ΔG^o = [(-743.6) + 0 - (-1582) - 0] kJ = $83\underline{8}.4$ = 838 kJ

The reaction is endothermic, absorbing 850 kJ of heat. The large positive value for ΔG^o indicates that the equilibrium composition is mainly reactants.

b.	$COCl_2(g)$ +	$H_2O(l)$ →	$CO_2(g)$ +	$2HCl(g)$
ΔH_f^o:	-220	-285.840	-393.5	2 x -92.31
ΔG_f^o:	-206	-237.192	-394.4	2 x -95.30

ΔH^o = [(-393.5 + (2)(-92.31)) - (-220) - (-285.840)] kJ = -72.28 = -72 kJ

ΔG^o = [(-394.4 + (2)(-95.30)) - (-206) - (-237.192)] kJ = $-14\underline{1}.808$ = -142 kJ

The reaction is exothermic; the ΔG^o value indicates mainly products at equilibrium.

18.41 Calculate ΔG^o, using the given ΔG_f^o values.

	$2H_2(g)$ +	$O_2(g)$ →	$2H_2O(l)$
ΔG_f^o:	0	0	2 x -237.2

ΔG^o = [(2 x -237.2) - (0)]kJ = $-474.\underline{4}$ kJ

Maximum work = ΔG^o = (-474.4 kJ) = -474.4 kJ. Since the maximum work is stipulated, no entropy is produced.

18.43 Calculate ΔG^o, per 1 mol Zn(s), using the given ΔG_f^o values.

	$Zn(s)$ +	$Cu^{2+}(aq)$ →	$Zn^{2+}(aq)$ +	$Cu(s)$
ΔG_f^o:	0	64.98	-147.21	0

ΔG^o = [(-147.21) + (0) - (64.98) - 0]kJ = -212.19 kJ/mol Zn

-212.19 kJ/mol Zn x (5.00 g ÷ 65.38 g/mol Zn) = $-16.\underline{2}27$ = -16.2 kJ

Maximum work = ΔG^o = (-16.3 kJ) = -16.3 kJ. Since maximum work is stipulated, no entropy is produced.

18.45 a. $K = K_p = \dfrac{P_{CO_2}P_{H_2}}{P_{CO}P_{H_2O}}$

b. $K = [Mg^{2+}][OH^-]^2$ (= K_{sp})

c. $K = [Li^+][OH^-]^2 P_{H2}$

18.47 First calculate ΔG^o, using the ΔG_f^o values from Table 18.2.

	$H_2(g)$	+	$Cl_2(g)$	→	$2HCl(g)$
ΔG_f^o =	0		0		-95.3

ΔG^o = 2(-95.3) kJ = -190.6 kJ

Use the equation ΔG^o = -2.303 RT log K to calculate K. Use 298 as an exact number, use R = 0.008314 kJ/K, and use the 4 sig figs for ΔG^o to assign 4 sig figs to the log K:

$$\log K = \frac{\Delta G^o}{-2.303\ RT} = \frac{-190.6\ kJ}{(-2.303)(0.008314\ kJ/K)(298\ K)} = 33.4\underline{2}04$$

$K = 2.\underline{6}3 \times 10^{33} = 2.6 \times 10^{33}$ (only 2 sig figs to the right of the decimal in log K)

18.49 First calculate ΔG^o, using the ΔG_f^o values from Table 18.2.

	$CO(g)$	+ $3H_2(g)$	→	$CH_4l(g)$	+	$H_2O(g)$
ΔG_f^o =	-137.2	0		-50.8		-228.6

Subtract ΔG_f^o of reactants from that of the products:

ΔG^o = [(-50.8 + -228.6) - (-137.2 + 0)]kJ = -142.2 kJ

Use the equation ΔG^o = -2.303 RT log K_p to calculate K_p, which is equal to K. Use 298 as an exact number, use R = 0.008314 kJ/K, and use the 4 sig figs for ΔG^o to assign 4 sig figs to the log K_p:

$$\log K_p = \frac{\Delta G^o}{-2.303\ RT} = \frac{-142.2\ kJ}{(-2.303)(0.008314\ kJ/K)(298\ K)} = 24.9\underline{2}1$$

$K_p = 8.\underline{3}3 \times 10^{24} = 8.3 \times 10^{24}$ (only 2 sig figs to the right of the decimal in log K)

18.51 First calculate ΔG^o, using the ΔG_f^o values from Table 18.2.

	$Mg(s)$	+ $Cu^{2+}(aq)$	→	$Mg^{2+}(aq)$	+	$Cu(s)$
ΔG_f^o =	0	64.98		-456.01		0

Subtract ΔG_f^o of reactants from that of the products:

ΔG^o = [(-456.01 + 0) - (64.98 + 0)]kJ = -520.99 kJ

Use the equation $\Delta G^o = -2.303\ RT \log K_c$ to calculate K_c, which is equal to K. Use 298 as an exact number, use R = 0.008314 kJ/K, and use the 5 sig figs for ΔG^o to assign 5 sig figs to the log K_c:

$$\log K_c = \frac{\Delta G^o}{-2.303\ RT} = \frac{-520.99\ kJ}{(-2.303)(0.008314\ kJ/K)(298\ K)} = 91.30\underline{8}07$$

$$K_c = 2.0\underline{3}2 \times 10^{91} = 2.03 \times 10^{91}$$ (only 3 sig figs to the right of the decimal in log K)

18.53 The sign of ΔS^o should be positive because there is an increase in moles of gas ($\Delta n_{gas} = +5$) as the solid reactant forms 5 moles of gas. The reaction is endothermic, implying a positive ΔH^o. The fact that the reaction is spontaneous implies that the product $T\Delta S^o$ is larger than ΔH^o, so that ΔG^o is negative, as required for a spontaneous reaction.

18.55 The ΔH value $\cong$ BE(H-H) + BE(Cl-Cl) - BE(H-Cl) $\cong$ [432 + 240 - 2(428)] kJ $\cong$ -184 kJ, and thus the reaction is exothermic. ΔS^o should be positive because there is an increase in disorder with the formation of unsymmetrical molecules from symmetrical H_2 and Cl_2. The reaction should be spontaneous because the contributions of both the ΔH term and the $-T\Delta S$ term are negative.

18.57 From Tables 6.2 and 18.1, we have:

	C(graphite)	+	$CO_2(g)$	$\rightleftharpoons$	$2CO(g)$
ΔH_f^o:	0		-393.5		2(-110.5)
S^o:	5.7		213.7		2(197.5)

Calculate ΔH^o and ΔS^o from these values.

$$\Delta H^o = [2(-110.5) - (0 + (-393.5))]\ kJ = 172.\underline{5}\ kJ$$

$$\Delta S^o = [2(197.5) - (5.7 + 213.7)]\ J/K = 175.\underline{6}\ J/K\ \ (0.1756\ kJ/K)$$

$$\Delta G^o = \Delta H^o - T\Delta S^o = 172.5\ kJ - (1273\ K)(0.1756\ kJ/K) = -51.\underline{0}4\ kJ$$

Substitute ΔG^o into the K_p equation at 1273 K, using R = 0.008314 kJ/K:

$$\log K_p = \frac{\Delta G^o}{-2.303\ RT} = \frac{-51.\underline{0}4\ kJ}{(-2.303)(0.008314\ kJ/K)(1273\ K)} = 2.0\underline{9}40$$

$$K_p = 1.\underline{2}41 \times 10^2 = 1.2 \times 10^2$$ (only 2 sig figs to right of decimal in log K_p)

Since K_p is greater than 1, it predicts that combustion of carbon should form significant amounts of CO product at equilibrium.

18.59 First calculate ΔH^o and ΔS^o, using the given ΔH_f^o and S^o values.

	$2NaHCO_3(s)$ →	$Na_2CO_3(s)$ +	$H_2O(g)$ +	$CO_2(g)$
ΔH_f^o =	2 x -947.7	-1130.8	-241.826	-393.5
ΔS^o =	2 x 102	139	188.72	213.7

ΔH^o = [(-1130.8 + -241.826 + -393.5) - 2(-947.7)] kJ = 129.$\underline{2}$74 kJ

ΔS^o = [(139 + 188.7 + 213.7) - 2(102)] J/K = 337.4 J/K (0.3374 kJ/K)

Substitute these values into $\Delta G^o = \Delta H^o - T\Delta S^o$, let $\Delta G^o = 0$, and rearrange to solve for T.

$$T = \frac{\Delta H^o}{\Delta S^o} = \frac{129.274 \text{ kJ}}{0.3374 \text{ kJ/K}} = 38\underline{3}.14 = 383 \text{ K}$$

18.61 When the liquid freezes, it releases heat: ΔH_{fus} = -69.0 J/g at 16.6°C (289.6 K) The entropy change is:

$$\Delta S = \frac{\Delta H_{fus}}{T} = \frac{-69.0 \text{ J/g}}{289.6 \text{ K}} \times \frac{60.05 \text{ g}}{1 \text{ mol}} = -14.\underline{2}9 = -14.3 \text{ J/(K•mol)}$$

18.63 a. ΔS^o is negative because there is a decrease in moles of gas (Δn_{gas} = -1) from 1 mole of gaseous reactant forming aqueous and liquid products. (Entropy decreases.)

b. ΔS^o is positive because there is an increase in moles of gas (Δn_{gas} = +5) from a solid reactant forming 5 moles of gas. (Entropy increases.)

c. ΔS^o is positive because there is an increase in moles of gas (Δn_{gas} = +3) from 2 moles of gaseous reactant forming 5 moles of gaseous products. (Entropy increases.)

d. ΔS^o is negative because there is a decrease in moles of gas (Δn_{gas} = -1) from 3 moles of gaseous reactants forming 2 moles of gaseous products. (Entropy decreases.)

18.65 ΔS^o is negative because there is a decrease in the moles of gas (Δn_{gas} = -2) from 3 moles of gaseous reactant forming 1 mole of gaseous product plus liquid product.

18.67 Calculate ΔS^o from the individual S^o values:

	$C_2H_5OH(l)$ +	$O_2(g)$ →	$CH_3COOH(l)$ +	$H_2O(l)$
S^o:	161	205.0	160	69.9

$\Delta S^o = \Sigma S^o(\text{prod}) - \Sigma S^o(\text{reac's})$

ΔS^o = [(160 + 69.9) · (161 + 205.0)] = -13$\underline{6}$.1 = -136 J/K

18.69 Calculate ΔG^o, using the ΔG_f^o values from Table 18.2.

	$H_2(g)$	+	$SO_2(g)$	→	$H_2S(g)$	+	$O_2(g)$
ΔG_f^o:	0		-300.2		-33		0

$$\Delta G^o = (-33 + 0) - (-300.2 + 0)\ kJ = 26\underline{7}.2 = 267\ kJ$$

Since ΔG^o is positive, the reaction is nonspontaneous as written, at 25°C.

18.71 At low (room) temperature ΔG^o or ($\Delta H^o - T\Delta S^o$) must be positive, but at higher temperatures ΔG^o or ($\Delta H^o - T\Delta S^o$) must be negative. Thus at the higher temperatures the $-T\Delta S^o$ term must become more negative than ΔH^o. Thus ΔS^o must be positive and so must ΔH^o be positive. If either were negative, ΔG^o would not become negative at higher temperatures.

18.73 First calculate ΔG^o, using the ΔG_f^o values in the exercise.

	$CaF_2(s)$ →	$Ca^{2+}(aq)$ +	$2F^-(aq)$
ΔG_f^o:	-1162	-553	2 x -276.5

Subtract ΔG_f^o of reactants from that of the product:

$$\Delta G_f^o = [-553 + 2 \times (-276.5) - (-1162)]kJ = 5\underline{6}.0\ kJ$$

Use the equation $\Delta G^o = -2.303\ RT \log K$ to calculate K_{sp}, equating K to K_{sp}. Use 298 as an exact number and use the 2 sig figs for ΔG^o to assign 2 sig figs to the log K_{sp}:

$$\log K_{sp} = \frac{\Delta G^o}{-2.303\ RT} = \frac{56.0\ kJ}{(-2.303)(0.008314\ kJ/K)(298\ K)} = -9.\underline{8}145$$

$K_{sp} = \underline{1}.53 \times 10^{-10} = 2 \times 10^{-10}$ (1 sig fig to the right of the decimal in log K_{sp})

18.75 From Table 18.2, we have:

	$COCl_2(g)$ →	$CO(g)$ +	$Cl_2(g)$
ΔH_f^o:	-220	-110.5	0
S^o:	284	197.5	223.0

Calculate ΔH^o and ΔS^o from these values.

$$\Delta H^o = [(-110.5) + 0 - (-220)]\ kJ = 10\underline{9}.5\ kJ$$

$$\Delta S^o = [197.5 + 223.0 - 284]\ J/K = 13\underline{6}.5\ J/K\ \ (0.13\underline{6}5\ kJ/K)$$

At 25°C : $\Delta G^o = \Delta H^o - T\Delta S^o = 109.5\ kJ - (298\ K)(0.1365\ kJ/K) = 6\underline{8}.8 = 69\ kJ$

At 800°C : $\Delta G^o = \Delta H^o - T\Delta S^o = 109.5\ kJ - (1073\ K)(0.1365\ kJ/K) = -3\underline{6}.9 = -37\ kJ$

Thus ΔG^o changes from a positive value, and a nonspontaneous reaction at 25°C to a negative value, and a spontaneous reaction at 800°C.

Cumulative-Skills Problems (require skills from previous chapters)

18.77 For the dissociation of HBr, assume that ΔH and ΔS are constant over the temperature range from 25°C to 375°C, and calculate the value of each to use to calculate K at 375°C. Start by calculating ΔH^o and ΔS^o at 25°C, using ΔH_f^o and S^o values.

	$2HBr(g)$ →	$H_2(g)$ +	$Br_2(g)$
ΔH_f^o:	2 x -36	0	30.91
ΔS^o:	2 x 198.59	130.6	245.38

$\Delta H^o = [(30.91 + 0) - 2(-36)]\text{ kJ} = 10\underline{2}.91\text{ kJ}$ ($\cong \Delta H$ at 375°C)

$\Delta S^o = [(245.38 + 130.6) - 2(198.59)]\text{ J/K} = -21.\underline{2}0\text{ J/K} = -0.02120\text{ kJ/K}$ ($\cong \Delta S$ at 375°C)

Substitute ΔH^o and ΔS^o into the Gibbs equation at 375°C (648 K):

$$\Delta G^o \cong \Delta H - T\Delta S \cong 102.91\text{ kJ} - (648\text{ K})(-0.02120\text{ kJ/K}) \cong 11\underline{6}.6\text{ kJ}$$

Substitute ΔG^o into the K equation at 648 K, using R = 0.008314 kJ/K:

$$\log K = \frac{\Delta G^o}{-2.303\,RT} = \frac{116.6\text{ kJ}}{(-2.303)(0.008314\text{ kJ/K})(648\text{ K})} = -9.4\underline{0}2$$

$K = 3.\underline{9}6 \times 10^{-10}$ (only 2 sig figs to right of decimal in log K)

Assuming $x = [H_2] = [Br_2]$, and assuming $[HBr] = (1.00 - 2x) \cong 1.00$ atm, substitute into the equilibrium expression:

$$\frac{[H_2][Br_2]}{[HBr]^2} \cong \frac{(x)(x)}{(1.00)^2} \cong 3.\underline{9}6 \times 10^{-10}$$

Solving for the approximate pressure of x:

$$x \cong \sqrt{(3.96 \times 10^{-10})(1.00)^2} \cong 1.\underline{9}89 \times 10^{-5}\text{ atm}$$

The percent dissociation at 1.00 atm is:

$$\%\text{ dissociated} = \frac{1.989 \times 10^{-5}\text{ atm}}{1.00\text{ atm}} \times 100\% = 0.0040\%\text{ (at 1.00 atm)}$$

Using a similar calculation at 10.0 atm, the % dissociation is $1.\underline{2}58 \times 10^{-3} = 0.0013\%$.

18.79 For the dissociation of NH_3, assume that ΔH and ΔS are constant over the temperature range from 25°C to 345°C and calculate values of each to calculate K at 345°C. First calculate ΔH^o and ΔS^o, using the given ΔH_f^o and S^o values.

	$2NH_3(g)$ →	$3H_2(g)$ +	$N_2(g)$
ΔH_f^o:	2 x -45.9	0	0
S^o:	2 x 193	3 x 130.6	191.5

ΔH^o = [0 + 0 - 2(-45.9)] kJ = 91.$\underline{8}$ kJ ≅ ΔH at 345°C

ΔS^o = [(3 x 130.6 + 191.5) - 2(193)] J/K = 19$\underline{7}$.3 J/K = 0.19$\underline{7}$3 kJ/K (≅ ΔS at 345°C)

Substitute ΔH^o and ΔS^o into the Gibbs equation at 345°C (618 K):

$\Delta G^o \cong \Delta H - T\Delta S \cong$ 91.8 kJ - (718 K)(0.1973 kJ/K) ≅ -30.$\underline{1}$3 kJ

Substitute ΔG^o into the K equation at 618 K, using R = 0.008314 kJ/K:

$$\log K = \frac{\Delta G^o}{-2.303\ RT} = \frac{-30.13\ \text{kJ}}{(-2.303)(0.008314\ \text{kJ/K})(618\ \text{K})} = 2.5\underline{4}28$$

$$K = 3.\underline{5}17 \times 10^2 = K_p$$

We now obtain K_c.

$$K_c = K_p(RT)^{-2}$$
$$= (3.517 \times 10^2)(0.0821 \times 618)^{-2} = 0.1\underline{3}66$$

The starting concentration of NH_3 is 1.00 mol/20.0L = 0.0500M. We obtain the following table.

	$2NH_3(g)$	→	$3H_2(g)$	+	$N_2(g)$
Starting	0.0500		0		0
Change	-2x		+3x		+x
Equilibrium	0.0500 - 2x		3x		x

The equilibrium equation is

$$\frac{[H_2]^3[N_2]}{[NH_3]^2} = K_c$$

or

$$\frac{(3x)^3 x}{(0.0500 - 2x)^2} = 0.1\underline{3}66$$

$$\frac{x^4}{(0.0500 - 2x)^2} = \frac{0.1366}{27} = 5.060 \times 10^{-3}$$

Taking the square root of both sides of this equation gives

$$\frac{x^2}{0.0500 - 2x} = 7.\underline{1}13 \times 10^{-2}$$

which rearranges to

$$x^2 + 1.423 \times 10^{-1}x - 3.556 \times 10^{-3} = 0$$

From the quadratic formula, we obtain

$$x = \frac{-0.1423 \pm \sqrt{(0.1423)^2 + 4 \times 3.556 \times 10^{-3}}}{2}$$

The positive root is

$$x = 0.02168$$

Hence,

$$[NH_3] = 0.0500 - 2(0.02168) = 6.64 \times 10^{-3}$$

$$\% \; NH_3 \text{ dissociated} = \left(1 - \frac{6.64 \times 10^{-3}}{0.0500}\right) \times 100$$

$$= 8\underline{6}.7\% = 87\%$$

18.81 First calculate ΔG^o at each temperature, using $\Delta G^o = -2.303RT \log K_a$:

$$25.0^oC: \; \Delta G^o = -(2.303)(0.008314 \text{ kJ/K})(298.2 \text{ K})(\log 1.754 \times 10^{-5}) = 27.15\underline{5}1 \text{ kJ}$$

$$50.0^oC: \; \Delta G^o = -(2.303)(0.008314 \text{ kJ/K})(323.2 \text{ K})(\log 1.633 \times 10^{-5}) = 29.62\underline{3}7 \text{ kJ}$$

Next solve 2 equations in 2 unknowns, assuming that ΔH^o and ΔS^o are constant over the range of 25°C to 50°C. Use 0.2982 K(kJ/J) and 0.3232 K(kJ/J) to convert ΔS^o in J to $T\Delta S^o$ in kJ.

1. $27.1552 \text{ kJ} = \Delta H^o - [298.2 \text{ K(kJ/J)} \; \Delta S^o]$

2. $29.6237 \text{ kJ} = \Delta H^o - [323.2 \text{ K(kJ/J)} \; \Delta S^o]$

Then rearrange equation 2 and substitute for ΔH^o into equation 2:

3a. $\Delta H^o = [(0.3232 \; \Delta S^o) \text{ K(kJ/J)} + 29.6237]$

3b. $27.1551 \text{ kJ} = [(0.3232 \; \Delta S^o) \text{ K(kJ/J)} + 29.6237] - [(0.2982 \; \Delta S^o) \text{ K(kJ/J)}]$

Solve for ΔS^o:

$$\Delta S^o = \frac{(29.6237 - 27.1552) \text{ kJ}}{(0.2982 - 0.3232) \text{ K(kJ/J)}} = -98.7\underline{4}4 = -98.74 \text{ J/K}$$

Substitute this value into equation 3a and solve for ΔH^o:

$$\Delta H^o = [(0.3232 \; \Delta S^o) \text{ K(kJ/J)} \times (-98.744 \text{ J/K}) + 29.6237 \text{ kJ}]$$

$$\Delta H^o = -2.2903 = -2.29 \text{ kJ}$$

CHAPTER 19

ELECTROCHEMISTRY

SOLUTIONS TO EXERCISES

Note on significant figures: The final answer to all mathematical solutions is given first with one nonsignificant figure (last significant figure underlined) and is then rounded to the correct number of figures. Intermediate answers usually also have at least one nonsignificant figure.

19.1 The electrode reactions are:

Cathode: $Ag^+(aq) + e^- \rightleftharpoons Ag(s)$

Anode: $Ni(s) \rightleftharpoons Ni^{2+}(aq) + 2e^-$

A sketch of the cell is given below:

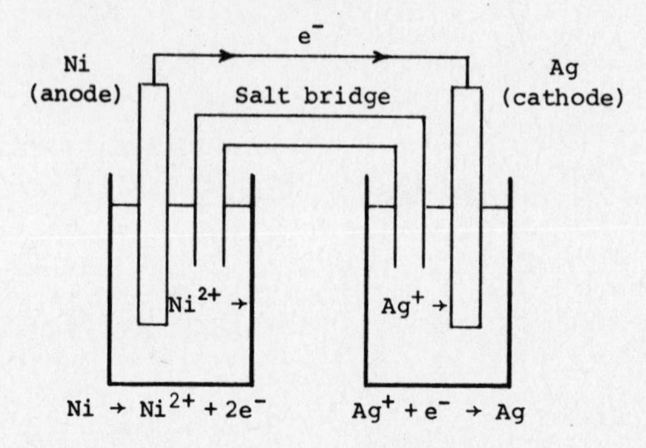

19.2 The notation for the cell is: $Zn(s)|Zn^{2+}(aq)||H^+(aq)|H_2(aq)|Pt(s)$

19.3 Given below are the half-cell reactions and their sum, the overall cell reaction:

$$Cd(s) \rightleftharpoons Cd^{2+}(aq) + 2e^-$$

$$2H^+(aq) + 2e^- \rightleftharpoons H_2(g)$$

$$Cd(s) + 2H^+(aq) \rightleftharpoons Cd^{2+}(aq) + H_2(g)$$

19.4 Since maximum work, $w_{max} = -nFE_{cell}$, write out the half-reactions to determine n:

$$Zn(s) \rightleftharpoons Zn^{2+}(aq) + 2e^-; \quad Cu^{2+}(aq) + 2e^- \rightleftharpoons Cu(s)$$

Since both are 2 electron half-reactions, n = 2. The maximum work per 1 mole of Zn(s) is thus:

$$w_{max} = -nFE_{cell} = -(2)(9.65 \times 10^4\ C)(1.10\ V) = -2.1\underline{2}3 \times 10^5\ V{\bullet}C = -2.12 \times 10^5\ J$$

The maximum work for 6.54 g of Zn(s) is:

$$6.54\ g\ Zn \times \frac{1\ mol\ Zn}{65.39\ g\ Zn} \times \frac{-2.123 \times 10^5\ J}{1\ mol\ Zn} = -2.1\underline{2}3 \times 10^4 = -2.12 \times 10^4\ J$$

19.5 The E° of the $NO_3^- \rightleftharpoons NO$ half-reaction is +0.96 V; the E° of the $Ag^+ \rightleftharpoons Ag(s)$ half reaction is +0.80 V. Thus the NO_3^- is the stronger oxidizing agent because it has the more positive E°.

19.6 The half-reactions and the corresponding E° values are:

$$I_2(s) + 2e^- \rightleftharpoons 2I^-(aq); \quad E^o = 0.54\ V$$

$$Cu^{2+}(aq) + 2e^- \rightleftharpoons Cu(s); \quad E^o = 0.34\ V$$

The stronger oxidizing agent is the one involved in the half reaction with the larger (more positive) standard electrode potential, so I_2 is the stronger oxidizing agent. Thus the reaction is nonspontaneous as written because I_2 is a product.

19.7 The half-reactions and the corresponding E° values are:

$$Zn^{2+}(aq) + 2e^- \rightleftharpoons Zn(s); \quad E^o = -0.76\ V$$

$$Cu^{2+}(aq) + 2e^- \rightleftharpoons Cu(s); \quad E^o = 0.34\ V$$

To obtain the standard cell emf, we must reverse the zinc half-reaction and reverse its half-cell potential, and then add the half-cell potentials:

$$Zn(s) \rightleftharpoons Zn^{2+}(aq) + 2e^- \qquad -E^o_{Zn} = 0.76\ V$$

$$Cu^{2+} + 2e- \rightleftharpoons Cu(s) \qquad E^o_{Cu} = 0.34\ V$$

$$Zn(s) + Cu^{2+}(aq) \rightleftharpoons Zn^{2+}(aq) + Cu(s) \qquad E^o_{cell} = 1.10\ V$$

19.8 The half-cell reactions, the corresponding E° values, and the addition of these to obtain the overall cell reaction and standard cell emf are as follows:

$$Sn^{2+}(aq) \rightleftharpoons Sn^{4+}(aq) + 2e^- \qquad -E^o_{Sn} = -0.15\ V$$

$$2Hg^{2+} + 2e- \rightleftharpoons Hg_2^{2+}(aq) \qquad E^o_{Hg} = 0.90\ V$$

$$Sn^{2+}(aq) + 2Hg^{2+}(aq) \rightleftharpoons Sn^{4+}(aq) + Hg_2^{2+}(aq) \qquad E^o_{cell} = 0.75\ V$$

Noting that n = 2 in the overall cell reaction, calculate ΔG^o:

$$\Delta G^o = -nFE^o_{cell} = -(2)(9.65 \times 10^4\ C)(0.75\ V) = -1.\underline{4}475 \times 10^5 = -1.4 \times 10^5\ J$$

$$\Delta G^o = -1.4 \times 10^2\ kJ$$

19.9 Write the equation with ΔG_f^o's beneath each substance:

	Mg(s) +	$Cu^{2+}(aq)$	⇌	$Mg^{2+}(aq)$ +	Cu(s); $n = 2e^-$
ΔG_f^o:	0	65.0		-456.0	0

$$\Delta G^o = -456.0 - (65.0) = -521.\underline{0}\ kJ\ (-521 \times 10^3\ C{\bullet}V)$$

$$\Delta G^o = -nFE^o_{cell}$$

$$-521 \times 10^3\ C{\bullet}V = -(2)(9.65 \times 10^4\ C)(E^o_{cell})$$

Rearrange and solve for E^o_{cell}. Decide the significant figures from the 4 sig figs of ΔG^o, assuming the F of 9.65×10^4 is an exact number, and not 3 significant figures:

$$E^o_{cell} = \frac{-521.0 \times 10^3\ C{\bullet}V}{-(2)(9.65 \times 10^4\ C)} = -2.69\underline{9}4 = -2.699\ V$$

19.10 Note that $K = K_c$; then rearrange the equation $E^o_{cell} = (0.0592\ V)/n \log K_c$ to solve for $\log K_c$. Begin by noting that the standard emf is 0.56 V. This is calculated from the $-E^o$ value of 0.41 V for $Fe(s) \rightleftharpoons Fe^{2+}(aq)$ and the E^o value of 0.15 V for $Sn^{4+}(aq) \rightleftharpoons Sn^{2+}(aq)$.

$$0.56\ V = \frac{0.0592\ V}{2} \log K_c$$

$$\log K_c = 1\underline{8}.91 \quad \text{(2 sig figs from the 2 sig figs of } E^o_{cell}\text{)}$$

$$K_c = 8.1 \times 10^{1\underline{8}} \quad \text{(neither of the digits in 8.1 is significant)}$$

19.11 The cell reaction is: $Zn(s) + 2Ag^+(aq) \rightleftharpoons Zn^{2+}(aq) + 2Ag(s)$; n = 2

The reaction quotient, Q, is:

$$\frac{[Zn^{2+}]}{[Ag^+]^2} = \frac{0.200}{(0.00200)^2} = 5.00 \times 10^4$$

E^o_{cell} is the sum of $-(-0.76\ V) + 0.80\ V$, or 1.56 V. Substitute this and Q into the Nernst equation to calculate E_{cell}:

$$E_{cell} = E^o_{cell} - \frac{0.0592}{n} \log Q = 1.56 - \frac{0.0592}{2} \log (5.00 \times 10^4)$$

$$E_{cell} = 1.5\underline{6}\ V - 0.139\underline{0}9\ V = 1.4\underline{2}09 = 1.42\ V$$

19.12 The half-reaction is: $Cu^{2+}(aq) + 2e^- \rightleftharpoons Cu(s)$

For this half-reaction, n = 2, and

$$Q = \frac{1}{[Cu^{2+}]} = \frac{1}{(0.0350)}; \; E^o = 0.34 \text{ V}$$

Substituting these values into the Nernst equation:

$$E_{cell} = E^o_{cell} - \frac{0.0592}{n} \log Q = 0.34 \text{ V} - \frac{0.0592}{2} \log\left(\frac{1}{0.0350}\right)$$

$$E_{cell} = 0.34 \text{ V} - 0.0430\underline{9}6 \text{ V} = 0.2\underline{9}69 = 0.30 \text{ V}$$

19.13 First calculate E^o_{cell}:

$$E^o_{cell} = -0.23 \text{ V} - (-0.76 \text{ V}) = 0.53 \text{ V}$$

Substitute this value and E = 0.34 V into the Nernst equation and solve for log Q:

$$E_{cell} = E^o_{cell} - \frac{0.0592}{n} \log Q; \; 0.34 \text{ V} = 0.53 \text{ V} - \frac{0.0592}{2} \log Q$$

$$\log Q = (0.53 \text{ V} - 0.34 \text{ V}) \times \frac{2}{0\ 0592 \text{ V}} = 6.\underline{4}19$$

$$Q = \frac{[Zn^{2+}]}{[Ni^{2+}]} = \frac{1}{[Ni^{2+}]} = \text{antilog } 6.\underline{4}19 = \underline{2}.624 \times 10^6$$

$$[Ni^{2+}] = \underline{3}.81 \times 10^{-7} = 4 \times 10^{-7} \text{ M}$$

19.14 a. The cathode reaction is: $K^+(l) + e^- \rightarrow K(l)$

The anode reaction is: $Cl^-(l) \rightarrow 1/2\ Cl_2(g) + e^-$

b. The cathode reaction is: $K^+(l) + e^- \rightarrow K(l)$

The anode reaction is: $4OH^-(l) \rightarrow O_2(g) + 2H_2O(g) + 4e^-$

19.15 Two possible cathode reactions are:

$Ag^+(aq) + e^- \rightarrow Ag(s); \; E^o = 0.80 \text{ V}$

$2H_2O(l) + 2e^- \rightarrow H_2(g) + 2OH^-(aq); \; E = -0.41 \text{ V}$ (at pH 7.00)

Since the electrode potential for silver is larger (more positive), it is easier to reduce; thus the cathode reaction is the first half-reaction above. The only possible anode reaction is:

$2H_2O(l) \rightarrow O_2(g) + 4H^+(aq) + 4e^-$

19.16 From the silver electrode equation ($Ag^+ + e^- \rightarrow Ag(s)$), we can write:

$$1 \text{ mol Ag} = 1 \text{ mol } e^-$$

Since 1 mol of e^- is equivalent to 9.65×10^4 C, the charge equivalent to 365 mg silver is:

$$0.365 \text{ g Ag} \times \frac{1 \text{ mol Ag}}{107.9 \text{ g}} \times \frac{1 \text{ mol } e^-}{1 \text{ mol Ag}} \times \frac{9.65 \times 10^4 \text{ C}}{1 \text{ mol } e^-} = 32\underline{6}.4 \text{ C}$$

Using the time of 216 min as 1.296×10^4 s, we calculate the current in amps:

$$\text{Current} = \frac{\text{charge}}{\text{time}} = \frac{326.4 \text{ C}}{1.296 \times 10^4 \text{ s}} = 2.5\underline{1}88 \times 10^{-2} = 2.52 \times 10^{-2} \text{ A}$$

19.17 When the current flows for 185 s, the charge is:

$$0.0565 \text{ A} \times 185 \text{ s} = 10.\underline{4}52 \text{ C}$$

The electrode reaction is:

$$2H_2O \rightarrow 4H^+ + O_2 + 4e^-$$

Thus 4 moles of electrons are equivalent to 1 mol of O_2. The mass of O_2 liberated is:

$$10.45 \text{ C} \times \frac{1 \text{ mol } e^-}{9.65 \times 10^4 \text{ C}} \times \frac{1 \text{ mol } O_2}{4 \text{ mol } e^-} \times \frac{32.00 \text{ g } O_2}{1 \text{ mol } O_2} = 8.6\underline{6}3 \times 10^{-4} = 8.66 \times 10^{-4} \text{ g } O_2$$

ANSWERS TO REVIEW QUESTIONS

19.1 A voltaic cell is an electrochemical cell in which a spontaneous reaction generates an electric current (energy). An electrolytic cell is an electrochemical cell that requires electrical current (energy) to drive a nonspontaneous reaction to the right.

19.2 In both the voltaic and electrolytic cells, the cathode is the electrode at which reduction occurs, and the anode is the electrode at which oxidation occurs. Cations move toward the cathode; anions move toward the anode.

19.3 The SI unit of electrical potential is the volt (V).

19.4 The faraday (F) is the magnitude of charge on one mole of electrons; it equals 9.65×10^4 C or 9.65×10^4 J/V.

19.5 It is necessary to measure the voltage of a voltaic cell when no current is flowing because the cell voltage exhibits its maximum value only when no current flows. Even if the current flows just for the time of measurement, the voltage drops enough so that what is measured is significantly less than the maximum.

19.6 Standard electrode potentials are defined relative to a standard electrode potential of zero volts (0.00, 0.000 V, etc.) for the $H^+/H_2(g)$ electrode. Because the cell emf is measured using the hydrogen electrode at standard conditions and a second electrode at standard conditions, the cell emf equals the E^o of the half-reaction at the second electrode.

19.7 The SI unit of energy = joules = coulombs x volts.

19.8 The mathematical relationships are as follows:

$$\Delta G^o = -nFE^o_{cell}$$

$$\Delta G^o = -2.303\ RT \log K$$

Combining these two equations gives:

$$\log K = nFE^o_{cell}/2.303\ RT$$

19.9 The first step in the corrosion of iron is: $2Fe(s) + O_2(g) + 2H_2O(l) \rightarrow 4OH^- + 2Fe^{2+}$

The Nernst equation for this reaction is:

$$E_{cell} = E^o_{cell} = -(0.0592/4) \log [OH^-]^4[Fe^{2+}]^2$$

If the pH increases, the $[OH^-]$ increases, and thus E_{cell} becomes more negative (this predicts that the reaction becomes less spontaneous). If the pH decreases, the $[OH^-]$ decreases, and thus E_{cell} becomes more positive (this predicts that the reaction becomes more spontaneous).

19.10 The zinc-carbon cell has a zinc can as the anode; the cathode is a graphite rod surrounded by a paste of manganese dioxide and carbon black. Around this is a second paste of ammonium and zinc chlorides. The electrode reactions involve oxidation of zinc metal to zinc(II) ion, and reduction at the cathode of $MnO_2(s)$ to $Mn_2O_3(s)$. The lead storage battery consists of of a spongy lead anode and a lead dioxide cathode, both immersed in aqueous sulfuric acid. At the anode, the lead is oxidized to lead sulfate, and at the cathode, lead dioxide is reduced to lead sulfate.

19.11 A fuel cell is essentially a battery which does not use up its electrodes. It instead operates with a continuous supply of reactants (fuel). An example is the hydrogen-oxygen fuel cell, in which oxygen is reduced at one electrode to the hydroxide ion and hydrogen is oxidized at the other electrode to water (H in the +1 oxidation state). Such a cell produces electrical energy in a spacecraft for long periods of time.

19.12 During the rusting of iron, one end of a drop of water exposed to air acts as one electrode of a voltaic cell; at this electrode, an oxygen molecule is reduced by four electrons to four hydroxide ions. Oxidation of metallic iron to iron(II) ion at the center of the drop of water supplies the electrons, and the center serves as the other electrode of the voltaic cell. Thus electrons flow from the center of the drop through the iron to the end of the drop.

19.13 When iron or steel is connected to an active metal such as zinc, a voltaic cell is formed with zinc as the anode and the iron as the cathode. Any type of moisture forms the electrolyte solution, and the zinc metal is then oxidized to zinc(II) ion in preference to the oxidation of iron metal. Oxygen is reduced at the cathode to hydroxide ions. If iron or steel is exposed to oxygen while connected to a less active metal such as tin, a voltaic cell is formed with iron as the anode and tin as the cathode, and iron is oxidized to iron(II) ion rather than tin being oxidized to tin(II) ion. Thus, exposed iron corrodes rapidly in a tin can. Fortunately, as long as the iron is covered by the tin it cannot corrode.

19.14 The addition of ionic species such as strong ionized sulfuric acid facilitates the passage of current through the solution.

19.15 Sodium metal can be prepared by electrolysis of molten sodium chloride.

19.16 The anode reaction in the electrolysis of molten potassium hydroxide is:

$$4OH^- \rightarrow O_2(g) + 2H_2O(g) + 4e^-$$

19.17 The reason different products are obtained from the electrolysis of dilute aqueous NaCl than from molten NaCl is that water, instead of Na^+, is reduced at the cathode during the electrolysis of aqueous NaCl. This is because the water has a more positive E^o (smaller decomposition voltage). At the anode, water, instead of chloride ion, is oxidized because water has a less positive E^o (smaller decomposition voltage).

19.18 The Nernst equation for the electrode reaction of $2Cl^-(aq) \rightarrow Cl_2(g) + 2e^-$ is:

$$E = -1.36 \text{ V} - (0.0592/2) \log (1/[Cl^-]^2) = -1.36 \text{ V} + 0.0592 \log [Cl^-]$$

This equation implies that E increases as $[Cl^-]$ increases. For a sufficiently large $[Cl^-]$, Cl^- will be more readily oxidized than the water solvent.

SOLUTIONS TO PRACTICE PROBLEMS

Note on significant figures: The final answer to all mathematical solutions is given first with one nonsignificant figure (last significant figure underlined) and is then rounded to the correct number of digits. Intermediate answers usually also have at least one nonsignificant figure.

19.19 Sketch of the cell:

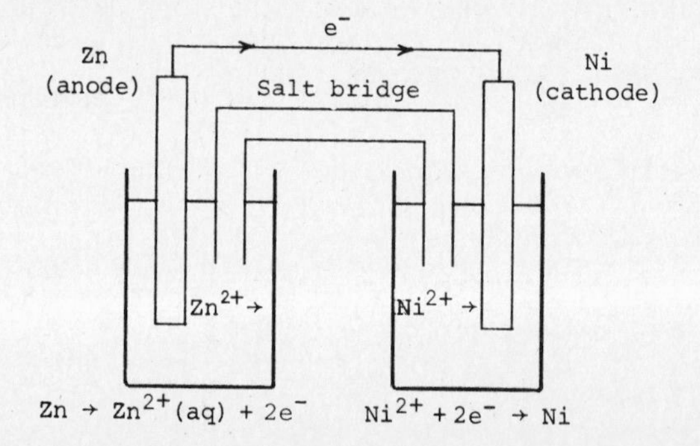

19.21 Sketch of the cell:

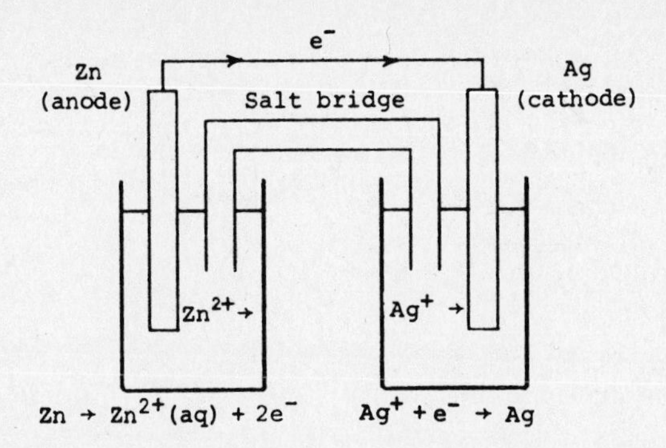

19.23 The electrode half-reactions and the overall cell reaction are:

anode: $Zn(s) + 2OH^-(aq) \rightarrow Zn(OH)_2(s) + 2e^-$

cathode: $Ag_2O(s) + H_2O(l) + 2e^- \rightarrow 2Ag(s) + 2OH^-(aq)$

overall: $Zn(s) + Ag_2O(s) + H_2O(l) \rightarrow Zn(OH)_2(s) + 2Ag(s)$

19.25 Because of its less negative E°, Pb^{2+} is reduced at the cathode and is written on the right; Cd(s) is oxidized at the anode and is written first, at the left, in the cell notation. The notation is: $Cd(s)|Cd^{2+}(aq)||Pb^{2+}(aq)|Pb(s)$.

19.27 Because of its less negative E°, H^+ is reduced at the cathode and is written on the right; Ni(s) is oxidized at the anode and is written first, at the left, in the cell notation. The notation is: $Ni(s)|Ni^{2+}(1\ \underline{M})||H^+(1\ \underline{M})|H_2(g)|Pt$.

19.29 The Fe(s), on the left, is the reducing agent. The Ag^+, on the right, is the oxidizing agent, gaining just one electron. Multiplying its half-reaction by two to equalize the numbers of electrons and writing both half-reactions gives the overall cell reaction:

$Fe(s) \rightarrow Fe^{2+}(aq) + 2e^-$

$2Ag^+(aq) + 2e^- \rightarrow 2Ag(s)$

$Fe(s) + 2Ag^+(aq) \rightarrow Fe^{2+}(aq) + 2Ag(s)$

19.31 The half-cell reactions, the overall cell reaction, and the sketch are:

$Cd(s) \rightarrow Cd^{2+}(aq) + 2e^-$

$Ni^{2+}(aq) + 2e^- \rightarrow Ni(s)$

$Cd(s) + Ni^{2+}(aq) \rightarrow Cd^{2+}(aq) + Ni(s)$

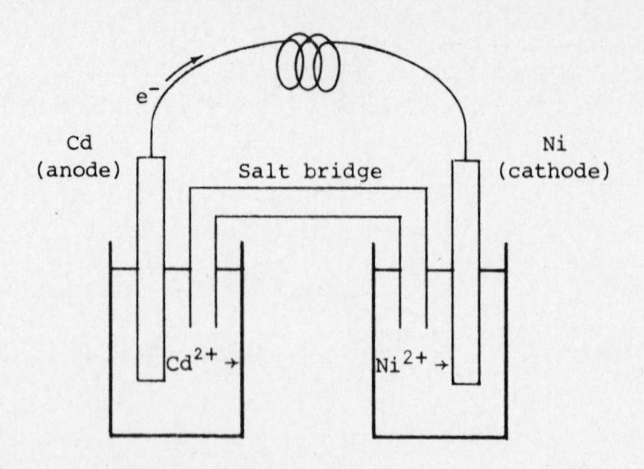

19.33 The half-cell reactions are:

$$2Fe^{3+}(aq) + 2e^- \rightarrow 2Fe^{2+}(aq) \text{ and } Zn(s) \rightarrow Zn^{2+}(aq) + 2e^-$$

Therefore for the cell, n = $2e^-$. So w, the maximum electrical work, is calculated:

$$w_{max} = -nFE_{cell} = -(2)(9.65 \times 10^4 C)(0.72\ V) = -1.\underline{3}89 \times 10^5\ C{\bullet}V = -1.\underline{3}89 \times 10^5\ J$$

Since this is the work obtained by reduction of 2 mol of Fe^{3+}, the work for 1 mol is:

$$w_{max} = -1.389 \times 10^5\ J/\ 2\ mol \div 2 = -6.\underline{9}45 \times 10^4 = -7.0 \times 10^4\ J/\ 1\ mol$$

19.35 The half-cell reactions are:

$$2Ag^+(aq) + 2e^- \rightarrow 2Ag(s) \text{ and } Ni(s) \rightarrow Ni^{2+}(aq) + 2e^-$$

Therefore for the cell, n = $2e^-$. So w, the maximum electrical work, is calculated:

$$w_{max} = -nFE_{cell} = -(2)(9.65 \times 10^4 C)(0.97\ V) = -1.\underline{8}7 \times 10^5\ C{\bullet}V = -1.\underline{8}7 \times 10^5\ J$$

Since this is the work obtained by reduction of 1 mol of Ni, the work for 15.0 g is:

$$15.0\ g\ Ni \times \frac{1\ mol\ Ni}{58.69\ g\ Ni} \times \frac{-1.87 \times 10^5\ J}{1\ mol\ Ni} = -4.\underline{7}7 \times 10^4 = -4.8 \times 10^4\ J$$

19.37 The species of interest all occur as reactants in Table 19.1. The order from top to bottom in which these species occur is the order of increasing oxidizing power of these species. The half-reactions, E^o values, and the order of increasing oxidizing power (increasing E^o) are:

$$NO_3^-(aq) + 4H^+(aq) + 3e^- \rightleftharpoons NO(g) + 2H_2O(l) \quad 0.96\ V$$

$$O_2(g) + 4H^+(aq) + 4e^- \rightleftharpoons 2H_2O(l) \quad 1.23\ V$$

$$H_2O_2(aq) + 2H^+(aq) + 2e^- \rightleftharpoons H_2O(aq) \quad 1.78\ V$$

In summary, the increasing order of strength is $NO_3^-(aq)$, $O_2(g)$, and $H_2O_2(aq)$.

19.39 The species of interest all occur as products in Table 19.1. The order from bottom to top in which these species occur is the order of increasing reducing power of these species. The half-reactions, their E^o values, and the order of increasing reducing power (decreasing value of E^o) are:

$Cu^{2+}(aq) + e^- \rightleftharpoons Cu^+(aq)$ 0.16 V (weakest reducing agent)

$Fe^{2+}(aq) + 2e^- \rightleftharpoons Fe(s)$ -0.41 V

$Zn^{2+}(aq) + 2e^- \rightleftharpoons Zn(s)$ -0.76 V (strongest reducing agent)

19.41 a. The reduction half-reactions and standard potentials are:

$Fe^{3+}(aq) + e^- \rightleftharpoons Fe^{2+}(aq)$ $E^o = 0.77$ V

$Sn^{4+}(aq) + 2e^- \rightleftharpoons Sn^{2+}(aq)$ $E^o = 0.15$ V

To calculate E^o_{cell}, the sign of E^o for the iron half-reaction will have to be reversed, giving $E^o_{cell} = -0.62$ V. Thus the reaction is not spontaneous.

b. The reduction half-reactions and standard potentials are:

$O_2(g) + 4H^+(aq) + 4e^- \rightleftharpoons 2H_2O(l)$ $E^o = 1.23$ V

$MnO_4^-(aq) + 5e^- \rightleftharpoons Mn^{2+}(aq)$ $E^o = 1.49$ V

To calculate E^o_{cell}, the sign of E^o for the oxygen half-reaction will have to be reversed, giving $E^o_{cell} = 0.26$ V. Thus the reaction is spontaneous.

19.43 The pertinent reduction half-reactions and their corresponding E^o values are:

$Br_2(l) + 2e^- \rightleftharpoons 2Br^-(aq)$ $E^o = 1.07$ V

$Cl_2(g) + 2e^- \rightleftharpoons 2Cl^-(aq)$ $E^o = 1.36$ V

$F_2(g) + 2e^- \rightleftharpoons 2F^-(aq)$ $E^o = 2.87$ V

From these we see that Cl_2 is a stronger oxidizing agent under standard conditions than is Br_2. To calculate E^o_{cell}, we would have to reverse the sign of the bromine half-reaction, giving $E^o_{cell} = 0.29$ V, a spontaneous reaction. The E^o_{cell} for the Cl_2 oxidation of F^- ion would be -1.51 V, a nonspontaneous reaction.

The possible reactions involve the oxidation of the anions by Cl_2 can be summarized as:

$Cl_2(g) + 2Br^-(aq) \rightarrow 2Cl^-(aq) + Br_2(l)$; spontaneous reaction

$Cl_2(g) + 2F^-(aq) \rightarrow 2Cl^-(aq) + F_2(g)$; nonspontaneous reaction

19.45 The pertinent reduction half-reactions and their corresponding E^o values are:

$Cr^{3+}(aq) + 3e^- \rightleftharpoons Cr(s)$ $E^o = -0.74$ V

$Hg_2^{2+}(aq) + 2e^- \rightleftharpoons 2Hg(l)$ $\quad E^o = 0.80\ V$

Reverse the first half-reaction, multiply by 2, and reverse the sign of its E_o. Then multiply the second equation by 3; this allows us to combine both half-reactions:

$$2Cr(s) \rightleftharpoons 2Cr^{3+}(aq) + 6e^- \qquad -E^o = 0.74\ V$$

$$3Hg_2^{2+}(aq) + 6e^- \rightleftharpoons 6Hg(l) \qquad E^o = 0.80\ V$$

$$2Cr(s) + 3Hg_2^{2+}(aq) \rightarrow 2Cr^{3+}(aq) + 6Hg(l) \qquad E^o_{cell} = 1.54\ V$$

19.47 The pertinent reduction half-reactions and their corresponding E^o values are:

$$Cr^{3+}(aq) + 3e^- \rightleftharpoons Cr(s) \qquad E^o = -0.74\ V$$

$$I_2(s) + 2e^- \rightleftharpoons 2I^-(aq) \qquad E^o = 0.54\ V$$

Reverse the first half-reaction, multiply by 2, and reverse the sign of its E_o. Then multiply the second equation by 3; this allows us to combine both half-reactions:

$$2Cr(s) \rightleftharpoons 2Cr^{3+}(aq) + 7H_2O(l) + 6e^- \qquad -E^o = 0.74\ V$$

$$3I_2(s) + 6e^- \rightleftharpoons 6I^-(aq) \qquad E^o = 0.54\ V$$

$$2Cr(s) + 3I(s) \rightarrow 2Cr^{3+}(aq) + 6I^-(aq) \qquad E^o_{cell} = 1.28\ V$$

19.49 First calculate E^o_{cell}; then find ΔG^o by multiplying by the electron change and the value of the faraday. The half-reactions and their respective E^os are:

$$Cu^{2+}(aq) + 2e^- \rightleftharpoons Cu(s) \qquad E^o = 0.34\ V$$

$$NO_3^-(aq) + 4H^+(aq) + 3e^- \rightleftharpoons NO(g) + 2H_2O(l) \qquad E^o = 0.96\ V$$

Reverse the first half-reaction, multiply by 3, and reverse the sign of its E_o. Then multiply the second equation by 2; this allows us to combine both half-reactions:

$$3Cu(s) \rightleftharpoons 3Cu^{2+}(aq) + 6e^- \qquad -E^o = -0.34\ V$$

$$2NO_3^-(aq) + 8H^+(aq) + 6e^- \rightleftharpoons 2NO(g) + 4H_2O(l) \qquad E^o = 0.96\ V$$

$$3Cu(s) + 2NO_3^-(aq) + 8H^+(aq) \rightarrow 3Cu^{2+}(aq) + 2NO(g) + 4H_2O(l) \qquad E^o_{cell} = 0.62\ V$$

$$\Delta G^o = -nFE^o_{cell} = -(6)(9.65 \times 10^4\ C)(0.62\ V) = -3.\underline{5}8 \times 10^5\ C{\bullet}V = -3.6 \times 10^5\ J$$

19.51 First calculate E^o_{cell}; then multiply by the electron change and the value of the faraday to find ΔG^o. The half-reactions and their respective E^o's are:

$I_2(s) + 2e^- \rightleftharpoons 2I^-(aq)$ $E^o = 0.54$ V

$Cl_2(g) + 2e^- \rightleftharpoons 2Cl^-(aq)$ $E^o = 1.36$ V

Reverse the first half-reaction, and reverse the sign of its E_o. Then combine both half-reactions:

$2I^-(aq) \rightleftharpoons I_2(s) + 2e^-$ $-E^o = -0.54$ V

$Cl_2(g) + 2e^- \rightleftharpoons 2Cl^-(aq)$ $E^o = 1.36$ V

$2I^-(aq) + Cl_2(g) \rightarrow I_2(s) + 2Cl^-(aq)$ $E^o_{cell} = 0.82$ V

$\Delta G^o = -nFE^o_{cell} = -(2)(9.65 \times 10^4 \text{ C})(0.81 \text{ V}) = -1.\underline{58} \times 10^5 \text{ C}\cdot\text{V} = -1.6 \times 10^5 \text{ J}$

19.53 Write the equation with ΔG_f^o's beneath each substance. Use ΔG^o to solve for E^o_{cell}.

$Al(s) + 3Ag^+(aq) \rightarrow Al^{3+}(aq) + 3Ag(s)$; $n = 3e^-$

ΔG_f^o: 0 3(77.111) -481.2 3(0)

$\Delta G^o = [(-481.1) - 3(77.111)] = -712.\underline{4}3 \text{ kJ} \; (-712.\underline{4}3 \times 10^3 \text{ C}\cdot\text{V})$

$\Delta G^o = -nFE^o_{cell}$

$-712.43 \times 10^3 \text{ C}\cdot\text{V} = -(3)(9.65 \times 10^4 \text{ C})(E^o_{cell})$

$$E^o_{cell} = \frac{-712.43 \times 10^3 \text{ C}\cdot\text{V}}{-(3)(9.65 \times 10^4 \text{ C})} = 2.4\underline{6}8 = 2.47 \text{ V}$$

19.55 Write the equation with ΔG_f^o's beneath each substance:

$PbO(s) + 2HSO_4^-(aq) + 2H^+(aq) + Pb(s) \rightarrow 2PbSO_4(s) + 2H_2O(l)$; $n = 2e^-$

ΔG_f^o: -219 2(-753) 0 0 2(-811) 2(-237.2)

$\Delta G^o = [2(-811) + 2(-237.2) - (-219) - 2(-753) = -37\underline{1}.4 \text{ kJ} \; (-371.4 \times 10^3 \text{ C}\cdot\text{V})$

$\Delta G^o = -nFE^o_{cell}$

$-37\underline{1}.4 \times 10^3 \text{ C}\cdot\text{V} = -(2)(9.65 \times 10^4 \text{ C})(E^o_{cell})$

Rearrange and solve for E^o_{cell}. Determine the significant figures from the 3 sig figs of ΔG^o.

$$E^o_{cell} = \frac{-371.4 \times 10^3 \text{ C}\cdot\text{V}}{-(2)(9.65 \times 10^4 \text{ C})} = 1.9\underline{2}47 = 1.92 \text{ V}$$

19.57 Rearrange the equation $E^o_{cell} = (0.0592\ V)/n \log K$ to solve for log K. Begin by calculating E^o_{cell}. Then substitute into the log K_c expression.

$Sn^{4+}(aq) + 2e^- \rightleftharpoons Sn^{2+}(aq)$ $\quad E^o = 0.15\ V$

$Fe^{3+}(aq) + e^- \rightleftharpoons Fe^{2+}(aq)$ $\quad E^o = 0.77\ V$

Reverse the first half-reaction and reverse the sign of its E_o. Then multiply the second equation by 2; this allows us to combine both half-reactions:

$Sn^{2+}(aq) \rightleftharpoons Sn^{4+}(aq) + 2e^-$ $\quad -E^o = -0.15\ V$

$2Fe^{2+}(aq) + 2e^- \rightleftharpoons 2Fe^{2+}(aq)$ $\quad E^o = 0.77\ V$

$Sn^{2+}(aq) + 2Fe^{3+}(aq) \rightleftharpoons Sn^{4+}(aq) + 2Fe^{2+}(aq)$ $\quad E^o_{cell} = 0.62\ V$

Now calculate log K.

$$0.62\ V = \frac{0.0592\ V}{2} \log K$$

$\log K = 2\underline{0}.94$ (2 sig figs from the 2 sig figs of E^o_{cell})

$K = 8.7 \times 10^{2\underline{0}}$ or rounding, $K = 10^{21}$ (neither of the digits in 8.7 is significant)

19.59 Rearrange the equation $E^o_{cell} = (0.0592\ V)/n \log K$ to solve for log K_c. Begin by calculating E^o_{cell}. Then substitute into the log K_c expression.

$Cu^{2+}(aq) + e^- \rightleftharpoons Cu^+(aq)$ $\quad E^o = 0.16\ V$

$Cu^+(aq) + e^- \rightleftharpoons Cu(s)$ $\quad E^o = 0.52\ V$

Reverse the first half-reaction and reverse the sign of its E_o. Then combine with the second equation:

$Cu^+(aq) \rightleftharpoons Cu^{2+}(aq) + e^-$ $\quad -E^o = -0.16\ V$

$Cu^+(aq) + e^- \rightleftharpoons Cu(s)$ $\quad E^o = 0.52\ V$

$2Cu^+(aq) \rightarrow Cu(s) + Cu^{2+}(aq)$ $\quad E^o_{cell} = 0.36\ V$

Now calculate log K.

$$0.36\ V = \frac{0.0592\ V}{1} \log K$$

$\log K_c = 6.\underline{0}81$ (2 sig figs from the 2 sig figs of E^o_{cell})

$K_c = \underline{1}.2 \times 10^6$ (1 sig fig because only 1 sig fig to the right of the decimal point)

19.61 First calculate E^o_{cell}:

$$E^o_{cell} = -(-0.74 \text{ V}) + (-0.23 \text{ V}) = 0.51 \text{ V}$$

The cell reaction is: $2Cr(s) + 3Ni^{2+}(aq) \rightarrow 2Cr^{3+}(aq) + 3Ni(s)$; n = 6

Substitute E^o_{cell} = 0.51 V, 1.0 x 10^{-2} M Cr^{3+}, and 2.0 M Ni^{2+} into the Nernst equation and solve for E_{cell}:

$$E_{cell} = E^o_{cell} - \frac{0.0592}{n}\log Q = 0.51 \text{ V} - \frac{0.0592}{6}\log\frac{(1.0 \times 10^{-2})^2}{(2.0)^3}$$

$$E_{cell} = 0.51 \text{ V} - (-0.048\underline{3}7 \text{ V}) = 0.5\underline{5}837 = 0.56 \text{ V}$$

19.63 First calculate E^o_{cell}:

$$E^o_{cell} = -(1.07 \text{ V}) + (1.49 \text{ V}) = 0.42 \text{ V}$$

The cell reaction is: $2MnO_4^-(aq) + 10Br^-(aq) + 16H^+(aq) \rightarrow$
$2Mn^{2+}(aq) + 5Br_2(l) + 8H_2O(l)$; n = 10

Substitute E^o_{cell} = 0.42 V, 0.010 M MnO_4^-, 0.010 M Br^-, 0.15 M Mn^{2+}, and 1.0 M H^+ into the Nernst equation and solve for E_{cell}:

$$E_{cell} = E^o_{cell} - \frac{0.0592}{n}\log Q = 0.42 \text{ V} - \frac{0.0592}{10}\log\frac{(0.15)^2}{(0.010)^2(0.010)^{10}(1.0)^{16}}$$

$$E_{cell} = 0.42 \text{ V} - (0.132\underline{3} \text{ V}) = 0.2\underline{8}77 = 0.29 \text{ V}$$

19.65 The half-reaction is: $Zn^{2+}(aq) + 2e^- \rightleftharpoons Zn(s)$

For this half-reaction, n = 2, and

$$Q = \frac{1}{[Zn^{2+}]} = \frac{1}{(2.0 \times 10^{-3})}; \quad E^o = -0.76 \text{ V}$$

Substituting these values into the Nernst equation:

$$E_{cell} = E^o_{cell} - \frac{0.0592}{n}\log Q = -0.76 \text{ V} - \frac{0.0592}{2}\log\left(\frac{1}{2.0 \times 10^{-3}}\right)$$

$$E_{cell} = -0.76 \text{ V} - 0.079\underline{8}8 \text{ V} = -0.8\underline{3}988 = -0.84 \text{ V}$$

19.67 Note that E^o_{cell} = 0.170 V. Substitute this value and E = 0.240 V into the Nernst equation and solve for log Q:

$$E_{cell} = E^o_{cell} - \frac{0.0592}{n}\log Q; \quad 0.240 \text{ V} = 0.170 \text{ V} - \frac{0.0592}{2}\log Q$$

$$\log Q = (0.240 \text{ V} - 0.170 \text{ V}) \times \left(-\frac{2}{0.0592 \text{ V}}\right) = -2.\underline{3}648$$

$$Q = \frac{[Cd^{2+}]}{[Ni^{2+}]} = \frac{[Cd^{2+}]}{1.0} = \text{antilog}(-2.\underline{3}648) = \underline{4}.31 \times 10^{-3}$$

$$[Cd^{2+}] = \underline{4}.31 \times 10^{-3} = 4 \times 10^{-3} \text{ M}$$

19.69 a. The cathode reaction is:

$$Ca^{2+}(l) + 2e^- \rightarrow Ca(l)$$

The anode reaction is:

$$2Cl^-(l) \rightarrow Cl_2(g) + 2e^-$$

b. The cathode reaction is:

$$Cs^+(l) + e^- \rightarrow Cs(l)$$

The anode reaction is:

$$4OH^-(l) \rightarrow O_2(g) + 2H_2O(g) + 4e^-$$

19.71 a. Two possible cathode reactions are:

$$Na^+(aq) + e^- \rightarrow Na(s);\ E^o = -2.71 \text{ V}$$

$$2H_2O(l) + 2e^- \rightarrow H_2(g) + 2OH^-(aq);\ E^o = -0.83 \text{ V}$$

Since the electrode potential for water is larger (less negative), it is easier to reduce; thus the cathode reaction is the second half-reaction above. The possible anode reactions are:

$$2H_2O(l) \rightarrow O_2(g) + 4H^+(aq) + 4e^-;\ E^o = -1.\underline{2}3 \text{ V}$$

$$2SO_4^{2-}(aq) \rightarrow S_2O_8^{2-}(aq) + 2e^-;\ E^o = -2.01 \text{ V}$$

Since the electrode potential for water is less negative, it is easier to oxidize; thus the anode reaction is the first half-reaction above.

The overall reaction is:

$$2H_2O(l) \rightarrow 2H_2(g) + O_2(g)$$

b. Two possible cathode reactions are:

$$K^+(aq) + e^- \rightarrow K(s);\ E^o = -2.92 \text{ V}$$

$$2H_2O(l) + 2e^- \rightarrow H_2(g) + 2OH^-(aq);\ E^o = -0.83 \text{ V}$$

Since the electrode potential for water is larger (less negative), it is easier to reduce; thus the cathode reaction is the second half-reaction above. Next we consider the possible anode reactions:

$2H_2O(l) \rightarrow O_2(g) + 4H^+(aq) + 4e^-;\ E^o = -1.23\text{ V}$

$2Br^-(aq) \rightarrow Br_2(l) + 2e^-;\ E^o = -1.07\text{ V}$

Since the electrode potential for bromide is less negative, it is easier to oxidize; thus the anode reaction is the second half-reaction above.

The overall reaction is:

$2Br^-(aq) + 2H_2O(l) \rightarrow Br_2(l) + H_2(g) + 2OH^-$

19.73 From the aluminum electrode equation ($Al^{3+}(l) + 3e^- \rightarrow Al(s)$): 1 mol Al = 3 mol e^-.

Since 1 mol of e^- is equivalent to 9.65×10^4 C, the charge equivalent to 5.12 kg aluminum is:

$$5.12 \times 10^3\text{ g} \times \frac{1\text{ mol Al}}{26.98\text{ g}} \times \frac{3\text{ mol e}^-}{1\text{ mol Al}} \times \frac{9.65 \times 10^4\text{ C}}{1\text{ mol e}^-} = 5.4\underline{9}38 \times 10^7 = 5.49 \times 10^7\text{ C}$$

19.75 From the lithium electrode equation ($Li^+(l) + e^- \rightarrow Li(s)$): 1 mol Li = 1 mol e^-.

Since 1 mol of e^- is equivalent to 9.65×10^4 C, the mass equivalent to 5.00×10^3 C is:

$$5.00 \times 10^3\text{ C} \times \frac{1\text{ mol e}^-}{9.65 \times 10^4\text{ C}} \times \frac{1\text{ mol Li}}{1\text{ mol e}^-} \times \frac{6.941\text{ g Li}}{1\text{ mol Li}} = 0.35\underline{9}6 = 0.360\text{ g Li}$$

19.77 The cell notation is: $Mg(s)|Mg^{2+}(aq)||Cl^-(aq)|Cl_2(g)|Pt(s)$. The reactions are:

Anode: $Mg(s) \rightarrow Mg^{2+}(aq) + 2e^-$

Cathode: $Cl_2(g) + 2e^- \rightarrow 2Cl^-(aq)$

$E^o_{cell} = 1.36\text{ V} - (-2.38\text{ V}) = 3.74\text{ V}$

19.79 In each case calculate the standard cell emf; if it is positive, the reaction is spontaneous.

a. The half-reactions and the corresponding E^o values are:

$Ni^{2+}(aq) + 2e^- \rightleftharpoons Ni(s);\quad E^o = -0.23\text{ V}$

$Fe^{3+}(aq) + e^- \rightleftharpoons Fe^{2+}(aq);\quad E^o = 0.77\text{ V}$

To obtain the standard cell emf, we must reverse the nickel half-reaction and reverse its half-cell potential, double the iron reaction, and then add the half-cell potentials:

$Ni(s) \rightleftharpoons Ni^{2+}(aq) + 2e^-\qquad -E^o_{Ni} = 0.23\text{ V}$

$2Fe^{3+} + 2e^- \rightleftharpoons 2Fe^{2+}(aq)\qquad E^o_{Fe} = 0.77\text{ V}$

$Ni(s) + 2Fe^{3+}(aq) \rightarrow Ni^{2+}(aq) + 2Fe^{2+}(aq)\qquad E^o_{cell} = 1.00\text{ V}$ (spontaneous)

b. The half-reactions and the corresponding E^o values are:

$$Sn^{4+}(aq) + 2e^- \rightleftharpoons Sn^{2+}(aq); \quad E^o = -0.23\ V$$

$$Fe^{3+}(aq) + e^- \rightleftharpoons Fe^{2+}(aq); \quad E^o = 0.77\ V$$

To obtain the standard cell emf, we must reverse the nickel half-reaction and reverse its half-cell potential, double the iron reaction, and then add the half-cell potentials:

$$Sn^{2+}(aq) \rightleftharpoons Sn^{4+}(aq) + 2e^- \qquad -E^o_{Sn} = -0.15\ V$$

$$2Fe^{3+} + 2e- \rightleftharpoons 2Fe^{2+}(aq) \qquad E^o_{Fe} = 0.77\ V$$

$$Sn^{2+}(aq) + 2Fe^{3+}(aq) \rightarrow Sn^{4+}(aq) + 2Fe^{3+}(aq) \quad E^o_{cell} = 0.62\ V\ \text{(spontaneous)}$$

19.81 First use K_{sp} to calculate $[Pb^{2+}]$:

$$[Pb^{2+}] = \frac{K_{sp}}{[SO_4^{2-}]} = \frac{1.7 \times 10^{-8}}{1.0} = 1.7 \times 10^{-8}\ M$$

Then calculate E^o_{cell}:

$$E^o_{cell} = -(-0.13\ V) + 0.00\ V = 0.13\ V$$

The cell reaction is: $Pb(s) + 2H^+(aq) \rightarrow Pb(s) + H_2(g);\ n = 2$

Substitute $E^o_{cell} = 0.13$ V, 1.7×10^{-8} M Pb^{2+}, and 1.0 M H^+ into the Nernst equation and solve for E_{cell}:

$$E_{cel\,l} = E^o_{cel\,l} - \frac{0.0592}{n} \log Q = 0.13\ V - \frac{0.0592}{2} \log \frac{(1.7 \times 10^{-8})}{(1.0)^2}$$

$$E_{cell} = 0.13\ V + (0.22\underline{9}9\ V) = 0.3\underline{5}99 = 0.36\ V$$

19.83 a. Rearrange the equation $E^o_{cell} = (0.0592\ V)/n \log K_c$ to solve for $\log K_c$. Note that $E^o_{cell} = 0.010$ V, and that n = 2. Now calculate log K.

$$0.010\ V = \frac{0.0592\ V}{2} \log K_c$$

$$\log K_c = 0.3\underline{3}78 \quad \text{(2 sig figs from the 2 sig figs of } E^o_{cell})$$

$$K_c = 2.\underline{1}76 = 2.2 \quad \text{(2 sig figs from the 2 sig digits to the right of the decimal)}$$

b. Now let x = $[Sn^{2+}]$, and 1.0 M - x = $[Pb^{2+}]$ and substitute into the equilibrium expression.

$$K_c = \frac{[Sn^{2+}]}{[Pb^{2+}]} = \frac{[x]}{[1.0 - x]} = 2.176$$

$x = 0.6\underline{8}51$

$[Pb^{2+}] = 1.0 - x = 0.\underline{3}149 = 0.3\ M$

19.85 The number of faradays is: a. 1 F b. 2 F c. 0.11 F d. 0.028 F
The number of coulombs is calculated below.

a. $1.0 \text{ mol e} \times \frac{1F}{1 \text{ mol e}} \times \frac{9.65 \times 10^4\ C}{1\ F} = 9.\underline{6}5 \times 10^4 = 9.6 \times 10^4\ C$

b. $2.0 \text{ mol e} \times \frac{1F}{1 \text{ mol e}} \times \frac{9.65 \times 10^4\ C}{1\ F} = 1.\underline{9}0 \times 10^5 = 1.9 \times 10^5\ C$

c. $1.0\ g \times \frac{1 \text{ mol}}{18.0\ g} \times \frac{2.0 \text{ mol e}}{1 \text{ mol}} \times \frac{1F}{1 \text{ mol e}} \times \frac{9.65 \times 10^4\ C}{1\ F} = 1.\underline{0}7 \times 10^4 = 1.1 \times 10^4\ C$

d. $1.0\ g \times \frac{1 \text{ mol}}{35.5\ g} \times \frac{2.0 \text{ mol e}}{2 \text{ mol}} \times \frac{1F}{1 \text{ mol e}} \times \frac{9.65 \times 10^4\ C}{1\ F} = 2.\underline{7}2 \times 10^3 = 2.7 \times 10^3\ C$

19.87 Find the moles of I_2 produced from the reaction $2I^- \rightarrow I_2 + 2e^-$ and the current and time.

$$65.3\ s \times (10.5 \times 10^{-3}\ A) \times \frac{1\ C}{1\ A\cdot s} \times \frac{1.0 \text{ mol e}}{9.65 \times 10^4\ C} \times \frac{1 \text{ mol } I_2}{2 \text{ mol e}} = 3.5\underline{5}3 \times 10^{-6} \text{ mol } I_2$$

From the equation in the problem, 1 mol I_2 reacts with 1 mol of H_3AsO_3 or 1 mol As, so:

$$3.5\underline{5}3 \times 10^{-6} \text{ mol } I_2 \times \frac{1 \text{ mol } H_3AsO_3}{1 \text{ mol } I_2} \times \frac{1 \text{ mol As}}{1 \text{ mol } H_3AsO_3} \times \frac{74.92 \text{ g As}}{1 \text{ mol As}} = 2.6\underline{6}1 \times 10^{-4} \text{ g As}$$

Cumulative-Skills Problems (require skills from previous chapters)

19.89 First calculate E^o_{cell}:

$$E^o_{cell} = -(1.07\ V) + (1.23\ V) = 0.16\ V$$

The cell reaction is: $O_2(g) + 4H^+(aq) + 4Br^-(aq) \rightarrow 2H_2O(l) + 2Br_2(l)$; $n = 4$

Convert pH to $[H^+]$: $[H^+] = \text{antilog}(-3.60) = 2.\underline{5}1 \times 10^{-4}\ M$

Substitute $E^o_{cell} = 0.16$ V, 1 M Br^-, 1 M Br_2, and 2.51×10^{-4} M H^+ into the Nernst equation and solve for E_{cell}:

$$E_{cell} = E^o_{cell} - \frac{0.0592}{n} \log Q = 0.16\ V - \frac{0.0592}{4} \log \frac{1}{(2.51 \times 10^{-4})^4}$$

$$E_{cell} = E^o_{cell} - \frac{0.0592}{n} \log Q = 0.16\ V - \frac{0.0592}{4} \log (2.519 \times 10^{14})$$

$$E_{cell} = E^o_{cell} - \frac{0.0592}{n} \log Q = 0.16\ V - \frac{0.0592}{4} (14.4\underline{0}1)$$

$E_{cell} = 0.16\ V - (0.213\underline{1}$ V, using 0.0592 as exact no) $= -0.0\underline{5}31 = -0.05$ V (nonspont)

19.91 In problem 19.89, E^o_{cell} was found to be 0.16 V. Neglecting ionization of HCNO, the $[H^+]$ of the buffer is:

$$[H^+] \cong 3.5 \times 10^{-4} \times \frac{(0.10\ M\ HCNO)}{(0.10\ M\ CNO^-)} \cong 3.5 \times 10^{-4}\ M$$

From the Nernst equation for the reaction of O_2 with $4Br^-$ and $4H^+$:

$$E_{cell} = E^o_{cell} - \frac{0.0592}{n} \log Q = 0.16\ V - \frac{0.0592}{4} \log \frac{1}{(3.5 \times 10^{-4})^4}$$

$$E_{cell} = E^o_{cell} - \frac{0.0592}{n} \log Q = 0.16\ V - \frac{0.0592}{4} \log (6.6639 \times 10^{13})$$

$$E_{cell} = E^o_{cell} - \frac{0.0592}{n} \log Q = 0.16\ V - \frac{0.0592}{4} (13.8\underline{2}37)$$

E_{cell} = 0.16 V - 0.204$\underline{5}$9 V (using 0.0592 as exact no) = -0.0$\underline{4}$459 = -0.04 V (nonspont)

19.93 The cell reaction is: $1/2H_2(g) + Ag^+(aq) \rightarrow Ag(s) + H^+(aq)$; n = 1

Then calculate E^o_{cell}:

$$E^o_{cell} = -0.00\ V + 0.80\ V = 0.80\ V$$

Substitute E_{cell} = 0.45 V, E^o_{cell} = 0.80 V, and 1.0 M H^+ into the Nernst equation and solve for log $[Ag^+]$:

$$E_{cell} = E^o_{cell} - \frac{0.0592}{n} \log Q;\ 0.45\ V = 0.80\ V - \frac{0.0592}{1} \log \frac{(1.0)}{[Ag^+]}$$

$$\log [Ag^+] = \frac{1(0.45 - 0.80)}{-0.0592} = -5.\underline{9}12\ V$$

$$[Ag^+] = 10^{-5.912} = \underline{1}.224 \times 10^{-6}\ M$$

$$K_{sp} = [Ag^+][SCN^-] = (1.224 \times 10^{-6})(0.10) = \underline{1}.224 \times 10^{-7} = 1 \times 10^{-7}$$

CHAPTER 20

NUCLEAR CHEMISTRY

SOLUTIONS TO EXERCISES

Note on significant figures: The final answer to all mathematical solutions is given first with one nonsignificant figure (last significant figure underlined) and is then rounded to the correct number of figures. Intermediate answers usually also have at least one nonsignificant figure.

20.1 The nuclide symbol for potassium-40 is $^{40}_{19}K$. Similarly, the nuclide symbol for calcium-40 is $^{40}_{20}Ca$. The equation for beta emission is

$$^{40}_{19}K \rightarrow ^{40}_{20}Ca + ^{0}_{-1}e$$

20.2 Write the nuclear equation for the decay of plutonium-239. Plutonium has atomic number 94. Thus, the nuclide symbol is $^{239}_{94}Pu$. An alpha particle has the symbol $^{4}_{2}He$. For the unknown product nucleus, write $^{A}_{Z}X$, where A is the mass number and Z is the atomic number. The nuclear equation is

$$^{239}_{94}Pu \rightarrow ^{A}_{Z}X + ^{4}_{2}He$$

From the superscripts, we can write

$$239 = A + 4; \quad A = 235$$

Similarly, for the subscripts we can write

$$94 = Z + 2; \quad Z = 92$$

Hence, A = 235 and Z = 92, so the product is $^{235}_{92}X$. Since element 92 is uranium, we write the product nucleus as $^{235}_{92}U$.

20.3 (a) $^{118}_{50}Sn$ has atomic number 50. It has 50 protons and 68 neutrons. Since its atomic number is less than 83 and it has an even number of protons and neutrons, it is expected to be stable.

(b) $^{76}_{33}As$ has atomic number 33. It has 33 protons and 43 neutrons. Since stable odd-odd nuclei are rare, we would expect $^{76}_{33}As$ to be one of the radioactive isotopes.

(c) $^{227}_{89}Ac$ has atomic number 89. Since its atomic number is greater than 83, $^{227}_{89}Ac$ is radioactive.

20.4 (a) $^{13}_{7}N$ has 7 protons and 6 neutrons (fewer neutrons than protons). It is expected to decay by electron capture or positron emission (more likely, since this is a light isotope).

(b) $^{26}_{11}Na$ has 11 protons and 15 neutrons. It is expected to decay by beta emission.

20.5 (a) The abbreviated notation is $^{40}_{20}Ca(d,p)\,^{41}_{20}Ca$.

(b) The nuclear equation is $^{12}_{6}C + ^{2}_{1}H \rightarrow ^{13}_{6}C + ^{1}_{1}H$.

20.6 We can write the nuclear equation as follows:

$$^{A}_{Z}X + ^{1}_{0}n \rightarrow ^{14}_{6}C + ^{1}_{1}H$$

To balance this equation in charge (subscripts) and mass number (superscripts), we write the equations

$A + 1 = 14 + 1$ (from superscripts)

$Z + 0 = 6 + 1$ (from subscripts)

Hence, A = 14 and Z = 7. Therefore, the nucleus that produces carbon-14 by this reaction is $^{14}_{7}N$.

20.7 Since an activity of 1.0 Ci is 3.7×10^{10} nuclei/s, the rate of decay in this sample is:

$$\text{Rate} = 13\ \text{Ci} \times \frac{3.7 \times 10^{10}\ \text{nuclei/s}}{1.0\ \text{Ci}} = 4.\underline{8}1 \times 10^{11}\ \text{nuclei/s}$$

The number of nuclei in this 2.5 μg sample of $^{99m}_{43}Tc$ can be calculated by noting that the molar mass in grams is approximately equal to the mass number.

$$2.5 \times 10^{-6}\ \text{g Tc–99m} \times \frac{1\ \text{mol Tc–99m}}{99\ \text{g Tc–99m}} \times \frac{6.02 \times 10^{23}\ \text{Tc–99m nuclei}}{1\ \text{mol Tc–99m}}$$

$$= 1.\underline{5}2 \times 10^{16}\ \text{Tc–99m nuclei}$$

Finally, solve the rate equation for k:

$$k = \frac{\text{rate}}{N_t} = \frac{4.81 \times 10^{11}\ \text{nuclei/s}}{1.52 \times 10^{16}\ \text{nuclei}} = 3.\underline{1}6 \times 10^{-5} = 3.2 \times 10^{-5}/\text{s}$$

20.8 First, substitute the value of k into the equation

$$t_{1/2} = \frac{0.693}{k} = \frac{0.693}{4.18 \times 10^{-9}/\text{s}} = 1.6\underline{5}8 \times 10^{8}\ \text{s}$$

Convert the half-life from seconds to years.

$$1.658 \times 10^{8}\ \text{s} \times \frac{1\ \text{min}}{60\ \text{s}} \times \frac{1\ \text{hour}}{60\ \text{min}} \times \frac{1\ \text{day}}{24\ \text{hours}} \times \frac{1\ \text{year}}{365\ \text{days}} = 5.2\underline{5}7 = 5.26\ \text{years}$$

20.9 Convert the half-life from years to seconds. Then calculate k. Next, use the rate equation to find the decay constant of the sample and finally the activity.

The conversion of the half-life to seconds gives

$$28.1 \text{ years} \times \frac{365 \text{ days}}{1 \text{ year}} \times \frac{24 \text{ hours}}{1 \text{ day}} \times \frac{60 \text{ min}}{1 \text{ hour}} \times \frac{60 \text{ s}}{1 \text{ min}} = 8.8\underline{6}1 \times 10^{8} \text{ s}$$

Since $t_{1/2} = 0.693/k$, solve this for k and substitute the half-life in seconds.

$$k = \frac{0.693}{t_{1/2}} = \frac{0.693}{8.861 \times 10^{8} \text{s}} = 7.8\underline{2}08 \times 10^{-10}/\text{s}$$

Before substituting into the rate equation, we need to know the number of nuclei in a sample containing 5.2×10^{-9} g of strontium-90.

$$5.2 \times 10^{-9} \text{g Sr–90} \times \frac{1 \text{ mol Sr–90}}{90 \text{ g}} \times \frac{6.02 \times 10^{23} \text{ Sr–90 nuclei}}{1 \text{ mol Sr–90}} = 3.\underline{4}78 \times 10^{13} \text{ Sr–90 nuclei}$$

Now, substitute into the rate equation:

$$\text{Rate} = K\,N_t = (7.8208 \times 10^{-10}/\text{s}) \times (3.478 \times 10^{13} \text{ nuclei})$$

$$= 2.\underline{7}2 \times 10^{4} \text{ nuclei/s}$$

Calculate the activity by dividing the rate in disintegrations of nuclei per second by 3.70×10^{10} disintegrations of nuclei per second per Curie.

$$\text{Activity} = \frac{2.72 \times 10^{4} \text{ nuclei/s}}{3.7 \times 10^{10} \text{ nuclei/(s} \cdot \text{Ci)}} = 7.\underline{3}51 \times 10^{-7} = 7.4 \times 10^{-7} \text{ Ci}$$

20.10 The decay constant, k, is $0.693/t_{1/2}$. If we substitute this into the equation

$$\log \frac{N_t}{N_0} = \frac{-kt}{2.303}$$

we get

$$\log \frac{N_0}{N_t} = \frac{0.693\,t}{2.303\,t_{1/2}}$$

Substitute the values t = 25.0 y and $t_{1/2}$ = 10.76 y into the equation and solve for N_0/N_t.

$$\log \frac{N_0}{N_t} = \frac{0.693 \times 25.0 \text{ y}}{2.303 \times 10.76 \text{ y}} = 0.69\underline{9}1$$

$$\frac{N_0}{N_t} = 5.0\underline{0}1$$

The fraction of krypton-85 remaining after 25.0 years is N_t/N_0.

$$\frac{N_t}{N_0} = \frac{1}{5.001} = 0.19\underline{9}9 = 0.200$$

20.11 Substitute $k = 0.693/t_{1/2}$ into the equation for the number of nuclei in a sample after time t:

$$\log \frac{N_t}{N_0} = \frac{-kt}{2.303} = \frac{2.303\,t_{1/2}}{0.693\,t}$$

Hence,

$$t = \frac{2.303\,t_{1/2}}{0.693} \times \log \frac{N_0}{N_t}$$

To obtain N_0/N_t assume that the ratio of $^{14}_{6}C$ to $^{12}_{6}C$ in the atmosphere has remained constant. Then we can say that 1.00 gram of total carbon from the jawbone gave 15.3 disintegrations per minute. The ratio of the number of $^{14}_{6}C$ originally present to the number that existed at the time of dating equals the ratio of rates of disintegration.

$$\frac{N_0}{N_t} = \frac{15.3}{4.5} = 3.\underline{4}00$$

Therefore, substituting this value of N_0/N_t and $t_{1/2}$ = 5730 y into the previous equation gives

$$t = \frac{2.303\ t_{1/2}}{0.693} \log \frac{N_0}{N_t} = \frac{2.303 \times 5730\ y}{0.693} \log 3.4 = 1.01 \times 10^4$$
$$= 1.0 \times 10^4 \text{ years}$$

20.12 Write the nuclear masses below each nuclide symbol. Then calculate Δm.

$$^{234}_{90}Th \rightarrow\ ^{234}_{91}Pa +\ ^{0}_{-1}e$$

233.9942 233.9931 0.000549 amu

$$\Delta m = (233.9931 + 0.000549 - 233.9942) \text{ amu} = -0.000551 \text{ amu}$$

(a) The mass change for molar amounts in this nuclear reaction is –0.000551 g or -5.51×10^{-7} kg. The energy change is

$$\Delta E = (\Delta m)c^2 = (-5.51 \times 10^{-7} \text{ kg})(3.00 \times 10^8 \text{ m/s})^2$$

For 1.00 g Th-234, the energy change is

$$1.00 \text{ g Th-234} \times \frac{1 \text{ mol Th-234}}{234 \text{ g Th-234}} \times \frac{-4.959 \times 10^{10} \text{ J}}{1 \text{ mol Th-234}} = -2.1\underline{1}9 \times 10^8 \text{ J}$$

(b) Convert the mass change for the reaction from amu to grams.

$$\Delta m = -5.51 \times 10^{-4} \times \frac{1 \text{ g}}{1 \text{ amu} \times 6.02 \times 10^{23}} = -9.1\underline{5}2 \times 10^{-28} \text{g} = 9.1\underline{5}2 \times 10^{-31} \text{ kg}$$

$$\Delta E = (\Delta m)c^2 = (-9.152 \times 10^{-31} \text{ kg})(3.00 \times 10^8 \text{ m/s})^2 = -8.2\underline{3}68 \times 10^{-14} \text{ J}$$

Convert this to MeV:

$$\Delta E = -8.2368 \times 10^{-14} \text{ J} \times \frac{1 \text{ MeV}}{1.602 \times 10^{-13} \text{ J}} = -0.51\underline{4}1 = -0.514 \text{ MeV}$$

ANSWERS TO REVIEW QUESTIONS

20.1 The two types of nuclear reactions and their equations are

Radioactive decay: $^{238}_{92}U \rightarrow\ ^{234}_{92}Th +\ ^{4}_{2}He$

Nuclear bombardment reactions: $^{27}_{13}Al +\ ^{4}_{2}He \rightarrow\ ^{30}_{15}P +\ ^{1}_{0}n$

20.2 Magic numbers are the numbers of nuclear particles in completed shells of protons or neutrons. Examples of nuclei with magic numbers of protons are $^{4}_{2}He$, $^{16}_{8}O$, and $^{40}_{20}Ca$.

20.3 To predict whether a nucleus will be stable, look for nuclei that have one of the magic numbers of protons and neutrons. Also look for nuclei that have an even number of protons and an even number of neutrons. Nuclei that fall in the band of stability are also very stable. There are no stable nuclei above atomic number 83.

20.4 The five common types of radioactive decay and the usual condition that leads to each type are

Alpha emission:	$Z > 83$
Beta emission:	N/Z is too large
Positron emission:	N/Z is too small
Electron capture:	N/Z is too small
Gamma emission:	the nucleus is in an excited state

20.5 The isotopes that begin each of the natural radioactive decay series are uranium-238, uranium-235, and thorium-232.

20.6 The equations are as follows:

(a) $^{14}_{7}N + ^{4}_{2}He \rightarrow ^{17}_{8}O + ^{1}_{1}H$

(b) $^{27}_{13}Al + ^{4}_{2}He \rightarrow ^{30}_{15}P + ^{1}_{0}n$

20.7 Particle accelerators are devices used to accelerate electrons, protons, and alpha particles and other ions to very high speeds. They operate by accelerating the charged particle toward a plate with charge opposite to that of the particle. Particle accelerators are required to accelerate alpha particles to speeds high enough to penetrate nuclei of large positive charge, which normally scatter alpha particles.

20.8 Before the discovery of transuranium elements, it was thought that americium ($Z = 95$) and curium ($Z = 96$) should be placed after actinium ($Z = 89$) in the periodic table as d-block transition elements. However, Seaborg and others discovered that these elements had properties similar to the lanthanides and placed them in a second series under the lanthanides.

20.9 The Geiger counter measures alpha particles by means of a tube filled with gas. When particles pass through the tube, they ionize the gas, freeing electrons, which creates a pulse of current that is detected by electronic equipment and is counted. A scintillation counter consists of phosphor, a substance that emits photons when struck by radiation; zinc sulfide is used for the detection of alpha particles; and sodium iodide containing thallium(I) iodide is used for gamma rays. The photons travel from the phosphor to a photoelectric detector, such as a photomultiplier, which magnifies the effect and gives a pulse of electric current that is measured.

20.10 A curie (Ci) equals 3.700×10^{10} nuclear disintegrations per second. A rad is the dosage of radiation that deposits 1×10^{-2} J of energy per kilogram of tissue. A rem is a unit of radiation dosage for biological destruction; it equals the rad multiplied by the relative biological effectiveness (RBE).

20.11 It will take cesium-137 three times its half-life of 30.2 y, or 90.6 y, to decay to 1/8 its original mass: 1/2 to 1/4 to 1/8 = 3 half-lives.

20.12 Since the $^{40}_{18}Ar$ was produced by radioactive decay of $^{40}_{19}K$, half of the initial amount of $^{40}_{19}K$ has decomposed. The age equals the half-life of 1.28×10^{9} y.

20.13 A radioactive tracer is a radioactive isotope added to a chemical, biological, or physical system to study it. For instance, $^{131}I^{-}$ is used as a tracer in the study of the dissolving of lead(II) iodide and its equilibrium in a saturated solution.

20.14 Isotope dilution is a technique designed to determine the quantity of a substance in a mixture, or to determine the total volume of a solution, by adding a known amount of an isotope to it. After removing a portion of the mixture, the fraction by which the isotope has been diluted provides a way of determining the quantity of substance, or the total volume of solution.

20.15 Neutron activation analysis is an analysis based on the conversion of stable isotopes to radioactive isotopes by bombarding a sample with neutrons. An unstable nucleus results, which then emits gamma rays or radioactive particles (such as beta particles). The amount of stable isotope is proportional to the measured emission.

20.16 The reason the deuteron, ${}^{2}_{1}H$, has a mass smaller than the sum of the masses of its constituents is that when nucleons come together to form a nucleus, energy is released. There must therefore be an equivalent decrease in mass since mass and energy are equivalent.

20.17 Iron-56 has a binding energy per nucleon that is near the maximum value. Two light nuclei, such as two C-12 nuclei, will undergo fusion (with the release of energy) as long as the product nuclei are lighter than iron-56 (which is the case with Na-23 and H-1).

20.18 (a) The nuclear fission reactor operates by means of a chain reaction of nuclear fissions, controlled to produce energy without explosion. (b) The breeder reactor is a nuclear reactor that obtains energy by means of nuclear fission but also produces plutonium-239 from uranium-238 for additional fuel. (c) A tokamak nuclear fusion reactor controls nuclear fusion by using a doughnut-shaped magnetic field to hold plasma away from material.

SOLUTIONS TO PRACTICE PROBLEMS

Note on significant figures: The final answer to all mathematical solutions is given first with one nonsignificant figure (last significant figure underlined) and is then rounded to the correct number of figures. Intermediate answers usually also have at least one nonsignificant figure.

20.19 ${}^{87}_{37}Rb \rightarrow {}^{87}_{38}Sr + {}^{0}_{-1}e$

20.21 ${}^{232}_{90}Th \rightarrow {}^{228}_{88}Ra + {}^{4}_{2}He$

20.23 Let X be the product nucleus. The nuclear equation is

$${}^{18}_{9}F \rightarrow {}^{A}_{Z}X + {}^{0}_{1}e$$

From the superscripts: $18 = A + 0$ or $A = 18$

From the subscripts: $9 = Z + 1$ or $Z = 8$

Thus, the product of the reaction is ${}^{18}_{8}X$, and since element 8 is oxygen, symbol O, the nuclear equation is

$${}^{18}_{9}F \rightarrow {}^{18}_{8}O + {}^{0}_{1}e$$

20.25 Let X be the product nucleus. The nuclear equation is

$${}^{210}_{84}Po \rightarrow {}^{A}_{Z}X + {}^{4}_{2}He$$

From the superscripts: $210 = A + 4$ or $A = 206$

From the subscripts: $84 = Z + 2$ or $Z = 82$

Thus, the product nucleus is $^{206}_{82}X$, and since element 82 is lead, symbol Pb, the nuclear equation is

$$^{210}_{84}Po \rightarrow \, ^{206}_{82}Pb + \, ^{4}_{2}He$$

20.27 (a) Neither nucleus has an atomic number that is a magic number of protons. Find how many neutrons are in each nucleus

Sb: # neutrons = A – Z = 122 – 51 = 71
Xe: # neutrons = A – Z = 136 – 54 = 82

82 is a magic number for neutrons (implying stability of nucleus) so we predict that $^{136}_{54}Xe$ is stable and $^{122}_{51}Sb$ is radioactive.

(b) $^{204}_{82}Pb$ has a magic number of protons (82), so it is expected to be the stable nucleus and $^{204}_{85}At$ is radioactive (atomic number greater than 83).

(c) Rb does not have an atomic number that is a magic number for protons. Find the numbers of neutrons in the two isotopes.

$^{87}_{37}Rb$: # of neutrons = 87 – 37 = 50

$^{80}_{37}Rb$: # of neutrons = 80 – 37 = 43

50 is a magic number for neutrons so we predict that $^{87}_{37}Rb$ is stable and $^{80}_{37}Rb$ is radioactive.

20.29 (a) α-emission is most likely for nuclei with $Z > 83$
(b) positron emission (more likely; $Z < 20$) or electron capture, since $N/Z = 3/5 < 1$
(c) β-emission

20.31 α-emission decreases the mass number by 4; β-emission does not affect the mass number.

$^{219}_{86}Rn$ belongs to the $^{235}_{92}U$ decay series, since the difference in mass numbers is 16, which is divisible by 4.

$^{220}_{86}Rn$ belongs to the $^{232}_{90}Th$ decay series, since the difference in mass numbers is 12, which is divisible by 4.

20.33 (a) $^{26}_{12}Mg(d, \alpha)\, ^{24}_{11}Na$

(b) $^{16}_{8}O(n, p)\, ^{16}_{7}N$

20.35 (a) $^{27}_{13}Al + \, ^{2}_{1}H \rightarrow \, ^{25}_{12}Mg + \, ^{4}_{2}He$

(b) $^{10}_{5}B + \, ^{4}_{2}He \rightarrow \, ^{13}_{6}C + \, ^{1}_{1}H$

20.37 $\frac{13.8\text{ MeV}}{1\text{ proton}} = \frac{13.8 \times 10^6\text{ eV}}{\text{proton}} \times \frac{1.602 \times 10^{-19}\text{ J}}{1\text{ eV}} \times \frac{1\text{ kJ}}{10^3\text{ J}} \times \frac{6.02 \times 10^{23}\text{ protons}}{1\text{ mol}}$

$= 1.3\underline{3}09 \times 10^9 = 1.33 \times 10^9\text{ kJ/mol}$

20.39 (a) $^{6}_{3}\text{Li} + ^{1}_{0}\text{n} \rightarrow ^{A}_{Z}\text{X} + ^{3}_{1}\text{H}$

From the superscripts: $6 + 1 = A + 3$ or $A = 6 + 1 - 3 = 4$

From the subscripts: $3 + 0 = Z + 1$ or $Z = 3 - 1 = 2$

The product nucleus is $^{4}_{2}\text{X}$. The element with $Z = 2$ is helium (He) so the missing nuclide is $^{4}_{2}\text{He}$.

(b) The reaction may be written

$^{232}_{90}\text{Th} + ^{A}_{Z}\text{X} \rightarrow ^{235}_{90}\text{U} + ^{1}_{0}\text{n}$

From the superscripts: $232 + A = 235 + 1$ or $A = 235 + 1 - 232 = 4$

From the subscripts: $90 + Z = 92 + 0$ or $Z = 92 - 90 = 2$

The projectile nucleus is $^{4}_{2}\text{X}$. The element with $Z = 2$ is helium, so the missing nuclide is $^{4}_{2}\text{He}$ or α.

The reaction is then written

$^{232}_{90}\text{Th}(\alpha,n) + ^{235}_{92}\text{U}$.

20.41 The reaction may be written

$^{A}_{Z}\text{X} + ^{4}_{2}\text{He} \rightarrow ^{242}_{96}\text{Cm} + ^{1}_{0}\text{n}$

From the superscripts: $A + 4 = 242 + 1$ or $A = 242 + 1 - 4 = 239$

From the subscripts: $Z + 2 = 96 + 0$ or $Z = 96 - 2 = 94$

The element with $Z = 94$ is plutonium (the target nucleus was $^{239}_{94}\text{Pu}$).

20.43 The rate of decay is 8.94×10^{10} nuclei/s. The number of nuclei in the sample is

$$0.250 \times 10^{-3}\text{ g H-3} \times \frac{1\text{ mol H-3}}{3.02\text{ g H-3}} \times \frac{6.02 \times 10^{23}\text{ nuclei}}{1\text{ mol H-3}} = 4.98 \times 10^{19}\text{ nuclei}$$

The rate equation is rate = kN_t. Solve for k.

$$k = \frac{\text{rate}}{N_t} = \frac{8.94 \times 10^{10}\text{ nuclei/s}}{4.98 \times 10^{19}\text{ nuclei}} = 1.7\underline{9}52 \times 10^{-9}\text{/s} = 1.80 \times 10^{-9}\text{/s}$$

20.45 Find the rate of decay from the activity

$$\text{Rate} = \frac{20.4\text{ Ci} \times 3.70 \times 10^{10}\text{ nuclei/s}}{1\text{ Ci}} = 7.5\underline{4}8 \times 10^{11}\text{ nuclei/s}$$

Convert the mass of S-35 to the number of nuclei. The molar mass in grams is approximately equal to the mass number.

$$0.48 \times 10^{-3} \text{ g S-35} \times \frac{1 \text{ mol S-35}}{35 \text{ g S-35}} \times \frac{6.02 \times 10^{23} \text{ nuclei}}{1 \text{ mol S-35}} = 8.\underline{2}56 \times 10^{18} \text{nuclei}$$

Solve the rate equation for k and substitute

$$k = \frac{\text{rate}}{N_t} = \frac{7.548 \times 10^{11} \text{ nuclei/s}}{8.256 \times 10^{18} \text{ nuclei}} = 9.\underline{1}4 \times 10^{-8}/\text{s} = 9.1 \times 10^{-8}/\text{s}$$

20.47 $$t_{1/2} = \frac{0.693}{k} = \frac{0.693}{4.6 \times 10^{-19}/\text{s}} \times \frac{1 \text{ min}}{60 \text{ s}} \times \frac{1 \text{ h}}{60 \text{ min}} \times \frac{1 \text{ d}}{24 \text{ h}} \times \frac{1 \text{ y}}{365 \text{ d}} = 4.\underline{7}77 \times 10^{10} \text{ y} = 4.8 \times 10^{10} \text{y}$$

20.49 $$k = \frac{0.693}{t_{1/2}} = \frac{0.693}{5.73 \times 10^{3} \text{y}} \times \frac{1 \text{ y}}{365 \text{ d}} \times \frac{1 \text{ d}}{24 \text{ h}} \times \frac{1 \text{ h}}{60 \text{ min}} \times \frac{1 \text{ min}}{60 \text{ s}} = 3.8\underline{3}505 \times 10^{-12}/\text{s} = 3.84 \times 10^{-12}/\text{s}$$

20.51 $$k = \frac{0.693}{t_{1/2}} = \frac{0.693}{2.69 \text{ d}} \times \frac{1 \text{d}}{24 \text{ h}} \times \frac{1 \text{ h}}{60 \text{ min}} \times \frac{1 \text{ min}}{60 \text{ s}} = 2.9\underline{8}2 \times 10^{-6}/\text{s}$$

Before substituting into the rate equation, find the number of gold nuclei from the mass.

$$0.43 \times 10^{-3} \text{ g Au-198} \times \frac{1 \text{ mol Au-198}}{198 \text{ g Au-198}} \times \frac{6.02 \times 10^{23} \text{ Au-198 nuclei}}{1 \text{ mol Au-198}}$$

$$= 1.\underline{3}1 \times 10^{18} \text{ Au-198 nuclei}$$

Now find the rate

$$\text{Rate} = kN_t = (2.982 \times 10^{-6}/\text{s})\ (1.31 \times 10^{18} \text{ nuclei}) = 3.\underline{9}06 \times 10^{12} \text{ nuclei/s}$$

$$\text{Activity} = \frac{3.906 \times 10^{12} \text{ nuclei/s}}{3.70 \times 10^{10} \text{ nuclei/s}\cdot\text{Ci}} = 1.\underline{0}55 \times 10^{2} \text{ Ci} = 1.1 \times 10^{2} \text{ Ci}$$

20.53 Find k from the half-life

$$k = \frac{0.693}{t_{1/2}} = \frac{0.693}{14.3 \text{ d}} \times \frac{1 \text{ d}}{24 \text{ h}} \times \frac{1 \text{ h}}{60 \text{ min}} \times \frac{1 \text{ min}}{60 \text{ s}} = 5.6\underline{0}9 \times 10^{-7}/\text{s}$$

Solve the rate equation for N_t

$$N_t = \frac{\text{rate}}{k} = \frac{6.0 \times 10^{12} \text{ nuclei/s}}{5.609 \times 10^{-7}/\text{s}} = 1.\underline{0}7 \times 10^{19} \text{ nuclei}$$

Convert N_t to the mass of P-32

$$1.\underline{0}7 \times 10^{19} \text{ P-32 nuclei} \times \frac{1 \text{ mol P-32}}{6.02 \times 10^{23} \text{ P-32 nuclei}} \times \frac{32 \text{ g P-32}}{1 \text{ mol P-32}}$$

$$= 5.\underline{6}8 \times 10^{-4} \text{ g} = 5.7 \times 10^{-4} \text{ g}$$

20.55 The decay constant, k, is equal to $0.693/t_{1/2}$. For a first-order reaction (like radioactive decay)

$$\log \frac{N_t}{N_0} = \frac{-kt}{2.303}$$

Substituting for k in the previous equation gives

$$\log \frac{N_0}{N_t} = \frac{0.693\ t}{2.303 t_{1/2}} = \frac{0.693(24.0\ h)}{2.303(15.0\ h)} = 0.48\underline{1}5$$

Taking the antilog of both sides of this equation gives

$$\frac{N_0}{N_t} = \text{antilog}\ (0.4815) = 3.0\underline{3}0$$

The fraction of Na-24 remaining at time t is N_t/N_0 so

$$\frac{N_t}{N_0} = \frac{1}{3.030} = 0.33\underline{0}03$$

After 24.0 h, 33.0% of the Na-24 remains.

$$5.0\ \mu g \times 0.33003 = 1.\underline{6}5015\ \mu g = 1.7\ \mu g$$

20.57 After 1.52 h, the amount of Ag-112 is $(1 - 0.280)N_0 = 0.720\ N_0$.

$$\log\left(\frac{N_0}{N_t}\right) = \frac{kt}{2.303} = \frac{0.693\ t}{2.303\ t_{1/2}}$$

Solve for $t_{1/2}$

$$t_{1/2} = \frac{0.693\ t}{2.303 \log\left(\frac{N_0}{N_t}\right)} = \frac{(0.693)\ (1.52\ h)}{2.303 \log\left(\frac{N_0}{0.720\ N_0}\right)} = 3.2\underline{0}59\ h = 3.21\ h$$

20.59 The rate equation for radioactive decay is

$$\text{rate} = kN_t$$

Rearrange this to give an expression for N_t

$$N_t = \frac{\text{rate}}{k}$$

At time t_0 rate = 125 nuclei/s, so $N_0 = 125/k$ nuclei.

At time t_1 (= 10.0 d) rate = 107 nuclei/s, so $N_1 = 107/k$ nuclei.

For radioactive decay

$$\log\left(\frac{N_0}{N_t}\right) = \frac{0.693\ t}{2.303\ t_{1/2}}$$

Solve for $t_{1/2}$

$$t_{1/2} = \frac{0.693\ t}{2.303 \log\left(\frac{N_0}{N_t}\right)} = \frac{0.693(10.0\ d)}{2.303 \log\left(\frac{125/k}{107/k}\right)} = 44.\underline{5}6\ d = 44.6\ d$$

20.61

$$\log\left(\frac{N_0}{N_t}\right) = \frac{0.693\ t}{2.303\ t_{1/2}}$$

Rearrange this to give an expression for t.

$$t = \frac{2.303\ t_{1/2} \log\left(\frac{N_0}{N_t}\right)}{0.693}$$

Since the number of C-14 nuclei present is directly proportional to the rate of decay (rate = kN_t), the ratio of the number of nuclei present at time 0 to the number of nuclei present at time t is equal to the ratio of the rates of disintegrations at those times.

$$\frac{N_0}{N_t} = \frac{15.3 \text{ nuclei/s}}{8.1 \text{ nuclei/s}} = 1.\underline{8}9$$

For C-14, $t_{1/2}$ = 5730 y, so

$$t = \frac{2.303(5730 \text{ y})(\log 1.89)}{0.693} = 5.\underline{2}6 \times 10^3 \text{ y} = 5.3 \times 10^3 \text{ y}$$

20.63 The half-life of C-14 is 5730 y, and the age of the sandals is 9.0×10^3 y. We substitute into the equation

$$\log \frac{N_t}{N_0} = \frac{-0.693\, t}{2.303\, t_{1/2}} = \frac{-0.693 \times (9.0 \times 10^3 \text{y})}{2.303 \times (5730 \text{ y})} = -0.4\underline{7}264$$

Taking the antilogarithm of both sides, we obtain

$$\frac{N_t}{N_0} = \frac{(\text{activity})_t}{(\text{activity})_0} = 0.3\underline{3}679$$

where $(\text{activity})_0$ is the original activity per gram sample, which was 15.3 disintegrations/min, and $(\text{activity})_t$ is the activity 9.0×10^3 y later.

$$\begin{aligned}(\text{activity})_t &= 15.3 \text{ disintegrations/(min} \cdot \text{g)} \times 0.3\underline{3}679 \\ &= 5.\underline{1}529 \text{ disintegrations/(min} \cdot \text{g)} \\ &= 5.2 \text{ disintegrations/(min} \cdot \text{g)}\end{aligned}$$

20.65 $$\Delta m = \frac{\Delta E}{c^2} = \frac{-185 \times 10^3 \text{ J}}{(3.00 \times 10^8 \text{ m/s})^2} = \frac{-1.85 \times 10^5 \text{ kg m}^2/\text{s}^2}{(3.00 \times 10^8 \text{ m/s})^2}$$

$$= -2.0\underline{5}55 \times 10^{-12} \text{ kg} = -2.06 \times 10^{-12} \text{ kg } (-2.06 \times 10^{-9} \text{ g})$$

20.67 $${}^{2}_{1}\text{H} + {}^{3}_{1}\text{H} \rightarrow {}^{4}_{2}\text{He} + {}^{1}_{0}\text{n}$$

masses: 2.01345 3.01550 4.00150 1.00867 amu

$$\Delta m = (4.00150 + 1.00867 - 2.01345 - 3.01550) \text{ amu} = -0.01878 \text{ amu}$$

The mass change is −0.01878 amu for 1 ${}^{2}_{1}\text{H}$ nucleus or −0.01878 g for 1 mol ${}^{2}_{1}\text{H}$

$$\begin{aligned}\Delta E = \Delta mc^2 &= (-0.01878 \times 10^{-3} \text{ kg})(2.998 \times 10^8 \text{ m/s})^2 \\ &= (-1.878 \times 10^{-5} \text{ kg})(2.998 \times 10^8 \text{ m/s})^2 \\ &= -1.68\underline{7}9 \times 10^{12} \text{ kg} \cdot \text{m}^2/\text{s}^2\end{aligned}$$

$$\Delta E = -1.68\underline{7}9 \times 10^{12} \text{ J} = -1.688 \times 10^{12} \text{ J (for 1 mol)}$$

Convert the mass change for one ${}^{2}_{1}\text{H}$ to kilograms

$$\Delta m = -0.01878 \text{ amu} \times \frac{1 \text{ g}}{6.022 \times 10^{23} \times 1 \text{ amu}} = -3.1186 \times 10^{-26} \text{ g } (-3.1186 \times 10^{-29} \text{ kg})$$

$$\Delta E = (-3.1186 \times 10^{-29} \text{ kg})(2.997 \times 10^8 \text{ m/s})^2 \left(\frac{1 \text{ MeV}}{1.602 \times 10^{-13} \text{ kg}\cdot\text{m}^2/\text{s}^2}\right)$$

$$= -17.4\underline{9}7 \text{ MeV} = -17.50 \text{ MeV}$$

20.69 Mass of 3 protons $= 3 \times 1.00728$ amu $= 3.02184$ amu
Mass of 3 neutrons $= 3 \times 1.00867$ amu $= \underline{3.02601 \text{ amu}}$
Total mass of nucleons $= 6.04785$ amu

Mass defect = Total nucleon mass − nuclear mass = (6.04785 − 6.01347) amu = 0.03438 amu

$$\Delta E = \Delta mc^2 = 0.03438 \text{ amu} \times \frac{1 \text{ g}}{1 \text{ amu} \times 6.022 \times 10^{23}} \times \frac{1 \text{ kg}}{10^3 \text{ g}} \times (2.998 \times 10^8 \text{ m/s})^2$$

$$\times \frac{1 \text{ MeV}}{1.602 \times 10^{-13} \text{ J}} = 32.03 \text{ MeV}$$

$$\text{Binding energy per nucleon} = \frac{32.03 \text{ MeV}}{6 \text{ nucleons}} = 5.33\underline{8}3 \text{ MeV/nucleon} = 5.338 \text{ MeV/nucleon}$$

20.71 Na-20, having fewer neutrons than the stable Na-23, is expected to decay to a nucleus with a lower atomic number (and hence a higher N/Z ratio) by electron capture or positron emission. Na-26, having more neutrons than the stable isotope, is expected to decay by beta emission to give a nucleus with a higher atomic number (and hence a lower N/Z ratio).

20.73 The overall reaction may be written

$$^{235}_{92}\text{U} \rightarrow {}^{207}_{82}\text{Pb} + n\,{}^{4}_{2}\text{He} + m\,{}^{0}_{-1}\text{e}$$

From the superscripts: $235 = 207 + n \times 4 + m \times 0$ or $n = \frac{235 - 207}{4} = 7$

From the subscripts: $92 = 82 + n \times 2 + m \times (-1)$ or $m = -(92 - 82 - 7 \times 2) = 4$

Therefore, there are 7 α emissions and 4 β emissions.

20.75 $$^{209}_{83}\text{Bi} + {}^{4}_{2}\text{He} \rightarrow {}^{A}_{85}\text{At} + 2\,{}^{1}_{0}\text{n}$$

From the superscripts: $209 + 4 = A + 2 \times 1$ or $A = 213 - 2 = 211$

The reaction is $^{209}_{83}\text{Bi} + {}^{4}_{2}\text{He} \rightarrow {}^{211}_{85}\text{At} + 2\,{}^{1}_{0}\text{n}$

20.77 $$^{238}_{92}\text{U} + {}^{12}_{6}\text{C} \rightarrow {}^{A}_{Z}\text{X} + 4\,{}^{1}_{0}\text{n}$$

From the superscripts: $238 + 12 = A + 4 \times 1$ or $A = 250 - 4 = 246$

From the subscripts: $92 + 6 = Z + 4 \times 0$ or $Z = 98$

The element with Z = 98 is Californium (Cf), so the equation is $^{238}_{92}\text{U} + {}^{12}_{6}\text{C} \rightarrow {}^{246}_{98}\text{Cf} + 4\,{}^{1}_{0}\text{n}$

20.79 $$\log\left(\frac{N_0}{N_t}\right) = \frac{0.693\ t}{2.303\ t_{1/2}}$$

Hence

$$t = \frac{2.303\ t_{1/2} \log\left(\frac{N_0}{N_t}\right)}{0.693}$$

The ratio N_0/N_t may be found from the ratio of the rates of decay.

$$\frac{N_0}{N_t} = \frac{\text{rate}_0}{\text{rate}_t} = \frac{\text{rate}_0}{0.78\ \text{rate}_0} = 1.\underline{2}8$$

$$t = \frac{2.303\ (12.3\ \text{y})\ (\log 1.28)}{0.693} = 4.\underline{3}8\ \text{y} = 4.4\ \text{y}$$

20.81 In the annihilation, the final mass is 0; that is, all mass is converted to energy. Therefore,

$$\Delta m = 0 - \text{mass of positron} - \text{mass of electron}$$

$$= -2(0.000549)\ \text{amu} \times \frac{1\ \text{g}}{1\ \text{amu} \times 6.02 \times 10^{23}} = \frac{1\ \text{kg}}{10^3\ \text{g}} = -1.824 \times 10^{-30}\ \text{kg}$$

The energy of each photon is $-\Delta E/2$.

$$E_{photon} = \frac{-\Delta E}{2} = \frac{-\Delta mc^2}{2} = \frac{-(-1.824 \times 10^{-30}\ \text{kg})\ (3.00 \times 10^8\ \text{m/s})^2}{2}$$

$$E = 8.2\underline{0}8 \times 10^{-14}\ \text{J}$$

The energy of a photon is related to its wavelength by the following equation

$$E = \frac{hc}{\lambda} \qquad \text{where h is Planck's constant}$$

Therefore

$$\lambda = \frac{hc}{E} = \frac{(6.626 \times 10^{-34}\ \text{J}\cdot\text{s})\ (3.00 \times 10^8\ \text{m/s})}{(8.208 \times 10^{-14}\ \text{J})} = 2.4\underline{2}1 \times 10^{-12}\ \text{m}$$

The wavelength of the photons is 2.42 pm

20.83

$$^{1}_{0}n \quad + \quad ^{235}_{92}U \quad \rightarrow \quad ^{136}_{53}I \quad + \quad ^{96}_{39}Y \quad + \quad 4\,^{1}_{0}n$$

masses 1.00867 234.9933 135.8401 95.8629 1.00867 amu

$$\Delta m = [135.8401 + 95.8629 + 4 \times 1.00867 - (1.00867 + 234.9934)]\ \text{amu} = -0.26439\ \text{amu}$$

When 1 mol of U-235 decays, the change in mass is –0.26439 g. Therefore for 1.00 kg of U-235, the change in mass is

$$1.00 \times 10^3\ \text{g U-235} \times \frac{1\ \text{mol U-235}}{234.9933\ \text{g U-235}} \times \frac{-0.26439\ \text{g}}{1\ \text{mol U-235}} = -1.125\ \text{g}$$

The change in mass when 1.00 kg of U-235 decays is –1.125 g

$$E = \Delta mc^2 = (-1.125 \times 10^{-3}\ \text{kg})\ (3.00 \times 10^8\ \text{m/s})^2$$
$$= -1.0\underline{1}2 \times 10^{14}\ \text{J (rounds to } -1.01 \times 10^{11}\ \text{kJ)}$$

For the combustion of carbon:

$$C(\text{graphite}) + O_2\ (g) \rightarrow CO_2\ (g)$$

ΔH°_f: 0 0 –394 kJ/mol

For the reaction, $\Delta H = \Delta E = -394$ kJ/mol

$$1.00 \times 10^3\ \text{g C} \times \frac{1\ \text{mol C}}{12.01\ \text{g C}} \times \frac{(-394\ \text{kJ})}{1\ \text{mol C}} = -3.2\underline{8}05 \times 10^4\ \text{kJ} = 3.28 \times 10^4\ \text{kJ}$$

The energy released in the fission of 1.00 kg of U-235 (1.01×10^{11} kJ) is larger by several orders of magnitude than the energy released by burning 1.00 kg of C (3.28×10^4 kJ).

Cumulative-Skills Problems (require skills from previous chapters)

20.85 Since the rate = kN_t, we first find the value of k in reciprocal seconds:

$$k = \frac{0.693}{t_{1/2}} = \frac{0.693}{14.3\ \text{d}} = 4.8\underline{4}6 \times 10^{-2}/\text{d}$$

$$\frac{4.846 \times 10^{-2}}{\text{d}} \times \frac{1\ \text{d}}{24\ \text{h}} \times \frac{1\ \text{h}}{60\ \text{min}} \times \frac{1\ \text{min}}{60\ \text{s}} = 5.6\underline{0}9 \times 10^{-7}/\text{s}$$

Now we calculate the number of P-32 nuclei in the sample (N_t). For this, we need the formula weight of Na_3PO_4 containing 15.6% of P-32 and (100.0 x 15.6)% = 84.4% naturally occurring P (we will see that this value differs only slightly from the formula weight of the naturally occurring isotopic mixture). The formula weight of Na_3PO_4 containing naturally occurring P is 163.9 amu; the formula weight of Na_3PO_4 when the P is 100% P-32 is 165.0 amu (we can assume that the atomic mass of P-32 is 32.0 amu). The formula weight of Na_3PO_4 containing 15.6% P-32 is obtained from the weighted average of the formula weights:

$$163.9\ \text{amu} \times 0.844 + 165.0\ \text{amu} \times 0.156 = 16\underline{4}.1\ \text{amu}$$

The moles of P in the sample equal

$$0.0545\ \text{g}\ Na_3PO_4 \times \frac{1\ \text{mol}\ Na_3PO_4}{164.1\ \text{g}\ Na_3PO_4} \times \frac{1\ \text{mol P}}{1\ \text{mol}\ Na_3PO_4} = 3.3\underline{2}1 \times 10^{-4}\ \text{mol P}$$

and the moles of P-32 equal

$$3.321 \times 10^{-4}\ \text{mol P} \times 0.156 = 5.1\underline{8}1 \times 10^{-5}\ \text{mol P-32}$$

Then, the number of P-32 nuclei is

$$5.181 \times 10^{-5}\ \text{mol P-32} \times \frac{6.022 \times 10^{23}\ \text{P-32 nuclei}}{1\ \text{mol P-32}} = 3.1\underline{2}0 \times 10^{19}\ \text{P-32 nuclei}$$

Finally, we calculate the rate of disintegrations.

$$\text{rate} = kN_t = (5.609 \times 10^{-7}/\text{s}) \times (3.120 \times 10^{19}\ \text{P-32 nuclei})$$
$$= 1.7\underline{5}0 \times 10^{13} = 1.75 \times 10^{13}\ \text{P-32 nuclei/s}$$

20.87 First calculate the value of the decay constant, k, for Po-210 to 4 significant figures using 2.303 log 2 = 0.6933:

$$k = \frac{0.6933}{t_{1/2}} = \frac{0.6933}{138.4\ \text{d}} = (5.00\underline{9}3 \times 10^{-3})/\text{d}$$

To calculate g PoO_2 decomposed, first calculate g PoO_2 left after 2 days; use the first-order relationship between time and concentration:

$$-\log = \frac{[A]_t}{[A]_o} = -\log \frac{[\text{g}\ PoO_2]_t}{[\text{g}\ PoO_2]_o} = -\log \frac{[\text{g}\ PoO_2]_{2\,d}}{[1.0000\ \text{g}]} = \frac{kt}{2.303}$$

$$-\log \frac{[\text{g}\ PoO_2]_{2\,d}}{[1.0000\ \text{g}]} = \frac{(5.0093 \times 10^{-3})/\text{d} \times 2.00\ \text{d}}{2.303}$$

$$-\log\ [\text{g}\ PoO_2]_{2\,d} = 4.3\underline{5}02 \times 10^{-3} - \log\ [1.0000\ \text{g}] = 4.3\underline{5}02 \times 10^{-3}$$

$$\log\ [\text{g}\ PoO_2]_{2\,d} = -(4.3\underline{5}02 \times 10^{-3})$$

$[g\ PoO_2]_{2\ d}$ = 0.99$\underline{0}$0333 g left after 2 days.

g PoO_2 decomposed after 2 d = 1.000$\underline{0}$ g – 0.99$\underline{0}$0333 = 0.00$\underline{9}$9667 g

$$\text{mol He} = \text{mol PoO}_2 \text{ decomposed} = 0.00\underline{9}9667 \text{ g Po} \times \frac{1 \text{ mol PoO}_2}{242 \text{ g PoO}_2} \times \frac{1 \text{ mol He}}{1 \text{ mol PoO}_2}$$

$$= 0.0000\underline{4}11 \text{ mol He}$$

Now calculate the volume of He at 25°C and 735 mmHg using R = 0.08205

$$V = \frac{nRT}{P} = \frac{(0.0000411 \text{ mol})\ (0.08205 \text{ L} \cdot \text{atm/(K} \cdot \text{mol}))\ (298 \text{ K})}{(735/760) \text{ atm}}$$

$$= 0.001039 \text{ L} = 0.001 \text{ L} \quad (1 \text{ mL})$$

20.89 $2p + 2n \rightarrow$ He-4

On a mole basis the mass difference, Δm, is:

2.01456 g/2 mol protons
+2.01734 g/2 mol neutrons
–4.00150 g/mol He

$$\Delta m = 0.03040 \text{ g/mol He} = -3.04\underline{0} \times 10^{-5} \text{ kg/mol He}$$

$$\Delta E = (\Delta m)c^2 = -3.040 \times 10^{-5} \text{ kg/mol He} \times (2.998 \times 10^8 \text{ m/s})^2 = -2.73\underline{2}3 \times 10^{12} \text{ kg} \cdot \text{m}^2/\text{s}^2$$

$$= -2.73\underline{2}3 \times 10^{12} \text{ J} = -2.73\underline{2}3 \times 10^9 \text{ kJ}$$

Next calculate $\Delta H°$ for burning of ethane:

	C_2H_6 (g)	+	$\frac{7}{2}O_2$ (g)	→	$2CO_2$ (g)	+	$3H_2O$ (g)
ΔH_f° =	–84.667 kJ		0		2(–393.5 kJ)		3(–241.826 kJ)

$$\Delta H° = [2(-393.\underline{5}) + 3(-241.826) - (-84.667) + 0] \text{ kJ} = 1427.\underline{8}1$$

$$= 1.427\underline{8}1 \times 10^3 \text{ kJ/mol ethane}$$

Now calculate the mol of ethane needed to obtain 2.736×10^9 kJ heat:

$$2.73\underline{2}3 \times 10^9 \text{ kJ} \times \frac{1 \text{ mol ethane}}{1.427\underline{8}1 \times 10^3 \text{ kJ}} = 1.91\underline{3}6 \times 10^6 \text{ mol ethane}$$

Finally convert moles to liters at 25°C and 725 mmHg using R = 0.08205

$$V = \frac{nRT}{P} = \frac{(1.9136 \times 10^6 \text{ mol})\ (0.08205 \text{ L} \cdot \text{atm/(K} \cdot \text{mol}))\ (298 \text{ K})}{(725/760) \text{ atm}}$$

$$= 4.9\underline{0}48 \times 10^7 = 4.90 \times 10^7 \text{ L}$$

CHAPTER 21

THE MAIN-GROUP ELEMENTS: GROUPS IA TO IIIA

SOLUTIONS TO EXERCISES

21.1 (a) Ga is more nonmetallic because it is farther to the right in the periodic table than Ca.
(b) B is more nonmetallic than Ga. They are both in Group IIIA and Ga is below B.
(c) Be is more nonmetallic than Ca because it is above and to the right of Cs.

21.2 Since thallium is a metal of the sixth period, the oxidation number corresponding to the group number is expected to be less stable than the oxidation number equal to the group number minus 2 $(3 - 2 = 1)$. Therefore, the substance that decomposes (is less stable) is expected to be $TlCl_3$ (Tl in oxidation state +3). This conclusion also agrees with the observed melting points. The compound with the lower melting point is more covalent (higher oxidation state). Thus, the compound that melts at 25°C is $TlCl_3$; the one that melts at 430°C is TlCl.

21.3 $Na\ (l) + KCl\ (l) \rightarrow NaCl\ (l) + K\ (g)$
$K\ (s) + O_2\ (g) \rightarrow KO_2\ (s)$

21.4 Add NaOH to a solution of the compound; the hydroxide, $Mg(OH)_2$ or $Be(OH)_2$, would precipitate. However, $Be(OH)_2$ would dissolve in more NaOH to give $Be(OH)_4^{2-}$ (aq).

21.5 $CaCO_3\ (s) \xrightarrow{\Delta} CaO\ (s) + CO_2\ (g)$
$CaO\ (s) + Mg^{2+}\ (aq) + H_2O\ (l) \rightarrow Mg(OH)_2\ (s) + Ca^{2+}\ (aq)$
$Mg(OH)_2\ (s) + H_2SO_4\ (aq) \rightarrow MgSO_4\ (aq) + 2H_2O\ (l)$

21.6 Make solutions of the compounds. If any two are mixed and give no precipitate, one is KOH and the other is $BaCl_2$. The one that gives a precipitate with an excess of the third solution, which is $Al_2(SO_4)_3$, must be $BaCl_2$ (the precipitate is $BaSO_4$).

21.7 (a) $Mg^{2+}\ (aq) + 2OH^-\ (aq) \rightarrow Mg(OH)_2\ (s)$
$Mg(OH)_2\ (s) + 2HCl\ (aq) \rightarrow MgCl_2\ (aq) + 2H_2O\ (l)$
(b) $MgCl_2\ (l) \rightarrow Mg\ (l) + Cl_2\ (g)$

ANSWERS TO REVIEW QUESTIONS

21.1 Four characteristics of a metal are its luster, its heat conductivity, its electrical conductivity, and its malleability or ductility.

21.2 The metallic characteristics of the elements in the periodic table decrease in going across a period from left to right. They also become more important as one goes down any column (group); this trend is most evident in Groups IIIA to VA.

21.3 In the Group VIA elements, the oxidation state rules are as follows. Oxygen: the most common oxidation state is equal to the group number minus 8, or –2. Sulfur: the most common positive oxidation state equals the group number, or +6. The oxidation state equal to the group number minus 8, or –2, is also important. Selenium, tellurium, and polonium: the most common oxidation state is equal to the group number minus 2, or +4.

21.4 The more acidic oxide in each pair is the one in which the element is in the higher oxidation state: (a) As_2O_5, (b) Tl_2O_3, (c) Sb_2O_5, (d) SO_3, and (e) SnO_2.

21.5 Some reasons for the difference in behavior of second-row elements from the remaining elements in the same column are that (1) the atomic size is relatively small, resulting in covalent rather than ionic bonding in many cases; (2) the bonding in these elements involves only s and p orbitals whereas the other elements may also use d orbitals; and (3) second-row elements often display strong multiple bonding (double or triple bonds) formed by overlap of the p orbitals.

21.6 The diagonal relationship between some elements occurs between the first three members of the second row (Li, Be, and B) and the three elements located diagonally below them (Mg, Al, and Si). The pairs that are diagonally located exhibit many similarities. Thus, Be and Al react with bases as well as acids. Also, $BeCl_2$ has a bridge Cl structure and so does $AlCl_3$.

21.7 Commercial sources for these elements are: Na, halite or rock salt; K, sylvite and carnallite; Mg, sea water, dolomite, and magnesite; Ca, sea shells, limestone, and gypsum; B, borax deposits; and Al, bauxite ores.

21.8 $^{227}_{89}Ac \rightarrow {}^{223}_{87}Fr + {}^{4}_{2}He$

21.9 Equations for the reactions of lithium and sodium with oxygen, water, and nitrogen are as follows:

$$4Li\,(s) + O_2\,(g) \rightarrow 2Li_2O\,(s)$$
$$2Na\,(s) + O_2\,(g) \rightarrow Na_2O_2\,(s)$$
$$2Li\,(s) + H_2O\,(l) \rightarrow H_2\,(g) + 2LiOH\,(aq)$$
$$2Na\,(s) + H_2O\,(l) \rightarrow H_2\,(g) + 2NaOH\,(aq)$$
$$6Li\,(s) + N_2\,(g) \rightarrow 2Li_3N(s)$$
$$Na\,(s) + N_2\,(g) \rightarrow NR$$

Note that Li forms the oxide, while Na forms mainly the peroxide. Also, Li forms the nitride, while Na does not react with nitrogen.

21.10 $2Na\,(s) + 2C_2H_5OH\,(l) \rightarrow H_2\,(g) + 2NaOC_2H_5\,(aq)$

21.11 (a) Cathode: $Li^+\,(l) + e^- \rightarrow Li\,(l)$
Anode: $2Cl^-\,(l) \rightarrow Cl_2\,(g) + 2e^-$
(b) Cathode: $Li^+\,(l) + e^- \rightarrow Li\,(l)$
Anode: $4OH^-\,(l) \rightarrow O_2\,(g) + 2H_2O\,(g) + 4e^-$

21.12 $2RbCl\,(l) + Ca\,(l) \rightarrow CaCl_2\,(l) + 2Rb\,(g)$

The reaction is forced in the direction of product Rb (g) because it is a gas and is removed from the equilibrium mixture.

21.13 The principal use of sodium is in the production of tetraethyllead. The principal use of potassium is for preparation of potassium superoxide for rebreathing gas masks.

21.14 Sodium hydroxide is manufactured by the electrolysis of aqueous sodium chloride; the other important product is chlorine gas.

21.15 Lithium carbonate is mixed with calcium hydroxide (lime), precipitating calcium carbonate and leaving lithium hydroxide in solution.

21.16 Radium exists only in the form of radioactive isotopes.

21.17 (a) $Ca\ (s) + 2H_2O\ (l) \rightarrow Ca^{2+}\ (aq) + 2OH^-\ (aq) + H_2\ (g)$
(b) $2Ca\ (s) + O_2\ (g) \rightarrow 2CaO\ (s)$
(c) $Ca\ (s) + H_2\ (g) \rightarrow CaH_2\ (s)$
(d) $3Ca\ (s) + N_2\ (g) \rightarrow Ca_3N_2\ (s)$

21.18 $CaCO_3\ (s) \rightarrow CaO\ (s) + CO_2\ (g)$
$Mg^{2+}\ (aq) + CaO\ (s) + H_2O\ (l) \rightarrow Mg(OH)_2\ (s) + Ca^{2+}\ (aq)$
$Mg(OH)_2\ (s) + 2HCl\ (aq) \rightarrow Mg^{2+}\ (aq) + 2Cl^-\ (aq) + 2H_2O\ (l)$ (evaporate)
$MgCl_2\ (l) \rightarrow Mg\ (l) + Cl_2\ (g)$

21.19 $3SrO\ (s) + 2Al\ (l) \rightarrow 3Sr\ (g) + Al_2O_3\ (s)$

21.20 $CaCO_3\ (s) + 2HCl\ (aq) \rightarrow Ca^{2+}\ (aq) + 2Cl^-\ (aq) + CO_2\ (g) + H_2O\ (l)$

21.21 $Ca(OH)_2\ (s) + CO_2\ (g) \rightarrow CaCO_3\ (s) + H_2O\ (g)$

21.22 Barium carbonate would dissolve in the hydrochloric acid in the stomach (see question 23.20) and produce soluble barium chloride, which is toxic.

21.23 (a) $Be(OH)_2\ (s) + 2HCl\ (aq) \rightarrow BeCl_2\ (aq) + 2H_2O\ (l)$
(b) $Be(OH)_2\ (s) + 2KOH\ (aq) \rightarrow K_2Be(OH)_4\ (aq)$

$Mg(OH)_2$ is similar to $Be(OH)_2$ in that it reacts with HCl; however, $Mg(OH)_2$ does not react with KOH.

21.24 Ruby contains primarily Al_2O_3, but the color comes from Cr_2O_3.

21.25 $2Al\ (s) + 6H^+\ (aq) \rightarrow 2Al^{3+}\ (aq) + 3H_2\ (g)$
$2Al\ (s) + 2NaOH\ (aq) + 6H_2O\ (l) \rightarrow 2Na^+\ (aq) + 2Al(OH)_4^-\ (aq) + 3H_2\ (g)$

21.26 Aluminum forms an adherent oxide coating that protects it against further reaction by oxidizing agents like nitric acid, or hydrogen ion.

21.27 The equations for the Hall-Héroult process are

Cathode: $4Al^{3+}\ (l) + 12e^- \rightarrow 4Al\ (s)$
Anode: $6O^{2-}\ (l) + 3C\ (g) \rightarrow 3CO_2\ (g) + 12e^-$

21.28 Boron trichloride vapor can be reduced with hydrogen on a hot tungsten wire to boron and hydrogen chloride gas.

21.29 Once a powdered aluminum-iron (III) oxide mixture is ignited, the heat evolved produces molten iron, which can be used for welding pieces of steel together.

21.30 Boric acid is obtained by adding sulfuric acid to borax solution and precipitating the boric acid.

$$Na_2B_4O_5(OH)_4\ (aq) + H_2SO_4\ (aq) + 3H_2O\ (l) \rightarrow Na_2SO_4\ (aq) + 4B(OH)_3\ (s)$$

21.31 Boric acid behaves like a monoprotic acid and functions as a Lewis acid by accepting an electron pair of OH^- in forming $B(OH)_4^-$ (aq) and H^+ (aq). It differs from HCN in that the proton comes from water; when HCN ionizes, the proton comes from HCN.

21.32 In tetraborane, the end B—H bonds are the usual two-center bonds. However, each interior B—H—B bond is a three-center bond formed by the overlapping of sp^3 hybrid boron orbitals with a 1s hydrogen orbital and containing just two electrons.

21.33 Aluminum(III) forms the $Al(H_2O)_6^{3+}$ ion, which loses a proton, and forms the $Al(H_2O)_5(OH)^-$ (aq) ion in water.

21.34 $BeCl_2$ and Al_2Cl_6 both contain chlorine bridges, but Al_2Cl_6 is a molecular substance whereas $BeCl_2$ is polymeric.

21.35 The basic steps in producing a pure metal from a natural source are (1) preliminary treatment of the ore (Bayer process for purifying Al_2O_3 from silicate and iron oxide impurities); (2) reduction to the impure metal (Al_2O_3 reduced to impure Al); and (3) refining of the impure metal to the pure metal (Hoopes process for electrolytically converting impure Al to pure Al).

21.36 Flotation is a physical separation process based on differences in the wettability of the mineral and the impurities (gangue).

21.37 The purpose of roasting zinc ore is to convert the zinc sulfide to zinc oxide.

21.38 To reduce WO_3 (s), hydrogen gas is used instead of carbon to avoid chemical combination of tungsten with carbon.

SOLUTIONS TO PRACTICE PROBLEMS

Note on significant figures: The final answer to all mathematical solutions is given first with one nonsignificant figure (last significant figure underlined) and is then rounded to the correct number of figures. Intermediate answers usually also have at least one nonsignificant figure.

21.39 (a) X is a nonmetal. The following properties of X lead to this conclusion:
1. X is nonlustrous.
2. X forms an acidic oxide (the oxide dissolves in water to give a solution that turns blue litmus red).
3. X forms oxoanions (XO_3^-).

(b) X is a metal, although it has some nonmetallic character. The following properties of X lead to this conclusion:
1. X is lustrous.
2. X is an electrical conductor.
3. X forms an oxide that reacts with HCl and with NaOH (indicating that the oxide is amphoteric).

21.41 (a) Na and Mg are in the same period and Mg is to the right of Na. Therefore Na is more metallic.

(b) Na is in the same column as Li and below it, so Na is more metallic than Li. Li is in the same period as Be and is to the left of it, so Li is more metallic than Be. Therefore Na is more metallic than Be.

(c) Ga is in the same period as K and to the right of it, so K is more metallic than Ga. Cs is in the same column as K and is below it, so Cs is more metallic than K. Therefore Cs is more metallic than Ga.

21.43 Pb is in the sixth period, so the oxidation state corresponding to the group number (IV) is expected to be less stable and more covalent. The brown solid, which has the lower melting point and decomposes, is lead(IV) oxide (PbO_2). The other is PbO.

21.45 Similarities

1. All show the same oxidation state (equal to group number) as other members of the same group.
2. Li forms halides and a hydride like other alkali metals.
3. Li and Be are metals, like other Group IA and IIA elements.
4. Li forms a basic oxide like other Group IA metals.

Differences

1. Be and B form compounds with covalent character (other elements of their respective groups are metals).
2. Li reacts with O_2 to form the normal oxide and with N_2 to form a nitride (other alkali metals form predominantly peroxides or superoxides; no nitride is formed with N_2).
3. Be forms ions such as $Be(OH)_4^{2-}$, indicating some nonmetallic character.
4. Be metal reacts with both acid and base, indicating nonmetallic as well as metallic character.
5. B is a semiconductor, and thus is a metalloid (other elements of Group IIIA are metals).
6. B forms an acidic oxide (other elements of Group IIIA give amphoteric or basic oxides).

21.47 $^{227}_{89}Ac \rightarrow {}^{227}_{90}Th + {}^{0}_{-1}e$

21.49 The reduction gives Ti metal and NaCl

$$TiCl_4\ (g) + 4Na\ (l) \rightarrow Ti\ (s) + 4NaCl\ (s)$$

21.51 $K_2SO_4\ (aq) + Ba(OH)_2\ (aq) \rightarrow 2KOH\ (aq) + BaSO_4\ (s)$

A barium sulfate precipitates, which drives the reaction to the right.

21.53 $CO_2\ (g) + NH_3\ (g) + NaCl\ (aq) + H_2O\ (l) \rightarrow NaHCO_3\ (s) + NH_4Cl\ (aq)$

$2NaHCO_3\ (s) \xrightarrow{\Delta} Na_2CO_3\ (s) + CO_2\ (g) + H_2O\ (g)$

$Ca(OH)_2\ (aq) + Na_2CO_3\ (aq) \rightarrow 2\ NaOH\ (aq) + CaCO_3\ (s)$

21.55 $^{230}_{90}Th \rightarrow {}^{226}_{88}Ra + {}^{4}_{2}He$

21.57 Place a small amount of the metal in strong base. If it is beryllium, it will react with evolution of hydrogen. If it is Mg, no reaction will occur.

21.59 The original solution contains Be^{2+}, Mg^{2+}, and Ba^{2+}. If Na_2SO_4 is added, $BaSO_4$ will precipitate and can be filtered off. If excess NaOH is added to the filtrate, $Mg(OH)_2$ will precipitate, leaving $Be(OH)_4^{2-}$ in the solution.

21.61 $Mg(OH)_2\ (s) + 2HCl\ (aq) \rightarrow MgCl_2\ (aq) + 2H_2O\ (l)$

$MgCl_2\ (l) \xrightarrow{\text{electrolysis}} Mg\ (l) + Cl_2\ (g)$

$BeF_2\ (l) + Mg\ (l) \xrightarrow[950^\circ C]{\Delta} Be(s) + MgF_2\ (l)$

21.63 $3Mg\ (s) + N_2\ (g) \rightarrow Mg_3N_2\ (s)$
$Mg_3N_2\ (s) + 6H_2O\ (l) \rightarrow 3Mg(OH)_2\ (aq) + 2NH_3\ (aq)$

21.65 $Fe_2O_3\ (s) + 2Al\ (l) \rightarrow Al_2O_3\ (l) + 2Fe\ (l)$

21.67 $2Al_2O_3\ (s) + 3C\ (s) + 6Cl_2\ (g) \xrightarrow{\Delta} 4AlCl_3\ (s) + 3CO_2\ (g)$
$AlCl_3\ (s) + 3K_{(amalgam)} \rightarrow 3KCl\ (s) + Al\ (s)$

21.69

$$H\text{—}\ddot{\underset{..}{O}}\text{—}\underset{\underset{\underset{H}{|}}{:\ddot{O}:}}{\overset{}{B}}\text{—}\ddot{\underset{..}{O}}\text{—}H + H\text{—}\ddot{\underset{..}{O}}\text{—}H \rightarrow \left[H\text{—}\ddot{\underset{..}{O}}\text{—}\underset{\underset{\underset{H}{|}}{:\ddot{O}:}}{\overset{\overset{H}{|}}{\overset{:\ddot{O}:}{|}}{B}}\text{—}\ddot{\underset{..}{O}}\text{—}H \right]^- + H^+$$

$$H^+ + H\text{—}\underset{\underset{H}{|}}{\ddot{N}}\text{—}H \rightarrow \left[H\text{—}\underset{\underset{H}{|}}{\overset{\overset{H}{|}}{N}}\text{—}H \right]^+$$

21.71 $Al(H_2O)_6^{3+}\ (aq) + HCO_3^-\ (aq) \rightarrow Al(H_2O)_5OH^{2+}\ (aq) + H_2O\ (l) + CO_2\ (g)$

21.73 Test portions of solutions of each compound with the others; the results can differentiate the compounds. $AlCl_3$ reacts only with KOH, giving a precipitate that dissolves in excess KOH. $BaCl_2$ reacts only with $BeSO_4$, giving a precipitate. $BeSO_4$ also gives a precipitate with KOH that dissolves in excess KOH.

21.75 (a) $2PbS\ (s) + 3O_2\ (g) \rightarrow 2PbO\ (s) + 2SO_2\ (g)$

(b) $PbO\ (s) + C\ (s) \xrightarrow{\Delta} Pb\ (s) + CO\ (g)$

(c) At the anode the impure lead is oxidized and goes into solution as Pb^{2+}. At the cathode the Pb^{2+} ions are reduced to Pb and deposited on the cathode, leaving the impurities in solution or at the impure anode.

21.77 $AlCl_3$ has a trigonal planar geometry.

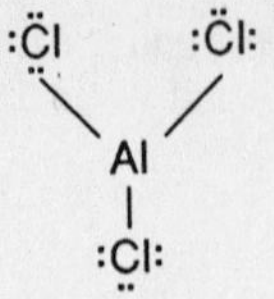

:C̈l :C̈l:
Al
:C̤l:

The geometry of Cl atoms about Al in Al_2Cl_6 is tetrahedral.

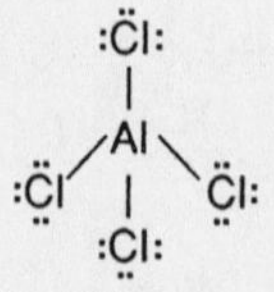

:C̈l:
Al
:C̈l C̈l:
:C̤l:

21.79 $k = \frac{0.693}{t_{1/2}} = \frac{0.693}{1.28 \times 10^{9} y} = 5.4\underline{1}4 \times 10^{-10}/y$

$\log \frac{N_0}{N_t} = \frac{kt}{2.303} = \frac{(5.414 \times 10^{-10}/y)\ (1.00 \times 10^{8} y)}{2.303} = 0.023\underline{5}08$

$\frac{N_0}{N_t}$ = antilog (0.023508) = 1.0$\underline{5}$56

The fraction remaining is N_t/N_0 or 1/1.0556 = 0.94$\underline{7}$32 = 0.947

21.81 $\frac{1}{2}Fe_2O_3\,(s) + Al\,(s) \rightarrow Fe\,(s) + \frac{1}{2}Al_2O_3\,(s)$

$\frac{1}{2}(-825.5)$ 0 0 $\frac{1}{2}(-1676)$ kJ

ΔH° = [1/2(-1676) - 1/2(-825.$\underline{5}$)] kJ = -42$\underline{5}$.25 = -425 kJ

The reaction is exothermic.

21.83 $SrCO_3\,(s) \rightarrow SrO\,(s) + CO_2\,(g)$

ΔH_f°: −1218 −592.0 −393.5 kJ

S°: 97.1 55.5 213.7 J/K

ΔH° = [−592.0 − 393.5 − (−1218)] kJ = 23$\underline{2}$.5 kJ

ΔS° = [55.5 + 213.7 − 97.1] J/K = 172.1 J/K (172.1 × 10^{-3} kJ/K)

$\Delta G^\circ = \Delta H^\circ - T\Delta S^\circ$

0 = 232.5 kJ − T(172.1 × 10^{-3} kJ/K)

$T = \frac{232.5}{172.1 \times 10^{-3}}$ K = 13$\underline{5}$0.1 = 1.35 × 10^{3} K

21.85 The overall disproportionation reaction can be considered as the sum of the following reactions

$2e^- + 2In^+\,(aq) \rightarrow 2In\,(s)$

$In^+\,(aq) \rightarrow In^{3+}\,(aq) + 2e^-$

$3In^+\,(aq) \rightarrow 2In\,(s) + In^{3+}$

$E^\circ_{cell} = E^\circ_{cathode} - E^\circ_{anode}$ = [−0.21 − (−0.40)] V = 0.19 V

$\Delta G° = -nFE°_{cell} = -(2)\ (9.65 \times 10^4\ C)\ (0.19\ J/C) = -3.\underline{6}67 \times 10^4\ J\ (-37\ kJ)$

$\Delta G°$ for the reaction as written is negative, so the disproportionation does occur spontaneously.

21.87 The body-centered cubic cell contains 2 atoms of Na (1 atom in the center and 1/8 atom at each of 8 corners). Let a be the cell dimension of the cubic cell. The cell diagonal has length $\sqrt{3}$ a. If the spheres touch along the diagonal, then the length of the diagonal is 4 times the radius of the spheres.

$$4r = \sqrt{3}\ a \quad \text{or} \quad a = \frac{4r}{\sqrt{3}}$$

Find the volume occupied by 2 atoms of Na.

$$2\ \text{atoms Na} \times \frac{1\ \text{mol Na}}{6.02 \times 10^{23}\ \text{atoms Na}} \times \frac{22.99\ \text{g Na}}{1\ \text{mol Na}} \times \frac{1\ \text{cm}^3}{0.97\ \text{g Na}} = 7.\underline{8}7 \times 10^{-23}\ \text{cm}^3$$

This volume is equal to the volume of the cubic cell, a^3

$$a = \left(\frac{4r}{\sqrt{3}}\right)^3 = 7.87 \times 10^{-23}\ \text{cm}$$

Solve for r

$$\frac{4r}{\sqrt{3}} = \sqrt[3]{7.87 \times 10^{-23}\ \text{cm}^3} = 4.\underline{2}9 \times 10^{-8}\ \text{cm}$$

$$r = \frac{4.29 \times 10^{-8}\left(\sqrt{3}\right)}{4}\ \text{cm} = 1.\underline{8}57 \times 10^{-8} = 1.9 \times 10^{-8}\ \text{cm}\ (1.9\ \text{Å})$$

21.89 $Mg^{2+} + Ca(OH)_2 \rightarrow Mg(OH)_2 + Ca^{2+}$

1 mol Mg^{2+} requires 1 mol $Ca(OH)_2$ so

$$1{,}272\ \text{g}\ Mg^{2+} \times \frac{1\ \text{mol}\ Mg^{2+}}{24.305\ \text{g}\ Mg^{2+}} \times \frac{1\ \text{mol}\ Ca(OH)_2}{1\ \text{mol}\ Mg^{2+}} \times \frac{74.10\ \text{g}\ Ca(OH)_2}{1\ \text{mol}\ Ca(OH)_2} = 3{,}87\underline{8}.02 = 3{,}878\ \text{g}$$

At least 3878 g of $Ca(OH)_2$ would be required to precipitate the Mg^{2+}.

21.91 $2LiOH\ (s) + CO_2\ (g) \rightarrow Li_2CO_3\ (s) + H_2O\ (g)$

Using the ideal gas law, find the amount of CO_2.

$$n = \frac{PV}{RT} = \frac{(30.0\ \text{mmHg})\ (1\ \text{atm}/760\ \text{mmHg})\ (1.00\ \text{L})}{0.08206\ \text{L} \cdot \text{atm/K} \cdot \text{mol}\ (298\ \text{k})} = 1.6\underline{1}4 \times 10^{3}\ \text{mol}$$

From the equation above

1 mol $CO_2 \triangleq 2$ mol LiOH

1 mol LiOH $\triangleq 23.95$ g LiOH

So

$$1.614 \times 10^{-3}\ \text{mol}\ CO_2 \times \frac{2\ \text{mol LiOH}}{1\ \text{mol}\ CO_2} \times \frac{23.95\ \text{g LiOH}}{1\ \text{mol LiOH}} = 7.7\underline{3}1 \times 10^{-2} = 7.73 \times 10^{-2}\ \text{g LiOH}$$

CHAPTER 22

THE MAIN-GROUP ELEMENTS: GROUPS IVA TO VIIA

SOLUTIONS TO EXERCISES

22.1 $PbO\ (s) + OH^-\ (aq) + H_2O\ (l) \rightarrow Pb(OH)_3^-\ (aq)$
$Pb(OH)_3^-\ (aq) + OCl^-\ (aq) \rightarrow PbO_2\ (s) + OH^-\ (aq) + H_2O\ (l) + Cl^-\ (aq)$

22.2 $2NaSn(OH)_3\ (aq) \rightarrow Sn\ (s) + 2Na^+\ (aq) + Sn(OH)_6^{2-}\ (aq)$

22.3 In the presence of excess oxygen, white phosphorus burns to phosphorus(V) oxide:

$$P_4\ (s) + 5O_2\ (g) \rightarrow P_4O_{10}\ (s)$$

Phosphorus(V) oxide reacts with water to give orthophosphoric acid, H_3PO_4:

$$P_4O_{10}\ (s) + 6H_2O\ (l) \rightarrow 4H_3PO_4\ (aq)$$

Finally, add base to form Na_3PO_4:

$$H_3PO_4\ (aq) + 3NaOH\ (aq) \rightarrow Na_3PO_4\ (aq) + 3H_2O\ (l)$$

22.4 $H_3AsO_3\ (aq) + 3Zn\ (s) + 6H^+\ (aq) \rightarrow AsH_3\ (g) + 3Zn^{2+}\ (aq) + 3H_2O\ (l)$

22.5 $S\ (s) + O_2\ (g) \rightarrow SO_2\ (g)$
$SO_2\ (g) + Na_2CO_3\ (aq) \rightarrow Na_2SO_3\ (g) + CO_2\ (g)$
$Na_2SO_3\ (aq) + S\ (s) \xrightarrow{\Delta} Na_2S_2O_3\ (aq)$

22.6 $2S_2O_3^{2-}\ (aq) + I_2\ (aq) \rightarrow 2I^-\ (aq) + S_4O_6^{2-}\ (aq)$

22.7 $2NaCl\ (aq) + 2H_2O\ (l) \xrightarrow{electrolysis} 2NaOH\ (aq) + H_2\ (g) + Cl_2\ (g)$
$3Cl_2\ (g) + 6NaOH\ (aq) \rightarrow NaClO_3\ (aq) + 5NaCl\ (aq) + 3H_2O\ (l)$
$2NaClO_3\ (aq) + SO_2\ (g) + H_2SO_4\ (aq) \rightarrow 2ClO_2\ (g) + 2NaHSO_4\ (aq)$

22.8 $NaBrO_3\ (aq) + F_2\ (g) + 2NaOH\ (aq) \rightarrow NaBrO_4\ (aq) + 2NaF\ (aq) + H_2O\ (l)$

22.9 The needed portion of Figure 22.23 is

+1.47 (from ClO_3^- to Cl_2)

$$ClO_3^- \qquad HClO \xrightarrow{+1.63} Cl_2 \xrightarrow{+1.36} Cl^-$$

Note that we should consider two possible disproportionations: (1) Cl_2 disproportionates to HClO and Cl^-, (2) Cl_2 disproportionates to ClO_3^- and Cl^-. The calculations of standard emf are as follows:

1. $E^\circ_{cell} = E^\circ_{cathode} - E^\circ_{anode}$ = [1.36 – (1.63)] V = –0.27 V
2. $E^\circ_{cell} = E^\circ_{cathode} - E^\circ_{anode}$ = [1.36 – (1.47)] V = –0.11 V

The smaller negative value is E°_{cell} = –0.11 V, but the reaction is nevertheless nonspontaneous.

22.10 The total number of valence electrons is 8 + 2 × 7 = 22. These are distributed to give the following Lewis formula:

:F̤̈ :K̤̈r: F̤̈:

Thus, there are five electron pairs about Kr, suggesting sp^3d hybridization. According to the VSEPR model, these electron pairs should have a trigonal bipyramidal arrangement. The three lone pairs occupy the equatorial positions, and the bonding pairs occupy the axial positions, giving a linear geometry.

22.11 From the problem statement, we have

$XeF_2 \rightarrow Xe + HF$ (acidic solution; not balanced)

If we balance the F atoms, H atoms, then charge, we get the following reduction half-reaction:

$XeF_2 + 2H^+ + 2e^- \rightarrow Xe + 2HF$

According to the problem statement, xenon difluoride will oxidize HCl to Cl_2. That is,

$HCl \rightarrow Cl_2$ (acidic solution; not balanced)

After balancing Cl and H atoms, then charge, we get the following oxidation half-reaction:

$2HCl \rightarrow 2H^+ + Cl_2 + 2e^-$

Adding these two half-reactions gives the balanced equation for the reaction:

$XeF_2 + 2HCl \rightarrow Xe + Cl_2 + 2HF$

ANSWERS TO REVIEW QUESTIONS

22.1 Silicon forms SiF_6^{2-} because it can utilize sp^3d^2 hybridization, but carbon cannot.

22.2 Carbon black is similar to graphite in that both have a layer structure, but it is different in that its stacking is disordered whereas graphite has ordered stacking.

22.3 Of the Group IVA elements, carbon, silicon, germanium, and tin have allotropes with a diamondlike structure.

22.4 According to LeChatelier's principle, high pressure would be expected to convert less dense graphite to more dense diamond because this would reduce the volume and partially reduce the pressure.

22.5 To prepare ultrapure silicon, quartz sand (SiO_2) is reduced with coke at 3000°C to silicon. Then the impure silicon is converted to $SiCl_4$, which is purified before being reduced back to pure silicon.

22.6 Catenation is the ability of an atom to bond covalently to like atoms, as in ethylene, $H_2C{=}CH_2$.

22.7 $$3Si(OH)_2O_2^{2-}\ (aq) \rightarrow \begin{array}{c} O^-\ \ O^-\ \ O^- \\ |\quad\ |\quad\ | \\ HOSiOSiOSiOH \\ |\quad\ |\quad\ | \\ O_-\ \ O_-\ \ O_- \end{array} + 2H_2O$$

22.8 $CO_2\ (g) + H_2O\ (l) \rightarrow H_2CO_3\ (aq)$
$SiO_2\ (s) + CaO\ (s) \xrightarrow{\Delta} CaSiO_3\ (l)$ This is glass.
$SnO_2\ (s) + 2OH^-\ (aq) + 2H_2O\ (l) \rightarrow Sn(OH)_6^{2-}\ (aq)$
$PbO_2\ (s) + 2OH^-\ (aq) + 2H_2O\ (l) \rightarrow Pb(OH)_6^{2-}\ (aq)$

22.9 (a) $Si\ (s) + 2Br_2\ (l) \rightarrow SiBr_4\ (g)$
$Sn\ (s) + 2Br_2\ (l) \rightarrow SnBr_4\ (s)$
$Pb\ (s) + Br_2\ (l) \rightarrow PbBr_2\ (s)$

(b) $Si\ (s) + O_2\ (g) \rightarrow SiO_2\ (s)$
$Sn\ (s) + O_2\ (g) \rightarrow SnO_2\ (s)$
$2Pb\ (s) + O_2\ (g) \rightarrow 2PbO\ (s)$

(c) $Si\ (s) + HCl\ (aq) \rightarrow NR$
$Sn\ (s) + 2[H^+\ (aq) + Cl^-\ (aq)] \rightarrow [Sn^{2+}\ (aq) + 2Cl^-\ (aq)] + H_2\ (g)$
$Pb(s) + 2[H^+\ (aq) + Cl^-\ (aq)] \rightarrow [Pb^{2+}\ (aq) + 2Cl^-\ (aq)] + H_2\ (g)$

(d) $Si\ (s) + HNO_3\ (aq) \rightarrow NR$
$3Sn\ (s) + 4HNO_3\ (aq) \rightarrow 3SnO_2\ (s) + 4NO\ (g) + 2H_2O\ (l)$
$3Pb\ (s) + 8HNO_3\ (aq) \rightarrow 3[Pb^{2+}\ (aq) + 2NO_3^-\ (aq)] + 2NO\ (g) + 4H_2O\ (l)$

22.10 Both of these ions have metal atoms in +2 oxidation states. As reducing agents, these metal atoms will be oxidized to +4 states. The +4 state of lead is normally strongly oxidizing; that is, lead(IV) is easily reduced to lead(II). Therefore, a lead(II) ion, such as $Pb(OH)_3^-$, would not be expected to be a reducing agent. In the case of tin, however, the +4 state tends to be more stable than the +4 state of lead. Thus, $Sn(OH)_3^-$ is expected to be the better reducing agent.

22.11 The Haber process is the catalytic preparation of ammonia from nitrogen and hydrogen at high pressure and temperature. It is important for the synthetic production of fertilizers and explosives.

22.12 The various nitrogen oxides are nitrous oxide (+1 oxidation state), nitric oxide (+2 oxidation state), nitrogen dioxide (+4 oxidation state), dinitrogen trioxide (+3 oxidation state), and dinitrogen pentoxide (+5 oxidation state).

22.13 The Ostwald process for nitric acid involves oxidizing ammonia to nitric oxide, oxidizing the nitric oxide to nitrogen dioxide, and reacting the nitrogen dioxide with water to give nitric acid.

22.14 White phosphorus is a molecular solid, P_4, with a tetrahedral structure having P—P—P bond angles of 60° (smaller than the usual 90°), which makes the bonds weaker than they would be otherwise and accounts for its high reactivity.

22.15 Oxygen reacts with phosphorus to give P_4O_{10}, with arsenic to give As_4O_6, with antimony to give Sb_4O_6, and with bismuth to give Bi_2O_3.

22.16 $Sb_4S_6 + 9O_2\ (g) \rightarrow Sb_4O_6\ (s) + 6SO_2\ (g)$
$Sb_4O_6\ (s) + 6C\ (s) \rightarrow 4Sb\ (s) + 6CO\ (g)$

22.17 The acids of the corresponding anhydrides are H_3PO_3 from P_4O_6, H_3PO_4 from P_4O_{10}, H_3AsO_3 from As_4O_6, and H_3AsO_4 from As_2O_5.

22.18 Since the hydrogens bonded to the phosphorus atom are not acidic, the acid will be monoprotic.

22.19 The first method is the treatment of $Ca_3(PO_4)_2$ with sulfuric acid, giving phosphoric acid and insoluble calcium sulfate. The second method is the treatment of $Ca_3(PO_4)_2$ with HF giving phosphoric acid and insoluble calcium fluoride.

22.20

$$
\begin{array}{ccccc}
 & \begin{array}{cc} H & H \\ O & O \\ | & | \end{array} & & \begin{array}{ccc} H & H & H \\ O & O & O \\ | & | & | \end{array} & \\
H_3PO_4 + & HOPOPOH & \rightarrow & HOPOPOPOH & + H_2O \\
 & \begin{array}{cc} | & | \\ O & O \end{array} & & \begin{array}{ccc} | & | & | \\ O & O & O \end{array} &
\end{array}
$$

22.21 Polyphosphates are added to detergents to form complexes with metal ions and thus prevent their precipitation onto clothes.

22.22 Priestley prepared oxygen by heating mercury(II) oxide:

$$2HgO\ (s) \xrightarrow{\Delta} 2Hg\ (l) + O_2\ (g)$$

22.23 The most important commercial means of producing oxygen is by distillation of liquid air. Air is filtered from dust particles, cooled to freeze out water and carbon dioxide, liquefied, and finally warmed until nitrogen and argon distill, leaving liquid oxygen behind.

22.24 According to molecular orbital theory, the configuration of O_2 is

$$(\sigma_{1s})^2(\sigma^*_{1s})^2(\sigma_{2s})^2(\sigma^*_{2s})^2(\pi_{2p})^4(\sigma_{2p})^2(\pi^*_{2p})^2$$

For O_2^-, it is the same except for $(\pi^*_{2p})^3$. For O_2^{2-}, it is the same except for $(\pi^*_{2p})^4$. The O_2 and O_2^- are paramagnetic; the O_2^{2-} is diamagnetic. The bond order is 2 for O_2; 1.5 for O_2^-; and 1.0 for O_2^{2-}.

22.25 Oxides are binary oxygen compounds where oxygen is in the –2 oxidation state whereas for peroxides the oxidation number of oxygen is –1 and the anion is O_2^{2-}. In the superoxides, the oxidation number of oxygen is $-\frac{1}{2}$ and the anion is O_2^-. An example of each is H_2O (oxide), H_2O_2 (peroxide), and KO_2 (superoxide).

22.26 Three natural sources of sulfur or sulfur compounds are sulfate minerals, sulfide minerals, and coal or petroleum compounds.

22.27 The structure of stable sulfur is the rhombic S_8 molecule.

22.28 As sulfur melts, it gives a straw-colored liquid. Upon further heating, this changes to a dark red-brown viscous liquid in which the S_8 molecules have opened up to join in long spiral chains of sulfur atoms. At about 200°C, the chains begin to break apart and the viscosity decreases.

22.29 Sulfur hexafluoride is prepared by treating solid sulfur with fluorine gas, yielding the gaseous SF_6.

22.30 The Frasch process involves melting underground sulfur deposits with superheated water, using air to force the melted sulfur upward to the surface, and cooling it to form solid sulfur.

22.31 H_2S is a reducing agent, giving S (s) as the product; Na_2SO_3 is a reducing agent, giving SO_4^{2-} as the product; hot concentrated H_2SO_4 is an oxidizing agent, giving SO_2 as the product; and $Na_2S_2O_3$ is a reducing agent, giving $S_4O_6^{2-}$ as the product.

22.32 (a) $2[H^+ (aq) + Cl^- (aq)] + ZnS (s) \rightarrow [Zn^{2+} (aq) + 2Cl^- (aq)] + H_2S (g)$

$CH_3CSNH_2 (aq) + 2H_2O (l) \xrightarrow{\Delta} NH_4^+ (aq) + CH_3COO^- (aq) + H_2S (aq)$

(b) $S (s) + O_2 (g) \rightarrow SO_2 (g)$

$2H^+ (aq) + SO_3^{2-} (aq) \rightarrow H_2O (l) + SO_2 (g)$

22.33 First step: $S (s) + O_2 (g) \rightarrow SO_2 (g)$

Second step: $2SO_2 (g) + O_2 (g) \xrightarrow[V_2O_5]{\Delta} 2SO_3 (g)$

Third step: $SO_3 (g) + H_2O (l) \rightarrow H_2SO_4 (aq)$

22.34

```
      O               O               O       O
      |               |               |       |
HO—S—OH  +  HO—S—OH ⇌ HO—S—O—S—OH  +  H2O
      |               |               |       |
      O               O               O       O
```

22.35 Sodium thiosulfate is prepared by heating a slurry of sulfur in sodium sulfite, Na_2SO_3, to give the thiosulfate anion, $S_2O_3^{2-}$, which can be isolated as the sodium salt.

22.36 Thiosulfuric acid cannot be prepared because the addition of acid to the thiosulfate anion causes it to decompose to sulfur dioxide gas and sulfur.

22.37 Sodium thiosulfate is used to fix the negative. In this process, thiosulfate ion dissolves silver halide by forming the thiosulfate complex of silver ion. If the negative were not fixed, the silver halide would darken over the entire negative when exposed to light, ruining the picture.

22.38 SO_2: H_2SO_3, weak; SeO_2: H_2SeO_3, weak; TeO_2: acid is unknown;
SO_3: H_2SO_4, strong; SeO_3: H_2SeO_4, strong; TeO_3: H_6TeO_6, weak.

22.39 Fluorine reacts with H_2O to give O_2 and F^-; Cl_2 gives HClO and Cl^-.

$2F_2 (g) + 2H_2O (l) \rightarrow 4[H^+ (aq) + F^- (aq)] + O_2 (g)$

$2Cl_2 (g) + 2H_2O (l) \rightarrow 2HClO (aq) + 2[H^+ (aq) + Cl^- (aq)]$

22.40 (a) NR
(b) $Cl_2 (aq) + 2Br^- (aq) \rightarrow Br_2 (l) + 2Cl^- (aq)$
(c) $Br_2 (aq) + 2I^- (aq) \rightarrow I_2 (s) + 2Br^- (aq)$
(d) NR

22.41 Add chlorine water and methylene chloride. For NaCl, there will be no reaction; for NaBr, an orange organic layer will appear; and for NaI, there will be a violet organic layer.

22.42 An interhalogen is a binary compound of one halogen with another: BrF.

22.43 Fluorine is obtained by electrolyzing a solution of KF in liquid HF. Chlorine is prepared by electrolysis of aqueous sodium chloride. Bromine can be obtained from sea water or brine by oxidizing the bromide ion present with chlorine. Iodine is prepared by oxidizing iodide ion in natural brines with chlorine.

22.44 HF has a higher boiling point because of strong hydrogen bonding among HF molecules; HCl has little or no hydrogen bonding.

22.45 HBr cannot be prepared by adding sulfuric acid to NaBr because the hot concentrated acid will oxidize the bromide ion to bromine.

22.46 $6HF\ (aq) + SiO_2\ (s) \rightarrow H_2SiF_6\ (s) + 2H_2O\ (l)$

22.47 $E° = 1.21\ V$; $ClO_3^-\ (aq) + 3H^+\ (aq) \rightarrow HClO_2\ (aq) + H_2O\ (l) + 2e^-$

22.48 Sodium hypochlorite is prepared by reaction of chlorine with NaOH:

$$Cl_2\ (g) + 2NaOH\ (aq) \rightarrow NaClO\ (aq) + NaCl\ (aq) + H_2O\ (l)$$

22.49 An aqueous solution of sodium hypochlorite should be basic because HClO is a weak acid. A solution of sodium perchlorate should be neutral because $HClO_4$ is a strong acid and NaOH is a strong base.

22.50 Bartlett found that PtF_6 reacted with molecular oxygen. Since the first ionization energy of xenon was slightly less than that of molecular oxygen, he reasoned that PtF_6 ought to react with xenon also.

SOLUTIONS TO PRACTICE EXERCISES

Note on significant figures: The final answer to all mathematical solutions is given first with one nonsignificant figure (last significant figure underlined) and is then rounded to the correct number of figures. Intermediate answers usually also have at least one nonsignificant figure.

22.51 $CH_4\ (g) + H_2O\ (g) \xrightarrow{\Delta} CO\ (g) + 3H_2\ (g)$

$$CO\ (g) + Na^+\ (aq) + OH^-\ (aq) \xrightarrow{\Delta} \underbrace{HCOO^-\ (aq) + Na^+\ (aq)}_{HCOONa\ \text{(sodium formate)}}$$

22.53 $SnO_2\ (s) + 2C\ (s) \xrightarrow{\Delta} Sn\ (l) + 2CO\ (g)$

$Sn\ (s) + 2HCl\ (aq) \rightarrow SnCl_2\ (aq) + H_2\ (g)$

22.55 The half-reactions and their sum are as follows:

$$3[Sn(OH)_3^- + 3OH^- \rightarrow Sn(OH)_6^{2-} + 2e^-]$$
$$2[Bi(OH)_3 + 3e^- \rightarrow Bi + 3OH^-]$$

$$3Sn(OH)_3^- + 2Bi(OH)_3 + \overset{3}{\cancel{9}}\ OH^- \rightarrow 3Sn(OH)_6^{2-} + 2Bi + \overset{3}{\cancel{6}}OH^-$$

Adding Na^+ ions gives

$$3NaSn(OH)_3 + 2Bi(OH)_3 + 3NaOH \rightarrow 3Na_2Sn(OH)_6 + 2Bi$$

22.57 $Sn(H_2O)_6^{2+}\ (aq) + H_2O\ (l) \rightarrow Sn(H_2O)_5(OH)^+\ (aq) + H_3O^+\ (aq)$

22.59 The electron-dot formula is

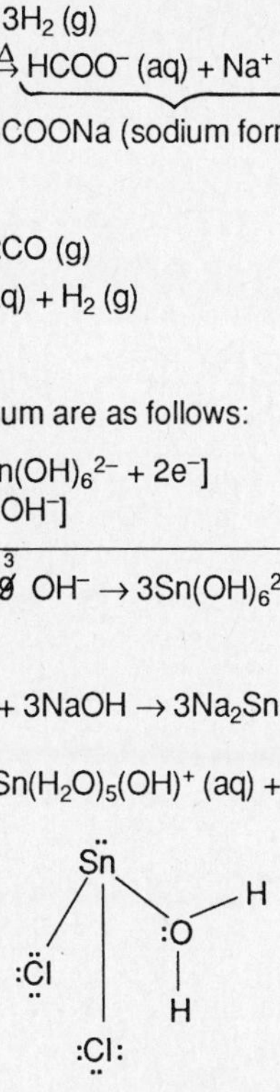

```
          ..
          Sn
         / | \
        /  |  :O—H
    :Cl:   |   |
     ..    |   H
          :Cl:
           ..
```

From the VSEPR model, the geometry is predicted to be trigonal pyramidal.

22.61 Prepare HNO_3 from NH_3:

$$4NH_3\ (g) + 5O_2\ (g) \xrightarrow{Pt} 4NO\ (g) + 6H_2O\ (g)$$
$$2NO\ (g) + O_2\ (g) \rightarrow 2NO_2\ (g)$$
$$3NO_2\ (g) + H_2O\ (l) \rightarrow 2HNO_3\ (aq) + NO\ (g)$$

To prepare N_2O, use HNO_3 just prepared:

$$NH_3\ (g) + HNO_3\ (aq) \rightarrow NH_4NO_3\ (aq)$$
$$NH_4NO_3\ (s) \xrightarrow{\Delta} N_2O\ (g) + 2H_2O\ (g)$$

22.63 (a) P_4O_6, H_3PO_3, PF_3 (b) P_4O_{10}, H_3PO_4, PF_5

22.65 $3As_4O_6\ (s) + 8HNO_3\ (aq) + 14H_2O\ (l) \rightarrow 12H_3AsO_4\ (aq) + 8NO\ (g)$
$2H_3AsO_4\ (s) \xrightarrow{\Delta} As_2O_5\ (s) + 3H_2O\ (g)$

22.67 $H_3PO_3 + H_2SO_4 \rightarrow H_3PO_4 + SO_2 + H_2O$

22.69

$-2 \times 2e^-$

$$\overset{+5}{AsO_4^{3-}} + \overset{-2}{N_2H_5^+} \rightarrow \overset{+3}{AsO_3^{3-}} + \overset{0}{N_2}$$

$+2e^-$

$$2AsO_4^{3-} + N_2H_5^+ \rightarrow 2AsO_3^{3-} + N_2$$

Balance O: $2AsO_4^{3-} + N_2H_5^+ \rightarrow 2AsO_3^{3-} + N_2 + 2H_2O$

Balance H^+: $2AsO_4^{3-} + N_2H_5^+ \rightarrow 2AsO_3^{3-} + N_2 + 2H_2O + H^+$

Solutions of N_2H_5Cl are acidic due to hydrolysis of $N_2H_5^+$:

$$N_2H_5^+\ (aq) + H_2O\ (l) \rightarrow N_2H_4\ (aq) + H_3O^+\ (aq)$$

22.71 PBr_4^+ has four pairs of electrons around the P atom, arranged in a tetrahedral fashion. The hybridization of P is sp^3. Each sp^3 hybrid orbital is used in the formation of a P—Br bond.

22.73 (a) $4Li\ (s) + 2O_2\ (g) \rightarrow 2Li_2O(s)$

(b) Organic materials burn in excess O_2 to give CO_2 and H_2O. The nitrogen becomes N_2.

$$4CH_3NH_2\ (g) + 9O_2\ (g) \rightarrow 4CO_2\ (g) + 2N_2\ (g) + 10H_2O\ (g)$$

(c) $P_4\ (s) + 5O_2\ (g) \rightarrow P_4O_{10}\ (s)$
$P_4O_{10}\ (s) + 6H_2O\ (l) \rightarrow 4H_3PO_4\ (aq)$

22.75 $2H_2S\ (g) + 3O_2\ (g) \rightarrow 2H_2O\ (g) + 2SO_2\ (g)$
$16H_2S\ (g) + 8SO_2\ (g) \rightarrow 16H_2O\ (l) + 3S_8\ (s)$

22.77

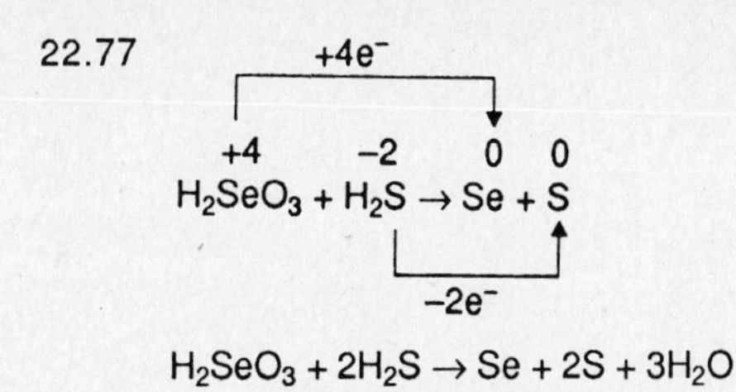

22.79 (a) The Lewis electron-dot formula of H_2Se is

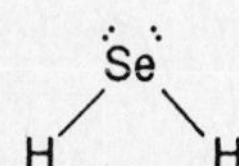

From the VSEPR model, the predicted geometry is angular, or bent. If the four pairs about Se are described in terms of equivalent hybrid orbitals, we would use sp^3 orbitals. We could diagram the bonding as follows:

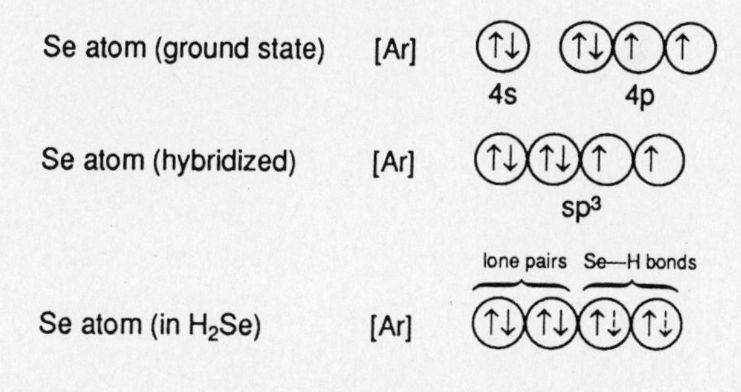

(b) The Lewis electron-dot formula of SeF_4 is

F
:Se F F
F

From the VSEPR model, the predicted geometry is distorted tetrahedral. If the five pairs about Se are described in terms of symmetrical hybrid orbitals, we would use sp^3d. We could diagram the bonding as follows:

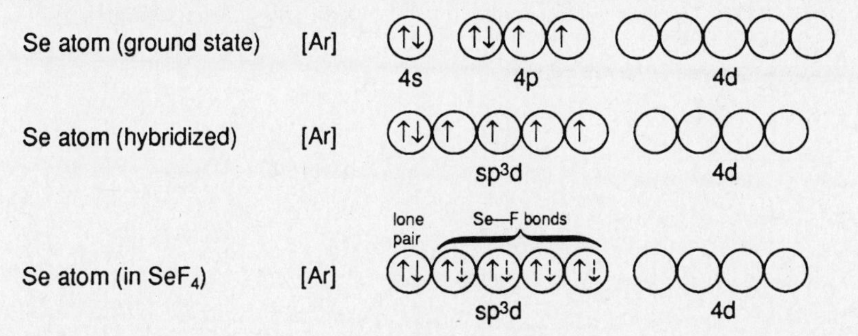

22.81 (a) $x_S + 6x_F = 0$
The oxidation number of F in compounds is always −1.
$x_S = -6x_F = -6(-1) = +6$

(b) $x_S + 3x_O = 0$
The oxidation number of O in most compounds is −2.
$x_S = -3x_O = -3(-2) = +6$

(c) $x_S + 2x_H = 0$
The oxidation number of H in most compounds is +1.
$x_S = -2x_H = -2(+1) = -2$

(d) $x_{Ca} + x_S + 3x_O = 0$
The oxidation number of Ca in compounds is +2; the oxidation number of O in most compounds is −2.
$x_S = -x_{Ca} - 3x_O = -(+2) - 3(-2) = +4$

22.83 $SiO_2 (s) + 4HF (g) \rightarrow 2H_2O (l) + SiF_4 (g)$

22.85 $CaF_2 (s) + H_2SO_4 (l) \xrightarrow{\Delta} CaSO_4 (s) + 2HF (g)$

$2HF (l) \xrightarrow[KF]{electrolysis} H_2(g) + F_2(g)$

$U (s) + 3F_2 (g) \rightarrow UF_6 (s)$

22.87

2 x (−1 e⁻)

$$\overset{-1}{2HCl} + K_2\overset{+6}{Cr_2}O_7 \rightarrow \overset{0}{Cl_2} + 2\overset{+3}{Cr^{3+}}$$

2 x (+3 e⁻)

$6HCl + K_2Cr_2O_7 \rightarrow 3Cl_2 + 2Cr^{3+}$

Balance O: $6HCl + K_2Cr_2O_7 \rightarrow 3Cl_2 + 2Cr^{3+} + 7H_2O$

Balance H: $8H^+ + 6HCl + K_2Cr_2O_7 \rightarrow 3Cl_2 + 2Cr^{3+} + 7H_2O$

Balance K: $8H^+ + 6HCl + K_2Cr_2O_7 \rightarrow 3Cl_2 + 2Cr^{3+} + 7H_2O + 2K^+$

(note that charge is balanced)

22.89 We can balance the net ionic equation first, then convert to the molecular equation if we wish:

+5 e⁻

$IO_3^- + HSO_3^- \rightarrow I_2 + SO_4^{2-}$

−2 e⁻

$2IO_3^- + 5HSO_3^- \rightarrow I_2 + 5SO_4^{2-}$

Balance O: $2IO_3^- + 5HSO_3^- \rightarrow I_2 + 5SO_4^{2-} + H_2O$

Balance H: $2IO_3^- + 5HSO_3^- \rightarrow I_2 + 5SO_4^{2-} + 3H^+ + H_2O$

The molecular equation is

$$2NaIO_3 + 5NaHSO_3 \rightarrow I_2 + 2Na_2SO_4 + 3NaHSO_4 + H_2O$$

22.91 The half-reactions are

oxidation: $Fe^{2+} \rightarrow Fe^{3+} + e^-$

reduction: $2e^- + 2HClO + 2H^+ \rightarrow 2H_2O + Cl_2$

$E^\circ_{cell} = E^\circ_{cathode} - E^\circ_{anode} = [1.63 - (0.77)]\ V = +0.86\ V$

E°_{cell} is positive for the reaction, $2H^+ + 2HClO + 2Fe^{2+} \rightarrow Cl_2 + 2Fe^{3+} + 2H_2O$

So ClO^- as HClO will oxidize Fe^{2+} to Fe^{3+} under standard conditions.

22.93 (a) The electron-dot formula of Cl_2O is

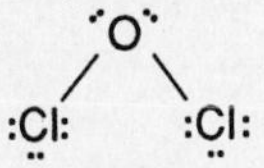

```
    ˙O˙
   /   \
:Cl:   :Cl:
```

The VSEPR model predicts a bent (angular) molecular geometry. We can describe the four electron pairs on O using sp^3 hybrid orbitals. The diagramming for the bond formation follows:

O atom (ground state) [He]

2s 2p

O atom (hybridized) [He]

sp^3

lone pairs Cl—O bonds

O atom (in Cl_2O) [He]

(b) An electron-dot formula of BrO_3^- is

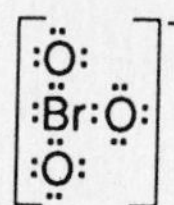

The VSEPR model predicts a trigonal pyramidal geometry. We can describe the four electron pairs on Br using sp^3 hybrid orbitals. The diagramming for the bond formation follows:

Br atom (ground state) [Ar]

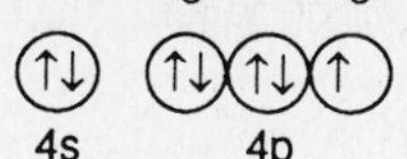

4s 4p

Br atom (hybridized) [Ar]

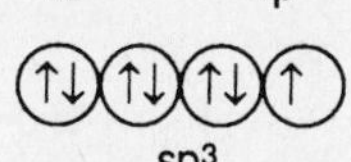

sp^3

lone pair Br—O bonds

Br atom (in BrO_3^-) [Ar]

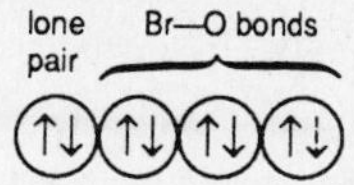

(Note that the additional electron accounts for the −1 charge of the ion; the bonds to O atoms are coordinate covalent.)

(c) The electron-dot formula of BrF_3 is

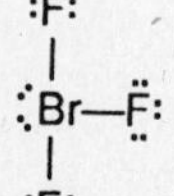

The five electron pairs on Br would be arranged trigonal bipyramidal. Putting the lone pairs in equatorial positions to reduce repulsions gives a T-shaped geometry for BrF_3. We can describe the five electron pairs on Br in terms of sp^3d hybrid orbitals. The diagramming for the bond formation follows:

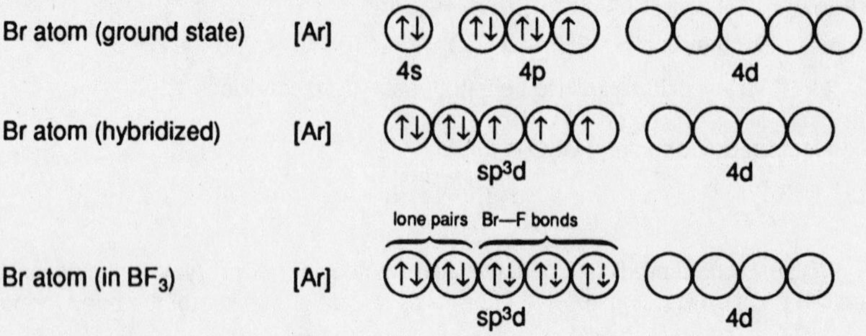

Br atom (ground state) [Ar] (↑↓) 4s (↑↓)(↑↓)(↑) 4p ( )( )( )( )( ) 4d

Br atom (hybridized) [Ar] (↑↓)(↑↓)(↑)(↑)(↑) sp^3d ( )( )( )( ) 4d

lone pairs Br—F bonds

Br atom (in BF_3) [Ar] (↑↓)(↑↓)(↑↓)(↑↓)(↑↓) sp^3d ( )( )( )( ) 4d

22.95 The total number of valence electrons is $8 + (4 \times 7) = 36$. These are distributed to give the following Lewis formula:

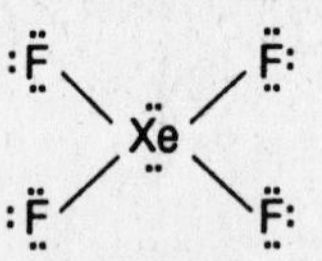

$$XeF_4 \text{ (Lewis formula: Xe with two lone pairs, bonded to four F atoms each with three lone pairs)}$$

The six electron pairs on Xe would have an octahedral arrangement, suggesting sp^3d^2 hybridization. The lone pairs on Xe would be directed above and below the molecule, which has a square planar geometry.

22.97 From the information given:

$XeF_2 \rightarrow Xe + O_2 + F^-$ (basic solution; not balanced)

The oxygen (and hydrogen) will be balanced by OH^- and H_2O from the basic solution. Balance half-reactions

Reduction: $XeF_2 \rightarrow Xe + F^-$

balance F atoms: $XeF_2 \rightarrow Xe + 2F^-$

balance charge: $2e^- + XeF_2 \rightarrow Xe + 2F^-$

Oxidation: $H_2O \rightarrow O_2$

balance O atoms: $2H_2O \rightarrow O_2$

balance H atoms: $2H_2O \rightarrow 4H^+ + O_2$

balance charge: $2H_2O \rightarrow 4H^+ + O_2 + 4e^-$

covert to base: $\cancel{2H_2O} + 4OH^- \rightarrow \overset{2}{\cancel{4}}H_2O + O_2 + 4e^-$

Add half-reactions to get overall reaction

$$4e^- + 2XeF_2 \rightarrow 2Xe + 4F^-$$
$$4OH^- \rightarrow O_2 + 2H_2O + 4e^-$$

$$2XeF_2 + 4OH^- \rightarrow 2Xe + 4F^- + O_2 + 2H_2O$$

22.99 (a) S_8 (b) P_4O_{10} (c) Br_2 (d) Sn

22.101 Test the pH. Na_2SO_4 (aq) is neutral; other solutions are acidic. Add HCl to acidic solutions. $NaHSO_3$ (aq) will evolve SO_2 (characteristic odor).

22.103 Salts of weak acids are hydrolyzed in aqueous solution.

$$H_2AsO_3^- (aq) + H_2O (l) \rightleftharpoons H_3AsO_3 (aq) + OH^- (aq) \qquad K_h = \frac{K_w}{K_a}$$

The equilibrium constant, K_h, is

$$K_h = \frac{K_w}{K_a} = \frac{1.0 \times 10^{-14}}{6 \times 10^{-10}} = \underline{1}.66 \times 10^{-5}$$

Concentration (M)	$H_2AsO_3^-$ →	H_3AsO_3 +	OH^-
Starting	0.050	0	0
Change	−x	+x	+x
Equilibrium	0.050 − x	x	x

At equilibrium

$$1.66 \times 10^{-5} = \frac{[H_3AsO_3][OH^-]}{[H_2AsO_3^-]} = \frac{(x)(x)}{(0.050 - x)}$$

Assume x is small compared to 0.050. Then

$$1.66 \times 10^{-5} \simeq \frac{x^2}{0.050}$$

$$x = \sqrt{1.66 \times 10^{-5} \times 0.050} \simeq \underline{9}.1 \times 10^{-4}$$

(The assumption was marginal. If 9.1×10^{-4} is used for x in 0.050 − x, and a new value of x is calculated, $x = 9.0 \times 10^{-4}$.) Thus at equilibrium $[OH^-] = \underline{9}.0 \times 10^{-4}$ M. Using K_w, find $[H^+]$.

$$K_w = [H^+][OH^-] \text{ so } [H^+] = \frac{K_w}{[OH^-]} = \frac{1.0 \times 10^{-14}}{9.0 \times 10^{-4}} = \underline{1}.11 \times 10^{-11} \text{ M}$$

$$pH = -\log [H^+] = -\log (1.1 \times 10^{-11}) = 10.\underline{9}54 = 11.0$$

22.105

$$\overset{+7}{KMnO_4} + 2H\overset{-1}{Cl} \rightarrow \overset{0}{Cl_2} + \overset{+2}{Mn^{2+}}$$

(Cl: −1 → 0, −2x(1e⁻); Mn: +7 → +2, +5e⁻)

$$2KMnO_4 + 10HCl \rightarrow 5Cl_2 + 2Mn^{2+}$$

Balance O: $2KMnO_4 + 10HCl \rightarrow 5Cl_2 + 2Mn^{2+} + 8H_2O$

Balance H: $6H^+ + 2KMnO_4 + 10HCl \rightarrow 5Cl_2 + 2Mn^{2+} + 8H_2O$

Balance K: $6H^+ + 2KMnO_4 + 10HCl \rightarrow 5Cl_2 + 2Mn^{2+} + 2K^+ + 8H_2O$

The source of H^+ is HCl, so the overall equation is

$$2KMnO_4 + 16HCl \rightarrow 5Cl_2 + 2MnCl_2 + 2KCl + 8H_2O$$

$$12.0 \text{ g } KMnO_4 \times \frac{1 \text{ mol } KMnO_4}{158.0 \text{ g } KMnO_4} \times \frac{16 \text{ mol HCl}}{2 \text{ mol } KMnO_4} \times \frac{1 \text{ L}}{1.50 \text{ mol HCl}} = 0.40\underline{5}06 \text{ L} = 0.405 \text{ L}$$

22.107 The molecular weight of $Ca(H_2PO_4)_2 \cdot H_2O$ is 252.07 amu.

$$\text{mass \% P} = \frac{\text{mass P}}{\text{mass } Ca(H_2PO_4)_2 \cdot H_2O} \times 100\% = \frac{2 \times 30.97 \text{ g}}{252.07 \text{ g}} \times 100\% = 24.57\% \text{ P}$$

$$\text{mass \% } Ca(H_2PO_4)_2 \cdot H_2O = \frac{\text{mass } Ca(H_2PO_4)_2 \cdot H_2O}{\text{mass fertilizer}} \times 100\%$$

Assume a sample of 100.0 g fertilizer. This contains 15.5 g P. Convert this to a mass of $Ca(H_2PO_4)_2 \cdot H_2O$:

$$15.5 \text{ g P} \times = \underbrace{\frac{100.00 \text{ g } Ca(H_2PO_4)_2 \cdot H_2O}{24.57 \text{ g P}}}_{\text{from mass \% P}} = 63.085 \text{ g } Ca(H_2PO_4)_2 \cdot H_2O$$

$$\text{mass \% } Ca(H_2PO_4)_2 \cdot H_2O = \frac{63.085 \text{ g } Ca(H_2PO_4)_2 \cdot H_2O}{100.0 \text{ g fertilizer}} \times 100\% = 63.\underline{0}85\% = 63.1\%$$

22.109 The reaction is $NaCl\ (aq) + H_2O\ (l) \rightarrow NaOCl\ (aq) + H_2\ (g)$

Note: This reaction is the sum of the following:

$2NaCl\ (aq) + 2H_2O\ (l) \rightarrow 2NaOH\ (aq) + Cl_2\ (g) + H_2\ (g)$

$2NaOH\ (aq) + Cl_2\ (g) \rightarrow NaOCl\ (aq) + NaCl\ (aq) + H_2O\ (l)$

$$1.00 \times 10^3 \text{ L} \times \frac{1 \text{ mL}}{10^{-3} \text{ L}} \times \frac{1.00 \text{ g soln}}{1 \text{ mL}} \times \frac{5.25 \text{ g NaOCl}}{100 \text{ g soln}} \times \frac{1 \text{ mol NaOCl}}{74.44 \text{ g NaOCl}} \times \frac{2 \text{ mol e}^-}{1 \text{ mol NaOCl}}$$

$$\times \frac{9.65 \times 10^4 \text{ C}}{1 \text{ mol e}^-} \times \frac{1 \text{ s}}{2.50 \times 10^3 \text{ C}} = 5.4\underline{44} \times 10^4 \text{ s} \times \frac{1 \text{ h}}{3.6 \times 10^3 \text{ s}} = 15.\underline{12} \text{ h} = 15.1\text{h}$$

22.111 $H_2O + NaOCl + 2I^- \rightarrow I_2 + NaCl + 2OH^-$

$I_2 + 2Na_2S_2O_3 \rightarrow 2NaI + Na_2S_4O_6$

$$34.6 \text{ mL } Na_2S_2O_3 \text{ soln} \times \frac{10^{-3} \text{ L}}{1 \text{ mL}} \times \frac{0.100 \text{ mol } Na_2S_2O_3}{1 \text{ L soln}} \times \frac{1 \text{ mol } I_2}{2 \text{ mol } Na_2S_2O_3} \times$$

$$\frac{1 \text{ mol NaOCl}}{1 \text{ mol } I_2} \times \frac{74.44 \text{ g NaOCl}}{1 \text{ mol NaOCl}} = 0.1288 \text{ g NaOCl}$$

If the density of bleach is taken as 1.00 g/mL, then 5.00 mL = 5.00 g

$$\text{mass \% NaOCl} = \frac{\text{mass NaOCl}}{\text{mass bleach}} \times 100\% = \frac{0.1288 \text{ g}}{5.00 \text{ g}} \times 100\% = 2.5\underline{7}56\% = 2.58\%$$

22.113 The disproportionation reaction has the following half-reactions:

reduction: $Sn(OH)_3^-\ (aq) + 2e^- \rightarrow Sn\ (s) + 3OH^-\ (aq)$

oxidation: $Sn(OH)_3^-\ (aq) + 3OH^-\ (aq) \rightarrow Sn(OH)_6^{2-}\ (aq) + 2e^-$

$E^\circ_{cell} = E_{cathode} - E_{anode} = [-0.79 - (-0.96)] \text{ V} = +0.17 \text{ V}$

E°_{cell} is positive, so the disproportionation is expected to occur.

22.115 The reduction half-reaction corresponding to the unknown E° value is

$$H_3PO_2\,(aq) + 4H^+\,(aq) + 4e^- \rightarrow PH_3\,(g) + 2H_2O\,(l)$$

If we add twice the oxidation half-reaction

$$H_2\,(g) \rightarrow 2H^+\,(aq) + 2e^-$$

to this, we get

$$H_3PO_2\,(aq) + 2H_2\,(g) \rightarrow PH_3\,(g) + 2H_2O\,(l) \qquad (1)$$

Reaction can be accomplished in two stages corresponding to the following reductions:

$$H_3PO_2\,(aq) \xrightarrow{-0.51} P_4 \xrightarrow{-0.04} PH_3$$

The reduction half-reactions are

$$H_3PO_2\,(aq) + H^+(ag) + e^- \rightarrow \frac{1}{4}P_4\,(s) + 2H_2O\,(l),\ E^\circ = -0.51\text{ V}$$

$$\frac{1}{4}P_4\,(s) + 3H^+\,(aq) + 3e^- \rightarrow PH_3\,(g),\ E^\circ = -0.04\text{ V}$$

Combining with the oxidation half-reaction $H_2\,(g) \rightarrow 2H^+\,(aq) + 2e^-$ gives

$$H_3PO_2\,(aq) + \frac{1}{2}H_2\,(g) \rightarrow \frac{1}{4}P_4\,(s) + 2H_2O\,(l) \qquad (2)$$

$$\frac{1}{4}P_4\,(s) + \frac{3}{2}H_2\,(g) \rightarrow PH_3\,(g) \qquad (3)$$

Note that reaction (1) equals the sum of reactions (2) and (3). Therefore, the standard free-energy change for (1) equals the sum of the standard free-energy changes for (2) and (3). The ΔG° values are in turn related to E° values. That is,

$$\Delta G_1^\circ = \Delta G_2^\circ + \Delta G_3^\circ$$

$$n_1FE_1^\circ = n_2FE_2^\circ + n_3FE_3^\circ$$

or

$$n_1E_1^\circ = n_2E_2^\circ + n_3E_3^\circ$$

$$4E_1^\circ = E_2^\circ + 3E_3^\circ = [(-0.51) + 3(-0.04)]\text{ V} = -0.63\text{ V}$$

$$E_1^\circ = -0.1\underline{5}75 = -0.16\text{ V}$$

CHAPTER 23

THE TRANSITION ELEMENTS

SOLUTIONS TO EXERCISES

23.1 Since the transition elements become less characteristically metallic in higher oxidation states, VF_5 is compound A, the colorless liquid.

23.2 Mn and Re are transition elements in the same column, and in $KMnO_4$ and $KReO_4$ these elements are in the same oxidation state. Since Mn is in the fourth period and Re is in the sixth period, Mn is expected to be more oxidizing than Re. Thus, $KMnO_4$ is a stronger oxidizing agent.

23.3 $Cu (s) + 2H_2SO_4 (l) \rightarrow [Cu^{2+} (aq) + SO_4^{2-} (aq)] + SO_2 (g) + 2H_2O (l)$
$2Cu^{2+} (aq) + 4I^- (aq) \rightarrow 2CuI (s) + I_2 (aq)$

23.4 Calculate the oxidation number, x_W, from the following:

$$x_{Ca} + x_W + 4x_O = 0$$

Note that the oxidation number of Ca in compounds is +2 and the usual oxidation number of O in compounds is −2. Therefore,

$$x_W = -x_{Ca} - 4x_O = -(+2) - 4(-2) = +6$$

23.5 The dichromate ion is a strong oxidizing agent in acidic solution. The two half-reactions are

$$Cr_2O_7^{2-} + 14H^+ + 6e^- \rightarrow 2Cr^{3+} + 7H_2O$$
$$6I^- \rightarrow 3I_2 + 6e^-$$

Adding the two half-reactions gives the balanced equation for the reaction of dichromate ion and iodide ion in acidic solution.

$$Cr_2O_7^{2-} (aq) + 14H^+ (aq) + 6I^- (aq) \rightarrow 2Cr^{3+} (aq) + 3I_2 (aq) + 7H_2O (l)$$

23.6 Since silver nitrate did not give a precipitate of AgCl, all of the chlorine atoms must be attached to the platinum. There are two potassium ions left over. These correspond to the second and third ions in the three-ion formula unit. This complex has the formula

$$K_2[PtCl_6]$$

23.7 (a) pentaamminechlorocobalt(III) chloride
(b) potassium aquapentacyanocobaltate(III)
(c) pentaaquahydroxoiron(III) ion

23.8 (a) $K_4[Fe(CN)_6]$ (b) $[Co(NH_3)_4Cl_2]Cl$ (c) $PtCl_4^{2-}$

23.9 (a) These two compounds display coordination isomerism because they differ in how the ligands are distributed to two metal atoms.

(b) These are linkage isomers because they differ in that the SCN^- ligand is attached to Mn by the S atom in one case and the N atom in the other.
(c) These are ionization isomers because they differ in the anion that is coordinated to the metal.
(d) These two compounds are hydrate isomers because they differ in the placement of water molecules in the complex.

23.10 The structural formula of the compound is

$[Co(NH_3)_4(H_2O)Cl]Cl_2$

A possible structural isomer of this compound is

$[Co(NH_3)_4Cl]Cl_2 \cdot H_2O$

This would be an example of a hydrate isomer.

23.11 (a) No geometric isomers.

(b)

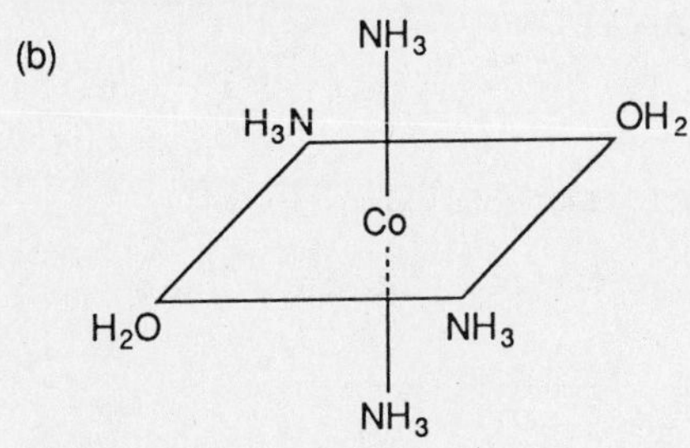

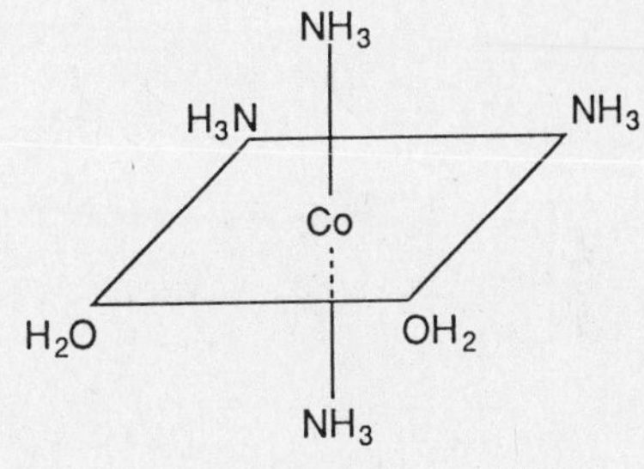

(c)

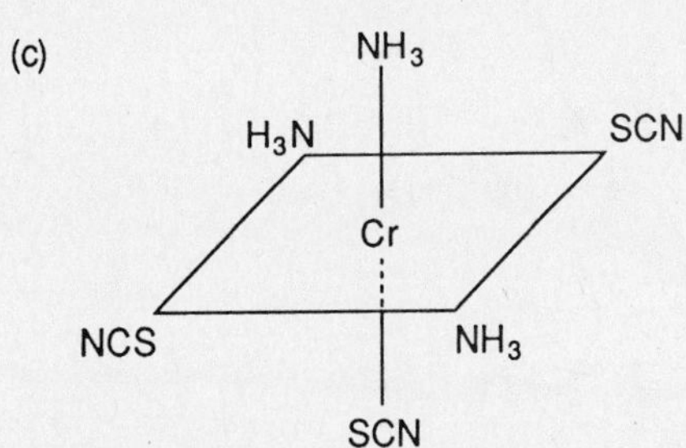

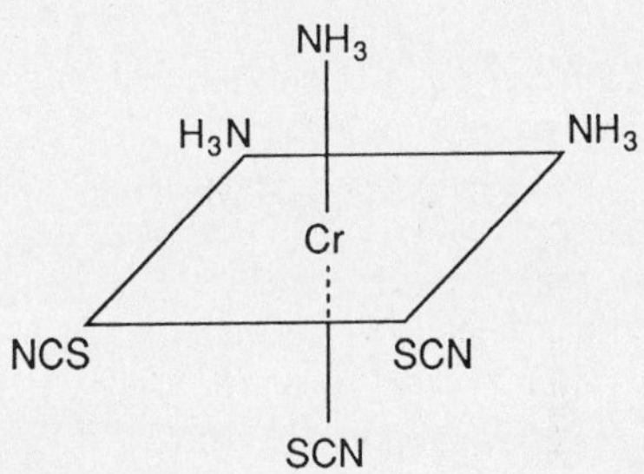

(d) No geometric isomers.

23.12 (a) No optical isomers.

(b)

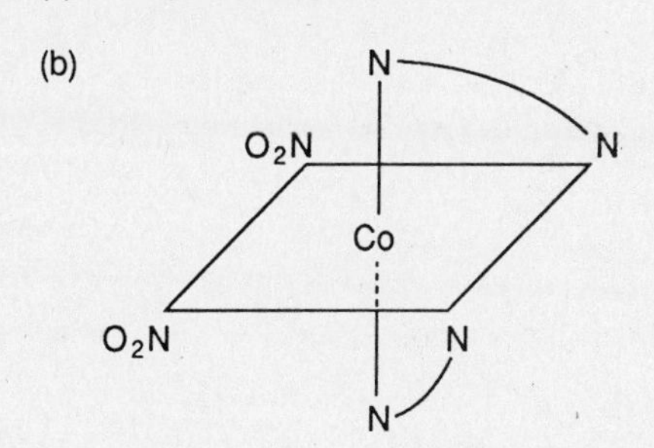

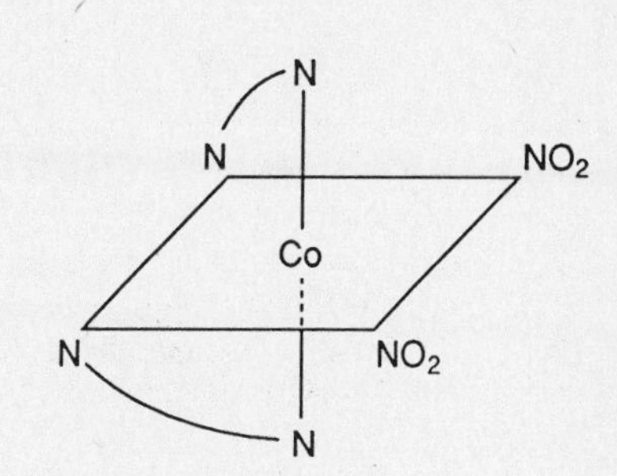

(c)

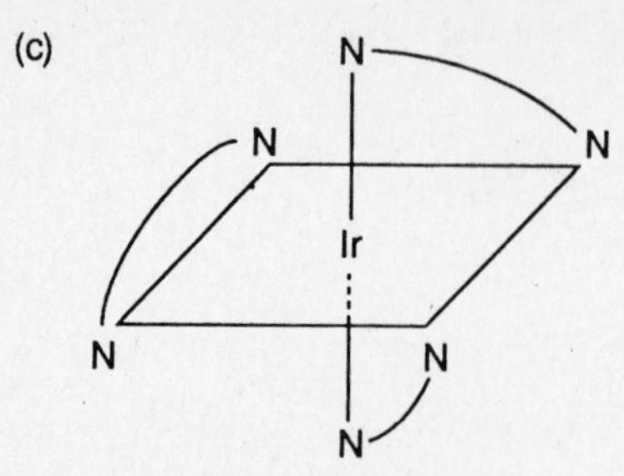

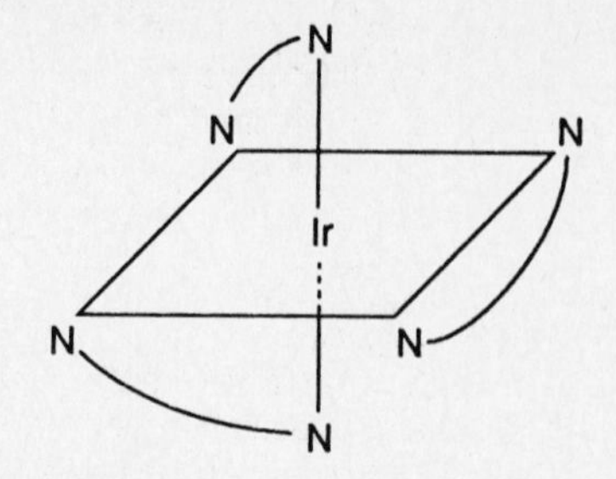

(d) No optical isomers.

23.13 The electronic configuration of cobalt is $[Ar]3d^74s^2$, and that of the cobalt(III) ion is $[Ar]3d^6$. The orbital diagram of a paramagnetic ion, such as CoF_6^{3-}, is

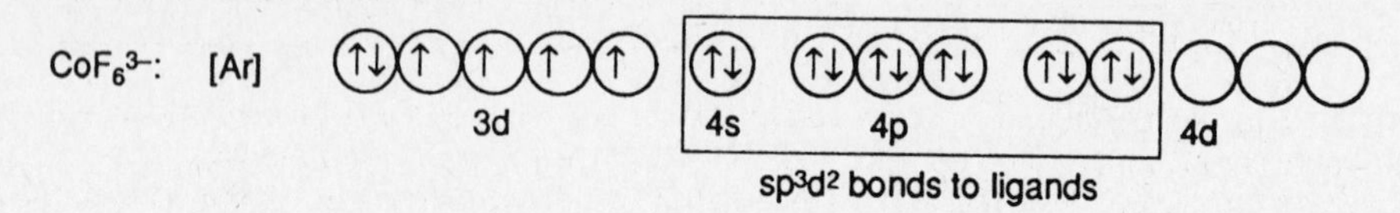

Bonding uses sp^3d^2 hybrid orbitals on Co; there are four unpaired electrons. The orbital diagram of a diamagnetic ion, such as $Co(NH_3)_6^{3+}$, is

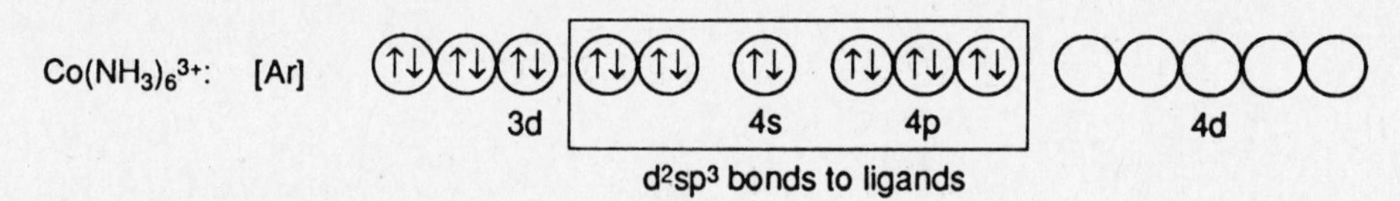

Bonding uses d^2sp^3 hybrid orbitals on Co; there are no unpaired electrons.

23.14 The Co^{2+} ion as the configuration $[Ar]3d^7$. The orbital diagram using dsp^2 bonds would give only one unpaired electron.

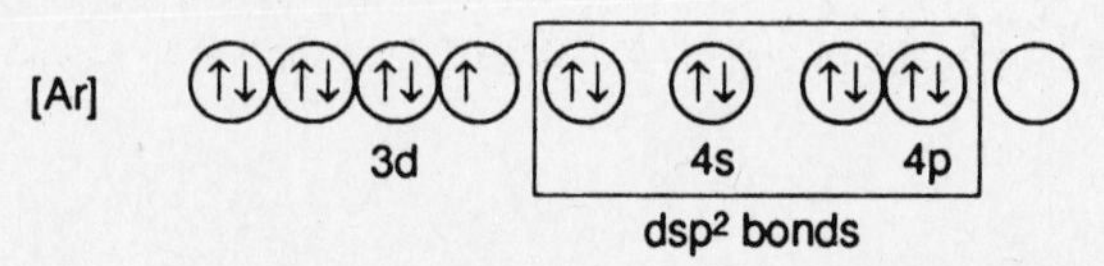

However, using sp^3 bonds would give the observed three unpaired electrons:

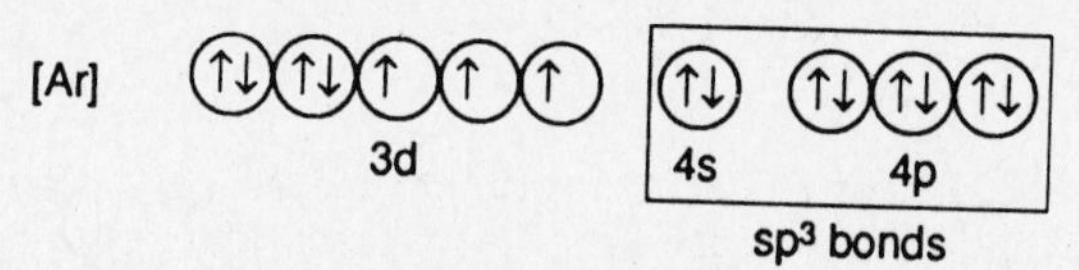

Therefore, the geometry is expected to be tetrahedral.

23.15 The electron configuration of Ni^{2+} is $[Ar]3d^8$. The distribution of electrons among the d orbitals of Ni in $Ni(H_2O)_6^{2+}$ is as follows:

$\underline{\uparrow}$ $\underline{\uparrow}$

$\underline{\uparrow\downarrow}$ $\underline{\uparrow\downarrow}$ $\underline{\uparrow\downarrow}$

Note that there is only one possible distribution of electrons, giving two unpaired electrons.

23.16 The electronic configuration of the Co^{2+} ion is [Ar]$3d^7$. The distribution of the d electrons in $CoCl_4^{2-}$ is as follows:

$$\underline{\uparrow} \quad \underline{\uparrow} \quad \underline{\uparrow}$$

$$\underline{\uparrow\downarrow} \quad \underline{\uparrow\downarrow}$$

23.17 The approximate wavelength of the maximum absorption for $Fe(H_2O)_6^{3+}$, which is pale purple, is 530 nm. The approximate wavelength of the maximum absorption for $Fe(CN)_6^{3-}$, which is red, is 500 nm. The shift is in the expected direction because CN^- is a more strongly bonding ligand than H_2O. As a result, Δ should increase and the wavelength of the absorption should decrease when H_2O is replaced by CN^-.

ANSWERS TO REVIEW QUESTIONS

23.1 Characteristics of the transition elements that set them apart from the main-group elements are: (1) The transition elements are metals with high melting points (only the IIB elements have low melting points). Most main-group elements have low melting points. (2) Each of the transition metals have several oxidation states (except for the IIIB and IIB elements). Most main-group metals have only one oxidation state in addition to 0. (3) Transition-metal compounds are often colored and many are paramagnetic. Most main-group compounds are colorless and diamagnetic.

23.2 Technetium has the electron configuration [Kr] $4d^5 5s^2$.

23.3 Molybdenum has the highest melting point of any element in the fifth period because it has the maximum number of unpaired electrons, which contribute to the strength of the metal bonding.

23.4 One reason iron, cobalt, and nickel are similar in properties is because these elements have similar covalent radii.

23.5 Nickel falls in the fourth period, which has much smaller covalent radii than the corresponding metals, palladium and platinum, in the fifth and sixth periods. However, palladium and platinum have very similar covalent radii.

23.6 $Cr(s) + 2HCl(aq) + 6H_2O(l) \rightarrow Cr(H_2O)_6^{2+} + 2Cl^-(aq) + H_2(g)$

$Cu(s) + HCl(aq) \rightarrow NR$

23.7 $Cr_2O_3(s) + 6HCl(aq) + 9H_2O(l) \rightarrow 2Cr(H_2O)_6^{3+} + 6Cl^-(aq)$

$Cr_2O_3(s) + 7H_2O(l) + 2OH^-(aq) \rightarrow 2Cr(H_2O)_2(OH)_4^-(aq)$

23.8 Four of the water molecules are associated with the copper(II) ion, and the fifth is hydrogen bonded to the sulfate ion as well as to water molecules on the copper(II). Heating changes the blue color to a white color, the color of anhydrous $CuSO_4$. The blue color is associated with $Cu(H_2O)_4^{2+}$. When the water molecules leave the copper ion, the blue color is lost.

23.9 $2Cu(H_2O)_6^+(aq) \rightarrow Cu(H_2O)_6^{2+}(aq) + Cu(s) + 6H_2O(l)$

Thus the other product is copper metal.

23.10 Werner showed that the electrical conductance of a solution of $[Pt(NH_3)_4Cl_2]Cl_2$ corresponded to that of three ions in solution and that two of the chloride ions could be precipitated as AgCl whereas the other two could not.

23.11 A complex ion is a metal atom or ion with Lewis bases attached to it through coordinate covalent bonds. A ligand is a Lewis base attached to a metal ion in a complex; it may be either a molecule or an anion, rarely a cation. The coordination number of a metal atom in a complex ion is the total number of bonds the metal forms with ligands. An example of a complex ion is $Fe(CN)_6^{4-}$; an example of a ligand is CN^-; and the coordination number of the preceding complex ion is six.

23.12 A bidentate ligand is a ligand that bonds to a metal ion through two atoms. Two examples are ethylenediamine, $H_2N—C_2H_4—NH_2$, and the oxalate ion, $^-O_2C—CO_2^-$.

23.13

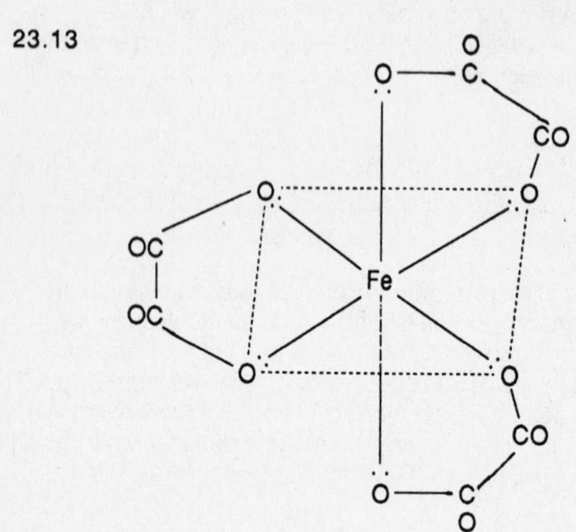

23.14 "Hexacyano" means that there are six CN^- ligands bonded to the iron cation. The Roman numeral "II" means that the oxidation state of the iron cation is two, so that the overall charge of the complex ion is –4. This requires four potassium ions to counterbalance the –4 charge.

23.15 The three properties are isomerism, paramagnetism, and color (or absorption of visible and ultraviolet radiation).

23.16 (a) Ionization isomerism involves isomers that are alike in that the same anions are present in the formula but different anions are coordinated to the metal ion. For example, the sulfate ion is coordinated to cobalt in $[Co(NH_3)_5(SO_4)]Br$, but the bromide ion is coordinated to cobalt in $[Co(NH_3)_5Br]SO_4$.

(b) Hydrate isomerism involves differences in the placement of water molecules in the complex ion. For example, $CrCl_3 \cdot 6H_2O$ exists as $[Cr(H_2O)_6]Cl_3$, $[Cr(H_2O)_4Cl_2]Cl \cdot 2H_2O$, and one other isomer.

(c) Coordination isomers are those in which both the cation and anion are complex and the ligands are distributed in different ways between the two metal atoms. For example,

$$[Cu(NH_3)_4]\,[PtCl_6] \quad \text{and} \quad [Pt(NH_3)_4]\,[CuCl_4].$$

(d) Linkage isomers are those in which two different donor atoms on the ligand may bond to the metal ion. For example, the SCN^- can bond to a metal ion through the sulfur atom or through the nitrogen atom.

23.17 Geometric isomers are isomers in which the atoms are joined to one another in the same way but differ because some atoms occupy different relative positions in space. In $[Pt(NH_3)_2Cl_2]$, the two NH_3's (or Cl's) can be arranged trans or cis to one another. Optical isomers are isomers that are nonsuperimposable mirror images of one another. See Figure 23.15 for an example of two cobalt optical isomers.

23.18 Compounds A and B are geometric isomers because these isomers have different physical properties whereas optical isomers do not.

23.19 A d optical isomer rotates the plane of polarized light to the right (dextrorotatory), and an *l* optical isomer rotates the plane to the left (levorotatory).

23.20 A racemic mixture is a mixture of 50% of the d isomer and 50% of the *l* isomer. One method of resolving a racemic mixture is to prepare a salt with an optically active ion of the opposite charge and crystallize the salts. They will no longer be optical isomers and will have different solubilities, and so one can be precipitated before the other.

23.21 According to the valence bond theory, a ligand orbital containing two electrons overlaps an unoccupied orbital on the metal ion.

23.22 (a) In the high-spin complex ion, all 3d orbitals of Fe^{2+} are occupied (four of the 3d orbitals each contain only one electron). Because they are occupied, those orbitals cannot be used for ligand bonding. Instead, sp^3d^2 hybrid orbitals form from the 4s, the three 4p, and two of the 4d orbitals. Each of six ligands donates a pair of electrons to one of these sp^3d^2 hybrid orbitals.

(b) In the low-spin complex ion, the six electrons in the 3d orbitals are paired so they occupy only three of the 3d orbitals. Then sp^3d^2 hybrid orbitals form from two of the 3d, the 4s, and the three 4p orbitals. Each of six ligands donates a pair of electrons to one of these d^2sp^3 hybrid orbitals.

23.23 The d orbitals of a transition metal atom may have different energies in the octahedral field of six negative charges because the electron pairs of the ligands point directly at the d_{z^2} and $d_{x^2-y^2}$ orbitals. These orbitals are raised much more in energy than the other three d orbitals because they occupy space between the ligands. Thus, there is a crystal field splitting between the first two d orbitals mentioned and the d_{xy}, d_{xz}, and d_{yz} orbitals.

23.24 The crystal field splitting is the difference in energy between the two sets of d orbitals for a given structure (such as octahedral) in complex ions. It is determined experimentally by measuring the energy of light absorbed by complex ions.

23.25 The pairing energy, P, is the energy required to place two electrons in the same orbital. If the crystal field splitting (Δ) is small because of weak-bonding ligands, then the pairing energy will be larger and the complex will be high-spin. If the crystal field splitting (Δ) is large because of strong-bonding ligands, then the pairing energy will be smaller and the complex will be low-spin.

23.26 (a) A high-spin Fe(II) octahedral complex is

↑ ↑

↑↓ ↑ ↑

(b) A low-spin Fe(II) octahedral complex is

___ ___

↑↓ ↑↓ ↑↓

23.27 The spectrochemical series is the arrangement of ligands in order of the relative size of the crystal field splittings (Δ) they induce in the d orbitals of a given oxidation state of a given metal ion. The order is the same, no matter what metal or oxidation state is involved. For Cl^-, H_2O, NH_3, and CN^-, the order of increasing crystal field splitting is

$$Cl^- < H_2O < NH_3 < CN^-,$$

where CN^- always acts as a strong-bonding ligand.

23.28 The complex absorbing red light would appear as a mixture of blue and green (approximately).

SOLUTIONS TO PRACTICE PROBLEMS

Note on significant figures: The final answer to all mathematical solutions is given first with one nonsignificant figure (last significant figure underlined) and is then rounded to the correct number of figures. Intermediate answers usually also have at least one nonsignificant figure.

23.29 The oxidation state of Cr in CrO_3 is +6; that in Cr_2O_3 is +3. The oxide of Cr in the higher oxidation state should be more acidic. Thus, the dark red oxide, which is acidic, should be CrO_3. This conclusion also agrees with the observed melting points. Bonding to Cr in the higher oxidation state is expected to be more covalent, and therefore, the melting point lower. The dark red oxide melts at 197°C compared to 2435°C for the green oxide, which must be Cr_2O_3.

23.31 W. The heavier element is expected to be more stable in the high oxidation state (+6).

23.33 $Cr(s) + 2HCl(aq) \rightarrow CrCl_2(aq) + H_2(g)$

$4CrCl_2(aq) + O_2(g) + 4HCl(aq) \rightarrow 4CrCl_3(aq) + 2H_2O(l)$

$CrCl_3(aq) + 3NaOH(aq) \rightarrow Cr(OH)_3(s) + 3NaCl(aq)$

23.35 (a) The charge on carbonate ion is −2. In order for $FeCO_3$ to be neutral the oxidation number of iron must be +2.

(b) The oxidation number of oxygen is −2. In order for the sum of the oxidation numbers of all atoms to be 0, manganese must be in the +4 oxidation state.

(c) The oxidation number of the chlorine is −1, so the oxidation number of copper must be +1.

(d) The oxidation number of oxygen is −2. The oxidation number of chlorine is −1. In order for the sum of the oxidation numbers to be zero, the oxidation number of chromium must be +6.

$$+6 + 2(-2) + 2(-1) = 0$$

23.37 The half-reactions are

$$Fe^{2+} \rightarrow Fe^{3+} + e^-$$

$$4H^+ + NO_3^- + 3e^- \rightarrow NO + 2H_2O$$

The balanced equation is

$$3Fe^{2+} + NO_3^- + 4H^+ \rightarrow 3Fe^{3+} + NO + 2H_2O$$

23.39 Since one mole of chloride ion is precipitated per formula unit of the complex, the chlorine atoms must be present as chloride ion. All the other ligands are coordinated to the cobalt. An appropriate formula is

$$[Co(NH_3)_4(NO_2)_2]Cl$$

23.41 (a) 4. There are four cyanide groups coordinated to the gold atom.
(b) 6. There are four ammonia molecules and two water molecules coordinated to the cobalt.
(c) 4. Each of the ethylenediamine molecules bonds to the gold atom through two nitrogen atoms.
(d) 6. Each ethylenediamine molecule bonds to the chromium atom through two nitrogen atoms and the oxalate ion bonds to the chromium through two oxygen atoms.

23.43 (a) The charge on the $Ni(CN)_4^{2-}$ ion is –2, to balance the charge of +2 from the $2K^+$ ions. Each cyanide ion has a charge of –1. The sum of the oxidation number of nickel and the charge on the cyanide ions must equal the charge, so

$$-2 = 4 \times (-1) + 1 \times [\text{Ox\# (Ni)}]$$
$$\text{Ox\# (Ni)} = -2 - (-4) = +2$$

(b) The charge on ethylenediamine is 0, so the oxidation of Mo is equal to the charge on the complex ion.

$$\text{Ox\# (Mo)} = +3$$

(c) Oxalate ion has a charge of –2, so

$$\text{Ox\# of Cr} = \text{charge of complex ion} - 3 \times (\text{charge of oxalate ion})$$
$$= (-3) - 3 \times (-2) = +3$$

(d) Chloride ion has a charge of –1, so the charge on the complex ion is +2. The NH_3 ligands are neutral and contribute nothing to the charge of the complex ion.

$$\text{Ox\# of Co} = \text{charge of complex ion} - \text{charge of nitrite ligand}$$
$$= +2 - (-1) = +3$$

23.45 (a) The charge of each chloride ligand is –1; the charge on the oxalate ligand is –2. The ammonia ligands are neutral.

$$\text{Ox\# of Cr} = \text{charge of complex ion} - \text{charge of oxalate} - 2 \times (\text{charge of chloride})$$
$$= -1 - (-2) - 2(-1) = +3$$

(b)

Formula	Name
NH_3	Ammine
Cl^-	Chloro
$C_2O_4^{2-}$	Oxalato

(c) 6. The chromium atom has one bond to each of the NH_3 ligands and Cl^- ligands and two bonds are to the $C_2O_4^{2-}$ ligand.

(d) If each NH_3 were replaced by one Cl^- and the $C_2O_4^{2-}$ ligand were replaced by 2 Cl^- ligands, there would be a total of 6 Cl^- ligands bonded to a chromium atom in the +3 oxidation state.

$$\text{charge on complex ion} = \text{Ox\# of Cr} + 6 \times (\text{charge on chloride ligand})$$
$$= +3 + 6 \times (-1) = -3$$

23.47 (a) Potassium hexafluoroferrate(III)
(b) Diamminediaquacopper(II) ion
(c) Ammonium aquapentafluoroferrate(III)
(d) Dicyanoargentate(I) ion

23.49 (a) Pentacarbonyliron(0)
(b) dicyanobis(ethylenediamine)rhodium(III)
(c) tetraamminesulfatochromium(III) chloride
(d) tetraoxomanganate(VII) ion [permanganate is the usual name]

23.51 (a) The charge on the complex ion equals

Ox# of Mn + 6 × (charge on cyanide ion) = +3 + 6(−1) = −3.

Hence, the formula is $K_3[Mn(CN)_6]$.

(b) The charge on the complex ion equals

Ox# of Zn + 4 × (charge on cyanide ion) = 2 + 4(−1) = −2.

Hence, the formula is $Na_2[Zn(CN)_4]$.

(c) The charge on the complex ion equals

Ox of Co + 2 × (charge on chloride ion).

Note that the ammine ligand (NH_3) is neutral. We get +3 + 2(−1) = +1.
The formula is $[Co(NH_3)_4Cl_2]NO_3$.

(d) The charge on the cation equals the Ox# of Cr (+3). The charge on the anion equals

Ox# of Cu + 4 × (charge on chloride ion) = +2 + 4(−1) = −2.

The formula is $[Cr(NH_3)_6]_2[CuCl_4]_3$.

23.53 (a) Linkage isomerism. In one case, the SCN^- ligand is attached to Co at the S; in the other, it is attached at the N.
(b) Coordination isomerism. In one case, the NH_3 ligands are attached to Co; in the other, they are attached to Cr.
(c) Hydrate isomerism. In the first compound, both H_2O molecules are coordinated to the Co atom; in the other, only one H_2O is directly attached to Co.
(d) Ionization isomerism. In the first compound, a chloride ion is produced upon ionization; in the other, a nitrite ion is produced.

23.55 The given compound must consist of a chloride ion (which can be precipitated with $AgNO_3$ soln) and a $[Co(NH_3)_4Br_2]^+$ ion, giving $[Co(NH_3)_4Br_2]Cl$. An ionization isomer is $[Co(NH_3)_4BrCl]Br$.

23.57 (a)

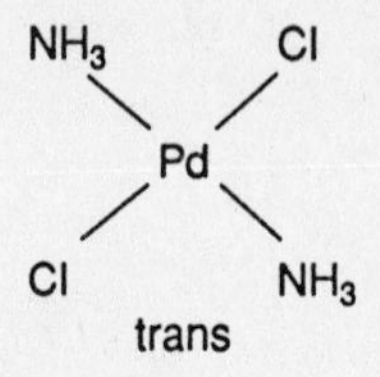

(b) No geometric isomerism.
(c) No geometric isomerism.

(d)

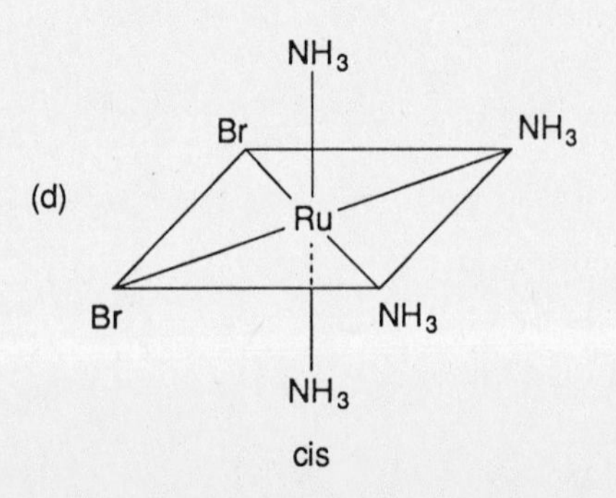

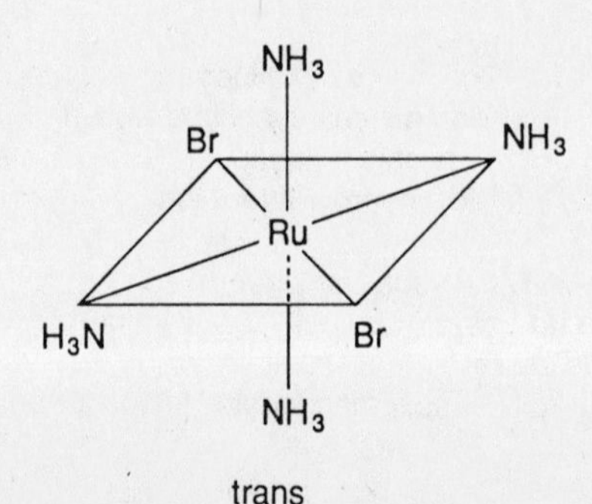

23.59 (a)

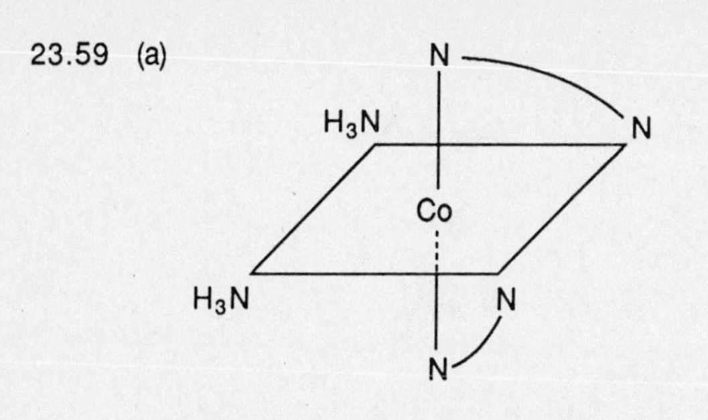

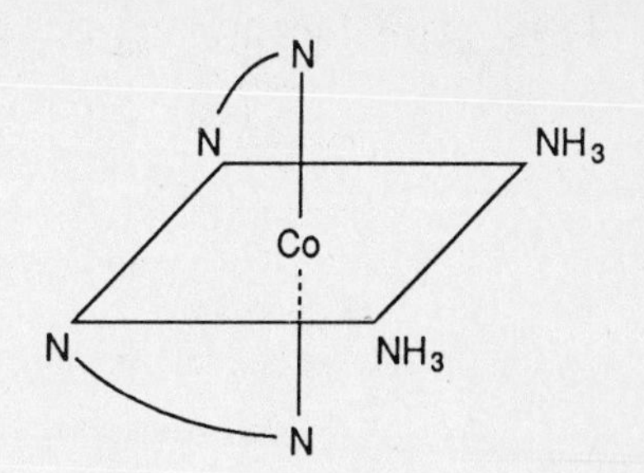

(b) No optical isomers.

23.61

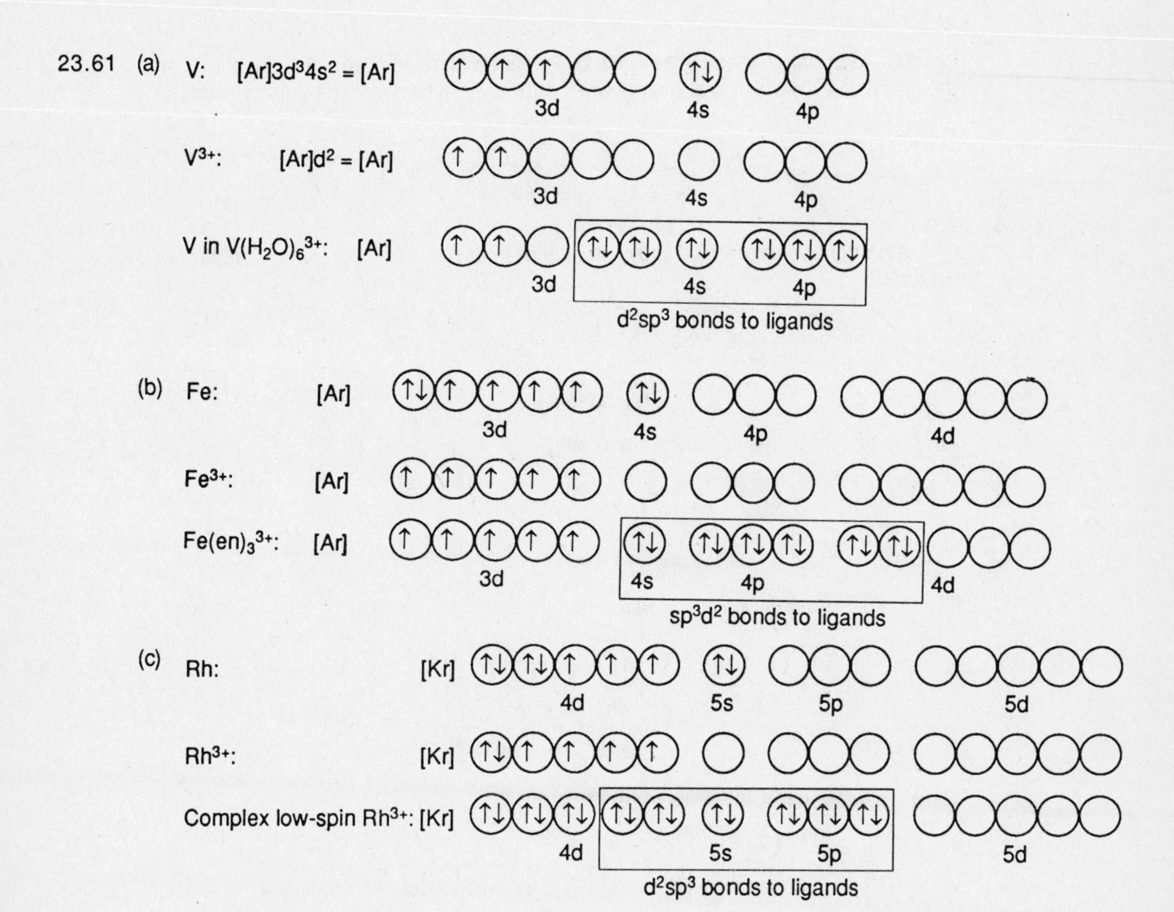

In the low-spin complex, two of the unpaired electrons in 4d orbitals are paired with two others to provide two empty d orbitals for cyanide ligands.

23.63 (a) Pt: [Xe] $4f^{14}5d^{8}6s^{2}$

= [Xe] $4f^{14}$ (↑↓)(↑↓)(↑↓)(↑)(↑) (↑↓) ( )( )( )

5d 6s 6p

Pt^{2+}: [Xe] $4f^{14}$ (↑↓)(↑↓)(↑↓)(↑)(↑) ( ) ( )( )( )

If there are no unpaired electrons in the complex, the configuration of the complexed Pt must be

[Xe] $4f^{14}$ (↑↓)(↑↓)(↑↓)(↑↓)(↑↓) (↑↓) (↑↓)(↑↓)( )

5d 6s 6p

dsp^2 bonds to ligand

in which the unpaired d electrons have been paired up to provide a 5d orbital for bonding to a ligand in a dsp^2 hybrid orbital. The geometry for dsp^2 hybridization is square planar.

(b) Co^{2+}: [Ar] (↑↓)(↑↓)(↑)(↑)(↑) ( ) ( )( )( )

3d 4s 4p

If there is only one unpaired electron in the complex, the configuration of the complexed Co must be

[Ar] (↑↓)(↑↓)(↑↓)(↑)(↑↓) (↑↓) (↑↓)(↑↓)( )

3d 4s 4p

dsp^2 bonds to ligands

dsp^2 hybridization = square planar geometry

(c) Fe^{3+}: [Ar] (↑)(↑)(↑)(↑)(↑) ( ) ( )( )( )

3d 4s 4p

If there are 5 unpaired electrons in the complex, the configuration of the Fe must be

[Ar] (↑)(↑)(↑)(↑)(↑) (↑↓) (↑↓)(↑↓)(↑↓)

3d 4s 4p

sp^3 bonds to ligands

sp^3 hybridization = tetrahedral geometry

(d) Co^{2+}: [Ar] (↑↓)(↑↓)(↑)(↑)(↑) ( ) ( )( )( )

3d 4s 4p

If there are 3 unpaired electrons in the complex, the configuration of the cobalt must be

[Ar] (↑↓)(↑↓)(↑)(↑)(↑) (↑↓) (↑↓)(↑↓)(↑↓)

3d 4s 4p

sp^3 bonds to ligands

sp^3 hybrid = tetrahedral

23.65 In an octahedral field the d orbitals are split so that three of them are of a lower energy than the remaining two.

increasing E ↑   __ __
__ __ __

(a) V^{3+} has 2 d electrons arranged as shown:

__ __
↑ ↑ __

There are 2 unpaired electrons.

(b) Co^{2+} has 7 d electrons. In the high-spin case they are arranged as follows:

↑ ↑
↑↓ ↑↓ ↑

There are 3 unpaired e^-.

(c) Mn^{3+} has 4 d electrons. In the high-spin case they are arranged as follows:

__ __
↑↓ ↑ ↑

There are 2 unpaired e^-.

23.67 (a) Pt^{2+} has 8 electrons in the 5d subshell. Since the complex is diamagnetic, the crystal field felt by the d orbitals is most likely square planar (no low-spin tetrahedral complexes are known). The arrangement of the d electrons is

__
↑↓
↑↓
↑↓ ↑↓

(b) Co^{2+} has 7 d electrons. One unpaired electron implies a low-spin complex. A square planar field will lead to low-spin complexes, so the geometry is probably square planar, with the d electrons arranged as follows:

__
↑
↑↓
↑↓ ↑↓

(c) Fe^{3+} has 5 d electrons. If they are all unpaired, that is, a high-spin complex, the field felt by the metal ion is most likely tetrahedral with the d electrons arranged as follows:

↑ ↑ ↑

↑ ↑

(d) Co^{2+} has 7 d electrons. Three unpaired electrons imply a high-spin complex. A tetrahedral field will lead to high-spin complexes, so the geometry is probably tetrahedral with the d electrons arranged as follows:

↑ ↑ ↑

↑↓ ↑↓

23.69 purple (from Table 23.9)

23.71 Yes. According to the spectrochemical series, H_2O is a more weakly bonding ligand than NH_3, so Δ should decrease. The wavelength of the absorption should increase ($\lambda = hc/\Delta$). The light absorbed by $Co(NH_3)_6{}^{3+}$ is violet-blue and the replacement of one NH_3 by H_2O shifts this toward blue. Thus the observed (complementary) color is shifted toward red, as is observed.

23.73 $$\Delta = \frac{hc}{\lambda} = \frac{(6.626 \times 10^{-34}\ J \cdot s)\ (2.998 \times 10^{8}\ m/s)}{(500 \times 10^{-9} m)} \times 6.02 \times 10^{23}/mol$$

$$= 2.3\underline{9}17 \times 10^5 = 2.39 \times 10^5\ J/mol\ \ (239\ kJ/mol)$$

23.75 The color in transition metal complexes is due to absorption of light when a d electron moves to a higher energy level. Since Sc^{3+} has no d electrons, it is expected to be colorless.

23.77 If $[Co(NH_3)_4Cl_2]^+$ had a regular planar hexagonal geometry, three geometric isomers would be expected.

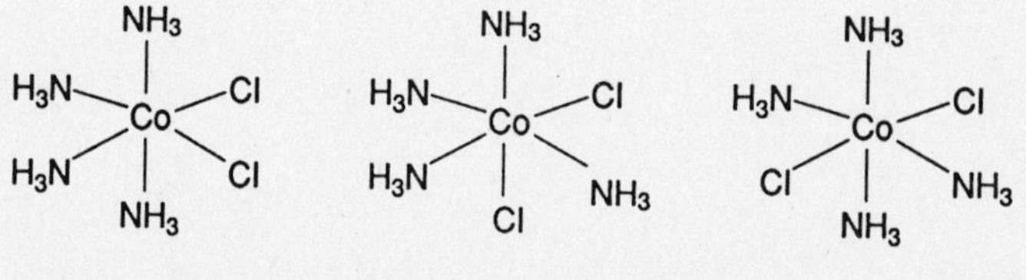

NH3, H3N, Cl, Co, H3N, Cl, NH3   NH3, H3N, Cl, Co, H3N, NH3, Cl   NH3, H3N, Cl, Co, Cl, NH3, NH3

The known existence of only two isomers of the complex, by itself, is not sufficient to rule out hexagonal geometry. There is the possibility that the third isomer has simply not been made. Other information is required to eliminate hexagonal geometry as a possibility.

23.79 $$k_d = 2.1 \times 10^{-13} = \frac{[Cu^{2+}]\ [NH_3]^4}{[Cu(NH_3)_4{}^{2+}]} \text{ for } Cu(NH_3)_4{}^{2+} \rightleftharpoons Cu^{2+}\ (aq) + 4NH_3\ (aq)$$

Concentration (M)	Cu^{2+} (aq)	+	$4NH_3$ (aq) $\rightleftharpoons$	$Cu(NH_3)_4{}^{2+}$ (aq)
Start	0.10		0.40	0
Change	$-x$		$-4x$	$+x$
Equilibrium	$0.10 - x$		$0.40 - 4x$	x

Substituting into the equilibrium expression

$$2.1 \times 10^{-13} = \frac{(0.10 - x)\,[4(0.10 - x)]^4}{x}$$

To simplify, let y = 0.10 – x. Then, x = 0.10 – y. The above equation becomes

$$2.1 \times 10^{-13} = \frac{y\,(4y)^4}{(0.10 - y)}$$

Assume that y is small compared to 0.1. Then, 0.10 – y $\simeq$ 0.10. Then,

$$2.1 \times 10^{-13} \simeq \frac{4^4 y^5}{0.10}$$

$$y^5 \simeq \frac{(2.1 \times 10^{-13})\,(0.10)}{4^4} \simeq 8.\underline{2}03 \times 10^{-17}$$

$$y \simeq \sqrt[5]{1.953 \times 10^{-17}} \simeq 6.\underline{0}64 \times 10^{-4}$$

Going back to the assumption above, we find that to 2 significant figures, 0.10 – 0.00046 = 0.10. The assumption was valid.

At equilibrium:

$[Cu^{2+}] = (0.10 - x)\ M = y = 6.\underline{0}64 \times 10^{-4} = 6.1 \times 10^{-4}\ M$

$[NH_3] = 4(0.10 - x)\ M = 4y = 4 \times 6.\underline{0}64 \times 10^{-4} = 2.\underline{4}3 \times 10^{-3} = 2.4 \times 10^{-3}\ M$

$[Cu(NH_3)_4{}^{2+}] = x = (0.10 - y) = 0.10\ M$

CHAPTER 24

ORGANIC CHEMISTRY

SOLUTIONS TO EXERCISES

24.1 (a) The longest continuous chain is numbered as follows:

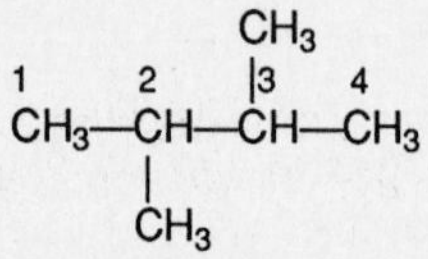

$$\overset{1}{CH_3}-\overset{2}{CH}(CH_3)-\overset{3}{CH}(CH_3)-\overset{4}{CH_3}$$

The name of the compound is 2,3-dimethylbutane.

(b) The longest continuous chain is numbered as follows:

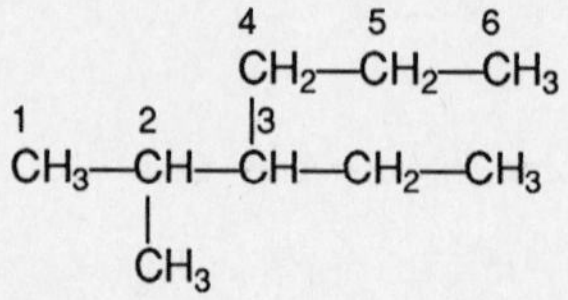

$$\overset{1}{CH_3}-\overset{2}{CH}(CH_3)-\overset{3}{CH}(\overset{4}{CH_2}-\overset{5}{CH_2}-\overset{6}{CH_3})-CH_2-CH_3$$

The name of the compound is 3-ethyl-2-methylhexane.

24.2 First write out the carbon skeleton for octane

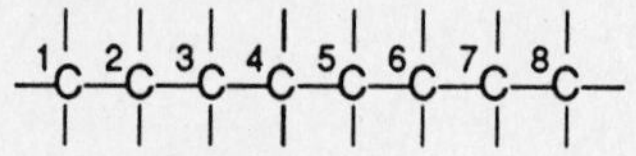

$$\overset{1}{C}-\overset{2}{C}-\overset{3}{C}-\overset{4}{C}-\overset{5}{C}-\overset{6}{C}-\overset{7}{C}-\overset{8}{C}$$

Then attach the alkyl groups

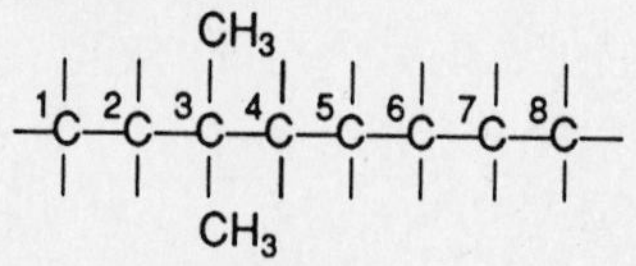

$$\overset{1}{C}-\overset{2}{C}-\overset{3}{C}(CH_3)_2-\overset{4}{C}-\overset{5}{C}-\overset{6}{C}-\overset{7}{C}-\overset{8}{C}$$

Finally, fill out the structure with H atoms

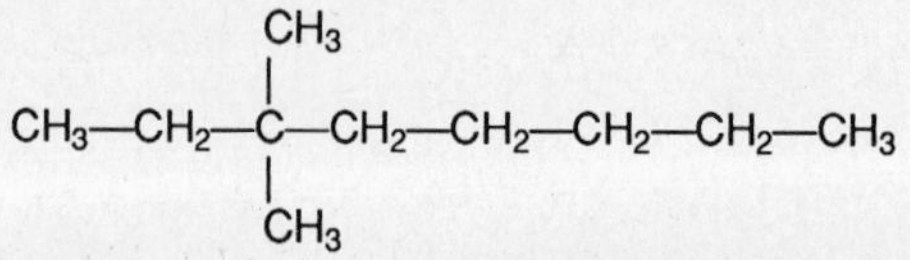

$$CH_3-CH_2-C(CH_3)_2-CH_2-CH_2-CH_2-CH_2-CH_3$$

24.3 (a) The numbering of the carbon chain is

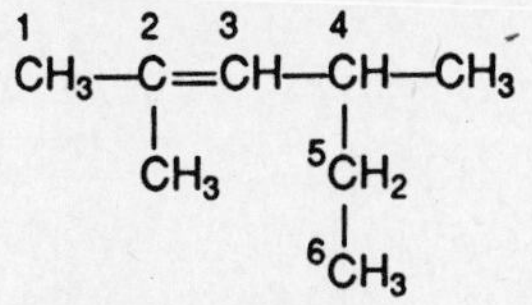

```
1     2  3     4
CH3—C=CH—CH—CH3
     |        |
     CH3     5CH2
              |
             6CH3
```

Since the longest chain containing a double bond has six carbons, this is a hexene. It is a 2-hexene because the double bond is between carbons 2 and 3. The name of the compound is 2,4-dimethyl-2-hexene.

(b) The numbering of the longest chain with a double bond is

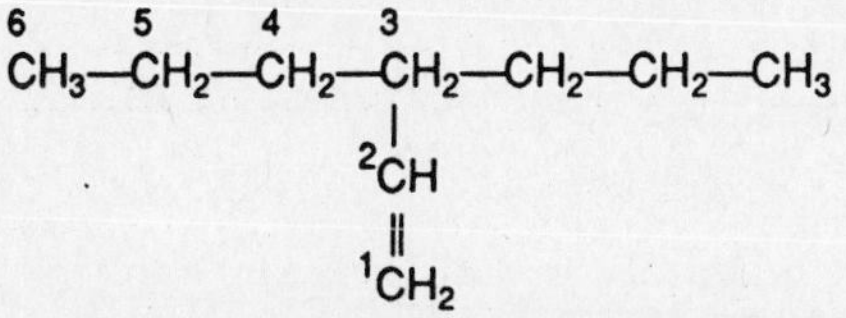

```
6     5     4     3
CH3—CH2—CH2—CH2—CH2—CH2—CH3
                 |
                2CH
                 ||
                1CH2
```

The longest chain containing the double bond has six carbon atoms; therefore, this is a hexene. It is a 1-hexene because the double bond is between carbons 1 and 2. The name of the compound is 3-propyl-1-hexene.

24.4 First, write out the carbon skeleton for heptene

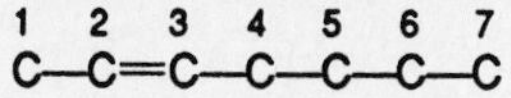

```
1  2  3  4  5  6  7
C—C=C—C—C—C—C
```

Then add the alkyl groups

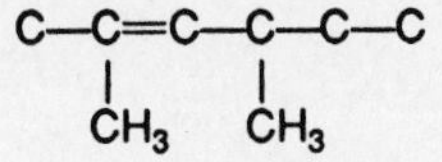

```
C—C=C—C—C—C
   |     |
   CH3   CH3
```

Finally, add the H atoms

```
CH3—C=CH—CH2—CH—CH2—CH3
     |         |
     CH3       CH3
```

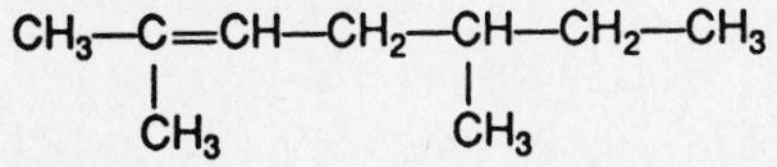

26.5 (a) Geometric isomers are possible.

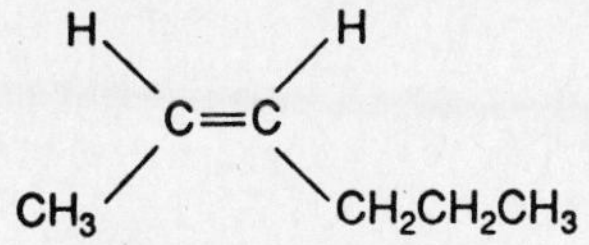

```
H          H                 H          CH2CH2CH3
  \       /                    \       /
    C=C                          C=C
  /       \                    /       \
CH3        CH2CH2CH3         CH3        H

   cis-2-hexene                trans-2-hexene
```

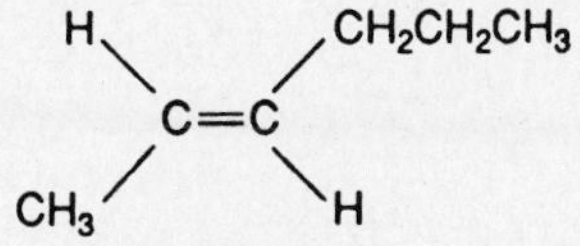

(b) No geometric isomers are possible because there are two H atoms attached to the second carbon of the double bond.

24.6 (a) There are only 3 carbons in the chain. The compound is propyne.
(b) The longest continuous chain containing the double bond has 5 carbon atoms. The compound is 3-methyl-1-pentyne.

24.7 (a) This compound has an ethyl group attached to a benzene ring.

C_6H_5—CH_2CH_3

(b) This compound has one phenyl group attached to each carbon atom in the ethane molecule.

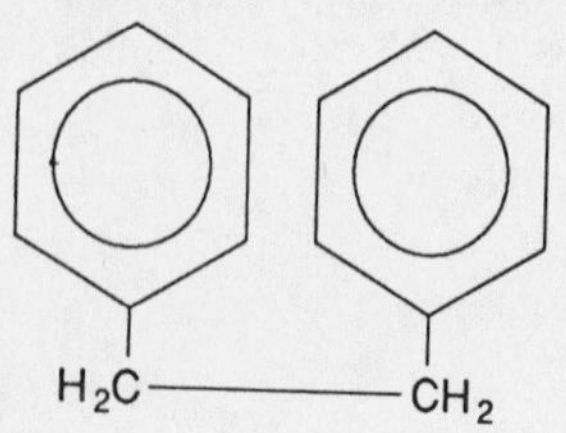

24.8 According to Markownikoff's rule, when HBr is added across the double bond in 1-butene, the H will add to carbon 1 (the C atom with the most bonds to H atoms) and the Br will add to carbon 2.

$$H_2C{=}CH{-}CH_2{-}CH_3 + HBr \rightarrow H_3C{-}\underset{\displaystyle Br}{\underset{|}{CH}}{-}CH_2{-}CH_3$$

The product is 2-bromobutane.

24.9 Since the compound has an —OH group, it is an alcohol. The longest carbon chain in the molecule has 6 carbons.

$$\overset{6}{CH_3}\overset{5}{CH_2}\overset{4}{CH_2}\overset{\displaystyle OH}{\overset{3|}{\underset{\displaystyle CH_2CH_3}{\underset{|}{C}}}}\overset{2}{CH_2}\overset{1}{CH_3}$$

The name of the compound is 3-ethyl-3-hexanol.

24.10 (a) Dimethyl ether (b) Methyl ethyl ether

24.11 (a) There are two alkyl groups attached to the carbonyl; therefore, the compound is a ketone. There are 5 carbon atoms in the chain. The name of the compound is 2-pentanone.
(b) There is a hydrogen atom attached to the carbonyl, so the compound is an aldehyde. The numbering of the stem carbon chain is

$$H{-}\overset{\displaystyle O}{\overset{\|1}{C}}{-}\overset{2}{CH_2}\overset{3}{CH_2}\overset{4}{CH_3}$$

The name of the compound is butanal.

24.12 The oxidation half-reaction is

$$CH_3{-}\underset{\displaystyle OH}{\underset{|}{CH_2}} \rightarrow CH_3{-}\underset{\displaystyle O}{\underset{\|}{C}}{-}H + 2H^+ + 2e^-$$

The half-reaction for the reduction of permanganate is

$$MnO_4^- + 8H^+ + 5e^- \rightarrow Mn^{2+} + 4H_2O$$

Multiplying the first reaction by 5 and the second by 2 and adding them gives the final balanced equation for the oxidation of ethanol to acetaldehyde.

$$5CH_3\text{—}\underset{\displaystyle OH}{\underset{|}{CH_2}} + 2MnO_4^- + 6H^+ \rightarrow 5CH_3\underset{\displaystyle O}{\underset{\|}{CH}} + 2Mn^{2+} + 8H_2O$$

24.13 (a) This is a tertiary alcohol, which is unreactive with most oxidizing agents.

(b) Aldehydes are easily oxidized to carboxylic acids.

$$CH_3\text{—}\underset{\displaystyle CHO}{\underset{|}{CH}}\text{—}CH_2\text{—}CH_3 + (O) \rightarrow CH_3\text{—}\underset{\displaystyle \underset{\displaystyle OH}{\underset{|}{C{=}O}}}{\underset{|}{CH}}\text{—}CH_2\text{—}CH_3$$

2-methyl-butanoic acid

(c) Aldehydes can also be reduced to alcohols.

$$CH_3\underset{\displaystyle CHO}{\underset{|}{CH}}CH_2CH_3 + 2(H) \rightarrow CH_3\underset{\displaystyle \underset{\displaystyle OH}{\underset{|}{CH_2}}}{\underset{|}{CH}}CH_2CH_3$$

2-methyl-1-butanol

(d) Secondary alcohols can be oxidized to form ketones.

$$CH_3\underset{\displaystyle OH}{\underset{|}{CH}}CH_2CH_3 + (O) \rightarrow CH_3\text{—}\underset{\displaystyle O}{\underset{\|}{C}}\text{—}CH_2CH_3 + H_2O$$

2-butanone

24.14 Esters are often prepared by heating an alcohol and a carboxylic acid in the presence of an inorganic acid. The choice of acid and alcohol depend on the ester that is desired. The R groups on the acid and alcohol must correspond to those in the ester.

$$CH_3\text{—}CH_2\text{—}\overset{\displaystyle O}{\overset{\|}{C}}\text{—}OH + HO\text{—}CH_3 \xrightarrow{H^+} CH_3CH_2\text{—}\overset{\displaystyle O}{\overset{\|}{C}}\text{—}O\text{—}CH_3 + H_2O$$

24.15 This addition polymer is formed when vinylidene chloride adds to itself across the double bond.

$$\cdots + CH_2{=}CCl_2 + CH_2{=}CCl_2 + CH_2{=}CCl_2 \rightarrow \text{—}CH_2\text{—}CCl_2\text{—}CH_2\text{—}CCl_2\text{—}CH_2\text{—}CCl_2\text{—}$$

ANSWERS TO REVIEW QUESTIONS

Note: All carbon atoms have been omitted in ring structures for simplicity.

24.1 The formula of an alkane with 30 carbon atoms is $C_{30}H_{62}$.

24.2 The molecules increase regularly in molecular weight. We, therefore, expect their intermolecular forces and thus their melting points to increase.

24.3 $H_3C—CH_2—CH_2—CH_2—CH_2—CH_3$

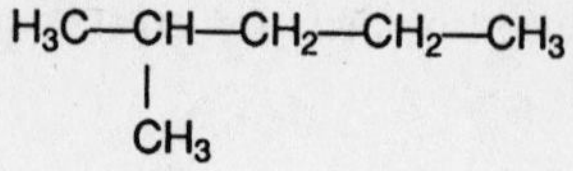

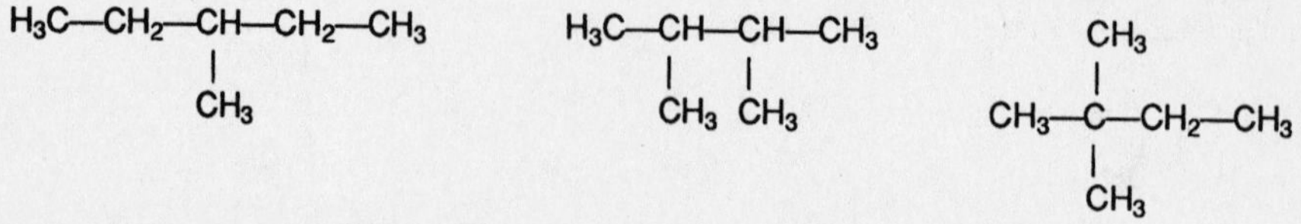

24.4 The structures of a seven-carbon alkane, cycloalkane, alkene, and aromatic hydrocarbon are

$CH_3—CH_2—CH_2—CH_2—CH_2—CH_2—CH_3$

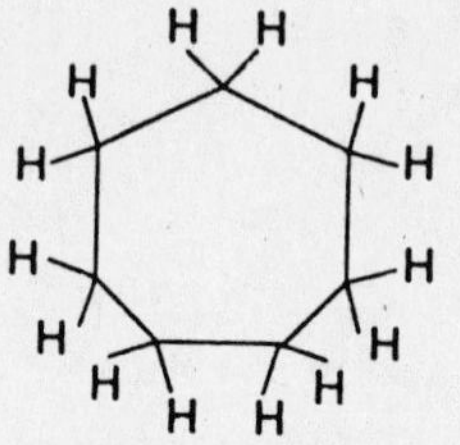

$CH_2{=}CH—CH_2—CH_2—CH_2—CH_2—CH_3$

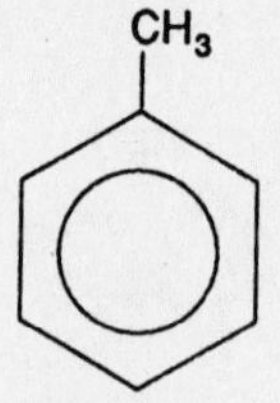

(toluene)

24.5 The two isomers of 2-butene are the cis and trans geometric isomers:

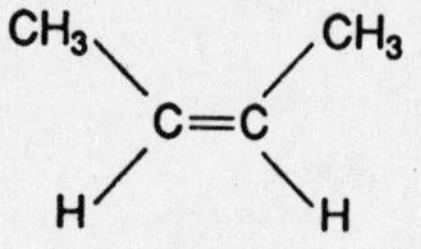

cis-2-butene

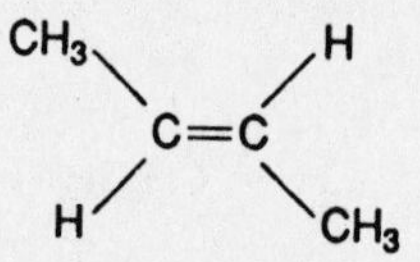

trans-2-butene

In the cis-2-butene, the two methyl groups are on the same side of the double bond; in the trans isomer, they are on opposite sides.

24.6 The structural formulas for the isomers of ethyl-methylbenzene are

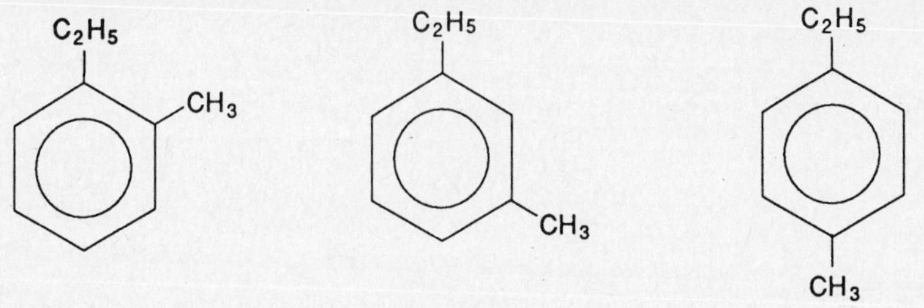

24.7 Methane: source, natural gas; use, home fuel.
Octane: source, petroleum; use, auto fuel.
Ethylene: source, petroleum refining; use, chemical industry raw material.
Acetylene: source, methane; use, acetylene torch.

24.8 CH_3CH_2Cl CH_2ClCH_2Cl $CHCl_2CHCl_2$ CCl_3CCl_2
CH_3CHCl_2 $CH_2ClCHCl_2$ $CHCl_2CCl_3$
CH_3CCl_3 CH_2ClCCl_3

24.9 A substitution reaction is a reaction in which part of the reagent molecule is substituted for a hydrogen atom on a hydrocarbon or hydrocarbon group. For example,

$$CH_4 + Cl_2 \rightarrow CH_3Cl + HCl$$

An addition reaction is a reaction in which parts of the reagent are added to each carbon atom of a carbon-carbon multiple bond, which then becomes a C—C single bond. For example,

$$CH_2{=}CH_2 + Br_2 \rightarrow CH_2Br{-}CH_2Br$$

24.10 The major product of HCl plus acetylene should be $HCCl_2{-}CH_3$ since Markownikoff's rule predicts this.

24.11 The octane number scale gives the "antiknock" characteristics of a gasoline. If a gasoline begins to burn before the sparkplug is ignited, the engine "knocks." The octane number scale is based on heptane and 2,2,4-trimethylpentane, given octane numbers of 0 and 100, respectively. The higher the octane number, the better the antiknock characteristics of the gasoline.

24.12 A functional group is a reactive portion of a molecule that undergoes predictable reactions no matter what the rest of the molecule is like. An example is a C=C bond, which always reacts with bromine or other addition reagents to add part of each reagent to each carbon atom.

24.13 An aldehyde is different from a ketone, carboxylic acid, and ester in that a hydrogen atom is always attached to the carbonyl group in addition to a hydrocarbon group.

24.14 Methanol: source, $CO + H_2$; use, solvent.
Ethanol: source, fermentation of glucose; use, solvent.
Ethylene glycol: source, ethylene; use, antifreeze.
Glycerol: source, from soap making; use, foods.
Formaldehyde: source, oxidation of methanol; use, plastics and resins.

24.15 (a) CO is a carbonyl group.
(b) $CH_3O—C$ is an ether group.
(c) C=C is a double bond.
(d) COOH is a carboxylic acid group.
(e) CHO is an aldehyde (carbonyl).
(f) CH_2OH, or –OH is a primary alcohol.

24.16 A primary alcohol is oxidized to an acid in two steps. An overall example reaction is

$$CH_3CH_2OH + 2[O] \rightarrow CH_3CO_2H + H_2O$$

A secondary alcohol is oxidized to a ketone. A reaction is

$$CH_3CHOHCH_3 + [O] + H^+ \rightarrow CH_3COCH_3 + H_2O$$

Tertiary alcohols are not oxidized.

An aldehyde is oxidized to an acid. The reaction is

$$CH_3CHO + [O] \rightarrow CH_3CO_2H$$

A ketone is not oxidized.

24.17 Ethyl ethanoate (acetate) is $CH_3CH_2OOC—CH_3$; methyl propanoate is $CH_3CH_2CO_2CH_3$.

24.18 $CH_3CO_2C_2H_5\,(l) + NaOH\,(aq) \rightarrow CH_3CO_2^-\,(aq) + Na^+\,(aq) + C_2H_5OH\,(l)$

24.19 The source of basicity of an amine is the pair of electrons on the nitrogen. An example of amine-acid reaction is

$$(CH_3CH_2)_3N\!: + CH_3CO_2H \rightarrow CH_3CO_2^- + (CH_3CH_2)_3NH^+$$

24.20 A condensation polymer is formed by splitting out a small molecule such as water between two molecules (or monomers) whereas addition polymers form when molecules add to one another, giving a chain.

Example of addition: $2nH_2C{=}CH_2 \rightarrow [–H_2C—CH_2—CH_2—CH_2–]_n$

Example of condensation:

$$2nHOROH + nHOOC—R—COOH \rightarrow [—O—R—O—CO—R—CO—O—R—O]_n + 2nH_2O$$

SOLUTIONS TO PRACTICE PROBLEMS

24.21 (a)

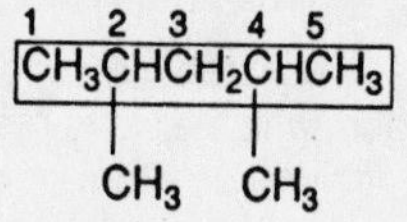

```
 1  2  3   4  5
[CH3CHCH2CHCH3]   longest chain     2,4-dimethylpentane
     |     |
    CH3   CH3
```

(b)

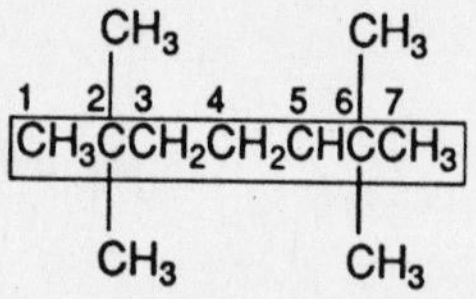

```
    CH3       CH3
     |         |
 1  2 3   4   5 6 7
[CH3CCH2CH2CHCCH3]   longest chain     2,2,6,6-tetramethylheptane
     |         |
    CH3       CH3
```

(c) 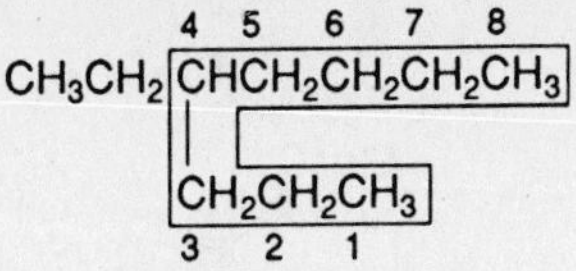 $CH_3CH_2\overset{4}{C}H\overset{5}{C}H_2\overset{6}{C}H_2\overset{7}{C}H_2\overset{8}{C}H_3$ longest chain 4-ethyloctane
with $\overset{3}{C}H_2\overset{2}{C}H_2\overset{1}{C}H_3$ on carbon 4

(d) 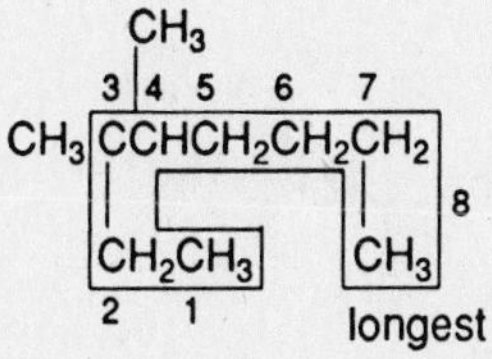 $CH_3\overset{3}{C}(CH_3)\overset{4}{C}H\overset{5}{C}H_2\overset{6}{C}H_2\overset{7}{C}H_2\overset{8}{C}H_3$ 3,4-dimethyloctane
with $\overset{2}{C}H_2\overset{1}{C}H_3$ on carbon 3; longest chain

24.23 (a) 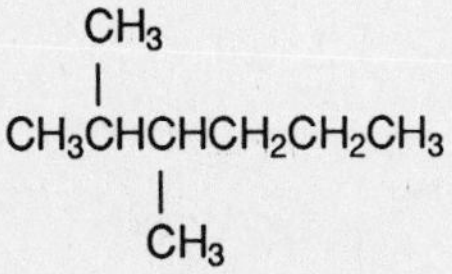 $CH_3CH(CH_3)CH(CH_3)CH_2CH_2CH_3$

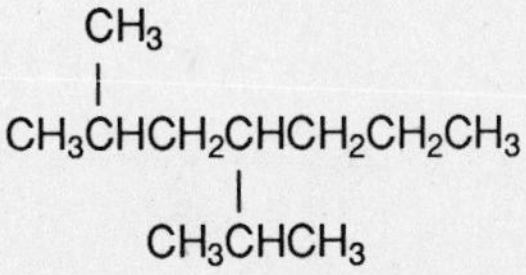

(b) 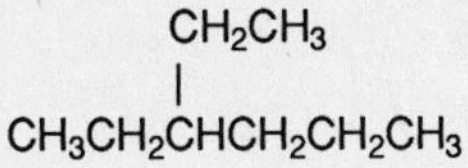 $CH_3CH_2CH(CH_2CH_3)CH_2CH_2CH_3$

(c) $CH_3CH(CH_3)CH_2CH(CH(CH_3)CH_3)CH_2CH_2CH_3$

(d) 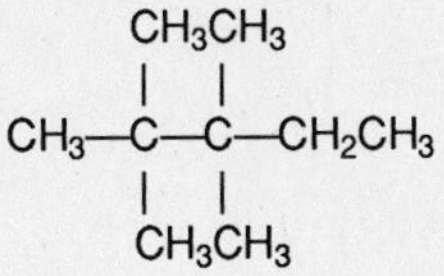 $CH_3—C(CH_3)_2—C(CH_3)_2—CH_2CH_3$

24.25 (a) 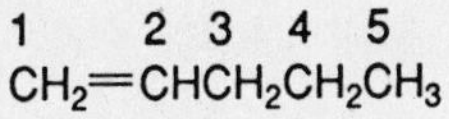 $\overset{1}{C}H_2{=}\overset{2}{C}H\overset{3}{C}H_2\overset{4}{C}H_2\overset{5}{C}H_3$ 1-pentene

(b) 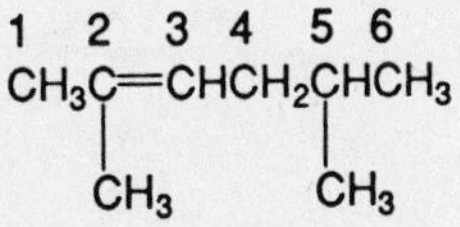 $\overset{1}{C}H_3\overset{2}{C}(CH_3){=}\overset{3}{C}H\overset{4}{C}H_2\overset{5}{C}H(CH_3)\overset{6}{C}H_3$ 2,5-dimethyl-2-hexene

24.27 (a) 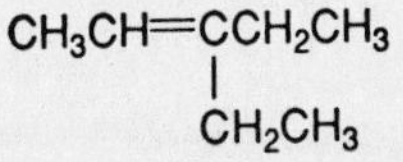 $CH_3CH{=}C(CH_2CH_3)CH_2CH_3$

(b) 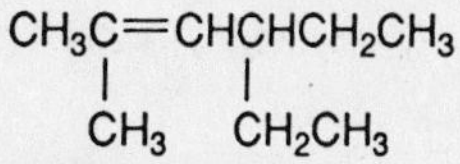 $CH_3C(CH_3){=}CHCH(CH_2CH_3)CH_2CH_3$

24.29 (a) 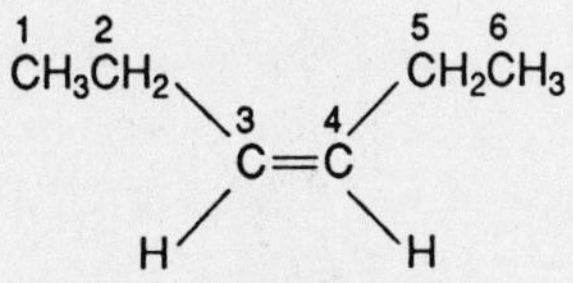 $\overset{1}{C}H_3\overset{2}{C}H_2(H)\overset{3}{C}{=}\overset{4}{C}(H)\overset{5}{C}H_2\overset{6}{C}H_3$ (both CH_2CH_3 groups on the same side)

cis-3-hexene

$CH_3CH_2(H)C{=}C(H)CH_2CH_3$ (the CH_2CH_3 groups on opposite sides)

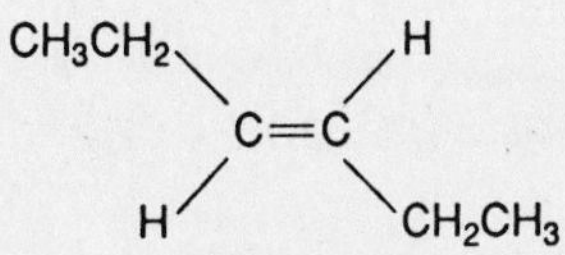

trans-3-hexene

(b) $CH_3(3)$, $H(4)$; $C{=}C$; CH_3CH_2 (1, 2), CH_2CH_3 (5, 6)

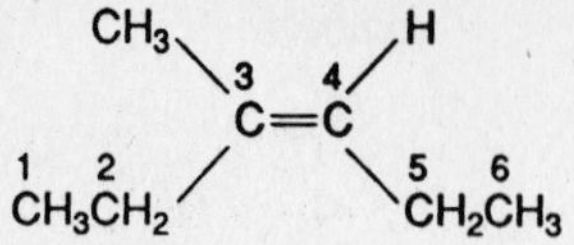

cis-3-methyl-3-hexene

CH_3, CH_2CH_3; $C{=}C$; CH_3CH_2, H

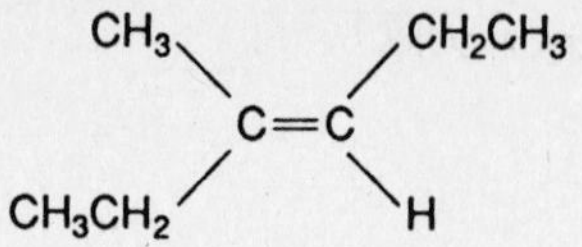

trans-3-methyl-3-hexene

24.31 (a) $\overset{1}{C}H_3\overset{2}{C}{\equiv}\overset{3}{C}\overset{4}{C}H_3$

2-butyne

(b) $\overset{1}{C}H{\equiv}\overset{2}{C}\overset{3}{C}H\overset{4}{C}H_3$ (with CH_3 on C-3)

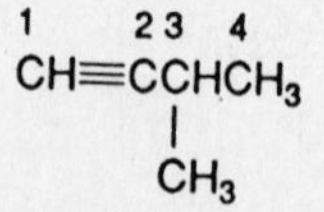

3-methyl-1-butyne

24.33 (a) $(C_6H_5)_3C{-}CH_3$

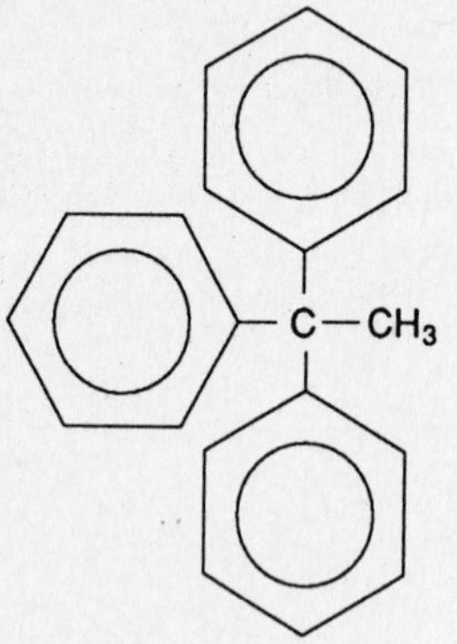

(b) CH_3, CH_2CH_3 (on benzene ring, adjacent positions)

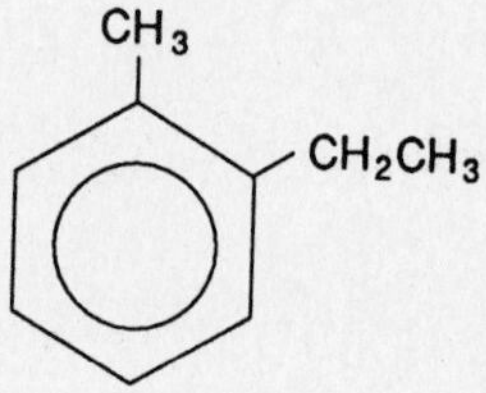

24.35 (a) $C_3H_6 + \frac{9}{2} O_2 \rightarrow 3CO_2 + 3H_2O$

Remove fraction:

$2C_3H_6 + 9O_2 \rightarrow 6CO_2 + 6H_2O$

(b)

$CH_2{=}CH_2 + MnO_4^- + H_2O \rightarrow \underset{OH}{CH_2}{-}\underset{OH}{CH_2} + MnO_2$

Oxidation: $CH_2{=}CH_2 \rightarrow \underset{OH}{CH_2}{-}\underset{OH}{CH_2}$

Balance O: $2OH^- + CH_2{=}CH_2 \rightarrow \underset{OH}{CH_2}{-}\underset{OH}{CH_2}$

Balance e⁻: $2OH^- + CH_2{=}CH_2 \rightarrow \underset{OH}{CH_2}{-}\underset{OH}{CH_2} + 2e^-$

Reduction: $MnO_4^- \rightarrow MnO_2$

Balance O: $MnO_4^- + 2H_2O \rightarrow MnO_2 + 4OH^-$

Balance e⁻: $MnO_4^- + 2H_2O + 3e^- \rightarrow MnO_2 + 4OH^-$

Add half reactions

$$3\left(2OH^- + CH_2{=}CH_2 \rightarrow \underset{\substack{|\\OH}}{CH_2}{-}\underset{\substack{|\\OH}}{CH_2} + 2e^-\right)$$

$$2(MnO_4^- + 2H_2O + 3e^- \rightarrow MnO_2 + 4OH^-)$$

$$\cancel{6}OH^- + 3CH_2{=}CH_2 + 2MnO_4^- + 4H_2O \rightarrow 3\,CH_2(OH){-}CH_2(OH) + 2MnO_2 + \overset{2}{\cancel{8}}OH^-$$

$$3CH_2{=}CH_2 + 2MnO_4^- + 4H_2O \rightarrow 3\,CH_2(OH){-}CH_2(OH) + 2MnO_2 + 2OH^-$$

(c) $CH_2{=}CH_2 + Br_2 \rightarrow CH_2(Br){-}CH_2(Br)$

(d) toluene ($C_6H_5CH_3$) $+ Br_2 \xrightarrow{FeBr_3}$ p-bromotoluene ($CH_3C_6H_4Br$, Br para to CH_3) $+ HBr$

(e) toluene ($C_6H_5CH_3$) $+ HNO_3 \xrightarrow{H_2SO_4}$ p-nitrotoluene ($CH_3C_6H_4NO_2$, NO_2 para to CH_3) $+ H_2O$

24.37 According to Markownikoff's rule, the major product is the one obtained when the H atom adds to the carbon atom of the double bond that already has the more hydrogen atoms attached to it. Therefore, 3-bromo-2-methylpropane is the major product.

$$CH_3{-}\underset{\substack{|\\CH_3}}{C}{=}CH_2 + HBr \rightarrow CH_3{-}\overset{\substack{Br\\|}}{\underset{\substack{|\\CH_3}}{C}}{-}CH_3$$

24.39 $CH_3CH_2CH_2CH_2CH_3 \xrightarrow[\Delta]{Al_2O_3 + SiO_2} CH_4 + CH_2{=}CHCH_2CH_3$

$CH_3CH_2CH_2CH_2CH_3 \xrightarrow[\Delta]{Al_2O_3 + SiO_2} CH_2{=}CH_2 + CH_3CH_2CH_3$

24.41 (a) $CH_3-\overset{O}{\overset{\|}{C}}-CH_2CH_2CH_3$ (C=O circled)

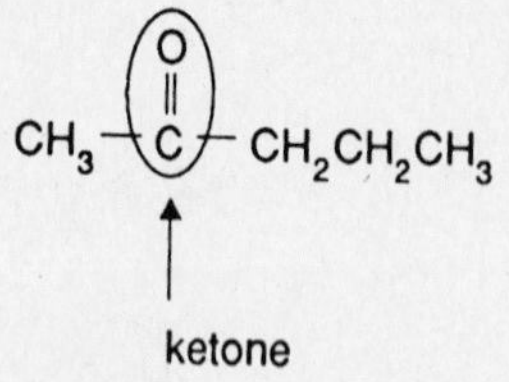

ketone

(b) $CH_3-\underset{}{\overset{OH}{\overset{|}{C}}}HCH_2CH_3$ ← alcohol (OH circled)

(c) $HO\overset{O}{\overset{\|}{C}}-CH_2CH_3$ (HOC=O circled)

carboxylic acid

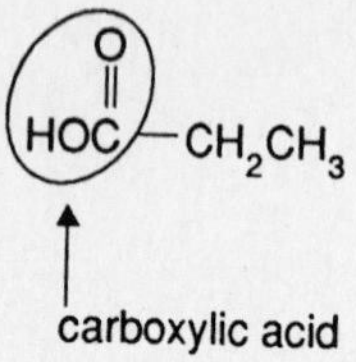

(d) $H-\overset{O}{\overset{\|}{C}}-CH_2CH_3$ (H–C=O circled)

aldehyde

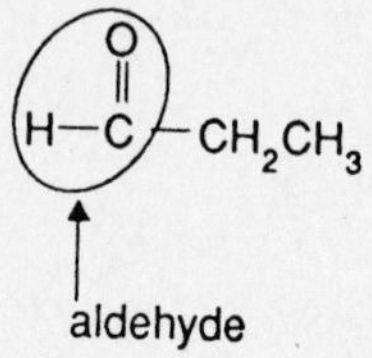

24.43 (a) (HO)– $\overset{1}{C}H_2\overset{2}{C}H_2\overset{3}{C}H_2\overset{4}{C}H_2\overset{5}{C}H_3$ 1-pentanol

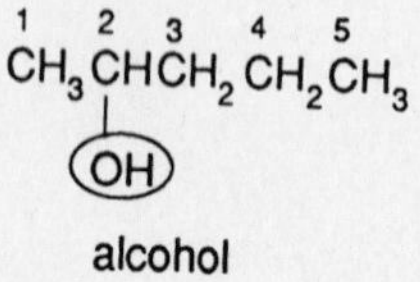

alcohol

(b) $\overset{1}{C}H_3\overset{2}{C}H\overset{3}{C}H_2\overset{4}{C}H_2\overset{5}{C}H_3$ 2-pentanol, with (OH) on C-2

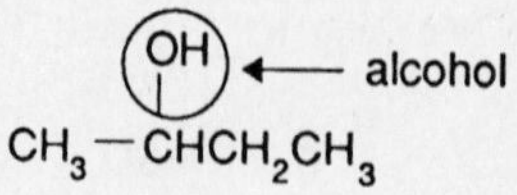

alcohol

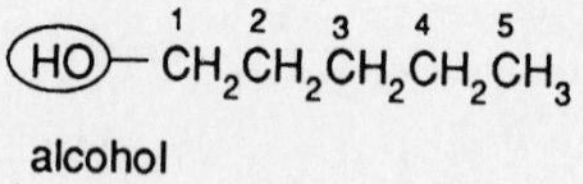

(c) $\overset{5}{C}H_3\overset{4}{C}H_2\overset{3}{C}H_2\overset{2}{C}HCH_2CH_2CH_3$ 2-propyl-1-pentanol, with $\overset{1}{C}H_2$(OH) on C-2 ← alcohol

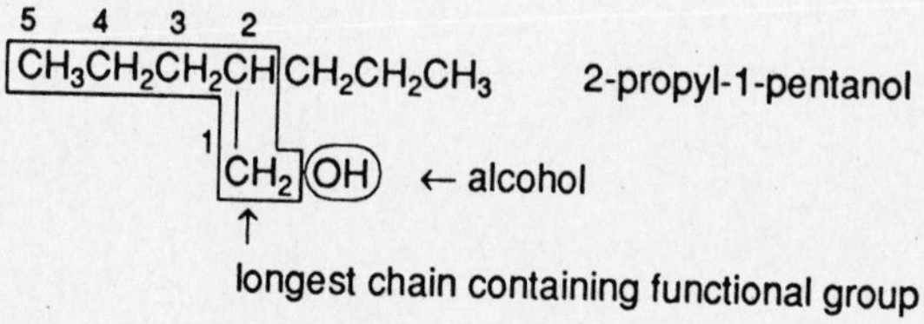

↑ longest chain containing functional group

(d) $\overset{7}{C}H_3\overset{6}{C}H_2\overset{5}{C}H_2\overset{4}{C}H\overset{3}{C}H_2\overset{2}{C}H_2\overset{1}{C}H_3$ 4-heptanol, with (OH) on C-4 — alcohol

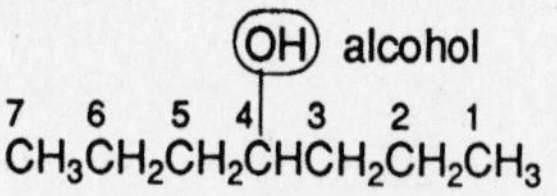

24.45 (a) ©H_3–CH–©H_2CH_3, with OH on the central CH

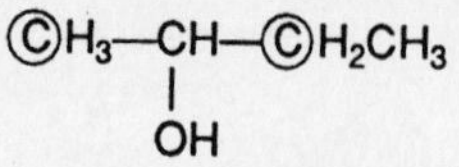

secondary alcohol

(b) ©H_3–C–©H_2CH_3, with ©H_2CH_3 above and OH below the central C

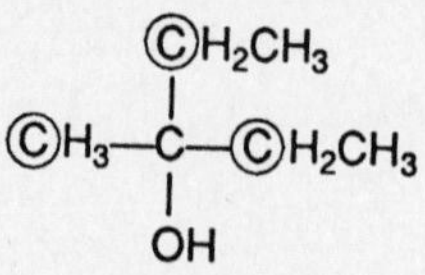

tertiary alcohol

(c) $HOCH_2—\text{Ⓒ}H$ with CH_3 above and CH_3 below the circled C

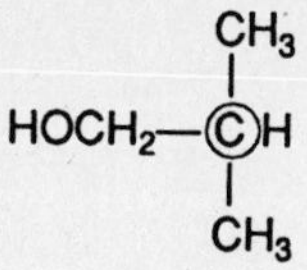

primary alcohol

(d) $HO—CH_2—\text{Ⓒ}—CH_3$ with CH_3 above and CH_3 below the circled C

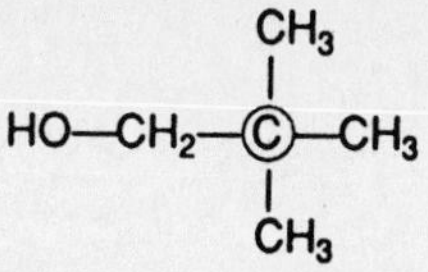

primary alcohol

24.47 (a) $CH_3CH_2—(O)—CH_2CH_2CH_3$ ethyl propyl ether

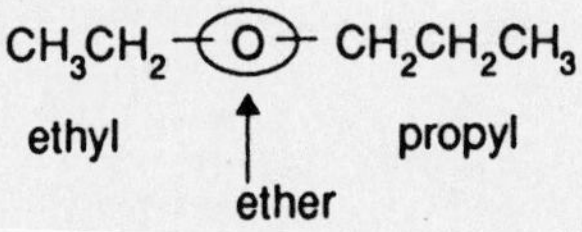

ethyl ↑ ether propyl

(b) $HC—(O)—CH_3$, with CH_3 above and CH_3 below the HC; methyl isopropyl ether

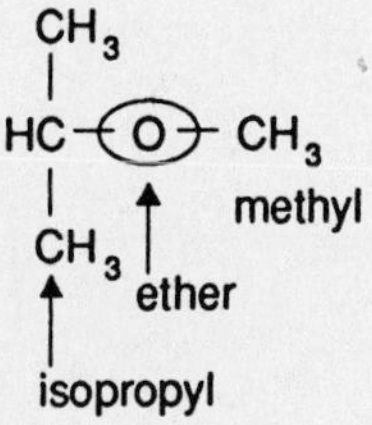

methyl

↑ ether

↑ isopropyl

24.49 (a) $\overset{1}{CH_3}\overset{2}{C}(=O)\overset{3}{CH_2}\overset{4}{CH_3}$ ketone

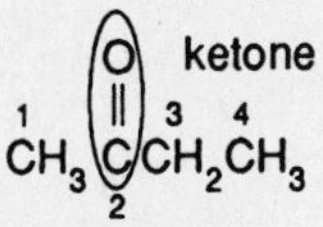

2-butanone

(b) $\overset{4}{CH_3}\overset{3}{CH_2}\overset{2}{CH_2}\overset{1}{CH}(=O)$ aldehyde

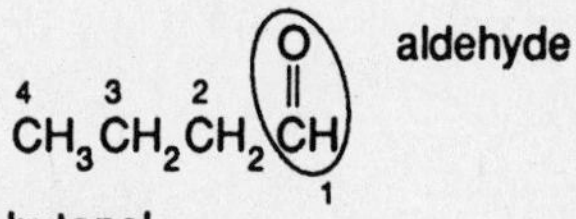

butanal

(c) $\overset{1}{HC}(=O)—\overset{2}{CH_2}\overset{3}{CH_2}\overset{4}{C}(CH_3)_2—\overset{5}{CH_3}$ aldehyde

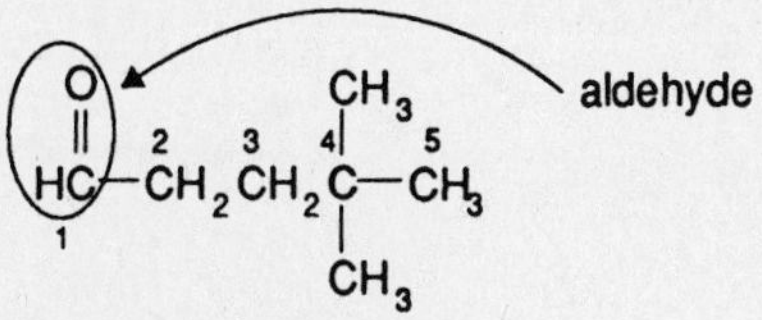

4,4-dimethylpentanal

(d) $CH_3—\overset{3}{CH}(\overset{4}{CH_2}\overset{5}{CH_3})\overset{2}{C}(=O)\overset{1}{CH_3}$ ketone

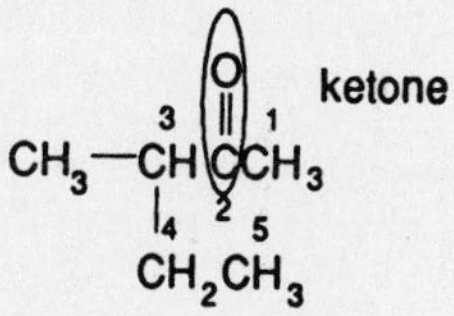

3-methyl-2-pentanone

24.51 $C_6H_5CHO + MnO_4^- \longrightarrow C_6H_5COO^- + MnO_2$

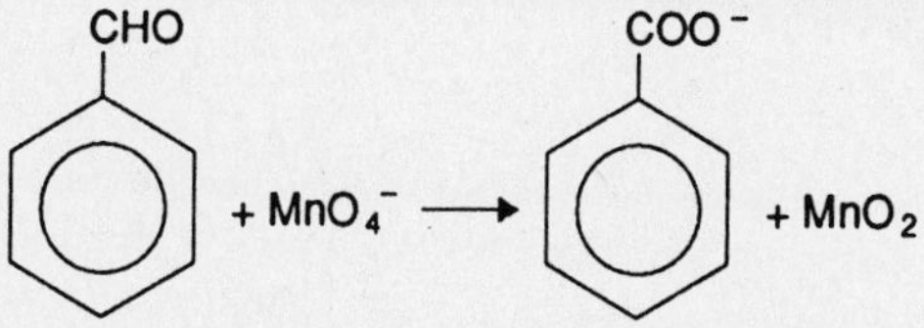

Oxidation: $C_6H_5CHO \longrightarrow C_6H_5COO^-$

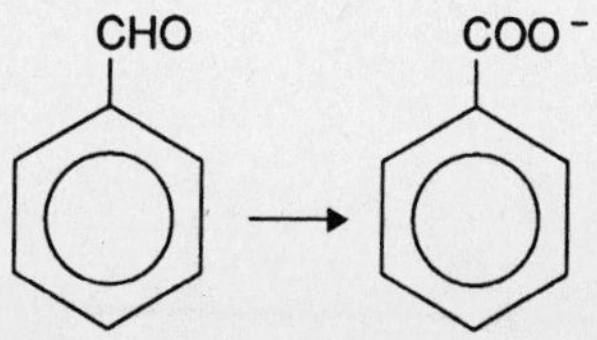

Balance O: $2OH^- + C_6H_5CHO \longrightarrow C_6H_5COO^- + H_2O$

Balance H: $3OH^- + C_6H_5CHO \longrightarrow C_6H_5COO^- + 2H_2O$

Balance charge: $3OH^- + C_6H_5CHO \longrightarrow C_6H_5COO^- + 2H_2O + 2e^-$

Reduction: $MnO_4^- \rightarrow MnO_2$

Balance O: $2H_2O + MnO_4^- \rightarrow MnO_2 + 4OH^-$

Balance charge: $3e^- + 2H_2O + MnO_4^- \rightarrow MnO_2 + 4OH^-$

Add half-reactions:

$$3\left(3OH^- + C_6H_5CHO \longrightarrow C_6H_5COO^- + 2H_2O + 2e^-\right)$$

$$2(3e^- + 2H_2O + MnO_4^- \rightarrow MnO_2 + 4OH^-)$$

$$OH^- + 3\,C_6H_5CHO + 2MnO_4^- \longrightarrow 3\,C_6H_5COO^- + 2H_2O + 2MnO_2$$

24.53 (a) $CH_3CH(CH_2CHO)CH_3 + (O) \rightarrow CH_3CH(CH_2COOH)CH_3$ (carbons numbered: $\overset{4}{C}H_3\overset{3}{C}H(\overset{2}{C}H_2\overset{1}{C}OOH)CH_3$)

aldehyde — an acid (3-methylbutanoic acid)

(b) No reaction. Tertiary alcohols are not easily oxidized.

(c) $C_6H_5COOH + 4(H) \longrightarrow C_6H_5CH_2OH + H_2O$

acid — an alcohol (phenylmethanol)

(d) $CH_3CH(OH)CH_2CH_3 + (O) \rightarrow CH_3C(=O)CH_2CH_3 + H_2O$

secondary alcohol — a ketone (2-butanone)

24.55 (a) $CH_3CH_2CH_2COOH + CH_3CH_2OH \overset{H^+}{\rightleftharpoons} CH_3CH_2CH_2COOCH_2CH_3 + H_2O$

butyric acid — ethyl butyrate

(b) $HCOOCH_3 + NaOH \rightarrow HCOO^-Na^+ + CH_3OH$

methyl formate — sodium formate — methanol

24.57 (a) $C_6H_5NH_2$ primary amine

(b) $CH_3CH_2NHCH_2CH_3$ secondary amine

24.59 $nCF_2{-}CF_2 \rightarrow {-}CF_2{-}CF_2{-}CF_2{-}CF_2{-}CF_2{-}CF_2{-}$

24.61 (a) 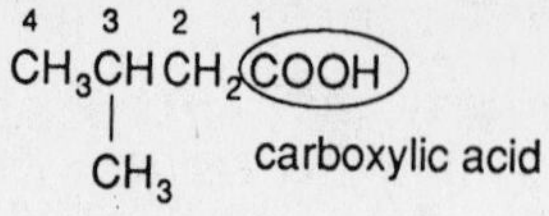 $\overset{4}{C}H_3\overset{3}{C}H\overset{2}{C}H_2\overset{1}{C}OOH$ (carboxylic acid circled), with CH_3 on C-3

carboxylic acid

3-methylbutanoic acid

(b) 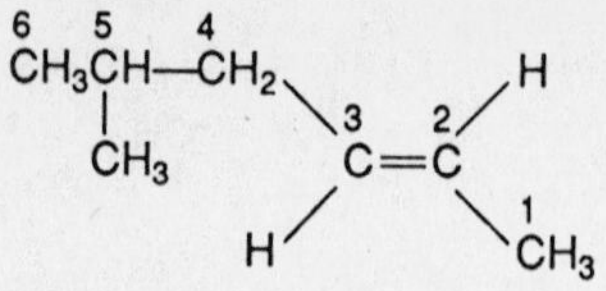 $\overset{6}{C}H_3\overset{5}{C}H—\overset{4}{C}H_2$, with CH_3 on C-5, attached to $\overset{3}{C}=\overset{2}{C}$; H on C-3, H on C-2, $\overset{1}{C}H_3$ on C-2

trans-5-methyl-2-hexene

(c) 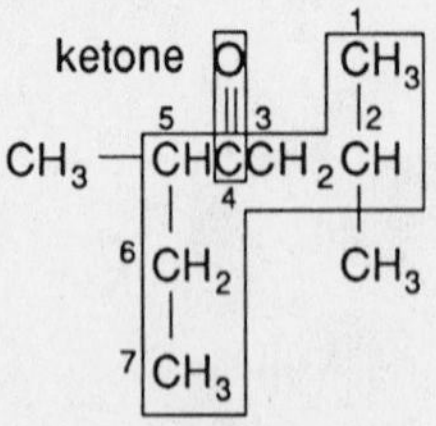 $CH_3—\overset{5}{C}H\overset{3}{C}(=O)\overset{4}{C}H_2\overset{2}{C}H(\overset{1}{C}H_3)$, with CH_3 on C-2 and $\overset{6}{C}H_2—\overset{7}{C}H_3$ on C-5

ketone

longest chain

2,5-dimethyl-4-heptanone

(d) 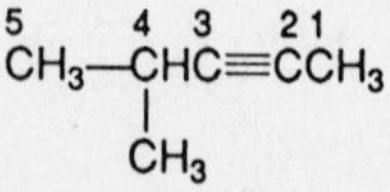 $\overset{5}{C}H_3—\overset{4}{C}H\overset{3}{C}≡\overset{2}{C}\overset{1}{C}H_3$, with CH_3 on C-4

4-methyl-2-pentyne

24.63 (a) 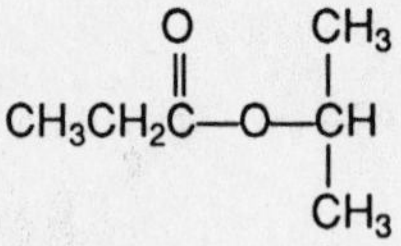 $CH_3CH_2C(=O)—O—CH(CH_3)_2$

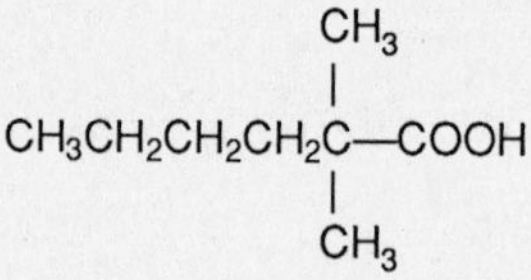

(b) 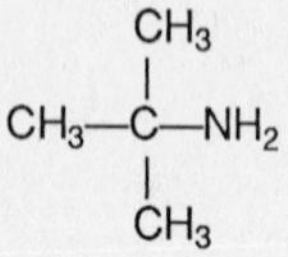 $CH_3—C(CH_3)_2—NH_2$

(c) $CH_3CH_2CH_2CH_2C(CH_3)_2—COOH$

(d) 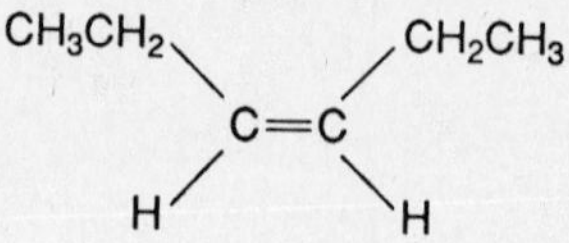 $CH_3CH_2(H)C=C(H)CH_2CH_3$ (the two CH_2CH_3 groups drawn on opposite sides: CH_3CH_2 upper left, CH_2CH_3 upper right, H lower left, H lower right)

24.65 (a) Addition of dichromate ion in acidic solution to propionaldehyde will cause the reagent to change from orange to green as the aldehyde is oxidized. Under similar conditions, acetone (a ketone) would not react.

(b) Addition of $CH_2{=}CH—C{\equiv}C—CH{=}CH_2$ to a solution of Br_2 in CCl_4 would cause the bromine color to disappear as the Br_2 was added to the double bonds. Addition of benzene to Br_2 in CCl_4 results in no reaction. Aromatic rings are not susceptible to attack by Br_2 in the absence of a catalyst.

24.67 (a) $CH_2—CH_2$, ethylene

(b) There must be an aromatic ring in the compound or the double bonds would react with Br_2

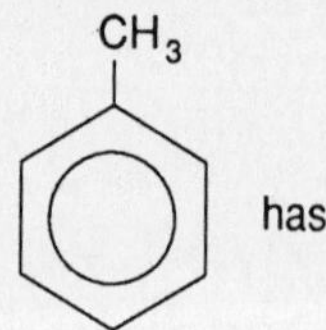

has the correct formula.

(c) CH_3NH_2 (d) CH_3OH

24.69 Assume 100.0 g of the unknown. This contains 85.6 g C and 14.4 g H. Convert these amounts to moles

$$85.6 \text{ g C} \times \frac{1 \text{ mol C}}{12.01 \text{ g C}} = 7.1\underline{2}7 \text{ mol C}$$

$$14.4 \text{ g H} \times \frac{1 \text{ mol H}}{1.008 \text{ g H}} = 14.\underline{2}857 \text{ mol H}$$

The molar ratio of H to C is 14.2857 : 7.12, or 2.00 : 1. The empirical formula is, therefore, CH_2. This formula unit has a mass of 12.011 + 2 × (1.008) amu = 14.027 amu.

$$\frac{56.1 \text{ amu}}{1 \text{ molecule}} \times \frac{1 \text{ formula unit}}{14.03 \text{ amu}} = \frac{4.00 \text{ formula units}}{1 \text{ molecule}}$$

The molecular formula is $(CH_2)_4$ or C_4H_8. The formula (C_nH_{2n}) indicates that the compound is either an alkene or a cycloalkane. Since it reacts with water and H_2SO_4, it must be an alkene. The product of the addition of H_2O to a double bond is an alcohol. Since the alcohol produced can be oxidized to a ketone it must be a secondary alcohol. The only secondary alcohol with 4 carbon atoms is 2-butanol.

$CH_3CH_2CHCH_3$ (with OH on the third carbon)

The original hydrocarbon from which it was produced is either 1-butene or 2-butene

$$CH_3CH{=}CHCH_3 + H_2O \xrightarrow{H_2SO_4} CH_3CH_2\underset{\displaystyle OH}{\underset{|}{C}}HCH_3$$

or

$$CH_3CH_2CH{=}CH_2 + H_2O \xrightarrow{H_2SO_4} CH_3CH_2\underset{\displaystyle OH}{\underset{|}{C}}HCH_3$$

CHAPTER 25

BIOCHEMISTRY

SOLUTIONS TO EXERCISES

25.1 The zwitterion has both amino groups and the carboxyl group in ionized form.

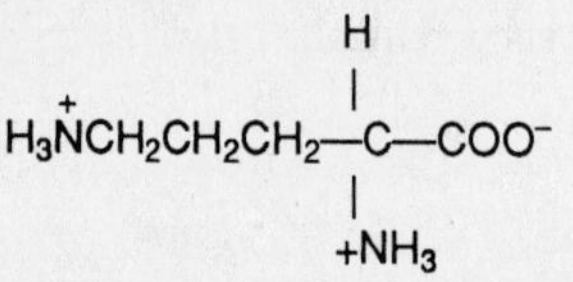

$$H_3\overset{+}{N}CH_2CH_2CH_2-\underset{\underset{+NH_3}{|}}{\overset{\overset{H}{|}}{C}}-COO^-$$

25.2 The possible dipeptides are gly-gly-ser, gly-ser-gly, and ser-gly-gly.

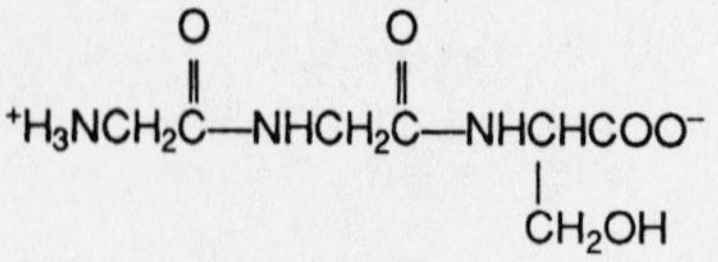

$$^+H_3NCH_2\overset{\overset{O}{\|}}{C}-NHCH_2\overset{\overset{O}{\|}}{C}-NH\underset{\underset{CH_2OH}{|}}{C}HCOO^-$$

gly-gly-ser

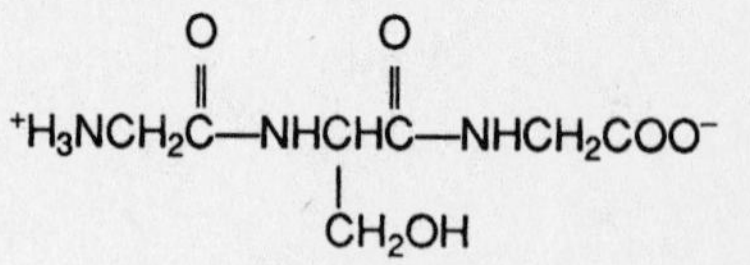

$$^+H_3NCH_2\overset{\overset{O}{\|}}{C}-NH\underset{\underset{CH_2OH}{|}}{C}H\overset{\overset{O}{\|}}{C}-NHCH_2COO^-$$

gly-ser-gly

$$^+H_3N\underset{\underset{CH_2OH}{|}}{C}H\overset{\overset{O}{\|}}{C}-NHCH_2\overset{\overset{O}{\|}}{C}-NHCH_2COO^-$$

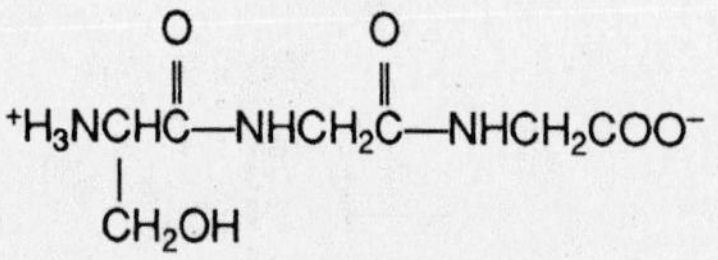

ser-gly-gly

25.3 The oxygen in the carbonyl of the glutamine side chain can hydrogen bond to the hydrogen of the hydroxyl group of the threonine side chain.

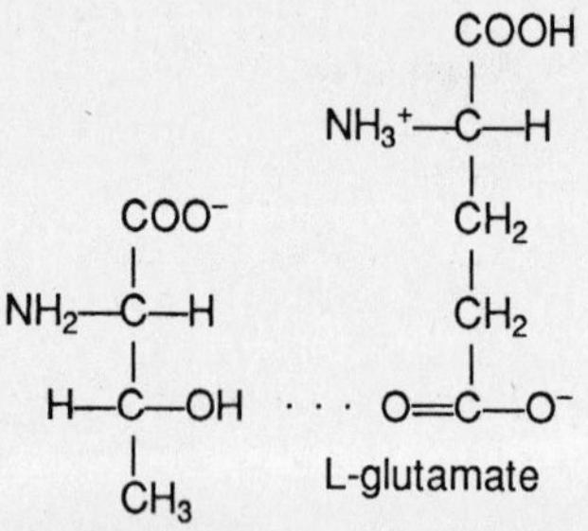

L-threonine: COO^- | NH_2-C-H | $H-C-OH$ | CH_3

L-glutamate: $COOH$ | NH_3^+-C-H | CH_2 | CH_2 | $O{=}C-O^-$

$$H-C-OH \cdots O{=}C-O^-$$

L-glutamate

L-threonine

25.4

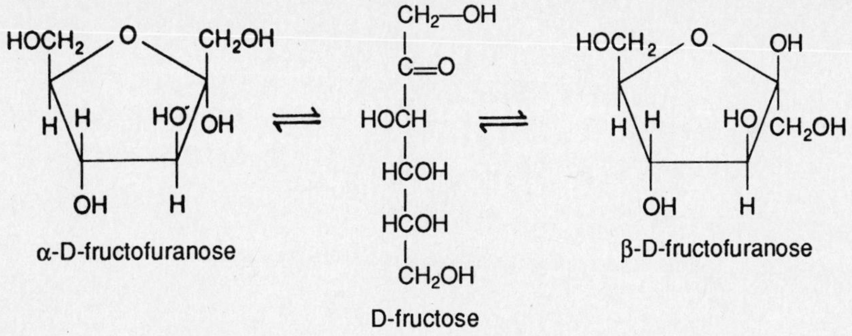

α-D-fructofuranose $\rightleftharpoons$ D-fructose $\rightleftharpoons$ β-D-fructofuranose

D-fructose: CH_2—OH / C=O / HOCH / HCOH / HCOH / CH_2OH

25.5 Maltose

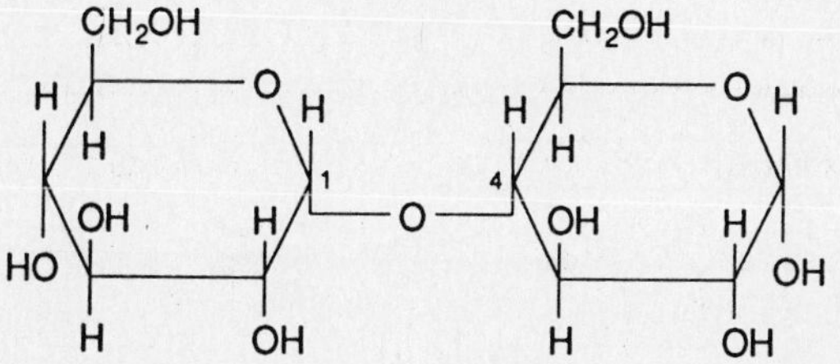

25.6 Cytidine 5'-monophosphate

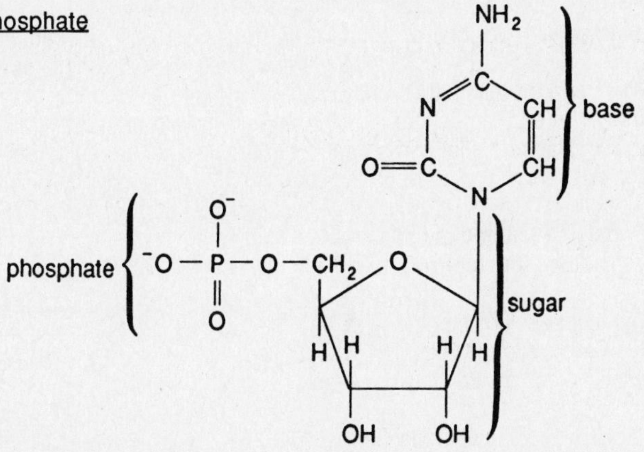

phosphate, sugar, base

25.7 RNA contains uracil, not thymine. Uracil is the complementary base for adenine.

ATGCTACGGATTCAA	Sequence given in Example 25.7
UACGAUGCCUAAGUU	RNA sequence

25.8 Consult Table 25.2. One possible nucleotide sequence is

AAA	CCU	GCU	UUU	UGG	GAG	CAU	GGU
(lys)	(pro)	(ala)	(phe)	(trp)	(glu)	(his)	(gly)

25.9 A triacylglycerol contains glycerol in an ester linkage with 3 fatty acids. In this case they are to be unsaturated fatty acids. One possibility is

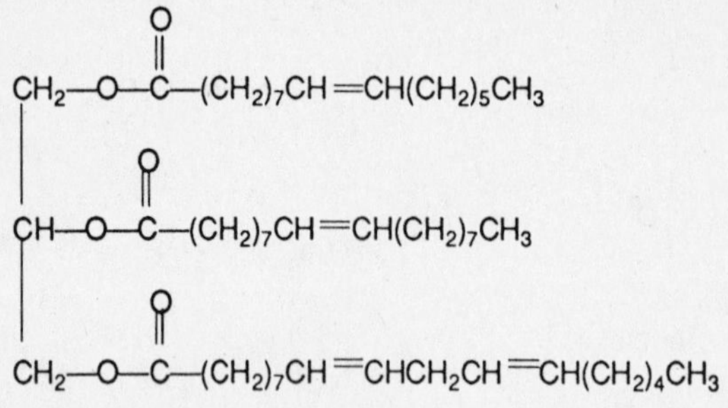

$$\begin{array}{l} CH_2-O-\overset{O}{\overset{\|}{C}}-(CH_2)_7CH=CH(CH_2)_5CH_3 \\ | \\ CH-O-\overset{O}{\overset{\|}{C}}-(CH_2)_7CH=CH(CH_2)_7CH_3 \\ | \\ CH_2-O-\overset{O}{\overset{\|}{C}}-(CH_2)_7CH=CHCH_2CH=CH(CH_2)_4CH_3 \end{array}$$

ANSWERS TO REVIEW QUESTIONS

25.1 Biological systems with positive free-energy changes proceed by using body energy supplied through coupling with a chemical reaction having negative free-energy change.

25.2 The primary structure of a protein refers to the order, or sequence, of the amino-acid units in the protein polymer. What makes one protein different from another of the same size is the arrangement of the various possible amino acids in the sequence. For example, there are 120 different sequences possible for a polypeptide with just five different amino acids. The basis of the unique conformation of a protein is the folding and coiling into a three-dimensional conformation in aqueous solution on the basis of the different side-chain amino-acid units. In stable conformations, the nonpolar side chains are buried within the structure away from water and the polar groups are on the surface of the conformation, where they can hydrogen bond with water.

25.3 The secondary structure of a protein is the simpler coiled or parallel arrangement of the protein chain. The tertiary structure refers to the folded nature of the structure.

25.4 An enzyme is a specific body catalyst, usually a globular protein, that possesses active sites. The specificity of an enzyme is explained by the active sites at which substrates will bind. Substrates fit into the active sites as a key fits into a lock.

25.5 A monosaccharide is a molecule with three to nine carbon atoms, all of which but one bear an OH group. The remaining carbon is always a carbonyl carbon, either an aldehyde or ketone. The four major monosaccharides are D-glucose, D-fructose, D-ribose, and 2-deoxy-D-ribose. D-glucose is blood sugar, an important energy source for cell function. D-fructose is a common sugar in fruits and a food source. D-ribose and 2-deoxy-D-ribose are parts of nucleic acids, carriers of species inheritance. An oligosaccharide is a molecule formed from monosaccharides in which the hemiacetal carbon atom of one is attached to the alcohol oxygen atom of another by a condensation reaction. The most common oligosaccharide is sucrose, a disaccharide. A polysaccharide is a long-chain molecular carbohydrate built of only one type of monomer or sometimes two alternating monomers. Some are made of glucose molecules only.

25.6 Cellulose forms the plant cell walls and is a linear polymer of β-D-glucopyranose units. Amylose is the major energy-storage substance of plants and is a linear polymer of α-D-glucopyranose units.

25.7 It is possible for glucose to exist in three forms in any solution such as blood: the straight-chain form, the α-D-glucopyranose form, and the β-D-glucopyranose form.

25.8 The complementary base pairs are the nucleotide bases that form strong hydrogen bonds with one another: adenine and thymine, adenine and uracil, and guanine and cytosine. A DNA molecule consists of two polynucleotide chains with base pairing along their entire lengths. The two chains are coiled about each other to form a double helix.

25.9 The only difference between ribonucleotides and deoxyribonucleotides is the sugar: ribonucleotides contain β-D-ribose, and deoxyribonucleotides contain 2-deoxy-β-D-ribose, with both sugars having furanose rings. Both form polymers by condensation (loss of H_2O).

25.10 The genetic code is the relationship between the nucleotide sequence in DNA and the amino-acid sequence in proteins. Genetic information is coded into the linear sequence of nucleotides in the DNA molecule, and this coding then directs the synthesis of the specific proteins that make a cell unique.

25.11 A codon is a code structure in messenger RNA; each such structure has a particular sequence of three nucleotides, usually denoted simply by their bases. An anticodon is a triplet sequence in transfer RNA complementary to the codon in DNA; the transfer RNA uses the anticodon to carry an amino acid to a ribosome. The messenger RNA and transfer RNA bond to each other through the codon and anticodon, respectively.

25.12 There are four RNA bases and three of these are arranged in a definite order in each codon. There are 16 different arrangements starting with each base, and since there are 4 bases, there are 64 different triplet codons.

25.13 A polypeptide is produced as follows: Imagine that we have a ribosome with messenger RNA attached in the proper way for translation and the first codon is in position to be read. (The messenger RNA has been synthesized with a sequence of bases complementary to that of the gene before attachment.) Transfer RNAs bring up various amino acids to be bonded to each other until a termination codon appears, to signal the end of the chain, which is then released from the ribosome.

25.14 Fats and oils have the same basic structure, but fats are solids and oils are liquids at room temperature.

25.15 A triacylglycerol is a triester formed from glycerol and three fatty acids bonded to the oxygens of the glycerol.

25.16 A biological membrane is composed of proteins inserted into a phospholipid matrix. A phospholipid resembles a triacylglycerol, but only two fatty acids form ester bonds; the third —OH group of glycerol is bonded to a phosphate group that in turn is bonded to an alcohol through an oxygen attached to phosphorus. Because the phospholipids form a bilayer, with the interior of the lipid bilayer being hydrophobic hydrocarbon chains of the fatty acids of the phospholipids, it presents a barrier to charged or polar substances (hydrophilic). The proteins in the membrane do catalyze the transport of particular substances across the membrane, some of which may be hydrophilic.

SOLUTIONS TO PRACTICE PROBLEMS

25.17 The free energy change for coupled reactions is the sum of the free energy changes for the individual reactions.

$$\Delta G_{overall} = +7.32 \text{ kcal/mol} + (-10.07) \text{ kcal/mol}$$
$$= -2.75 \text{ kcal/mol}$$

25.19 In general, reactions that break complex molecules into simpler molecules are energy-releasing, so the breakdown of a fat to give CO_2 and H_2O is expected to release free energy.

25.21

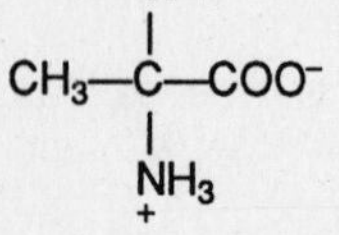

$$CH_3-\overset{\displaystyle H}{\underset{\displaystyle \overset{+}{N}H_3}{C}}-COO^-$$

25.23 There are two possible dipeptides containing one molecule each of L-alanine and L-histidine

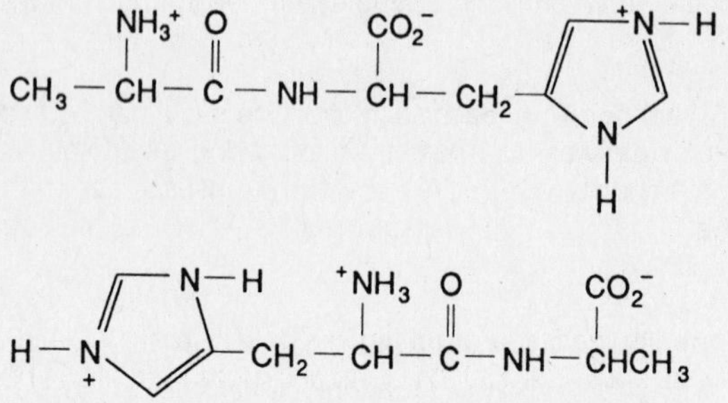

$$CH_3-\overset{NH_3^+}{CH}-\overset{O}{\overset{\|}{C}}-NH-\overset{CO_2^-}{CH}-CH_2-(\text{imidazolium ring: } {}^+N-H,\ N-H)$$

$$H-\overset{+}{N}(\text{imidazolium ring, } N-H)-CH_2-\overset{{}^+NH_3}{CH}-\overset{O}{\overset{\|}{C}}-NH-\overset{CO_2^-}{CHCH_3}$$

25.25

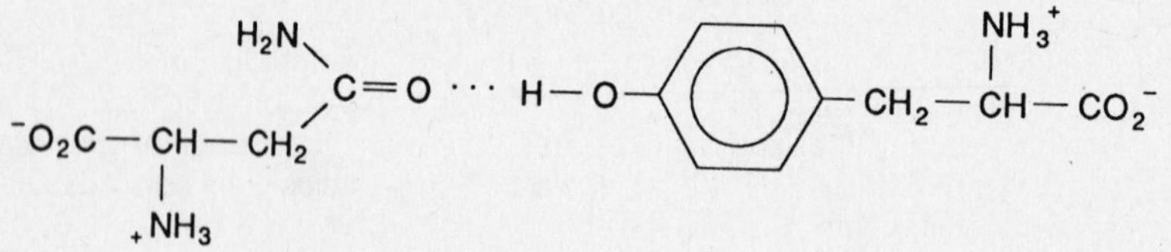

$$^-O_2C-\underset{{}_+NH_3}{CH}-CH_2-\overset{H_2N}{C}{=}O\cdots H-O-C_6H_4-CH_2-\overset{NH_3^+}{CH}-CO_2^-$$

25.27

(pyranose ring: CH_2OH, HO, H, OH, H, H, O, H, OH, OH) $\rightleftharpoons$ open-chain form:

$$\begin{array}{c} O{=}C-H \\ H-C-OH \\ HO-C-H \\ HO-C-H \\ H-C-OH \\ CH_2OH \end{array}$$

$\rightleftharpoons$ (pyranose ring: CH_2OH, HO, H, OH, H, H, O, OH, H, OH)

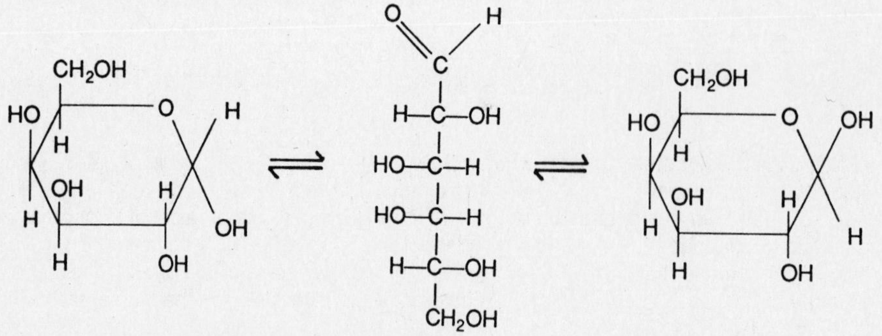

25.29

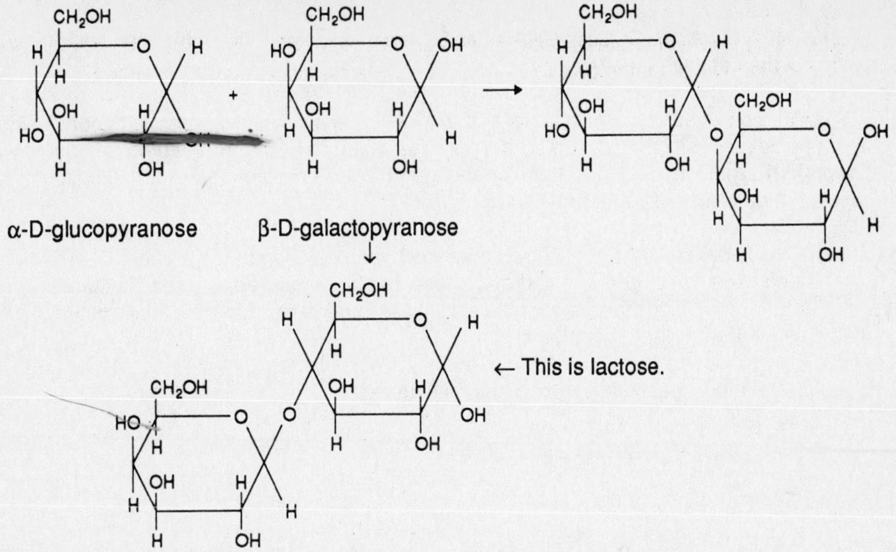

α-D-glucopyranose β-D-galactopyranose

← This is lactose.

25.31 DNA consists of two strands of polynucleotides with the adenine units in one strand paired to thymine units in the other. Similarly guanine units are paired with cytosine units. Therefore the ratios A : T and G : C are both 1:1.

25.33 Adenosine consists of ribose and adenine.

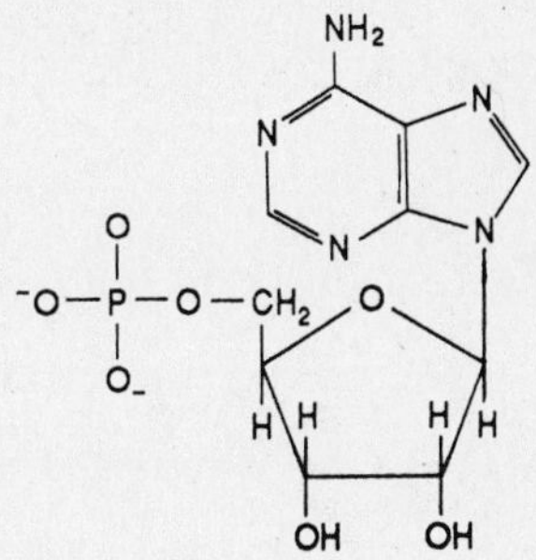

25.35 3 hydrogen bonds link a guanine-cytosine base pair.

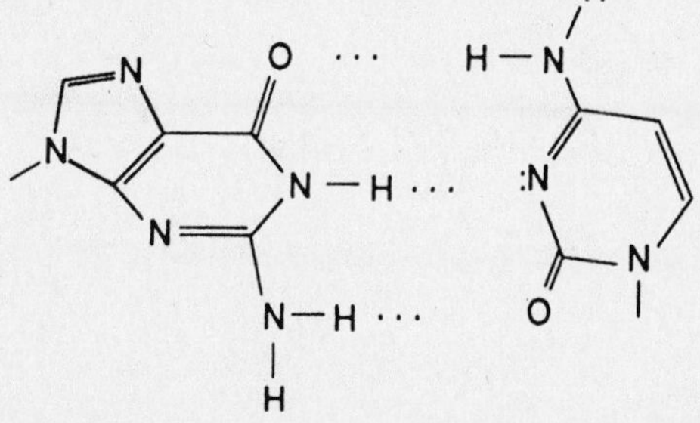

Since only two hydrogen bonds link an adenine-uracil base pair, the bonding would be expected to be stronger in the guanine-cytosine pair.

25.37 TGACTGCGTTAACTGGCG In DNA, adenine and thymine are complementary and guanine and cytosine are complementary.

25.39 If a codon consisted of two nucleotides, there would be 4×4 or 16 possible codons using the four nucleotides. Since there are 20 amino acids which must be represented uniquely by a codon for protein synthesis, the codon must be longer than two nucleotides. A 2 nucleotide codon would not be workable as an amino acid code.

25.41 When DNA is denatured the hydrogen bonds between base pairs are broken. DNA with a greater percent composition of guanine and cytosine would be denatured less readily than DNA with greater percent composition of adenine and thymine because there are more hydrogen bonds between the former than the latter.

25.43 Mark off the message in triplets, beginning at the left.

```
GGA | UCC | CGC | UUU | GGG | CUG | AAA | UAG
Gly-Ser-Arg-Phe-Gly-Leu-Lys
```

Note that the UAG at the right codes for the end of the sequence.

25.45 The codons have bases that are complementary to those in the anticodon.

Anticodons	GAC	UGA	GGG	ACC
Codons	CUG	ACU	CCC	UGG

25.47 Consult Table 27.2 to find what nucleotides correspond to the amino acids in the sequence

```
leu-ala-val-glu-asp-cys-met-trp-lys
CUU GCU GUU GAA GAU UGU AUG UGG AAA
```

25.49

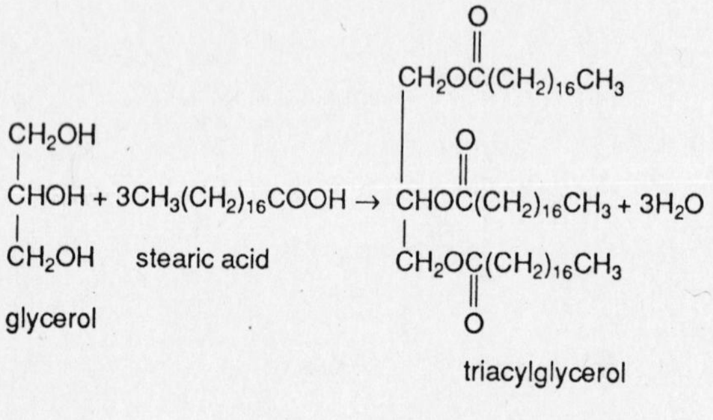

$$\begin{matrix} CH_2OH \\ | \\ CHOH \\ | \\ CH_2OH \end{matrix} + 3CH_3(CH_2)_{16}COOH \rightarrow \begin{matrix} CH_2OC(=O)(CH_2)_{16}CH_3 \\ | \\ CHOC(=O)(CH_2)_{16}CH_3 \\ | \\ CH_2OC(=O)(CH_2)_{16}CH_3 \end{matrix} + 3H_2O$$

glycerol, stearic acid, triacylglycerol

25.51

$$\begin{matrix} CH_2OC(=O)(CH_2)_{16}CH_3 \\ | \\ CHOC(=O)(CH_2)_{16}CH_3 \\ | \\ CH_2OC(=O)(CH_2)_{16}CH_3 \end{matrix} + 3NaOH \rightarrow \begin{matrix} CH_2OH \\ | \\ CHOH \\ | \\ CH_2OH \end{matrix} + 3CH_3(CH_2)_{16}CO_2^-Na^+$$

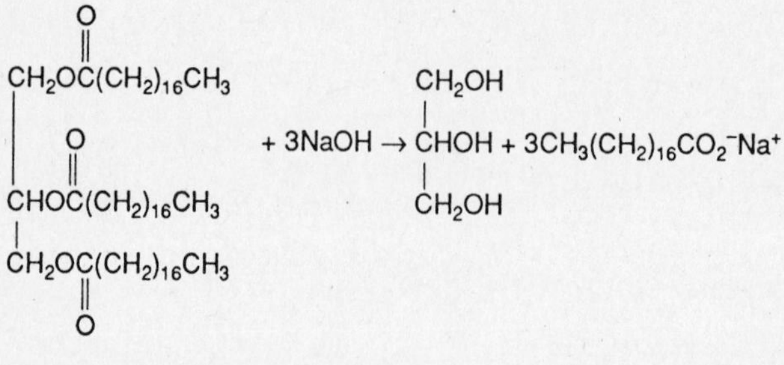

25.53 Since the interior of a cell membrane is hydrophobic, non-polar molecules such as 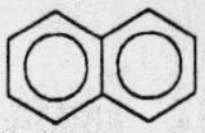 diffuse more readily across it. The remaining species in the list are ionic or polar, and so would diffuse slowly across the membrane.

25.55 `AGTCGACCGTTAAT`

Complement `TCAGCTGGCAATTA`

25.57 The number of possible sequences is 6 • 5 • 4 • 3 • 2 • 1 or 720.

27.59

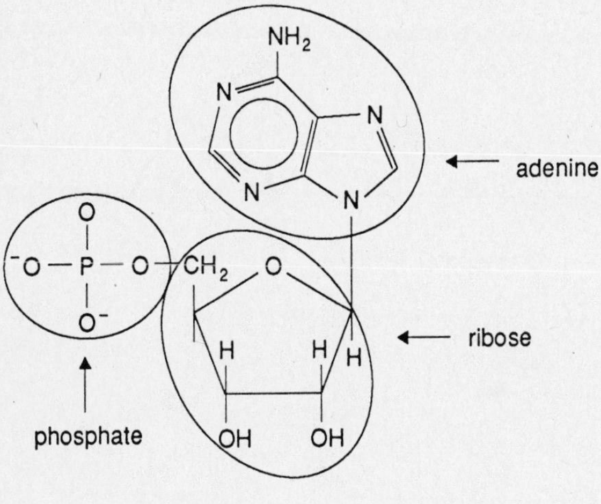

25.61 $CH_3SCH_2CH_2CH(^+NH_3)—CO_2^-$ has a nonpolar side chain. The other amino acid has a polar SH group in the side chain.

25.63 The two possibilities are

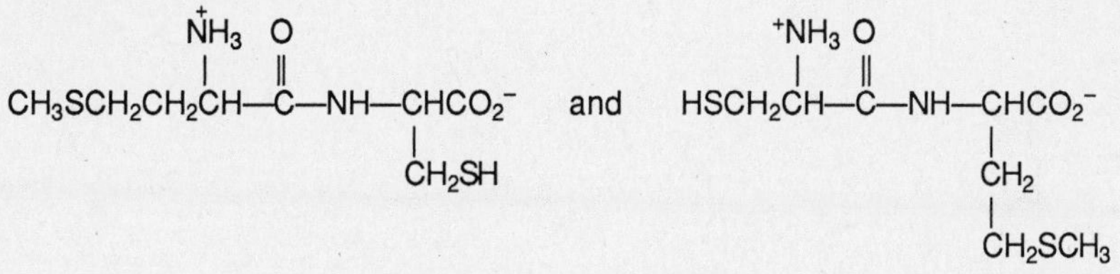

25.65

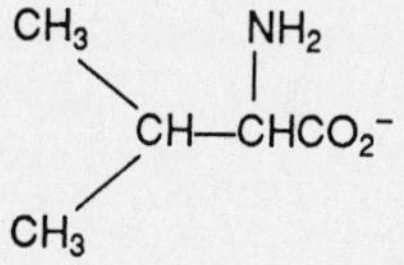

At high pH the amino group is not protonated and the carboxyl group has lost its hydrogen.

25.67 The triacylglycerol is made up of glycerol and three fatty acids

glycerol	fatty acids
CH_2OH	$CH_3(CH_2)_{14}CO_2H$ palmitic acid
|	
$CHOH$	$CH_3(CH_2)_7CH—CH(CH_2)_7CO_2H$ oleic acid
|	
CH_2OH	$CH_3(CH_2)_5CH—CH(CH_2)_7CO_2H$ palmitoleic acid
glycerol	

APPENDIX A

MATHEMATICAL SKILLS

SOLUTIONS TO EXERCISES

1. (a) Either leave as 4.38 or write it as 4.38×10^0.
   (b) Shift the decimal point left and count the number of positions shifted (3). The answer is 4.380×10^3.
   (c) Shift the decimal point right and count the number of positions shifted (4). The answer is 4.83×10^{-4}.

2. (a) Shift the decimal point right 3 places. The answer is 7025.
   (b) Shift the decimal point left 4 places. The answer is 0.000897.

3. Express 2.8×10^{-6} as 0.028×10^{-4}. Then the sum can be written $(3.142 \times 10^{-4}) + (0.028 \times 10^{-4})$ or $(3.142 + 0.028) \times 10^{-4} = 3.170 \times 10^{-4}$.

4. (a) $(5.4 \times 10^{-7}) \times (1.8 \times 10^8) = (5.4 \times 1.8) \times 10^{-7} \times 10^8 = 9.72 \times 10^1$. This rounds to 9.7×10^1.

   (b) $\dfrac{5.4 \times 10^{-7}}{6.0 \times 10^{-5}} = \dfrac{5.4}{6.0} \times 10^{-7} \times 10^5 = 0.90 \times 10^{-2} = 9.0 \times 10^{-3}$

5. (a) $(3.56 \times 10^3)^4 = (3.56)^4 \times (10^3)^4 = 161 \times 10^{12} = 1.61 \times 10^{14}$

   (b) $\sqrt[3]{4.81 \times 10^2} = \sqrt[3]{0.481 \times 10^3} = \sqrt[3]{0.481} \times \sqrt[3]{10^3} = 0.784 \times 10^1 = 7.84 \times 10^0$

6. (a) $\log 0.00582 = -2.235$
   (b) $\log 689 = 2.838$

7. (a) antilog $5.728 = 5.35 \times 10^5$
   (b) antilog $(-5.728) = 1.87 \times 10^{-6}$

8. $$x = \frac{-0.850 \pm \sqrt{(0.850)^2 - 4(1.80)(-9.50)}}{2(1.80)} = \frac{-0.850 \pm \sqrt{69.12}}{3.60} = \frac{-0.850 \pm 8.314}{3.60}$$

   The positive root is $\dfrac{7.46}{3.60} = 2.07$.